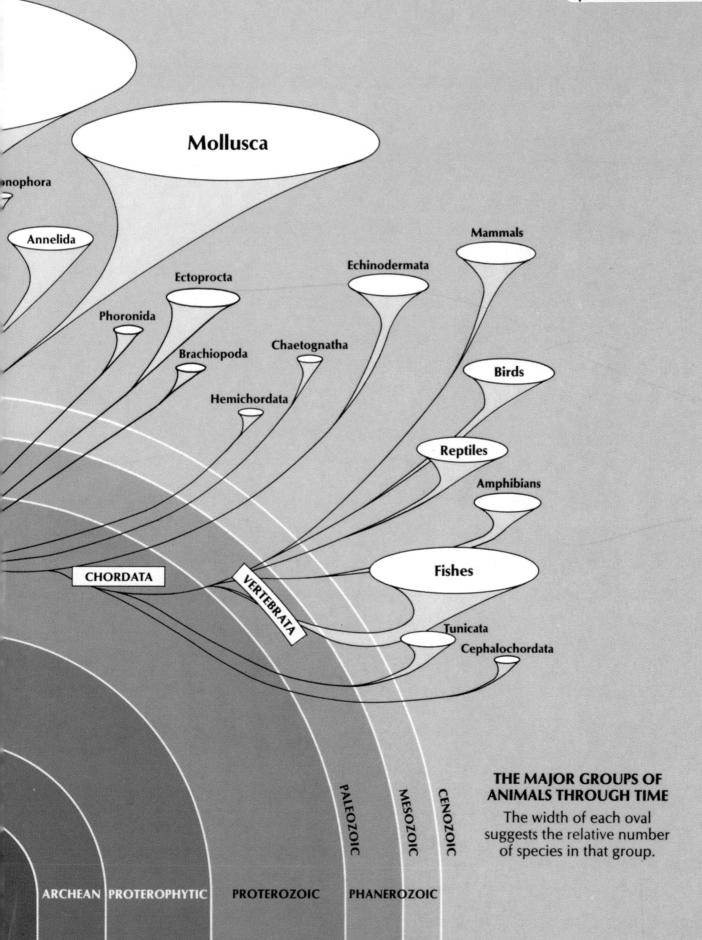

Mollusca

...onophora

Annelida

Ectoprocta

Phoronida

Brachiopoda

Chaetognatha

Hemichordata

Echinodermata

Mammals

Birds

Reptiles

Amphibians

Fishes

CHORDATA

VERTEBRATA

Tunicata

Cephalochordata

W9-CPD-677

PALEOZOIC

MESOZOIC

CENOZOIC

ARCHEAN PROTEROPHYTIC   PROTEROZOIC   PHANEROZOIC

THE MAJOR GROUPS OF ANIMALS THROUGH TIME

The width of each oval suggests the relative number of species in that group.

ANIMAL | diversity

ANIMAL | diversity
Second Edition

CLEVELAND P. HICKMAN, JR.
Washington and Lee University

LARRY S. ROBERTS
University of Miami

ALLAN LARSON
Washington University

Original Artwork by
WILLIAM C. OBER, M.D. *and* CLAIRE W. GARRISON

Boston   Burr Ridge, IL   Dubuque, IA   Madison, WI   New York   San Francisco   St. Louis
Bangkok   Bogotá   Caracas   Lisbon   London   Madrid
Mexico City   Milan   New Delhi   Seoul   Singapore   Sydney   Taipei   Toronto

McGraw-Hill Higher Education

A Division of The **McGraw-Hill** *Companies*

ANIMAL DIVERSITY, SECOND EDITION

Copyright © 2000, 1995 by The McGraw-Hill Companies, Inc. All rights reserved. Printed in the United States of America. Except as permitted under the United States Copyright Act of 1976, no part of this publication may be reproduced or distributed in any form or by any means, or stored in a data base or retrieval system, without the prior written permission of the publisher.

 This book is printed on recycled, acid-free paper containing 10% postconsumer waste.

1 2 3 4 5 6 7 8 9 0 VNH/VNH 0 9 8 7 6 5 4 3 2 1 0

ISBN 0-07-012200-8

Vice president and editorial director: *Kevin T. Kane*
Publisher: *Michael D. Lange*
Sponsoring editor: *Margaret J. Kemp*
Developmental editor: *Donna Nemmers*
Marketing manager: *Michelle Watnick*
Senior project manager: *Gloria G. Schiesl*
Senior production supervisor: *Mary E. Haas*
Design manager: *Stuart D. Paterson*
Photo research coordinator: *John C. Leland*
Supplement coordinator: *Sandra M. Schnee*
Compositor: *Precision Graphics*
Typeface: *10/12 Garamond Book*
Printer: *Von Hoffmann Press, Inc.*

Cover/interior design: *Mary Sailor*
Cover image: *Minden Pictures*
Photo research: *Connie Mueller Photo Research*

The credits section for this book begins on page 412 and is considered an extension of the copyright page.

Library of Congress Catalog Number: 99—62705

www.mhhe.com

Brief Contents

Contents

Preface

Students find the study of zoology both fascinating and challenging. The fascination is generated by a natural inquisitiveness toward animals. The challenge is being introduced to the vast amount of information in a general zoology text. This second edition of *Animal Diversity* presents a survey of the animal kingdom with emphasis on diversity, evolutionary relationships, functional adaptations, and environmental interactions. It is tailored for the restrictive requirements of a one-semester or one-quarter course and is appropriate for both non-science and science majors.

Organization and Coverage

The sixteen survey chapters, drawn from a nearly identical treatment in the seventh edition of *Biology of Animals,* are prefaced by discussions of the principles of evolution, animal architecture, and classification.

Chapter 1 begins with a brief explanation of the scientific method—what science is (and what it is not) and then moves to a discussion of evolutionary principles, slightly condensed from a similar treatment in *Biology of Animals.* Following a historical account of Charles Darwin's life and discoveries, the five major components of Darwin's evolutionary theory are presented, together with important challenges and revisions to his theory and an assessment of its current scientific status. This approach reflects our current understanding that Darwinism is not a single, simple statement that is easily confirmed or refuted. It also prepares the student to dismiss the arguments of creationists who misconstrue scientific challenges to Darwinism as contradictions to the validity of organic evolution. The chapter ends with explanations of micro- and macroevolution.

Chapter 2 on animal architecture is a short but important chapter that defines the organization and development of body plans distinguishing the major groups of animals. This chapter includes a picture essay of tissue types and a section explaining important developmental features associated with the evolutionary diversification of the bilateral metazoa.

Chapter 3 treats the classification and phylogeny of animals. We present a brief history of how animal diversity has been organized for systematic study, emphasizing the current use of Darwin's theory of common descent as the major principle underlying animal taxonomy. Continuing controversies between the schools of evolutionary taxonomy and phylogenetic systematics (cladistics) are presented, including a discussion of how these alternative taxonomic philosophies affect our study of evolution. Chapter 3 also emphasizes that current issues in ecology, evolution, and conservation biology all depend on our taxonomic system.

The 16 survey chapters are a comprehensive, modern, and thoroughly researched coverage of the animal phyla. We emphasize the unifying architectural and functional theme of each group. Structure and function of representative forms are described, together with their ecological, behavioral, and evolutionary relationships. The distinctive themes and features of each group assist the student's approach to each chapter.

New with this edition are succinct statements of "Position in the Animal Kingdom" and "Biological Contributions" at the beginning of each survey chapter. All survey chapters are updated for this edition and many new illustrations are added. Also new with this edition, are a selection of Internet web descriptions dealing with the chapter's topics, which are found at the end of each survey chapter. The text's web site contains chapter-sorted line Internet links for each end-of-chapter description.

The classifications in each chapter follow coverage of the particular group, in most cases immediately preceding the summary at the end of the chapter. The discussions of phylogenetic relationships are written from a cladistic viewpoint, and cladograms have been presented where possible. These show the inferred branching events in each group's history and the origin of some of the principal shared derived characters. Traditional phylogenetic trees have been drawn to agree with cladistic analysis as closely as possible. Because cladistics is still not embraced by all teachers, we have presented cladograms as supplemental to the conventional Linnaean classifications.

Teaching and Learning Aids

Vocabulary Development

Key words are boldfaced and the derivations of generic names of animals are given where they first appear in the text. In addition, the derivations of many technical and zoological terms are provided in the text; in this way students gradually become familiar with the more common roots that recur in many technical terms. An extensive glossary provides the pronunciation, derivation, and definition of each term.

Chapter Prologues

A distinctive feature of this text is an opening essay placed in a panel at the beginning of each chapter. Each essay presents a theme or topic relating to the subject of the chapter. Some present biological, particularly evolutionary, principles; others illuminate distinguishing characteristics of the group treated in the chapter. Each is intended to present an important concept drawn from the chapter in an interesting manner that will facilitate learning by students, as well as engage their interest and pique their curiosity.

Boxed Notes

Boxed notes, which appear throughout the book, augment the text material and offer interesting sidelights without interrupting the narrative.

For Review

Each chapter ends with a concise summary, a list of review questions, and annotated selected references. The review questions enable students to test themselves for retention and understanding of the more important chapter material.

Art Program

The appearance and usefulness of this text are much enhanced by numerous full-color paintings by William C. Ober and Claire W. Garrison. Bill's artistic skills, knowledge of biology, and experience gained from an earlier career as a practicing physician have enriched the authors' other zoology texts through several editions. Claire practiced pediatric and obstetric nursing before turning to scientific illustration as a full-time career. Texts illustrated by Bill and Claire have received national recognition and won awards from the Association of Medical Illustrators, American Institute of Graphic Arts, Chicago Book Clinic, Printing Industries of America, and Bookbuilders West. Bill and Claire also are recipients of the Art Directors Award.

Internet Links

At the end of each survey chapter is a selection of web pages and their internet links dealing with the chapter's topics. The hot links for the pages are found at the text's web site at *http://www.mhhe.com/zoology.*

Supplements

Instructor's Manual/Test Item File

The Instructor's Manual provides a chapter outline, test bank, commentary and lesson plan, and a listing of resource references for each chapter. We trust that this material will be particularly helpful for first-time users of the text, although experienced teachers also may find much of value.

Laboratory Manual

The laboratory manual by Cleveland P. Hickman, Jr., Frances M. Hickman, and Lee B. Kats, *Laboratory Studies in Integrated Zoology,* is designed specifically for a survey course in zoology. The popular redesigned wall chart, "Chief taxonomic subdivisions and organ systems of animals," is available on request.

Computerized Test Bank

Test questions contained in the Instructor's Manual/Test Item File are available as a computerized test generation system for IBM-compatible and Macintosh computers. Using this MicroTest system, instructors can create tests and quizzes quickly and easily. Instructors can sort questions by type or level of difficulty, and can add their own questions to the bank of questions provided.

Transparency Acetates

A set of full-color transparency acetates of important textual illustrations is available with this edition of *Animal Diversity.* Labeling is clear, dark, and bold for easy reading.

Animal Diversity Slides

A set of animal diversity slides, photographed by the authors (CPH and LSR) and Bill Ober on their various excursions, are offered in this unique textbook supplement. Both invertebrates and vertebrates are represented. Descriptions, including specific names of each animal and brief overview of the animal's ecology and/or behavior, accompany the slides.

Student Study Guide

The *General Zoology Student Study Guide* by Jane Aloi and Gina Erickson is a useful tool for student review and study. It provides self-testing, valuable study tips, and chapter summary activities, including critical thinking exercises.

McGraw-Hill Zoology Web Site

The Internet provides a new route for learning and studying. McGraw-Hill has designed a web site to support the Zoology field of study. This site provides live links to related Internet sites that are described in *Animal Diversity*'s end-of-chapter pedagogy. In addition, you will find on-line quizzing, information about careers in Zoology, flash cards, and much more. You can find this site at http://www.mbhe.com/zoology. Click on the coverage of *Animal Diversity* to access the book-specific Intenet Links.

Life Science Animation Video Series

Difficult concepts like DNA replication, oxidation, and respiration can be learned though animation in this series of five videotapes. Over 65 animations present difficult life science processes in a method that fosters easier learning and review.

Life Science Living Lexicon CD-ROM

This powerful, interactive CD-ROM contains a complete lexicon of life science terminology, including a glossary of more than 4,000 terms, complete with examples and definitions. It also provides a glossary of common biological roots, prefixes, and suffixes; audio pronunciations of 300 terms; nearly 500 terms accompanied by labeled illustrations; and interactive quizzing.

Acknowledgments

We wish to thank the following zoologists for reviewing the second edition of this text:

Jane Aloi, *Saddleback College*
Sneed B. Collard, *University of West Florida*
John W. Fleeger, *Louisiana State University*
James Horwitz, *Palm Beach Community College*
Gwilym S. Jones, *Northeastern University*
Harold E. Klaassen, *Kansas State University*
Frank Pezold, *Northeast Louisiana University*
Robert J. Reinsvold, *University of Northern Colorado*
Ronald G. Wolff, *University of Florida*

Also, we are grateful to the reviewers of the first edition of this text:

Wade L. Collier, M.D., *Manatee Community College*
Richard C. Funk, *Eastern Illinois University*
Davis W. Pritchett, *Northeast Louisiana University*
Jeff Wooters, *Pensacola Jr. College*

The authors express their gratitude to the able and conscientious staff of McGraw-Hill who brought this book to its present form. We extend special thanks to Sponsoring Editor Marge Kemp, Developmental Editor Donna Nemmers, and Senior Project Manager Gloria Schiesl. All played essential roles in shaping this second edition.

Cleveland P. Hickman, Jr.
Larry S. Roberts
Allan Larson

Science of Zoology and Evolution of Animal Diversity

CHAPTER | one

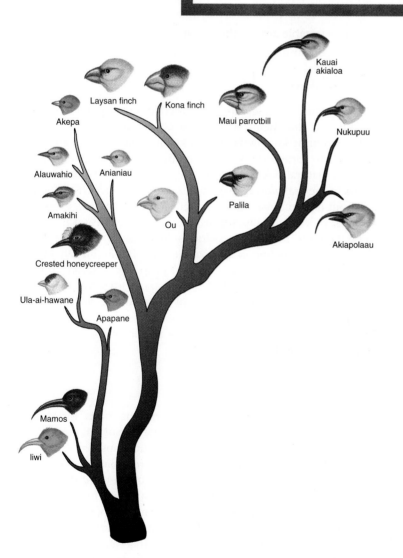

Kauai akialoa

Laysan finch

Kona finch

Akepa

Maui parrotbill

Nukupuu

Alauwahio

Anianiau

Amakihi

Ou

Palila

Crested honeycreeper

Akiapolaau

Ula-ai-hawane

Apapane

Mamos

Iiwi

A Legacy of Change

The major feature of life's history is the legacy of perpetual change. Despite the apparent permanence of the natural world, change characterizes all things on Earth and in the universe. Countless kinds of animals and plants have flourished and disappeared, leaving behind an imperfect fossil record of their existence. Many, but not all, have left living descendants that bear a partial resemblance to them.

Life's changes are perceived and measured in many ways. On a short evolutionary timescale, we see changes in the frequencies of different genetic variants within populations. Evolutionary changes in the relative frequencies of light- and dark-colored moths were observed within a single human lifetime in the polluted countryside of industrial England. The formation of new species and dramatic changes in the appearances of organisms, as seen in the evolutionary diversification of Hawaiian birds, requires longer timescales covering 100,000 to 1 million years. Major evolutionary trends and periodic mass extinctions occur on even larger timescales covering tens of millions of years. The fossil record of horses through the past 50 million years shows a series of different species replacing older ones through time and ending with the familiar horses that we know today. The fossil record of marine invertebrates shows us a series of mass extinctions separated by intervals of approximately 26 million years.

The earth bears its own record of the irreversible, historical change that we call organic evolution. Because every feature of life as we know it today is a product of the evolutionary process, biologists consider organic evolution the keystone of all biological knowledge.

Zoology (Gr. *zōon,* animal, + *logos,* discourse on, study of) is the scientific study of animals. It is a subdivision of biology (Gr. *bios,* life, + *logos,* discourse on, study of), the study of all life. The panorama of animal diversity—how animals function, live, reproduce, and interact with their environment—is exciting, fascinating, and awe inspiring. A complete understanding of all phenomena included in zoology is beyond the ability of any single person, perhaps of all humanity, but the satisfaction of knowing as much as possible is worth the effort.

To understand the diversity of animal life, we must study its long history, which began more than 600 million years ago. From the earliest animals to the millions of animal species living today, this history demonstrates perpetual change, which we call **evolution.** We depict the history of animal life as a branching genealogical tree, called a **phylogeny.** We place the earliest species ancestral to all animals at the trunk; then all living animal species fall at the growing tips of the branches. Each successive branching event represents the formation of new species from an ancestral one. The newly formed species inherit many characteristics from their immediate ancestor, but they also evolve new features that appear for the first time in the history of life. Each branch therefore has its own unique combination of characteristics and contributes a new dimension to the spectrum of animal diversity.

The study of animal diversity has two major goals. The first is to reconstruct the phylogeny of animal life and to find where in evolutionary history we can locate the origins of multicellularity, the **coelom, spiral cleavage,** vertebrae, **homeothermy,** and the many other features that comprise animal diversity as we know it today. The second major goal is to understand the historical processes that have generated and maintained diverse species and adaptations throughout evolutionary history. Darwin's theory of evolution makes possible the application of scientific principles to attain both of these goals.

Principles of Science

A basic understanding of zoology requires an understanding of what science is, what it is not, and how knowledge is gained using the scientific method. In this section we examine the methodology that zoology shares with science as a whole. These features distinguish the sciences from those activities, such as art and religion, that we exclude from science.

Despite the enormous impact that science has had on our lives, many people have only a minimal understanding of science. Public misunderstanding of scientific principles as applied to the study of animal diversity was evident on March 19, 1981, when the governor of Arkansas signed into law the Balanced Treatment for Creation-Science and Evolution-Science Act (Act 590 of 1981). This act falsely presented creation-science as a valid scientific endeavor. Creation-science is a religious position advocated by a minority of the American religious community, and it does not qualify as science.

Enactment of this law led to a historic lawsuit tried in December 1981 in the court of Judge William R. Overton, U.S. District Court, Eastern District of Arkansas. The suit was brought by the American Civil Liberties Union on behalf of 23 plaintiffs, including religious leaders and groups representing several denominations, individual parents, and educational associations. The plaintiffs contended that the law was a violation of the First Amendment to the U.S. Constitution, which prohibits establishment of religion by the government. This prohibition includes passing a law that would aid one religion or prefer one religion over another. On January 5, 1982, Judge Overton permanently prohibited the state of Arkansas from enforcing Act 590.

Considerable testimony during the trial addressed the nature of science. On the basis of testimony by scientists, Judge Overton stated explicitly these essential characteristics of science:

1. It is guided by natural law.
2. It has to be explanatory by reference to natural law.
3. It is testable against the empirical world.
4. Its conclusions are tentative, that is, are not necessarily the final word.
5. It is falsifiable.

The pursuit of scientific knowledge must be guided by the physical and chemical laws that govern the state of existence. Scientific knowledge must explain what is observed by reference to natural law without requiring the intervention of any supernatural being or force. We must be able to observe events in the real world, directly or indirectly, to test hypotheses about nature. If we draw a conclusion relative to an event, we must be ready to discard or modify our conclusion if it is contradicted by further observations. As Judge Overton stated, "While anybody is free to approach a scientific inquiry in any fashion they choose, they cannot properly describe the methodology used as scientific, if they start with a conclusion and refuse to change it regardless of the evidence developed during the course of the investigation." Science is neutral on religion, and the results of science do not favor one religious position over another.

Scientific Method

These essential criteria of science form the basis for the **hypothetico-deductive method.** The first step of this method is the generation of **hypotheses,** or potential answers to the question being asked. These hypotheses are usually based on prior observations of nature (Figure 1-1), or they are derived from theories based on such observations. Scientific hypotheses often constitute general statements that may explain a large number of diverse observations about nature. Natural selection, for example, explains the observations that many different species have properties that adapt them to their environments. On the basis of the hypothesis, the scientist must say, "If my hypothesis is a valid explanation of past observations, then future observations ought to have certain characteristics."

massive evidence that most biologists view repudiation of evolution as tantamount to repudiation of reality. Nonetheless, evolution, along with all other theories in science, has not been proven in a mathematical sense, but it is testable, tentative, and falsifiable.

Experimental and Evolutionary Sciences

The many questions that people have asked about the animal world since the time of Aristotle can be grouped into two major categories. The first category seeks to understand the **proximate causes** (also called immediate causes) that underlie the functioning of biological systems at all levels of complexity. It includes the problems of explaining how animals perform their metabolic, physiological, and behavioral functions at the molecular, cellular, organismal, and even population levels. For example, how is genetic information expressed to guide the synthesis of proteins? What causes cells to divide to produce new cells? How does population density affect the physiology and behavior of organisms?

The biological sciences that address proximate causes are called **experimental sciences** because they use the **experimental method.** This method consists of three steps: (1) predicting how a system being studied will respond to a disturbance, (2) making the disturbance, and then (3) comparing the observed results with the predicted ones. Experimental conditions are repeated to eliminate chance occurrences that might produce errors. **Controls** (repetitions of the experimental procedure that lack the disturbance) are

FIGURE 1-1

A few of the many dimensions of zoological research. **A,** Observing coral growth in the Caribbean Sea. **B,** Studying insect larvae collected from an arctic pond on Canada's Baffin Island. **C,** Separating growth stages of crab larvae at a marine laboratory. **D,** Observing nematocyst discharge from hydrozoan tentacles (**E**).

If a hypothesis is very powerful in explaining a large variety of related phenomena, it attains the status of a **theory.** Natural selection is a good example. Natural selection provides a potential explanation for the occurrence of many different traits distributed among animal species. Each of these instances constitutes a specific hypothesis generated from the theory of natural selection. The most useful theories are those that can explain the largest array of different natural phenomena.

We emphasize that the word "theory," when used by scientists, is not just speculation as it is in ordinary English usage. The failure to make this distinction has been prominent in criticism of evolution by creationists, who have called evolution "only a theory" to imply that it is little better than a random guess. In fact, the theory of evolution is supported by such

established to eliminate any unperceived factors that may bias the outcome of the experiment.

The processes by which animals maintain a body temperature under different environmental conditions, digest their food, migrate to new habitats, or store energy are some additional examples of phenomena that are studied using the experimental method. Subfields of biology that qualify as experimental sciences include molecular biology, cell biology, endocrinology, developmental biology, and community ecology.

In contrast to proximate causes, the **evolutionary sciences** address questions of **ultimate causes** that have produced biological systems and their properties through evolutionary time. For example, what are the evolutionary factors that caused some birds to acquire complex patterns of

The Animal Rights Controversy

In recent years, the debate surrounding the use of animals to serve human needs has intensified. Most controversial of all is the issue of animal use in biomedical and behavioral research and in the testing of commercial products.

A few years ago, Congress passed a series of amendments to the Federal Animal Welfare Act, a body of laws covering animal care in laboratories and other facilities. These amendments have become known as the three R's: *Reduction* in the number of animals needed for research; *Refinement* of techniques that might cause stress or suffering; *Replacement* of live animals with simulations or cell cultures whenever possible. As a result, the total number of animals used each year in research and in testing of commercial products has declined. Developments in cellular and molecular biology also have contributed to a decreased use of animals for research and testing. The animal rights movement, composed largely of vocal antivivisectionists, has created an awareness of the needs of animals used in research and has stimulated researchers to discover cheaper, more efficient, and more humane alternatives.

However, computers and culturing of cells can simulate the effects on organismal systems of, for instance, drugs, only when the basic principles involved are well known.

When the principles themselves are being scrutinized and tested, computer modeling is not sufficient. A recent report by the National Research Council concedes that although the search for alternatives to the use of animals in research and testing will continue, "the chance that alternatives will completely replace animals in the foreseeable future is nil." Realistic immediate goals, however, include reduction in number of animals used, replacement of mammals with other vertebrates, and refinement of experimental procedures to reduce discomfort of the animals being tested.

Medical and veterinary progress depends on research using animals. Every drug and every vaccine developed to improve the human condition has been tested first on animals. Research using animals has enabled medical science to eliminate smallpox and polio, and to immunize against diseases previously common and often deadly, including diphtheria, mumps, and rubella. It also has helped to create treatments for cancer, diabetes, heart disease, and manic-depressive psychoses, and to develop surgical procedures including heart surgery, blood transfusions, and cataract removal. AIDS research is wholly dependent on studies using animals. The similarity of simian AIDS, identified in rhesus monkeys, to human AIDS has permitted the disease in

According to the U.S. Department of Health and Human Services, animal research has helped extend our life expectancy by 20.8 years.

monkeys to serve as a model for the human disease. Recent work indicates that cats, too, may prove to be useful models for the development of an AIDS vaccine. Skin grafting experiments, first done with cattle and later

---

seasonal migration between temperate and tropical regions? Why do different species of animals have different numbers of chromosomes in their cells? Why do some animal species maintain complex social systems, whereas animals of other species remain largely solitary?

The evolutionary sciences proceed largely using the **comparative method** rather than experimentation. Characteristics of molecular biology, cell biology, organismal structure, development, and ecology are compared among related species to identify their patterns of variation. The patterns of similarity and dissimilarity then can be used to test hypotheses of relatedness and thereby to reconstruct the evolutionary tree that relates the species being compared. Comparative studies also serve to test hypotheses of the evolutionary processes that have generated animal diversity. Clearly, the evolutionary sciences use results of the experimental sciences as a starting point. Evolutionary sciences include comparative biochemistry, molecular evolution, comparative cell biology, comparative anatomy, comparative physiology, and phylogenetic systematics.

Origins of Darwinian Evolutionary Theory

Charles Robert Darwin and Alfred Russel Wallace (Figure 1-2) were the first to establish evolution as a powerful scientific theory. Today the reality of organic evolution can be denied only by abandoning reason. As the noted English biologist Sir Julian Huxley wrote, "Charles Darwin effected the greatest of all revolutions in human thought, greater than Einstein's or Freud's or even Newton's, by simultaneously establishing the fact and discovering the mechanism of organic evolution." Darwinian theory allows us to understand both the genetics of populations and long-term trends in the fossil record. Darwin and Wallace were not the first, however, to consider the basic idea of organic evolution, which has an ancient history. We review the history of evolutionary thinking as it led to Darwin's theory and then discuss evidence supporting it.

with other animals, opened a new era in immunological research with vast ramifications for treatment of disease in humans and other animals.

Research using animals also has benefited *other animals* through the development of veterinary cures. The vaccines for feline leukemia and canine parvovirus were first introduced to other cats and dogs. Many other vaccinations for serious diseases of animals were developed through research on animals: for example, rabies, distemper, anthrax, hepatitis, and tetanus. No endangered species is used in general research (except to protect that species from total extinction). Thus, research using animals has provided enormous benefits to humans and other animals. Still, much remains to be learned about treatment of diseases such as cancer, AIDS, diabetes, and heart disease, and research with animals will be required for this purpose.

Despite the remarkable benefits produced by research on animals, advocates of animal rights often present an inaccurate and emotionally distorted picture of this research. The ultimate goal of most animal rights activists, who have focused specifically on the use of animals in science rather than on the treatment of animals in all contexts, remains the total abolition of all forms of research using animals. The scientific community is deeply concerned about the impact of these attacks on the ability of scientists to conduct important experiments that will benefit people and animals. They argue that if we are justified to use animals for food and fiber and as pets, we are justified in experimentation to benefit human welfare when these studies are conducted humanely and ethically.

References on Animal Rights Controversy

Commission on Life Sciences, National Research Council. 1988. Use of laboratory animals in biomedical and behavioral research. Washington, D.C., National Academy Press. *Statement of national policy on guidelines for the use of animals in biomedical research. Includes a chapter on the benefits derived from the use of animals.*

Goldberg, A. M., and J. M. Frazier. 1989. Alternatives to animals in toxicity testing. Sci. Am. **261**:24–30 (Aug.). *Describes alternatives that are being developed for the costly and time-consuming use of animals in the testing of thousands of chemicals that each year must be evaluated for potential toxicity to humans.*

Pringle, L. 1989. The animal rights controversy. San Diego, California, Harcourt Brace Jovanovich, Publishers. *Although no one writing about the animal rights movement can honestly claim to be totally objective and impartial on such an emotionally charged issue, this book comes as close as any to presenting a balanced treatment.*

Rowan, A. N. 1984. Of mice, models, and men: a critical evaluation of animal research. Albany, New York, State University of New York Press. *Good review of the issues. Chapter 7 deals with the use of animals in education, and notes that our educational system provides little help in resolving the contradiction of teaching kindness to animals while using animals in experimentation in biology classes.*

Sperling, S. 1988. Animal liberators: research and morality. Berkeley, University of California Press. *Thoughtful and carefully researched study of the animal rights movement, its ideological roots, and the passionate idealism of animal rights activists.*

A   **B**

Pre-Darwinian Evolutionary Ideas

Before the eighteenth century, speculation on the origin of species rested on myth and superstition, not on anything resembling a testable scientific theory. Creation myths viewed the world as a constant entity that did not change after its creation. Nevertheless, some thinkers approached the idea that nature has a long history of perpetual and irreversible change.

FIGURE 1-2

Founders of the theory of natural selection. **A,** Charles Robert Darwin (1809-1882), as he appeared in 1881, the year before his death. **B,** Alfred Russel Wallace (1823-1913) in 1895. Darwin and Wallace independently developed the same theory. A letter and essay from Wallace written to Darwin in 1858 spurred Darwin into writing *On the Origin of Species,* published in 1859.

Early Greek philosophers, notably Xenophanes, Empedocles, and Aristotle, developed a primitive idea of evolutionary change. They recognized fossils as evidence for a former life that they believed had been destroyed by natural catastrophe. Despite their spirit of intellectual inquiry, the Greeks failed to establish an evolutionary concept, and the issue declined well before the rise of Christianity. The opportunity for evolutionary thinking became even more restricted as the biblical account of the earth's creation became accepted as a tenet of faith. The year 4004 B.C. was fixed by Archbishop James Ussher (mid-seventeenth century) as the time of life's creation. Evolutionary views were considered rebellious and heretical. Still, some speculation continued. The French naturalist Georges Louis Buffon (1707–1788) stressed the influence of environment on the modifications of animal type. He also extended the age of the earth to 70,000 years.

FIGURE 1-3

Jean Baptiste de Lamarck (1744–1829), French naturalist who offered the first scientific explanation of evolution. Lamarck's hypothesis that evolution proceeds by the inheritance of acquired characteristics has been discredited.

FIGURE 1-4

Sir Charles Lyell (1797–1875), English geologist and friend of Darwin. His book *Principles of Geology* greatly influenced Darwin during Darwin's formative period. This photograph was made about 1856.

Lamarckism: The First Scientific Explanation of Evolution

The first complete explanation of evolution was authored by the French biologist Jean Baptiste de Lamarck (1744–1829) (Figure 1-3) in 1809, the year of Darwin's birth. He made the first convincing case for the idea that fossils were the remains of extinct animals. Lamarck's evolutionary mechanism, **inheritance of acquired characteristics,** was engagingly simple: organisms, by striving to meet the demands of their environments, acquire adaptations and pass them by heredity to their offspring. According to Lamarck, the giraffe evolved its long neck because its ancestors lengthened their necks by stretching to obtain food and then passed the lengthened neck to their offspring. Over many generations, these changes accumulated to produce the long neck of the modern giraffe.

We call Lamarck's concept of evolution *transformational,* because it claims that individual organisms transform their appearance to produce evolution. We now reject transformational theories because genetic studies show that traits acquired by an organism during its lifetime, such as strengthened muscles, are not inherited by offspring. Darwin's evolutionary theory differs from Lamarck's in being a *variational* theory. Evolutionary change is caused by differential survival and reproduction among organisms that differ in hereditary traits, not by inheritance of acquired characteristics.

Charles Lyell and Uniformitarianism

The geologist Sir Charles Lyell (1797–1875) (Figure 1-4) established in his *Principles of Geology* (1830–1833) the principle of **uniformitarianism.** Uniformitarianism encompasses two important principles that guide the scientific study of the history of nature. These principles are (1) that the laws of physics and chemistry remain the same throughout the history of the earth, and (2) that past geological events occurred by natural processes similar to those that we observe in action today. Lyell showed that natural forces, acting over long periods of time, could explain the formation of fossil-bearing rocks. Lyell's geological studies led him to conclude that the earth's age must be reckoned in millions of years. These principles were important for discrediting miraculous and supernatural explanations of the history of nature and replacing them with scientific explanations. Lyell also stressed the gradual nature of geological changes that occur through time, and he argued further that such changes have no inherent directionality. We will see that both of these claims left important marks on Darwin's evolutionary theory.

Darwin's Great Voyage of Discovery

"After having been twice driven back by heavy southwestern gales, Her Majesty's ship *Beagle,* a ten-gun brig, under the command of Captain Robert FitzRoy, R.N., sailed from Devonport on the 27th of December, 1831." Thus began Charles Darwin's account of the historic five-year voyage of the *Beagle* around the world (Figure 1-5). Darwin, not quite 23 years old, had been asked to accompany Captain FitzRoy on the *Beagle,* a small vessel only 90 feet in length, which was about to depart on an extensive surveying voyage to South America and

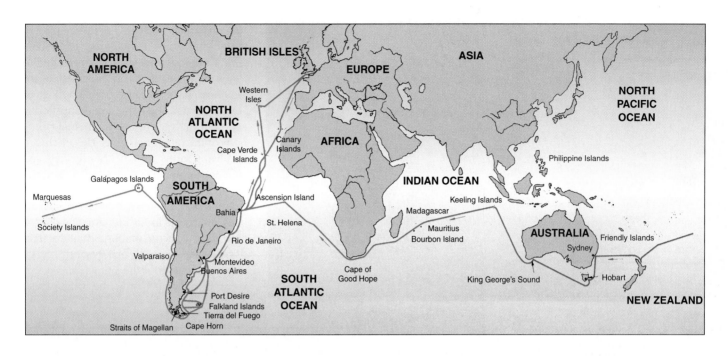

FIGURE 1-5

Five-year voyage of the H.M.S. *Beagle.*

A   **B**

FIGURE 1-6

Charles Darwin and the H.M.S. *Beagle.* **A,** Darwin in 1840, four years after the *Beagle* returned to England, and a year after his marriage to his cousin, Emma Wedgwood. **B,** The H.M.S. *Beagle* sails in Beagle Channel, Tierra del Fuego, on the southern tip of South America in 1833. This watercolor was painted by Conrad Martens, one of two official artists during the voyage of the *Beagle.*

the Pacific (Figure 1-6). It was the beginning of one of the most important voyages of the nineteenth century.

During the voyage (1831–1836), Darwin endured sea-sickness and the erratic companionship of the authoritarian Captain FitzRoy. But Darwin's youthful physical strength and early training as a naturalist equipped him for his work. The

Beagle made many stops along the harbors and coasts of South America and adjacent regions. Darwin made extensive collections and observations on the fauna and flora of these regions. He unearthed numerous fossils of animals long extinct and noted the resemblance between fossils of the South American pampas and the known fossils of North America. In the Andes

American mainland, yet differed from them in curious ways. Each island often contained a unique species that was related to forms on other islands. In short, Galápagos life must have originated in continental South America and then undergone modification in the various environmental conditions of the different islands. He concluded that living forms were neither divinely created nor immutable; they were, in fact, the products of evolution. Although Darwin devoted only a few pages to Galápagos animals and plants in his monumental *On the Origin of Species,* published more than two decades later, his observations on the unique character of the animals and plants were, in his own words, the "origin of all my views."

FIGURE 1-7

The Galápagos Islands viewed from the rim of a volcano.

> *"Whenever I have found that I have blundered, or that my work has been imperfect, and when I have been contemptuously criticized, and even when I have been overpraised, so that I have felt mortified, it has been my greatest comfort to say hundreds of times to myself that 'I have worked as hard and as well as I could, and no man can do more than this.'"*
> —Charles Darwin, in his autobiography, 1876.

he encountered seashells embedded in rocks at 13,000 feet. He experienced a severe earthquake and watched the mountain torrents that relentlessly wore away the earth. These observations strengthened his conviction that natural forces were responsible for the geological features of the earth.

In mid-September of 1835, the *Beagle* arrived at the Galápagos Islands, a volcanic archipelago straddling the equator 600 miles west of Ecuador (Figure 1-7). The fame of the islands stems from their infinite strangeness. They are unlike any other islands on Earth. Some visitors today are struck with awe and wonder, others with a sense of depression and dejection. Circled by capricious currents, surrounded by shores of twisted lava, bearing skeletal brushwood baked by the equatorial sun, almost devoid of vegetation, inhabited by strange reptiles and by convicts stranded by the Ecuadorian government, the islands indeed had few admirers among mariners. By the middle of the seventeenth century, the islands were already known to the Spaniards as "Las Islas Galápagos"—the tortoise islands. The giant tortoises, used for food first by buccaneers and later by American and British whalers, sealers, and ships of war, were the islands' principal attraction. At the time of Darwin's visit, the tortoises already were heavily exploited.

During the *Beagle*'s five-week visit to the Galápagos, Darwin began to develop his views of the evolution of life on Earth. His original observations of the giant tortoises, marine iguanas, mockingbirds, and ground finches, all contributed to the turning point in Darwin's thinking.

Darwin was struck by the fact that, although the Galápagos Islands and the Cape Verde Islands (visited earlier in this voyage of the *Beagle*) were similar in climate and topography, their fauna and flora were altogether different. He recognized that Galápagos plants and animals were related to those of the South

On October 2, 1836, the *Beagle* returned to England, where Darwin conducted the remainder of his scientific work (Figure 1-8). Most of Darwin's extensive collections had preceded him there, as had most of his notebooks and diaries kept during the cruise. Darwin's journal was published three years after the *Beagle*'s return to England. It was an instant success and required two additional printings within the first year. In later versions, Darwin made extensive changes and titled his book *The Voyage of the Beagle.* The fascinating account of his observations written in a simple, appealing style has made the book one of the most lasting and popular travel books of all time.

Curiously, the main product of Darwin's voyage, his theory of evolution, did not appear in print for more than 20 years after the *Beagle*'s return. In 1838, he "happened to read for amusement" an essay on populations by T. R. Malthus (1766–1834), who stated that animal and plant populations, including human populations, tend to increase beyond the capacity of the environment to support them. Darwin already had been gathering information on the artificial selection of animals under domestication by humans. After reading Malthus's article, Darwin realized that a process of selection in nature, a "struggle for existence" because of overpopulation, could be a powerful force for evolution of wild species.

He allowed the idea to develop in his own mind until it was presented in 1844 in a still-unpublished essay. Finally in 1856 he began to assemble his voluminous data into a work on

FIGURE 1-8

Darwin's study at Down House in Kent, England, is preserved today much as it was when Darwin wrote *On the Origin of Species.*

the origin of species. He expected to write four volumes, a very big book, "as perfect as I can make it." However, his plans were to take an unexpected turn.

In 1858, he received a manuscript from Alfred Russel Wallace (1823–1913), an English naturalist in Malaya with whom he was corresponding. Darwin was stunned to find that in a few pages, Wallace summarized the main points of the natural selection theory on which Darwin had been working for two decades. Rather than withhold his own work in favor of Wallace as he was inclined to do, Darwin was persuaded by two close friends, the geologist Lyell and the botanist Hooker, to publish his views in a brief statement that would appear together with Wallace's paper in the *Journal of the Linnean Society.* Portions of both papers were read before an unimpressed audience on July 1, 1858.

For the next year, Darwin worked urgently to prepare an "abstract" of the planned four-volume work. This book was published in November 1859, with the title *On the Origin of Species by Means of Natural Selection, or the Preservation of Favoured Races in the Struggle for Life.* The 1250 copies of the first printing were sold the first day! The book instantly generated a storm that has never completely abated. Darwin's views were to have extraordinary consequences on scientific and religious beliefs and remain among the greatest intellectual achievements of all time.

Once Darwin's caution had been swept away by the publication of *On the Origin of Species,* he entered an incredibly productive period of evolutionary thinking for the next 23 years, producing book after book. He died on April 19, 1882, and was buried in Westminster Abbey. The little *Beagle* had

already disappeared, having been retired in 1870 and presumably broken up for scrap.

Darwin's Theory of Evolution

Darwin's theory of evolution is now over 130 years old. Biologists today are frequently asked, "What is Darwinism?" and "Do biologists still accept Darwin's theory of evolution?" These questions cannot be given simple answers because Darwinism encompasses several different, although mutually compatible, theories. Professor Ernst Mayr of Harvard University has argued that Darwinism should be viewed as five major theories. These five theories have somewhat different origins and different fates and cannot be discussed accurately as if they were only a single statement. The theories are (1) **perpetual change,** (2) **common descent,** (3) **multiplication of species,** (4) **gradualism,** and (5) **natural selection.** The first three theories are generally accepted as having universal application throughout the living world. The theories of gradualism and natural selection are controversial among evolutionists. Gradualism and natural selection are clearly part of the evolutionary process, but they might not be as pervasive as Darwin thought. Legitimate controversies regarding gradualism and natural selection often are misrepresented by creationists as challenges to the first three theories, whose validity is strongly supported by all relevant facts.

1. **Perpetual change.** This is the basic theory of evolution on which the others are based. It states that the living world is neither constant nor perpetually cycling, but is always changing. The properties of organisms undergo modification across generations throughout time. This theory originated in antiquity but did not gain widespread acceptance until Darwin advocated it in the context of his other four theories. "Perpetual change" is documented by the fossil record, which clearly refutes creationists' claims for a recent origin of all living forms. Because it has withstood repeated testing and is supported by an overwhelming number of observations, we now regard "perpetual change" as a scientific fact.

2. **Common descent.** The second Darwinian theory, "common descent," states that all forms of life descended from a common ancestor through a branching of lineages (Figure 1-9). The opposing argument, that the different forms of life arose

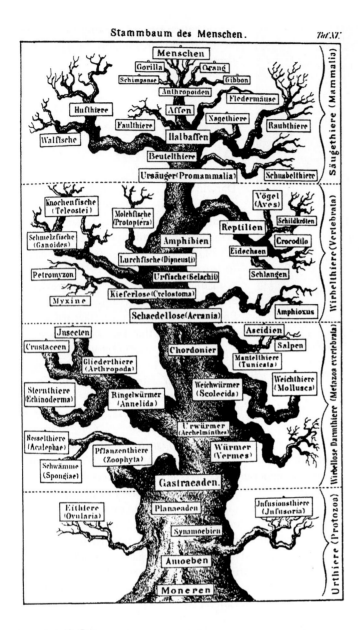

FIGURE 1-9

An early tree of life drawn in 1874 by the German biologist, Ernst Haeckel, who was strongly influenced by Darwin's theory of common descent. Many of the phylogenetic hypotheses shown in this tree, including the unilateral progression of evolution toward humans (= Menschen, *top*), subsequently have been refuted.

research is guided by Darwin's theory of common descent toward reconstructing life's phylogeny using the patterns of similarity and dissimilarity observed among species. The resulting phylogeny serves as the basis for our taxonomic classification of animals (Chapter 3).

3. **Multiplication of species.** Darwin's third theory states that the evolutionary process produces new species by the splitting and transformation of older ones. Species are now generally viewed as reproductively distinct populations of organisms that usually but not always differ from each other in organismal form. Once species are fully formed, interbreeding does not occur among members of different species. Evolutionists generally agree that the splitting and transformation of lineages produce new species, although much controversy remains concerning the details of this process and the precise meaning of the term "species" (Chapter 3). Biologists are actively studying the evolutionary processes that generate new species.

4. **Gradualism.** Darwin's theory of gradualism states that the large differences in anatomical traits that characterize different species originate by accumulation of many small incremental changes over very long periods of time. This theory opposes the notion that

Thomas Henry Huxley (1825–1895), one of England's greatest zoologists, on first reading the convincing evidence of natural selection in Darwin's On the Origin of Species *is said to have exclaimed, "How extremely stupid not to have thought of that!" He became Darwin's foremost advocate and engaged in often bitter debates with Darwin's critics. Darwin, who disliked publicly defending his own work, was glad to leave such encounters to his "bulldog," as Huxley called himself.*

independently and descended to the present in linear, unbranched genealogies, has been refuted by comparative studies of organismal form, cell structure, and macromolecular structures (including those of the genetic material, DNA). All of these studies confirm the theory that life's history has the structure of a branching evolutionary tree, known as a phylogeny. Species that share relatively recent common ancestry have more similar features at all levels than do species that have only an ancient common ancestry. Much current

Darwin's Explanatory Model of Evolution by Natural Selection

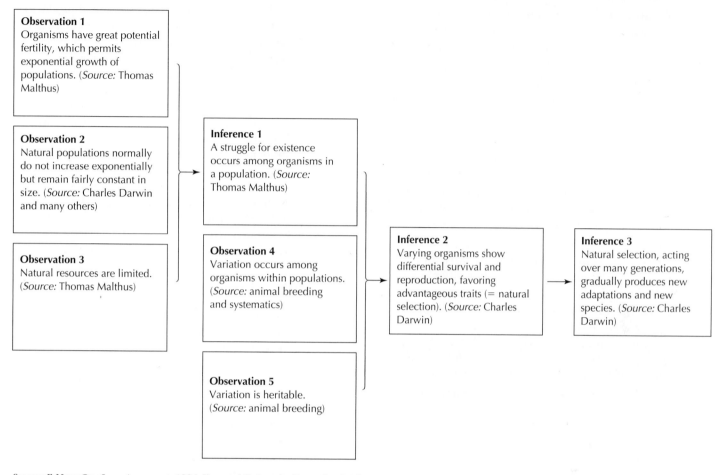

Observation 1
Organisms have great potential fertility, which permits exponential growth of populations. (*Source:* Thomas Malthus)

Observation 2
Natural populations normally do not increase exponentially but remain fairly constant in size. (*Source:* Charles Darwin and many others)

Observation 3
Natural resources are limited. (*Source:* Thomas Malthus)

Observation 4
Variation occurs among organisms within populations. (*Source:* animal breeding and systematics)

Observation 5
Variation is heritable. (*Source:* animal breeding)

Inference 1
A struggle for existence occurs among organisms in a population. (*Source:* Thomas Malthus)

Inference 2
Varying organisms show differential survival and reproduction, favoring advantageous traits (= natural selection). (*Source:* Charles Darwin)

Inference 3
Natural selection, acting over many generations, gradually produces new adaptations and new species. (*Source:* Charles Darwin)

Source: E. Mayr, One Long Argument, *1991, Harvard University Press, Cambridge, MA.*

large anatomical differences arise by sudden genetic changes. This theory is important because genetic changes having very large effects on organismal form are usually harmful to the organism. It is possible, however, that some genetic variants that have large effects on the organism are nonetheless sufficiently beneficial to be favored by natural selection. Therefore, although gradual evolution is known to occur, it may not explain the origin of all structural differences that we observe among species. Scientists are studying this question actively.

5. **Natural selection.** Natural selection explains why organisms are constructed to meet the demands of their environments, a phenomenon called **adaptation.** This theory describes a natural process by which populations accumulate favorable characteristics throughout long periods of evolutionary time. Adaptation was viewed previously as strong evidence against evolution. Darwin's theory of natural selection was therefore important for convincing people that a natural process, capable of being studied scientifically,

could produce new adaptations and new species. Demonstration that natural processes could produce adaptation was important to the eventual acceptance of all five Darwinian theories. Darwin developed his theory of natural selection as a series of five observations and three inferences from them:

Observation 1—Organisms have great potential fertility. All populations produce large numbers of gametes and potentially large numbers of offspring each generation. Population size would increase exponentially at an enormous rate if all individuals that were produced each generation survived and reproduced. Darwin calculated that, even in a slow-breeding species such as the elephant, a single pair breeding from age 30 to 90 and having only six young could produce 19 million descendants in 750 years.

Observation 2—Natural populations normally remain constant in size, except for minor fluctuations. Natural populations fluctuate in size across generations and sometimes go extinct, but no natural populations show the continued exponential growth that their reproductive capacity theoretically could sustain.

Observation 3—Natural resources are limited. Exponential growth of a natural population would require unlimited natural resources to provide food and habitat for the expanding population, but natural resources are finite.

Inference 1—There exists a continuing *struggle for existence* among members of a population. Survivors represent only a part, often a very small part, of the individuals produced each generation. Darwin wrote in *On the Origin of Species* that "it is the doctrine of Malthus applied with manifold force to the whole animal and vegetable kingdoms." The struggle for food, shelter, and space becomes increasingly severe as overpopulation develops.

Observation 4—All organisms show *variation.* No two individuals are exactly alike. They differ in size, color, physiology, behavior, and many other ways.

Observation 5—Variation is heritable. Darwin noted that offspring tend to resemble their parents, although he did not understand how. The hereditary mechanism discovered by Gregor Mendel would be applied to Darwin's theory many years later.

Inference 2—There is *differential survival and reproduction* among varying organisms in a population. Survival in the struggle for existence is not random with respect to hereditary variation present in the population. Some traits give their possessors an advantage in using the environment for effective survival and reproduction.

The popular phrase "survival of the fittest" was not originated by Darwin but was coined a few years earlier by the British philosopher Herbert Spencer, who anticipated some of Darwin's principles of evolution. Unfortunately the phrase later came to be coupled with unbridled aggression and violence in a bloody, competitive world. In fact, natural selection operates through many other characteristics of living things. The fittest animal may be the most helpful or the most caring. Fighting prowess is only one of several means toward successful reproductive advantage.

Inference 3—Over many generations, differential survival and reproduction generate new adaptations and new species. The differential reproduction of varying organisms gradually transforms species and results in the long-term "improvement" of types. Darwin

knew that people often use hereditary variation to produce useful new breeds of livestock and plants. *Natural* selection acting over millions of years should be even more effective in producing new types than the *artificial* selection imposed during a human lifetime. Natural selection acting independently on geographically separated populations would cause them to diverge from each other, thereby generating the reproductive barriers that lead to speciation.

Natural selection can be viewed as a two-step process with a random component and a nonrandom component. The production of variation among organisms is the random component. The mutational process has no inherent tendency to generate traits that are favorable to the organism; if anything, the reverse is probably true. The nonrandom component is the differential persistence of different traits, determined by the effectiveness of traits in permitting their possessors to use environmental resources to survive and to reproduce. The phenomenon of differential survival and reproduction among varying organisms is now called **sorting** and should not be equated with natural selection. We now know that even random processes (genetic drift, p. 28) can produce sorting. Darwin's theory of natural selection states that sorting occurs *because certain traits give their possessors advantages in survival and reproduction* relative to others that lack those traits. Selection is therefore a specific cause of sorting.

Evidence for Darwin's Five Theories of Evolution

Perpetual Change

Perpetual change in the form and diversity of animal life throughout its 600- to 700-million-year history is seen most directly in the fossil record. A **fossil** is a remnant of past life uncovered from the crust of the earth (Figure 1-10). Some fossils constitute complete remains (insects in amber and mammoths), actual hard parts (teeth and bones), or petrified skeletal parts that are infiltrated with silica or other minerals (ostracoderms and molluscs). Other fossils include molds, casts, impressions, and fossil excrement (coprolites). In addition to documenting organismal evolution, fossils reveal profound changes in the earth's environment, including major changes in the distributions of lands and seas. Because many organisms left no fossils, a complete record of the past is

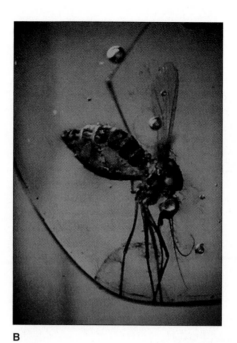

C

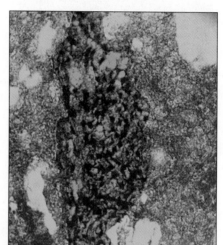

A

B

D

FIGURE 1-10

Four examples of fossil material. **A,** Stalked crinoids (sea lilies, class Crinoidea, phylum Echinodermata p. 248) from 85-million-year-old Cretaceous rocks. The fossil record of these echinoderms shows that they reached their peak millions of years earlier and began a slow decline to the present. **B,** The fossil of an insect that got stuck in the resin of a tree 40 million years ago and the resin hardened into amber. **C,** Fish fossil from rocks of the Green River Formation, Wyoming. Such fish swam here during the Eocene epoch of the Tertiary period, approximately 55 million years ago. **D,** Electron micrograph of tissue from a fly fossilized as shown in **B;** the nucleus of a cell is marked in red.

always beyond our reach; nonetheless, discovery of new fossils and reinterpretation of existing ones expand our knowledge of how the form and diversity of animals changed through geological time.

Fossil remains may on rare occasions include soft tissues preserved so well that recognizable cellular organelles can be viewed with the electron microscope! Insects are frequently found entombed in amber, the fossilized resin of trees. One study of a fly entombed in 40-million-year-old amber revealed structures corresponding to muscle fibers, nuclei, ribosomes, lipid droplets, endoplasmic reticulum, and mitochondria (Figure 1-10D). This extreme case of mummification probably occurred because chemicals in the plant sap diffused into the embalmed insect's tissues. The fictional extraction and cloning of DNA from embalmed insects that had bitten and then sucked the blood of dinosaurs was the technical basis of Michael Crichton's best-seller Jurassic Park.

Interpreting the Fossil Record

The fossil record is biased because preservation is selective. Vertebrate skeletal parts and invertebrates with shells and other hard structures left the best record (Figure 1-10). Soft-bodied animals, including the jellyfishes and most worms, are fossilized only under very unusual circumstances such as those that formed the Burgess Shale of British Columbia (Figure 1-11). Exceptionally favorable conditions for fossilization produced the Precambrian fossil bed of South Australia, the tar pits of Rancho La Brea (Hancock Park, Los Angeles), the great dinosaur beds (Alberta, Canada, and Jensen, Utah; Figure 1-12) and the Olduvai Gorge of Tanzania.

Fossils are deposited in stratified layers with new deposits forming on top of older ones. If left undisturbed, which is rare, a sequence is preserved with the ages of fossils being directly proportional to their depth in the stratified layers. Characteristic fossils often serve to identify particular layers. Certain widespread marine invertebrate fossils, including various foraminiferans (p. 78) and echinoderms (p. 248), are such good indicators of specific geological periods that they are called "index," or "guide," fossils. Unfortunately, the layers are usually tilted or

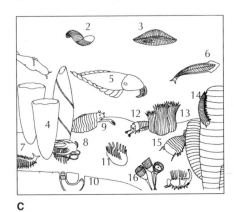

FIGURE 1-11

A, Fossil trilobites (p. 198) visible at the Burgess Shale Quarry, British Columbia. **B,** Animals of the Cambrian period, approximately 580 million years ago, as reconstructed from fossils preserved in the Burgess Shale of British Columbia, Canada. The main new body plans that appeared rather abruptly at this time established the body plans of animals familiar to us today. **C,** Key to Burgess Shale drawing. *Amiskwia* (1), from an extinct phylum; *Odontogriphus* (2), from an extinct phylum; *Eldonia* (3), a possible echinoderm (p. 248); *Halichondrites* (4), a sponge (p. 87); *Anomalocaris canadensis* (5), from an extinct phylum; *Pikaia* (6), an early chordate (p. 269); *Canadia* (7), a polychaete; *Marrella splendens* (8), a unique arthropod; *Opabinia* (9), from an extinct phylum; *Ottoia* (10), a priapulid (p. 142); *Wiwaxia* (11), from an extinct phylum; *Yohoia* (12), a unique arthropod; *Xianguangia* (13), an anemone-like animal; *Aysheaia* (14), an onychophoran (p. 240) or extinct phylum; *Sidneyia* (15), a unique arthropod (p. 196); *Dinomischus* (16), from an extinct phylum; *Hallucigenia* (17), from an extinct phylum.

folded or show faults (cracks). Old deposits exposed by erosion may be covered with new deposits in a different plane. When exposed to tremendous pressures or heat, stratified sedimentary rock metamorphoses into crystalline quartzite, slate, or marble, which destroys fossils.

Geological Time

Long before the earth's age was known, geologists divided its history into a table of succeeding events based on the ordered layers of sedimentary rock. The "law of stratigraphy" produced a relative dating with the oldest layers at the bottom and the youngest at the top of the sequence. Time was divided into eons, eras, periods, and epochs as shown on the endpaper inside the back cover of this book. Time during the last eon (Phanerozoic) is expressed in eras (for example, Cenozoic), periods (for example, Tertiary), epochs (for example, Paleocene), and sometimes smaller divisions of an epoch.

In the late 1940s, radiometric dating methods were developed for determining the absolute age in years of rock formations. Several independent methods are now used, all based on the radioactive decay of naturally occurring elements into other elements. These "radioactive clocks" are independent of pressure and temperature changes and therefore are not affected by often violent earth-building activities.

FIGURE 1-12

A hadrosaur skeleton from Dinosaur Provincial Park, Alberta, Canada.

The more well-known carbon-14 (^{14}C) dating method is of little help in estimating the age of geological formations because the short half-life of ^{14}C restricts its use to quite recent events (less than about 40,000 years). It is especially useful, however, for archaeological studies. This method is based on the production of radioactive ^{14}C (half-life of approximately 5570 years) in the upper atmosphere by bombardment of nitrogen-14 (^{14}N) with cosmic radiation. The radioactive ^{14}C enters the tissue of living animals and plants, and an equilibrium is established between atmospheric ^{14}C and the ^{14}N in the organism. At death, ^{14}C exchange with the atmosphere stops. In 5570 years, only half of the original ^{14}C remains in the preserved fossil. Its age is found by comparing the ^{14}C content of the fossil with that of living organisms.

One method, potassium-argon dating, depends on the decay of potassium-40 (^{40}K) to argon-40 (^{40}Ar) (12%) and calcium-40 (^{40}Ca) (88%). The half-life of potassium-40 is 1.3 billion years; half of the original atoms will decay in 1.3 billion years, and half of the remaining atoms will be gone at the end of the next 1.3 billion years. This decay continues until all radioactive potassium-40 atoms are gone. To measure the age of the rock, one calculates the ratio of remaining potassium-40 atoms to the amount of potassium-40 originally there (the remaining potassium-40 atoms plus the argon-40 and calcium-40 into which they have decayed). Several such isotopes exist for dating purposes, some for dating the age of the earth itself. One of the most useful radioactive clocks depends on the decay of uranium into lead. With this method, rocks over 2 billion years old can be dated with a probable error of less than 1%.

The fossil record of macroscopic organisms begins near the start of the Cambrian period of the Paleozoic era, approximately 600 million years BP. Geological time before the Cambrian is called the Precambrian era or Proterozoic eon. Although the Precambrian era occupies 85% of all geological time, it has received much less attention than later eras, partly because oil, which provides the commercial incentive for much geological work, seldom exists in Precambrian formations. The Precambrian era contains well-preserved fossils of bacteria and algae, and casts of jellyfishes, sponge spicules, soft corals, segmented flatworms, and worm trails. Most, but not all, are microscopic fossils.

Evolutionary Trends

The fossil record allows us to view evolutionary change across the broadest scale of time. Species arise and then become extinct repeatedly throughout the fossil record. Animal species typically survive approximately 1 million to 10 million years, although their duration is highly variable. When we study patterns of species or taxon replacement through time, we observe **trends.** Trends are directional changes in the characteristic features or patterns of diversity in a group of organisms. Fossil trends clearly demonstrate Darwin's principle of perpetual change.

A well-studied fossil trend is the evolution of horses from the Eocene epoch to the present (Figure 1-13). Looking back at the Eocene epoch, we see many different genera and species of horses that replaced each other through time (Figure 1-13). George Gaylord Simpson (p. 58) showed that this trend is compatible with Darwinian evolutionary theory. The three characteristics that show the clearest trends in horse evolution are body size, foot structure, and tooth structure. Compared to modern horses, the horses in extinct genera were small, their teeth had a relatively small grinding surface, and their feet had a relatively large number of toes (four). Throughout the subsequent Oligocene, Miocene, Pliocene, and Pleistocene epochs, there were continuing patterns of new genera arising and old ones becoming extinct. In each case, there was a net increase in body size, expansion of the grinding surface of the teeth, and reduction in the number of toes. As the number of toes was reduced, the central digit became increasingly more prominent in the foot, and eventually only this central digit remained.

The fossil record shows a net change not only in the characteristics of horses but also variation in the numbers of different horse genera (and numbers of species) that exist through time. The many horse genera of past epochs have been lost to extinction, leaving only a single survivor. Evolutionary trends in diversity are observed in fossils of many different groups of animals (Figure 1-14).

Trends in fossil diversity through time are produced by different rates of species formation versus extinction through time. Why do some lineages generate large numbers of new species whereas others generate relatively few? Why do different

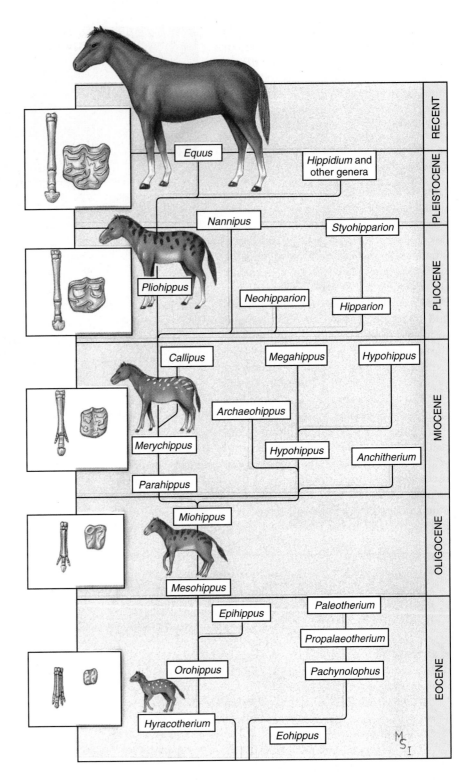

FIGURE 1-13

A reconstruction of the genera of horses from the Eocene to the present. Evolutionary trends toward increased size, elaboration of molars, and loss of toes are shown together with a hypothetical genealogy of extant and fossil genera.

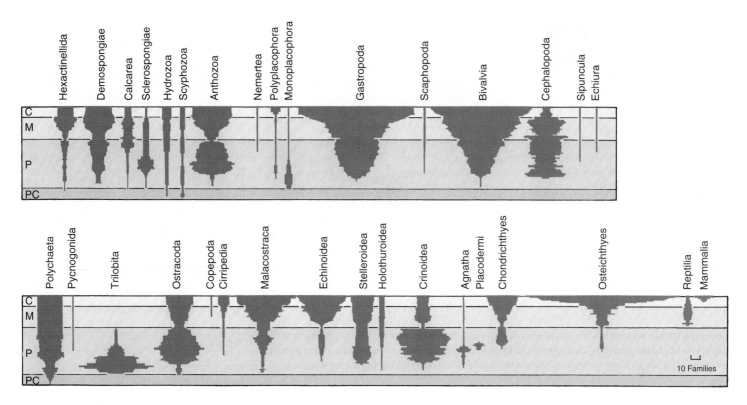

FIGURE 1-14

Diversity profiles of taxonomic families from different animal groups in the fossil record. The scale marks off the Precambrian (PC) and the Paleozoic (P), Mesozoic (M), and Cenozoic (C) eras. The number of families is indicated by the width of the profile.

lineages undergo higher or lower rates of extinction (of species, genera, or families) throughout evolutionary time? To answer these questions, we must turn to Darwin's other four theories of evolution. Regardless of how we answer these questions, however, the observed trends in animal diversity clearly illustrate Darwin's principle of perpetual change. Because the remaining four theories of Darwinism rely on the theory of perpetual change, evidence supporting these theories strengthens Darwin's theory of perpetual change.

Common Descent

Darwin proposed that all plants and animals have descended from "some one form into which life was first breathed." Life's history is depicted as a branching tree, called a phylogeny, that gives all of life a unified evolutionary history. Pre-Darwinian evolutionists, including Lamarck, advocated multiple independent origins of life, each of which gave rise to lineages that changed through time without extensive branching. Like all good scientific theories, common descent makes several important predictions that can be tested and potentially used

to reject it. According to this theory, we should be able to trace the genealogies of all modern species backward until they converge on ancestral lineages shared with other species, both living and extinct. We should be able to continue this process, moving farther backward through evolutionary time, until we reach the primordial ancestor of all life on earth. All forms of life, including many extinct forms that represent dead branches, will connect to this tree somewhere. Although reconstructing the history of life in this manner may seem almost impossible, it has in fact been extraordinarily successful. How has this difficult task been accomplished?

Homology and Reconstruction of Phylogeny

Darwin recognized the major source of evidence for common descent in the concept of **homology.** Darwin's contemporary, Richard Owen (1804–1892), used this term to denote "the same organ in different organisms under every variety of form and function." The classic example of homology is the limb skeleton of vertebrates. The bones of the vertebrate limb maintain characteristic structures and patterns of connection

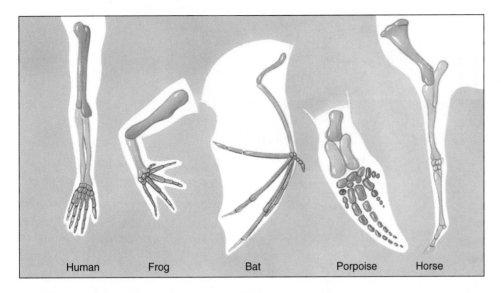

Human   Frog   Bat   Porpoise   Horse

FIGURE **1-15**

Forelimbs of five vertebrates show skeletal homologies: green, humerus; yellow, radius and ulna; purple, "hand" (carpals, metacarpals, and phalanges). Homologies of bones and patterns of connection are evident despite evolutionary modification for various particular functions.

FIGURE **1-16**

This 1873 advertisement for Merchant's Gargling Oil ridicules Darwin's theory of the common descent of humans and apes, which hardly received universal acceptance during Darwin's lifetime.

despite diverse modifications for different functions (Figure 1-15). According to Darwin's theory of common descent, the structures that we call homologies represent characteristics inherited with some modification from a corresponding feature in a common ancestor.

Darwin devoted an entire book, *The Descent of Man and Selection in Relation to Sex,* largely to the idea that humans share common descent with apes and other animals. This idea was repugnant to the Victorian world, which responded with predictable outrage (Figure 1-16). Darwin built his case mostly on anatomical comparisons revealing homology between humans and apes. To Darwin, the close resemblances between apes and humans could be explained only by common descent.

Throughout the history of all forms of life, evolutionary processes generate new characteristics that are transmitted across generations. Every time a new feature arises on a lineage destined to be ancestral to others, we see the origin of a new homology. The pattern formed by these homologies provides evidence for common descent and allows us to reconstruct the branching evolutionary history of life. We can illustrate such evidence using a phylogenetic tree of the ground-dwelling ratite birds (Figure 1-17). A new skeletal homology (see Figure 1-17) arises on each of the lineages shown (descriptions of the homologies are not included because they are highly technical). The different groups of species located at the tips of the branches contain different combinations of these homologies that reflect ancestry. For example, the ostriches show homologies 1 through 5 and 8, whereas the kiwis show homologies 1, 2, 13, and 15. The branches of the tree combine these species into a **nested hierarchy** of groups within groups (see Chapter 3). Smaller groups (species grouped near terminal branches) are contained within

larger ones (species grouped by basal branches, including the trunk of the tree). If we erase the tree structure but retain the patterns of homology observed in the terminal groups of species, we will be able to reconstruct the branching structure of the entire tree. Evolutionists test the theory of common descent by observing the patterns of homology present within all groups of organisms. The pattern formed by all homologies taken together should specify a single branching tree that represents the evolutionary genealogy of all living organisms.

The nested hierarchical structure of homology is so pervasive in the living world that it forms the basis for our systematic classification of all forms of life (genera grouped into families, families grouped into orders, and so on). Hierarchical classification even preceded Darwin's theory because this pattern is so evident, but it was not explained adequately before Darwin. Once the idea of common descent was accepted, biologists began investigating the structural, molecular, and/or chromosomal homologies of animal groups. Taken together, the nested hierarchical patterns uncovered by these studies have permitted us to reconstruct the evolutionary trees of many groups and to continue investigating others. The use of Darwin's theory of common descent to reconstruct the evolutionary history of life and to classify animals is the subject of Chapter 3.

Note that the earlier evolutionary hypothesis that life arose many times, forming unbranched lineages, predicts linear sequences of evolutionary change with no nested hierarchy of homologies among species. Because we do observe

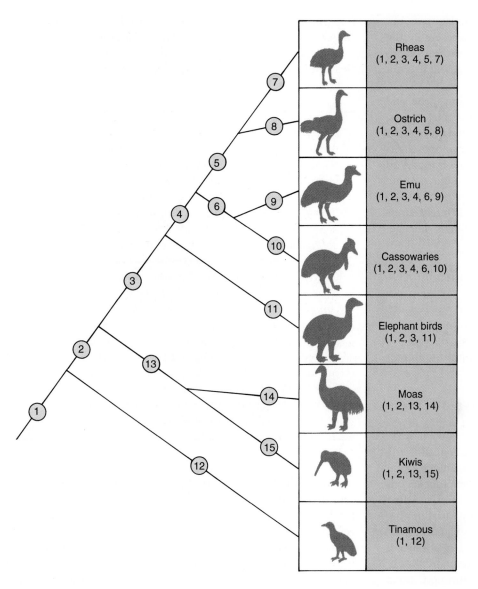

FIGURE **1-17**

The phylogenetic pattern specified by fifteen homologous structures in the skeletons of ratite birds. Homologous features are numbered one through fifteen and are marked both on the branches of the tree on which they arose and on the birds that have them. If you were to erase the tree structure, you would be able to reconstruct it without error from the distributions of homologous features shown for the birds at the terminal branches.

nested hierarchies of homologies, that hypothesis is rejected. Note also that the creationist argument, which is not a scientific hypothesis, can make no testable predictions about any pattern of homology.

Ontogeny, Phylogeny, and Recapitulation

Ontogeny is the history of the development of an organism through its entire life. Early developmental and embryological features contribute greatly to our knowledge of homology and common descent. Comparative studies of ontogeny show how the evolutionary alteration of developmental timing generates

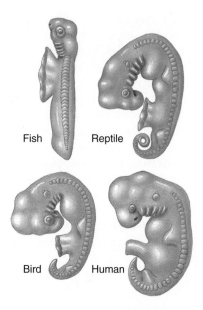

FIGURE **1-18**

Comparison of gill arches of different embryos. All are shown separated from the yolk sac. Note the remarkable similarity of the four embryos at this early stage in development.

new **phenotypes** (expressed characteristics or appearance of an organism), thereby causing evolutionary divergence among lineages.

The German zoologist Ernst Haeckel, a contemporary of Darwin, proposed that each successive stage in the development of an individual represented one of the adult forms present in its evolutionary history. For example, the human embryo with gill depressions in the neck was considered to signify a fish-like ancestor. On this basis Haeckel gave his generalization: *ontogeny (individual development) recapitulates (repeats) phylogeny (evolutionary descent).* This notion later became known simply as **recapitulation** or the **biogenetic law.** Haeckel based his biogenetic law on the flawed premise that evolutionary change occurs by successively adding stages onto the end of an unaltered ancestral ontogeny, condensing the ancestral ontogeny into earlier developmental stages. This notion was based on Lamarck's concept of the inheritance of acquired characteristics (p. 6).

The nineteenth-century embryologist, K. E. von Baer, gave a more satisfactory explanation of the relationship between ontogeny and phylogeny. He argued that early developmental features were simply more widely shared among different animal groups than later ones. For example, Figure 1-18 shows early embryological similarities of organisms whose adult forms are very different. The adults of animals with relatively short and simple ontogenies often resemble pre-adult stages of other animals whose ontogeny is more elaborate, but the embryos of descendants do not necessarily resemble the adults of their ancestors. Even early development undergoes evolutionary divergence among groups, however, and it is not as stable as von Baer believed.

We now know many parallels between ontogeny and phylogeny, but the features of an ancestral ontogeny can be shifted either to earlier or later stages in descendant ontogenies. Evolutionary change in the timing of development is called **heterochrony,** a term initially used by Haeckel to denote exceptions to recapitulation. Because the lengthening or shortening of ontogeny can change different parts of the body independently, we often see a mosaic of different kinds of developmental evolutionary change in a single lineage. Therefore, cases in which an entire ontogeny recapitulates phylogeny are rare.

Despite the many changes that have occurred throughout the years in scientific thinking about the relationship between ontogeny and phylogeny, one important fact remains clear. The theory of common descent is strengthened enormously by the many homologies found among the various developmental stages of organisms belonging to different species.

Multiplication of Species

The multiplication of species through time is a logical corollary to Darwin's theory of common descent. A branch point on the evolutionary tree means that an ancestral species has split into two different ones. Darwin's theory postulates that the variation present within a species, especially variation that occurs between geographically separated populations, provides the material from which new species are produced. Because evolution is a branching process, the total number of species produced by evolution increases through time, although most of these species eventually become extinct. A major challenge for evolutionists is to discover the process by which an ancestral species "branches" to form two or more descendant species.

Before we explore the multiplication of species, we must decide what we mean by "species." As we will see in Chapter 3, no consensus exists regarding the definition of species. Most biologists would agree, however, that important criteria for recognizing species include (1) descent from a common ancestral population, (2) reproductive compatibility (ability to interbreed) within and reproductive incompatibility between species, and (3) maintenance within species of genotypic and phenotypic cohesion (lack of abrupt differences among populations in allelic frequencies [see the following text] and organismal appearance). The criterion of reproductive compatibility has received the greatest attention in studies of species formation, also called **speciation.**

The biological factors that prevent different species from interbreeding are called **reproductive barriers.** The primary problem of speciation is to discover how two initially compat-ible populations evolve reproductive barriers that cause them to become distinct, separately evolving lineages. How do populations diverge from each other in their reproductive properties while maintaining complete reproductive compatibility within each population?

Geographic barriers between populations are not the same thing as reproductive barriers. Geographic barriers refer to the spatial separation of two populations. They prevent gene exchange and are usually a precondition for speciation. Reproductive barriers are the result of evolution and refer to the various physical, physiological, ecological, and behavioral factors that prevent interbreeding between different species. Geographic barriers do not guarantee that reproductive barriers will evolve. Reproductive barriers are most likely to evolve under conditions that include small population size, the right combination of selective factors, and long periods of geographic isolation. One or both of a pair of geographically isolated populations may become extinct prior to evolution of reproductive barriers between them. Over the vast span of geological time, however, the conditions required for speciation have been repeated millions of times.

Reproductive barriers between populations usually evolve gradually. Evolution of reproductive barriers requires that diverging populations must be kept physically separate for long periods of time. If the diverging populations were reunited before reproductive barriers were completely formed, interbreeding would occur between the populations and they would merge. Speciation by gradual divergence in animals usually requires long periods of time, perhaps 10,000 to 100,000 years or more. Geographical isolation followed by gradual divergence is the most effective way for reproductive barriers to evolve, and many evolutionists consider geographical separation a prerequisite for branching speciation. Speciation that results from the evolution of reproductive barriers between geographically separated populations is known as **allopatric speciation,** or geographic speciation.

Evidence for allopatric ("in another land") speciation occurs in many forms, but perhaps the most convincing is the occurrence of geographically separated but adjoining, closely related populations that illustrate the gradual origin of reproductive barriers. Populations of the salamander, *Ensatina eschscholtzii,* in California are a particularly clear example (Figure 1-19). These populations show evolutionary divergence in color pattern and collectively form a geographic ring around California's central valley. Genetic exchange between differentiated, geographically adjoining populations is evident through the formation of hybrids and occasionally regions of extensive genetic exchange (called zones of introgression).

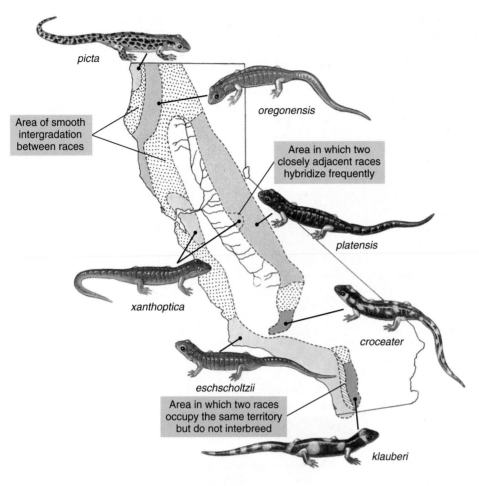

FIGURE 1-19

Speciation in progress: geographic variation of color patterns in the salamander genus *Ensatina*. Populations of *Ensatina eschscholtzii* form a geographic ring around the Central Valley of California. Adjacent, differentiated populations throughout the ring can exchange genes except at the bottom of the ring where the subspecies *E. e. eschscholtzii* and *E. e. klauberi* overlap without interbreeding. These two subspecies would be recognized as distinct species if the intermediate populations linking them across the ring were extinct. This example demonstrates that reproductive barriers between populations can evolve gradually.

the islands are effectively isolated geographically from their parental populations and can undergo divergent evolution, leading to reproductive barriers and speciation. Archipelagoes, such as the Galápagos Islands, greatly increase the opportunities for speciation in this manner.

The production of many ecologically diverse species from a common ancestral stock is called **adaptive radiation.** The Galápagos finches clearly illustrate adaptive radiation on an oceanic archipelago (Figures 1-20 and 1-21). The Galápagos finches (the name "Darwin's finches" was popularized in the 1940s by the British ornithologist David Lack) are close relatives, but each species differs from the others in the size and shape of the beak and in feeding habits. Darwin's finches descended from a single ancestral population that arrived from the mainland and subsequently colonized the different islands of the Galápagos archipelago. The finches underwent adaptive radiation, occupying habitats that on the mainland would have been denied to them by the presence of other species that are better able to exploit those habitats. The Galápagos finches thus assumed the characteristics of mainland families as diverse and unfinchlike as warblers and woodpeckers (Figure 1-21B). A fourteenth species of finch, found on isolated Cocos Island far north of the Galápagos archipelago, represents yet another speciation event in this impressive adaptive radiation.

Gradualism

Darwin's theory of gradualism opposed arguments for the sudden origin of species. Small differences, resembling those that we observe among organisms within populations today, are the raw material from which the different major forms of life evolved. This theory shares with Lyell's uniformitarianism the notion that we must not explain past changes by invoking unusual catastrophic events that are not observable today. If new species originated in single, catastrophic events, we should be able to see these events happening today and we do not. What we observe instead are small, continuous changes in the phenotypes that are present in natural populations. Such continuous changes can produce major differences among species only by accumulating over many thousands to millions of years. A simple statement of Darwin's theory of gradualism is that accumulation of quantitative changes leads to qualitative change.

Two populations at the southern tip of the geographical range (called *E. e. eschscholtzii* and *E. e. klauberi*) make contact but do not interbreed. A gradual accumulation of reproductive differences among contiguous populations around the ring is visible, with the two southernmost populations being separated by strong reproductive barriers.

Additional evidence for allopatric speciation comes from the observation of animal diversification on islands. Oceanic islands that were formed by volcanoes are initially devoid of life. They are gradually colonized by plants and animals from a continent or from other islands in separate invasions. The invaders often encounter situations ideal for evolutionary diversification, because environmental resources that were exploited heavily by other species on the mainland are free for colonization on the sparsely populated island. Because colonization of oceanic islands is rare, populations established on

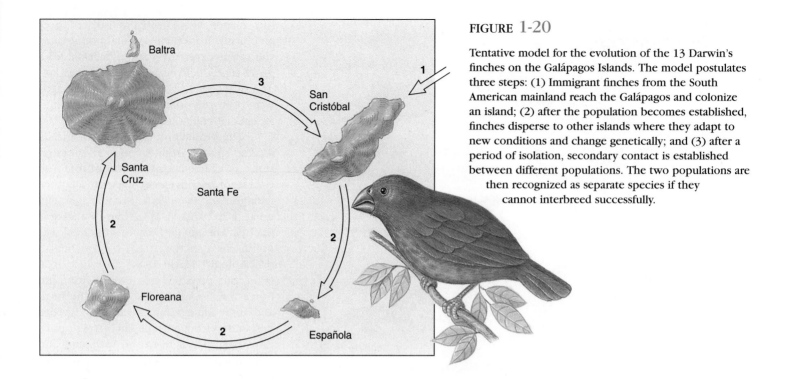

FIGURE 1-20

Tentative model for the evolution of the 13 Darwin's finches on the Galápagos Islands. The model postulates three steps: (1) Immigrant finches from the South American mainland reach the Galápagos and colonize an island; (2) after the population becomes established, finches disperse to other islands where they adapt to new conditions and change genetically; and (3) after a period of isolation, secondary contact is established between different populations. The two populations are then recognized as separate species if they cannot interbreed successfully.

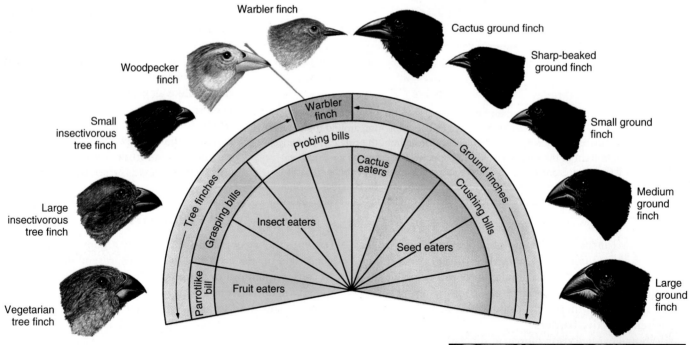

A

FIGURE 1-21

A, Adaptive radiation in ten species of Darwin's finches from Santa Cruz, one of the Galápagos Islands. Differences in bills and feeding habits are shown. All apparently descended from a single common ancestral finch from the South American continent. **B,** Woodpecker finch, one of the 13 species of Galápagos Islands finches, using a slender twig as a tool for feeding. This finch worked for about 15 minutes before spearing and removing a wood roach from a break in the tree.

FIGURE 1-22

The ancon breed of sheep arose from a "sporting mutation" that caused dwarfing of the legs. Many of his contemporaries criticized Darwin for his claim that such mutations are not important in the process of evolution by natural selection.

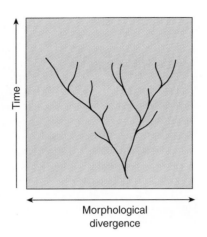

FIGURE 1-23

The gradualist model of evolutionary change in morphology, viewed as proceeding more or less steadily through geological time (millions of years). Bifurcations followed by gradual divergence led to speciation.

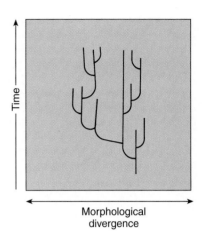

FIGURE 1-24

The punctuated equilibrium model sees evolutionary change being concentrated in relatively rapid bursts of branching speciation (*lateral lines*) followed by prolonged periods of little change throughout geological time (millions of years).

Phenotypic gradualism was controversial when Darwin first proposed it, and it is still controversial. Not all phenotypic changes are small, incremental ones. Some mutations that appear during artificial breeding, traditionally called 'sports,' change the phenotype substantially in a single mutational step. Sports that produce dwarfing are observed in many species, including humans, dogs, and sheep, and have been used by animal breeders to achieve desired results; for example, a sport that deforms the limbs was used to produce ancon sheep, which cannot jump hedges and are therefore easily contained (Figure 1-22). Many colleagues of Darwin who accepted his other theories considered phenotypic gradualism too extreme. If sporting mutations can be used in animal breeding, why must we exclude them from our evolutionary theory? In favor of gradualism, some have replied that sporting mutations always have negative side effects that would prevent them from surviving in natural populations. Indeed, it is questionable whether the ancon sheep, despite their attractiveness to farmers, would propagate successfully in the presence of their long-legged relatives without human intervention.

When we view Darwinian gradualism on a geological timescale, we may expect to find in the fossil record a long series of intermediate forms connecting the phenotypes of ancestral and descendant populations (Figure 1-23). This predicted pattern is called **phyletic gradualism.** Darwin recognized that phyletic gradualism is not often revealed by the fossil record. Studies conducted since Darwin's time likewise have failed to produce the continuous series of fossils predicted by phyletic gradualism. Is the theory of gradualism therefore refuted by the fossil record? Darwin and others claim that it is not, because the fossil record is too imperfect to preserve transitional series. Although evolution is a slow process by our standards, it is rapid relative to the rate at which good fossil deposits accumulate. Others have argued, however, that the abrupt origins and extinctions of species in the fossil record force us to conclude that phyletic gradualism is rare.

Niles Eldredge and Stephen Jay Gould proposed **punctuated equilibrium** in 1972 to explain the discontinuous evolutionary changes observed throughout geological time. Punctuated equilibrium states that phenotypic evolution is concentrated in relatively brief events of branching speciation, followed by much longer intervals of evolutionary stasis (Figure 1-24). Speciation is an episodic event, occurring over a period of approximately 10,000 to 100,000 years. Because species may survive for 5 million to 10 million years, the speciation event is a "geological instant," representing 1% or less of a species' existence. Ten thousand years is plenty of time, however, for Darwinian evolution to accomplish dramatic changes. A small fraction of the total evolutionary history of a group therefore accounts for most of the morphological evolutionary change that we observe.

Evolutionists who lamented the imperfect state of the fossil record were treated in 1981 to the opening of an uncensored page of fossil history in Africa. Peter Williamson, a British paleontologist working in fossil beds 400 m deep near Lake Turkana, documented a remarkably clear record of speciation in freshwater snails. The geology of the Lake Turkana basin reveals a history of instability. Earthquakes, volcanic eruptions, and climatic changes caused the waters to rise and fall periodically, sometimes by hundreds of feet. Thirteen lineages of snails show long periods of stability interrupted by relatively brief periods of rapid change in shell shape when

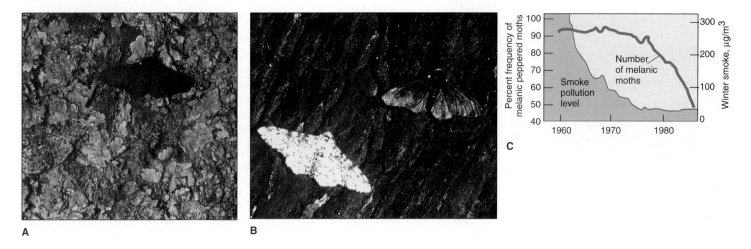

A   **B**   **C**

FIGURE 1-25

Light and melanic forms of the peppered moth, *Biston betularia* on, **A,** a lichen-covered tree in unpolluted countryside and, **B,** a soot-covered tree near industrial Birmingham, England. These color variants have a simple genetic basis. **C,** Recent decline in the frequency of the melanic form of the peppered moth with falling air pollution in industrial areas of England. The frequency of the melanic form still exceeded 90% in 1960, when smoke and sulfur dioxide emissions were still high. Later, as emissions fell and light-colored lichens began to grow again on the tree trunks, the melanic form became more conspicuous to predators. By 1986, only 50% of the moths were still of the melanic form, the rest having been replaced by the light form.

snail populations were fragmented by receding waters. These populations diverged to produce new species that then remained unchanged through thick deposits before becoming extinct and being replaced by descendant species. The transitions occurred within 5000 to 50,000 years. In the few meters of sediment where speciation occurred, transitional forms were visible. Williamson's study conforms well to the punctuated equilibrium model, which remains an important challenge to gradualism on a geological timescale.

Natural Selection

Many examples show how natural selection alters populations in nature. Sometimes selection can proceed very rapidly as, for example, in the evolution of high resistance to insecticides by insects, especially flies and mosquitoes. Doses that at first killed almost all pests later were ineffective in controlling them. As more insects were exposed to insecticides, those most sensitive were killed, leaving more space and less competition for resistant strains to multiply. Thus, as a result of selection, mutations bestowing high resistance, but previously rare in the population, increased in frequency.

Perhaps the most famous instance of rapid selection is that of industrial melanism (dark pigmentation) in the peppered moth of England (Figure 1-25). Before 1850, the peppered moth was white with black speckling in the wings and body. In 1849, a mutant black form of the species appeared. It became increasingly common, reaching frequencies of 98% in Manchester and other heavily industrialized areas by 1900. The peppered moth, like most moths, is active at night. It rests during the day in exposed places, depending upon its cryptic col-

oration for protection. Experimental studies have shown that, consistent with the hypothesis of natural selection, birds are able to locate and to eat moths that do not match their surroundings, but that birds in the same area frequently fail to find moths that match their surroundings. The mottled pattern of the normal white form blends well with lichen-covered tree trunks. With increasing industrialization, the soot from thousands of chimneys darkened the bark of trees for miles around centers such as Manchester. Against this dark background, the white moth was conspicuous to predatory birds, whereas the mutant black form was camouflaged. When pollution was diminished, the frequency of lightly pigmented individuals in moth populations increased (Figure 1-25C), consistent with the hypothesis of natural selection.

A recurring criticism of natural selection is that it cannot generate new structures or species but can only modify old ones. Most structures in their early evolutionary stages could not have performed the functions that the fully formed structures perform, and therefore it is unclear how natural selection could have favored them. What use is half a wing or the rudiment of a feather for a flying bird? To answer this criticism, evolutionists propose that many structures evolved initially for purposes different from the ones that they have today. Rudimentary feathers could have been useful in thermoregulation, for example. Their role in flying would have evolved later after they incidentally had acquired some aerodynamic properties that permitted them to be selected for improvement of flying. Because the anatomical differences observed among organisms from different, closely related species resemble variation observed within species, it is unreasonable to propose that selection will never lead beyond the species boundary.

Revisions of Darwinian Evolutionary Theory

Neo-Darwinism

The most serious weakness in Darwin's argument was his failure to identify correctly the mechanism of inheritance. Darwin saw heredity as a blending phenomenon in which the characteristics of the parents melded together in the offspring. Darwin also invoked the Lamarckian hypothesis that an organism could alter its heredity through the use and disuse of body parts and through the direct influence of the environment. Darwin did not realize that hereditary factors could be discrete and nonblending, and that a new genetic variant therefore could persist unaltered from one generation to the next. The German biologist August Weismann (1834–1914) rejected Lamarckian inheritance by showing experimentally that modifications of an organism during its lifetime do not change its heredity, and he revised Darwinian evolutionary theory accordingly. We now use the term **neo-Darwinism** to denote Darwinian evolutionary theory as revised by Weismann. The genetic basis of neo-Darwinism eventually became what is now called the **chromosomal theory of inheritance,** a synthesis of Mendelian genetics and cytological studies of the segregation of chromosomes into gametes.

Gregor Mendel published his theories of inheritance in 1868, 14 years before Darwin died. Darwin presumably never read the work, although it was found in Darwin's extensive library after his death. Had Mendel written Darwin about his results, it is possible that Darwin would have modified his theory accordingly. However, it is also possible that Darwin could not have seen the importance of the hereditary mechanisms to the continuous variations and gradual changes that represent the hub of Darwinian evolution. It took other scientists many years to establish the relationship of genetics to Darwin's theory of natural selection.

Emergence of Modern Darwinism: The Synthetic Theory

In the 1930s, a new breed of geneticists began to reevaluate Darwinian evolutionary theory from a different perspective. These were population geneticists, scientists who studied variation in natural populations of animals and plants and who had a sound knowledge of statistics and mathematics. Gradually, a new comprehensive theory emerged that brought together population genetics, paleontology, biogeography, embryology, systematics, and animal behavior in a Darwinian framework.

Population geneticists study evolution as a change in the genetic composition of populations. With the establishment of population genetics, evolutionary biology became divided into two different subfields. **Microevolution** pertains to evolutionary changes in the frequencies of variant forms of genes within populations. **Macroevolution** refers to evolution on a grand scale, encompassing the origin of new organismal structures and designs, evolutionary trends, adaptive radiation, phylogenetic relationships of species, and mass extinction. Macroevolutionary research is based in systematics and the comparative method (p. 4). Following the evolutionary synthesis, both macroevolution and microevolution have operated firmly within the tradition of neo-Darwinism, and both have expanded Darwinian theory in important ways.

Microevolution: Genetic Variation and Change within Species

Microevolution is the study of genetic change occurring within natural populations. The variant forms of a single gene are called **alleles.** The occurrence of different alleles of a gene in a population is called **polymorphism.** All of the alleles of all genes possessed by members of a population collectively form the **gene pool.** The amount of polymorphism present in large populations is potentially enormous, because at observed mutation rates, many different alleles are expected for all genes.

Population geneticists study polymorphism by identifying the different allelic forms of a gene that are present in a population and then measuring their relative frequencies in the population. The relative frequency of a particular allele of a gene in a population is known as its **allelic frequency.** For example, in the human population, there are three different allelic forms of the gene encoding the ABO blood types: I^A, I^B, and i. Because each individual **genotype** contains two copies of this gene, the total number of copies present in the population is twice the number of individuals. What fraction of this total is represented by each of the three different allelic forms? In France, we find the following allelic frequencies: $I^A = 0.46$, $I^B = 0.14$, and $i = 0.40$. In Russia, the corresponding allelic frequencies differ ($I^A = 0.38$, $I^B = 0.28$, and $i = 0.34$), demonstrating microevolutionary divergence between these populations (Figure 1-26). Genetically, alleles I^A and I^B are dominant to i, but i is nearly as frequent as I^A and exceeds the frequency of I^B in both populations. Dominance describes the *phenotypic effect* of an allele in **heterozygous** individuals, not its relative abundance in a population of individuals. We will demonstrate that Mendelian inheritance and dominance do not alter allelic frequencies directly or produce evolutionary change in a population.

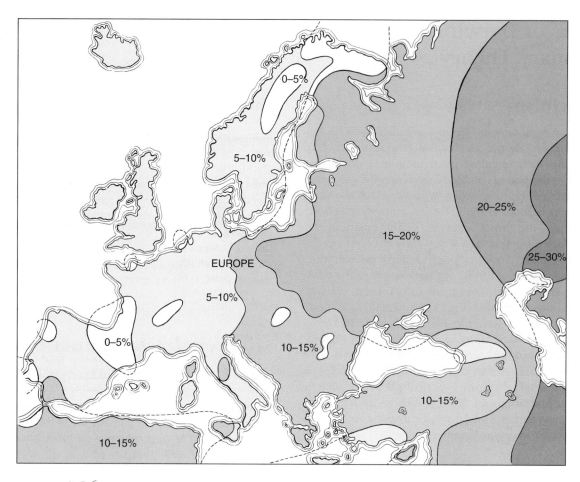

FIGURE 1-26

Frequencies of the blood-type B allele among humans in Europe. The allele is more common in the east and rarer in the west. The allele may have arisen in the east and gradually diffused westward through the genetic continuity of human populations. There is no known selective advantage of this allele, and its changing frequency probably represents the effects of random genetic drift.

Genetic Equilibrium

In many human populations, genetically recessive traits, including the O blood type, blond hair, and blue eyes, are very common. Why have not the genetically dominant alternatives gradually supplanted these recessive traits? It is a common misconception that a characteristic associated with a dominant allele increases in proportion because of its genetic dominance. This notion is not true because there is a tendency in *large* populations for allelic frequencies to remain in equilibrium generation after generation. This principle is based on the **Hardy-Weinberg equilibrium** (see box), which forms the foundation for population genetics. According to this theorem, the hereditary process alone does not produce evolutionary change. In large biparental populations, allelic frequencies and genotypic ratios attain an equilibrium in one generation and remain constant thereafter unless disturbed by recurring mutations, natural selection, migration, nonrandom mating, or genetic drift (random sorting). Such disturbances are the sources of microevolutionary change.

A rare allele, according to this principle, does not disappear from a large population merely because it is rare. That is why certain rare traits, such as albinism and cystic fibrosis, persist for endless generations. For example, albinism in humans is caused by a rare recessive allele a. Only one person in 20,000 is an albino, and this individual must be homozygous (a/a) for the recessive allele. Obviously the population contains many carriers, people with normal pigmentation who are heterozygous (A/a) for albinism. What is their frequency? A convenient way to calculate the frequencies of genotypes in a population is with the binomial expansion of $(p + q)^2$ (see box). We will let p represent the allelic frequency of A and q the allelic frequency of a.

Hardy-Weinberg Equilibrium: Why the Hereditary Process Does Not Change Allelic Frequencies

The Hardy-Weinberg law is a logical consequence of Mendel's first law of segregation and expresses the tendency toward equilibrium inherent in Mendelian heredity.

Let us select for our example a population having a single locus bearing just two alleles, *T* and *t*. The phenotypic expression of this gene might be, for example, the ability to taste a chemical compound called phenylthiocarbamide. Individuals in the population will be of three genotypes for this locus, *T/T*, *T/t* (both tasters), and *t/t* (nontasters). In a sample of 100 individuals, let us suppose we have determined that there are 20 of *T/T* genotype, 40 of *T/t* genotype, and 40 of *t/t* genotype. We could then set up a table showing the allelic frequencies as follows (remember that every individual's genotype has two copies of the gene):

Genotype	Number of Individuals	Copies of the *T* Allele	Copies of the *t* Allele
T/T	20	40	
T/t	40	40	40
t/t	40		80
TOTAL	100	80	120

Of the 200 copies, the proportion of the *T* allele is 80/200 = 0.4 (40%); and the proportion of the *t* allele is 120/200 = 0.6 (60%). It is customary in presenting this equilibrium to use *p* and *q* to represent the two allelic frequencies. The genetically dominant allele is represented by *p*, and the genetically recessive by *q*. Thus:

$$p = \text{frequency of } T = 0.4$$
$$q = \text{frequency of } t = 0.6$$
$$\text{Therefore } p + q = 1$$

Having calculated allelic frequencies in the sample, let us determine whether these frequencies will change spontaneously in a new generation of the population. Assuming that the mating is random (and this is important; all mating combinations of genotypes must be equally probable), each individual will contribute an equal number of gametes to the common pool from which the next generation is formed. The frequencies of gametes in the pool will be proportional to the allelic frequencies in the sample. That is, 40% of the gametes will be *T*, and 60% will be *t* (ratio of 0.4:0.6). Both ova and sperm will, of course, show the same frequencies. The next generation is formed as follows:

	Ova	
Sperm	*T* = 0.4	*t* = 0.6
T = 0.4	*T/T* = 0.16	*T/t* = 0.24
t = 0.6	*T/t* = 0.24	*t/t* = 0.36

Collecting the genotypes, we have:

$$\text{frequency of } T/T = 0.16$$
$$\text{frequency of } T/t = 0.48$$
$$\text{frequency of } t/t = 0.36$$

Next, we determine the values of *p* and *q* from the randomly mated populations. From the table, we see that the frequency of *T* will be the sum of genotypes *T/T*, which is 0.16, and one-half of the genotype *T/t*, which is 0.24:

$$T(p) = 0.16 + .5(0.48) = 0.4$$

Similarly, the frequency of *t* will be the sum of genotypes *t/t*, which is 0.36, and one-half the genotype *T/t*, which is 0.24:

$$t(p) = 0.36 + .5(0.48) = 0.6$$

The new generation bears exactly the same allelic frequencies as the parent population! Note that no increase has occurred in the frequency of the genetically dominant allele *T*. Thus *in a freely interbreeding, sexually reproducing population, the frequency of each allele would remain constant generation after generation in the absence of natural selection, migration, recurring mutation, and genetic drift* (see text). The more mathematically minded reader will recognize that the genotype frequencies *T/T*, *T/t*, and *t/t* are actually a binomial expansion of $(p + q)^2$:

$$(p + q)^2 = p^2 + 2pq + q^2 = 1$$

Assuming that mating is random (a questionable assumption, but one that we will accept for our example), the distribution of genotypic frequencies is $p^2 = A/A$, $2pq = A/a$, and $q^2 = a/a$. Only the frequency of genotype *a/a* is known with certainty, 1/20,000; therefore:

$$q^2 = \frac{1}{20,000}$$

$$q = \sqrt{\frac{1}{20,000}} = \frac{1}{141}$$

$$p = 1 - q$$

$$= \frac{141}{141} - \frac{1}{141} = \frac{140}{141}$$

The frequency of carriers is as follows:

$$A/a = 2pq = 2 \times \frac{140}{141} \times \frac{1}{141} = \frac{1}{70}$$

One person in every 70 is a carrier! Although a recessive trait may be rare, it is amazing how common a recessive allele may be in a population. There is a message here for anyone proposing to eliminate a "bad" recessive allele from a population by controlling reproduction. It is practically impossible. Because only the homozygous recessive individuals reveal the phenotype against which artificial selection could act (by sterilization, for example), the allele would continue to surface from heterozygous carriers. For a recessive allele present in 2 of every 100 persons (but homozygous in only 1 in 10,000 persons), 50 generations of complete selection against the homozygotes are required just to reduce its frequency to 1 in 100 persons.

> *Eugenics is the study that deals with the improvement of hereditary qualities in humans by social control of mating and reproduction. While the genetic argument against eugenics cited in the text is compelling in itself, a eugenics program lingered in the United States (and elsewhere) until finally dispatched following Adolph Hitler's attempt to "purify" races in Europe by genocide.*

Processes of Evolution: How Genetic Equilibrium Is Upset

Genetic equilibrium is disturbed in natural populations by (1) random genetic drift, (2) nonrandom mating, (3) recurring mutation, (4) migration, (5) natural selection, and interactions among these factors. Recurring mutation is the ultimate source of variability in all populations, but it usually requires interaction with one or more of the other factors to upset genetic equilibrium. We will look at these other factors individually.

Genetic Drift

Some species, like the cheetah (Figure 1-27), contain very little genetic variation, probably because their ancestral lineages sometimes were restricted to very small populations. A small population clearly cannot contain large amounts of genetic variation. Each individual organism has at most two different allelic forms of each gene, and a single breeding pair contains

FIGURE 1-27

The cheetah, a species whose genetic variability has been depleted to very low levels because of small population size in the past.

at most four different allelic forms of a gene. Suppose that we have such a breeding pair. We know from Mendelian genetics that chance decides which of the different allelic forms of a gene gets passed to offspring. It is therefore possible by chance alone that one or two of the parental alleles in this example will not be passed to any offspring. It is highly unlikely that the different alleles present in a small ancestral population are all passed to descendants without any change of allelic frequency. This chance fluctuation in allelic frequency from one generation to the next, including loss of alleles from the population, is called **genetic drift.**

Genetic drift occurs to some degree in all populations of finite size. Perfect constancy of allelic frequencies, as predicted by Hardy-Weinberg equilibrium, occurs only in infinitely large populations, and such populations occur only in mathematical models. All populations of animals are finite and therefore experience some effect of genetic drift, which becomes greater, on average, as population size declines. Genetic drift erodes the genetic variability of a population. If population size remains small for many generations in a row, genetic variation can be greatly depleted. This loss is harmful to the evolutionary success of a species because it restricts potential genetic responses to environmental change. Indeed, biologists are concerned that cheetah populations may have insufficient variation for continued survival.

> *During the 1960s, genetic drift was deemphasized as a factor of much importance in evolution because it would be strong enough to oppose the action of natural selection only in small populations. However, evolutionists now realize that most breeding populations of animals are small. A natural barrier, such as a stream or a ravine between two hilltops, may effectively separate a population into two separate and independent evolutionary units.*

Nonrandom Mating

If mating is nonrandom, genotypic frequencies will deviate from the Hardy-Weinberg expectations. For example, if two different alleles of a gene are equally frequent ($p = q = .5$), we expect half of the genotypes to be heterozygous ($2pq = 2 [.5] [.5] = .5$) and one-quarter to be homozygous for each of the respective alleles ($p^2 = q^2 = [.5]^2 = .25$). If we have **positive assortative mating,** individuals mate preferentially with others of the same genotype, such as albinos mating with other albinos. Matings among homozygous parents generate offspring that are homozygous like themselves. Matings among heterozygous parents produce on average 50% heterozygous offspring and 50% homozygous offspring (25% of each alternative type) each generation. Positive assortative mating increases the frequency of homozygous genotypes and decreases the frequency of heterozygous genotypes in the population but does not change allelic frequencies.

Preferential mating among close relatives also increases homozygosity and is called **inbreeding.** Whereas positive assortative mating usually affects one or a few traits, inbreeding simultaneously affects all variable traits. Strong inbreeding greatly increases the chances that rare recessive alleles will become homozygous and be expressed.

> *Inbreeding has surfaced as a serious problem in zoos holding small populations of rare mammals. Matings of close relatives tend to bring together genes from the same ancestor and increase the probability that two copies of a deleterious gene will come together in the same organism. The result is "inbreeding depression." Our management solution is to enlarge genetic diversity by bringing together captive animals from different zoos or by introducing new stock from the wild if possible. Paradoxically, where zoo populations are extremely small and no wild stock can be obtained, deliberate inbreeding is recommended. This procedure selects for genes that tolerate inbreeding; deleterious genes disappear because they kill the animals that bear them in the homozygous condition.*

Because inbreeding and genetic drift are both promoted by small population size, they are often confused with each other. Their effects are very different, however. Inbreeding alone cannot change allelic frequencies in the population, only the ways that alleles are combined into genotypes. Genetic drift changes allelic frequencies and consequently also changes genotypic frequencies. Even very large populations have the potential for being highly inbred if a behavioral preference for mating with close relatives exists, although this situation rarely occurs in nature. Genetic drift, however, will be relatively weak in very large populations.

Migration

Migration prevents different populations of a species from diverging. If a species is divided into many small populations, genetic drift and selection acting separately in the different populations can produce evolutionary divergence between them. A small amount of migration between populations each generation keeps the different populations from becoming too different. For example, the French and Russian populations whose ABO allele frequencies were discussed previously show some genetic divergence, but continuing migration between them prevents them from becoming completely distinct.

Natural Selection

Natural selection can change both allelic frequencies and genotypic frequencies in a population. Although the effects of selection are often reported for particular polymorphic genes, we must stress that natural selection acts on the whole animal, not on isolated traits. The organism that possesses the superior combination of traits will be favored. An animal may have some traits that confer no advantage or even a disadvantage, but it is successful overall if its combination of traits is favorable. When we claim that a genotype at a particular gene has a higher relative **fitness** than others, we state that on average that genotype confers an advantage in survival and reproduction in the population. If alternative genotypes have unequal probabilities of survival and reproduction, the Hardy-Weinberg equilibrium will be upset.

Some traits and combinations of traits are advantageous for certain aspects of the organism's survival or reproduction and disadvantageous for others. Darwin used the term **sexual selection** to denote the selection of traits that are advantageous for obtaining mates but may be harmful for survival. Bright colors and elaborate feathers may enhance a male bird's competitive ability in obtaining mates while simultaneously increasing his vulnerability to predators (Figure 1-28). Changes in the environment can alter the selective value of different traits. The action of selection on character variation is therefore very complex.

Selection is often studied using quantitative traits, those that show continuous variation with no obvious pattern of

FIGURE 1-28

A pair of wood ducks. Brightly colored feathers of male birds probably confer no survival advantage and might even be harmful by alerting predators. Such colors nonetheless confer advantage in attracting mates, which overcomes, on average, the negative consequences of these colors for survival. Darwin used the term "sexual selection" to denote traits that give an individual an advantage in attracting mates, even if the traits are neutral or harmful for survival.

Mendelian segregation in their inheritance. The values of the trait in offspring often are intermediate between the values in the parents. Such traits are influenced by variation at many genes, each of which follows Mendelian inheritance and contributes a small, incremental amount to the total phenotype. Traits that show quantitative variation include tail length in mice, length of a leg segment in grasshoppers, number of gill rakers in sunfishes, number of peas in pods, and height of adult humans. When the values are graphed with respect to frequency distribution, they often approximate a normal, bell-shaped curve (Figure 1-29A). Most individuals fall near the average, fewer fall well above or below the average, and the extremes form the "tails" of the frequency curve with increasing rarity.

Selection can act on quantitative traits to produce three different kinds of evolutionary response (see Figure 1-29B, C, and D). **Stabilizing selection** favors average values of the trait and disfavors extreme ones (Figure 1-29B). **Directional selection** favors an extreme value of the phenotype and causes the population average to shift toward it (Figure 1-29C). When we think about natural selection producing evolutionary change, it is usually directional selection that we have in mind, although we must remember that it is not the only possibility. A third alternative is **disruptive selection** in which two different extreme phenotypes are favored simultaneously, but the average is disfavored (Figure 1-29D). The population will become bimodal, meaning that two very different phenotypes will predominate.

Interactions of Selection, Drift, and Migration

Subdivision of a species into small populations that exchange migrants is an optimal one for promoting rapid adaptive evolution of the species as a whole. The interaction of genetic drift and selection in the different populations permits many different genetic combinations of many polymorphic genes to be tested against natural selection. The migration among the populations permits particularly favorable new genetic combinations to spread throughout the species as a whole. The interaction of selection, genetic drift, and migration in this example produces evolutionary change that is qualitatively different from what would result if any of these three factors acted alone. Natural selection, genetic drift, mutation, nonrandom mating, and migration interact in natural populations to create an enormous opportunity for evolutionary change; the perpetual stability predicted by Hardy-Weinberg equilibrium almost never lasts across any significant amount of evolutionary time.

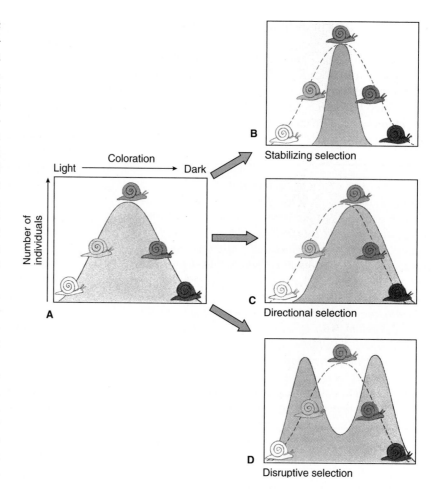

FIGURE 1-29

Responses to selection on a continuous (polygenic) character, coloration in a snail. **A,** The frequency distribution of coloration before selection. **B,** Stabilizing selection culls extreme variants from the population, in this case eliminating individuals that are unusually light or dark, thereby stabilizing the mean. **C,** Directional selection shifts the population mean, in this case by favoring darkly colored variants. **D,** Disruptive selection favors both extremes but not the mean; the mean is unchanged but the population no longer has a bell-shaped distribution of phenotypes.

Macroevolution: Major Evolutionary Events

Macroevolution describes large-scale events in organic evolution. The process of speciation links macroevolution and microevolution. The major trends in the fossil record described earlier (see Figures 1-13 and 1-14) fall clearly within the realm of macroevolution. The patterns and processes of macroevolutionary change emerge from those of microevolution, but they acquire some degree of autonomy in doing so. The emergence of new adaptations and species, and the varying rates of speciation and extinction observed in the fossil record go beyond the fluctuations of allelic frequencies within populations.

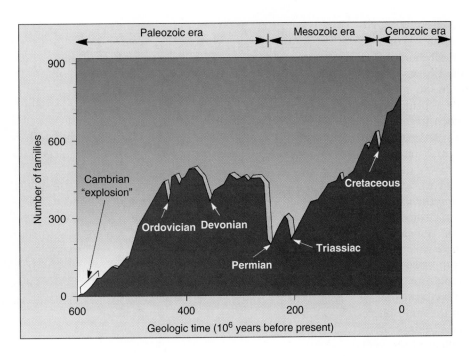

FIGURE 1-30

Changes in numbers of taxonomic families of marine animals through time from the Cambrian period to the present. Sharp drops represent five major extinctions of skeletonized marine animals. Note that despite the extinctions, the overall number of marine families has increased to the present.

Speciation and Extinction through Geological Time

Evolutionary change at the macroevolutionary level provides a new perspective on Darwin's theory of natural selection. A species has two possible evolutionary fates: it may give rise to new species or become extinct without leaving descendants. Rates of speciation and extinction vary among lineages, and the lineages that have the highest speciation rates and lowest extinction rates produce the greatest diversity of living forms. The characteristics of a species may make it more or less likely than others to undergo speciation or extinction events. Because many characteristics are passed from ancestral to descendant species (analogous to heredity at the organismal level), lineages whose properties enhance the probability of speciation and confer resistance to extinction should come to dominate the living world. This species-level process that produces differential rates of speciation and extinction among lineages is analogous in many ways to natural selection. It represents an expansion of Darwin's theory of natural selection.

Species selection is the differential survival and multiplication of species through geological time based on variation among lineages in species-level properties. These species-level properties include mating rituals, social structuring, migration patterns, and geographic distribution. Descendant species usually resemble their ancestors for these properties. For example, a "harem" system of mating in which a single male and several females compose a breeding unit characterizes some mammalian lineages but not others. We expect speciation rates to be enhanced by social systems that promote the founding of new populations by small numbers of individuals. Certain social systems may increase the likelihood that a species will survive environmental challenges through cooperative action. Such properties would be favored over geological time by species selection.

Mass Extinctions

When we study evolutionary change on an even larger timescale, we observe periodic events in which large numbers of taxa go extinct nearly simultaneously. These events are called **mass extinctions** (Figure 1-30). The most cataclysmic of these extinction episodes happened about 225 million years ago, when at least half of the families of shallow-water marine invertebrates, and fully 90% of marine invertebrate species disappeared within a few million years. This was the **Permian extinction.** The **Cretaceous extinction,** which occurred about 65 million years ago, marked the end of the dinosaurs, as well as numerous marine invertebrates and many small reptilian species.

The causes of mass extinctions and their occurrence at intervals of approximately 26 million years are difficult to explain. Some researchers have proposed biological explanations for these periodic mass extinctions and others consider them artifacts of our statistical and taxonomic analyses. Walter Alvarez proposed that the earth was periodically bombarded by asteroids, causing these mass extinctions (Figure 1-31). The drastic effects of such a bombardment of a planet were observed in July 1994 when fragments of Comet Shoemaker-Levy 9 bombarded Jupiter. The first fragment to hit Jupiter was estimated to have the force of 10 million hydrogen bombs. Twenty additional fragments hit Jupiter within the following week, one of which was 25 times more powerful than the first fragment. This bombardment was the most violent event in the recorded history of the solar system. Bombardments by asteroids or comets could change the earth's climate drastically, sending debris into the atmosphere and blocking sunlight. Temperature changes would have challenged the ecological tolerances of many species. This hypothesis is being tested in several ways, including a search for impact craters left by the asteroids and for altered mineral content of the rock strata where mass extinctions occurred. Unusual concentrations of the rare-earth element iridium in some strata imply that this element entered the earth's atmosphere through asteroid bombardment.

Sometimes, lineages favored by species selection are unusually susceptible to mass extinction. The climatic changes produced by the hypothesized asteroid bombardments could produce selective challenges very different from those encountered at other times in the earth's history. Selective discrimination of particular biological traits by events of mass extinction is termed **catastrophic species selection.** For example, mammals survived the end-Cretaceous mass extinction that destroyed the dinosaurs and other prominent vertebrate and invertebrate groups. Following this event, the mammals were able to use environmental resources that previously had been denied them, leading to their adaptive radiation.

Natural selection, species selection, and catastrophic species selection interact to produce the macroevolutionary trends that we see in the fossil record. The study of these interacting causal processes has made modern paleontology an active and exciting field.

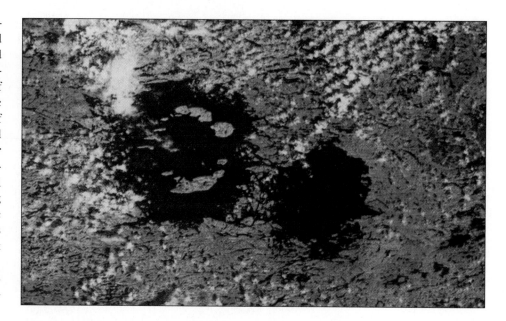

FIGURE 1-31

Twin craters of Clearwater Lakes in Canada show that multiple impacts on the earth are not as unlikely as they might seem. Evidence suggests that at least two impacts within a short time were responsible for the Cretaceous mass extinction.

Summary

Zoology is the scientific study of animals. Science is characterized by a particular approach to the acquisition of human knowledge. It is guided by, and is explanatory with reference to, natural law, and it is testable, tentative, and falsifiable. These criteria form the basis for the scientific method, which may be described as hypothetico-deductive in character. On the basis of prior observations, the scientist formulates an explanatory hypothesis. Predictions about future observations based on the hypothesis can support or falsify the hypothesis. A hypothesis for which there is a large amount of supporting data, particularly one that explains a very large number of observations, may be elevated to the status of a theory.

The experimental sciences seek to understand proximate or immediate causes in biological systems and use the experimental method. Tests of hypotheses are performed by means of experiments, which must include controls. The evolutionary sciences seek ultimate causes in biological systems by using the comparative method.

Organic evolution explains the diversity of living organisms as the historical outcome of gradual change from previously existing forms. Evolutionary theory is strongly identified with Charles Robert Darwin, who presented the first credible explanation for evolutionary change. Darwin derived much of the material used to construct his theory from his experiences on a five-year voyage around the world aboard the H.M.S. *Beagle.*

Darwin's evolutionary theory has five major components. Its most basic proposition is *perpetual change,* the theory that the world is neither constant nor perpetually cycling but is steadily undergoing irreversible change. The fossil record amply demonstrates perpetual change in the continuing fluctuation of animal form and diversity following the Cambrian explosion 600 million years ago. Darwin's theory of *common descent* states that all organisms descend from a common ancestor through a branching of genealogical lineages. This theory explains morphological homologies among organisms as characteristics inherited with modification from a corresponding feature in their common evolutionary ancestor. Patterns of homology formed by common descent with modification permit us to classify organisms according to their evolutionary relationships.

A corollary of common descent is the *multiplication of species* through evolutionary time. Allopatric speciation describes the evolution of reproductive barriers between geographically separated populations to generate new species. Adaptive radiation is the proliferation of many adaptively diverse species from a single ancestral lineage. Oceanic archipelagoes, such as the Galápagos Islands, are particularly conducive to the adaptive radiation of terrestrial organisms.

Darwin's theory of *gradualism* states that large phenotypic differences between species are produced by the accumulation through evolutionary time of many individually small changes. Gradualism is still controversial. Mutations that have large effects on the phenotype have been useful in animal breeding, leading some to dispute Darwin's claim that such mutations are not important in evolution. On a macroevolutionary perspective, punctuated equilibrium states that most evolutionary change occurs in relatively brief events of branching speciation, separated by long intervals in which little phenotypic change accumulates.

Darwin's fifth major statement is that *natural selection* is the guiding force of evolution. This principle is founded on observations that all species overproduce their kind, causing a struggle for the limited resources that support existence. Because no two organisms are exactly alike, and because variable traits are at least partially heritable, those whose hereditary endowment enhances their use of resources for survival and reproduction contribute disproportionately to the next generation. Over many generations, the sorting of variation by selection produces new species and new adaptations.

Population geneticists discovered the principles by which genetic properties of populations change through time. A particularly important discovery, known as Hardy-Weinberg equilibrium, showed that the hereditary process itself does not change the genetic composition of populations. The important sources of evolutionary change include mutation, genetic drift, nonrandom mating, migration, natural selection, and their interactions. Mutations are the ultimate source of all new variation on which selection acts.

Macroevolution comprises the study of evolutionary change on a geological timescale. Macroevolutionary studies measure rates of speciation, extinction, and changes of diversity through time. These studies have expanded Darwinian evolutionary theory to include higher-level processes that regulate rates of speciation and extinction among lineages, including species selection and catastrophic species selection.

Review Questions

1. What are the essential characteristics of science? Describe how evolutionary studies fit these characteristics whereas "scientific creationism" does not.

2. What is the relationship between a hypothesis and a theory?

3. Explain how biologists distinguish between experimental and evolutionary sciences.

4. Briefly summarize Lamarck's concept of the evolutionary process. What is wrong with this concept?

5. What is "uniformitarianism"? How did it influence Darwin's evolutionary theory?

6. Why was the *Beagle's* journey so important to Darwin's thinking?

7. What was the key idea contained in Malthus's essay on populations that was to help Darwin formulate his theory of natural selection?

8. Explain how each of the following contributes to Darwin's evolutionary theory: fossils, geographic distributions of closely related animals, homology, animal classification.

9. How do modern evolutionists view the relationship between ontogeny and phylogeny?

10. What is the main evolutionary lesson provided by Darwin's finches on the Galápagos Islands?

11. How is the observation of "sporting mutations" in animal breeding used to challenge Darwin's theory of gradualism? Why did Darwin reject such mutations as having little evolutionary importance?

12. What does the theory of punctuated equilibrium state about the occurrence of speciation throughout geological time?

13. Describe the observations and inferences that compose Darwin's theory of natural selection.

14. Identify the random and nonrandom components of Darwin's theory of natural selection.

15. Describe some recurring criticisms of Darwin's theory of natural selection. How can these criticisms be refuted?

16. It is a common but mistaken belief that because some alleles are genetically dominant and others are recessive, the dominants will eventually replace (drive out) all the recessives in a population. How does the Hardy-Weinberg equilibrium refute this notion?

17. Assume that you are sampling a trait in animal populations; the trait is controlled by a single allelic pair A and a, and you can distinguish all three phenotypes AA, Aa, and aa (intermediate inheritance). Your sample includes:

Population	AA	Aa	aa	TOTAL
I	300	500	200	1000
II	400	400	200	1000

Calculate the distribution of phenotypes in each population as expected under Hardy-Weinberg equilibrium. Is population I in equilibrium? Is population II in equilibrium?

18. If after studying a population for a trait determined by a single pair of alleles you find that the population is not in equilibrium, what possible reasons might explain the lack of equilibrium?

19. Explain why genetic drift is more powerful in small populations.

20. Is it easier for selection to remove a deleterious recessive allele from a randomly mating population or a highly inbred population? Why?

21. Distinguish between microevolution and macroevolution.

Selected References

See also general references on page 395.

Avise, J. C. 1994. Molecular markers, natural history and evolution. New York, Chapman & Hall. *An exciting and readable account of the evolutionary discoveries made using molecular studies, with particular attention to conservation.*

Buss, L. W. 1987. The evolution of individuality. Princeton, New Jersey, Princeton University Press. *An original and provocative thesis on the relationship between development and evolution, with examples drawn from many different animal phyla.*

Darwin, C. 1859. On the origin of species by means of natural selection, or the preservation of favoured races in the struggle for life. London, John Murray. *There were five subsequent editions by the author.*

Desmond, A., and J. Moore. 1991. Darwin. Warner Books, New York. *An interpretive biography of Charles Darwin.*

Freeman, S., and J. C. Herron. 1998. Evolutionary analysis. Upper Saddle River, New Jersey, Prentice-Hall. *An introductory textbook on evolutionary biology designed for undergraduate biology majors.*

Futuyma, D. J. 1998. Evolutionary biology, ed. 3. Sunderland, Massachusetts, Sinauer Associates. *A very thorough introductory textbook on evolution.*

Glen, W. 1994. The mass extinction debates: how science works in a crisis. Stanford, Stanford University Press. *A discussion of mass extinction presented in the form of a debate and panel discussion among concerned scientists.*

Gould, S. J. 1989. Wonderful life: the Burgess Shale and the nature of history. New York, W. W. Norton & Company. *An insightful discussion of what fossils tell us about the nature of life's evolutionary history.*

Hall, B. K. 1992. Evolutionary developmental biology. New York, Chapman & Hall. *A review of the interaction of genetics and development in evolving lineages, with particular attention to issues of heterochrony, homology, and developmental constraints on evolution.*

Hartl, D. L., and A. G. Clark. 1997. Principles of population genetics. Sunderland, Massachusetts, Sinauer Associates. *A current textbook on population genetics.*

Kitcher, P. 1982. Abusing science: the case against creationism. Cambridge, Massachusetts, MIT Press. *A treatise on how knowledge is gained in science and why creationism does not qualify as science.*

Kohn, D. 1985. The Darwinian heritage. Princeton, New Jersey, Princeton University Press. *An edited volume on Darwin and Darwinism with contributions from many leading historians of evolutionary theory.*

Mayr, E. 1988. Toward a new philosophy of biology. Cambridge, Massachusetts, Harvard University Press. *A collection of essays on many aspects of evolution by a leading evolutionary biologist.*

Raff, R. A. 1996. The shape of life: genes, development and the evolution of animal form. Chicago, Illinois, University of Chicago Press. *A provocative discussion of the genetic and developmental processes underlying evolution of animal diversity.*

Ruse, M. 1998. Philosophy of biology. Amherst, New York, Prometheus Books. *A collection of essays on evolutionary biology, including information on the Arkansas Balanced Treatment for Creation-Science and Evolution-Science Act.*

Somit, A., and S. A. Peterson. 1992. The dynamics of evolution: the punctuated equilibrium debate in the natural and social sciences. Ithaca, New York, Cornell University Press. *Numerous authors examine punctuated equilibrium and its implications for biological and social sciences.*

Links to the Internet

Visit this textbook's web site at http://www.mhhe.com/zoology to find live Internet links for each of the references listed below.

Science and Technology

1. New Scientist. Science and technology news is the focus of this web site. Internet Magazine called this the best web site in 1995. A must-see site for amateur and professional scientists.

Science and Biology

2. Biology. This site, maintained by California State University at Long Beach, provides a wide variety of biological images and activities. You can access numerous attractive photographs of animals and their environment. Click on Global Campus, Science, then Biology.

Evolution and Geological Time

3. Geological Time. This University of California Berkeley Museum of Paleontology site presents an excellent description of the geological timescale and the events that highlight major geological time periods.

4. Evolution and Behavior. This site provides answers to FAQs related to many topics on evolution. It provides book reviews, information on evolutionary scientists, and perspectives on evolution and religion.

5. Punctuated Equilibrium. This site provides a history of the observations and ideas that led to the punctuated equilibrium model of evolution, as well as a detailed description of the model.

Ecology, Biodiversity, and the Environment

6. Biodiversity and Biological Collections Web Server. This great site with lots of connections provides information on biodiversity studies from around the world.

7. Biodiversity. A fact sheet on biodiversity from the Ecological Society of America, with links to more information.

8. Virtual Library of Biodiversity, Ecology and the Environment. A clickable index, with lists of many endangered species, state issues, and legislation related to endangered species.

9. EE-Link. This site is a source of information on endangered species in the United States. It provides species lists sorted by region and taxonomic group, and information on natural history, laws, treaties, and periodicals.

10. Ecological Society of America. This site presents news, environmental policy updates, a newsletter, ESA press releases, membership information, and many links.

11. Endangered Species. The U.S. Fish and Wildlife Service maintains information on endangered species. This site provides answers to FAQs, information on the Endangered Species Act, and species lists.

12. National Wildlife Federation. This site, maintained by the National Wildlife Federation, provides an environmental hot line and presents information on pending legislation concerning environmental issues. It also provides activities and games for K–12 classrooms.

Careers in Biology and Zoology

13. Careers in Biology and Zoology. These sites contain information on careers in biology and zoology.

14. Careers in Medicine. Thinking of a career in health care? This terrific site is a must-see for anyone considering medicine. Much thought-provoking information, links, and lists of organizations with more information.

15. Careers in Health and Human Development. Another good site with information and links to more sites related to careers in the allied health professions.

Animal Architecture

CHAPTER two

New Designs for Living

Thirty-two phyla of multicellular animals are recognized by
zoologists today, each phylum characterized by a distinctive
body plan and array of biological properties that set it apart from
all other phyla. Nearly all are the survivors of perhaps 100 phyla
that were generated 600 million years ago during the Cambrian
explosion, the most important evolutionary event in the history
of animal life. Within the space of a few million years virtually all
of the major body plans that we see today, together with many
other novel plans that we know only from the fossil record, were
established. Entering a world sparse in species and mostly free of
competition, these new life forms began widespread
experimentation, producing new themes in animal architecture.
Nothing since then has equaled the Cambrian explosion. Later
bursts of speciation that followed major extinction events
produced only variations on established themes.

　　Once forged, a major body plan becomes a limiting
determinant of body form for descendants of that ancestral line.
Molluscs beget only molluscs and birds beget only birds, nothing
else. Despite the appearance of structural and functional
adaptations for distinctive ways of life, the evolution of new
forms always develops within the architectural constraints of the
phylum's ancestral pattern. This is why we shall never see
molluscs that fly or birds confined within a protective shell.

The English satirist Samuel Butler proclaimed that the human body was merely "a pair of pincers set over a bellows and a stewpan and the whole thing fixed upon stilts." While human attitudes toward the human body are distinctly ambivalent, most people less cynical than Butler would agree that the body is a triumph of intricate, living architecture. Less obvious, perhaps, is that humans and most other animals share an intrinsic material design and fundamental functional plan despite vast differences in structural complexity. This essential uniformity of biological organization derives from the common ancestry of animals and from their basic cellular construction. In this chapter, we will consider the limited number of body plans that underlie the diversity of animal form and examine some of the common architectural themes that animals share.

The Hierarchical Organization of Animal Complexity

Among the different metazoan groups, we can recognize five major grades of organization (Table 2-1). Each grade is more complex than the preceding one and builds upon it in a hierarchical manner.

TABLE 2-1
Levels of Organization in Organismal Complexity

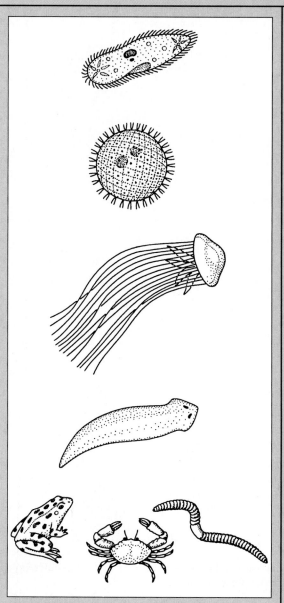

1. *Protoplasmic level of organization.* Protoplasmic organization is found in protozoa and other unicellular organisms. All life functions are confined within the boundaries of a single cell, the fundamental unit of life. Within the cell, the living substance is differentiated into organelles capable of carrying on specialized functions.

2. *Cellular level of organization.* Cellular organization is an aggregation of cells that are functionally differentiated. A division of labor is evident, so that some cells are concerned with, for example, reproduction, others with nutrition. Such cells have little tendency to become organized into tissues (a tissue is a group of similar cells organized to perform a common function). Some protozoan colonial forms that have distinct somatic and reproductive cells might be placed at the cellular level of organization. Many authorities also place sponges at this level.

3. *Cell-tissue level of organization.* A step beyond the preceding is the aggregation of similar cells into definite patterns or layers, thus becoming a **tissue.** Some authorities assign the sponges to this level, although the jellyfish and their relatives (Cnidaria) more clearly demonstrate the tissue plan. Both groups are still largely of the cellular grade of organization because most of the cells are scattered and not organized into tissues. An excellent example of a tissue in cnidarians is the **nerve net,** in which the nerve cells and their processes form a definite tissue structure, with the function of coordination.

4. *Tissue-organ level of organization.* The aggregation of tissues into organs is a further step in advancement. Organs are usually made up of more than one kind of tissue and have a more specialized function than tissues. The first appearance of this level is in the flatworms (Platyhelminthes), in which there are a number of well-defined organs such as eyespots, digestive tract, and reproductive organs. In fact, the reproductive organs are well organized into a reproductive system.

5. *Organ-system level of organization.* When organs work together to perform some function we have the highest level of organization—the organ system. The systems are associated with the basic body functions—circulation, respiration, digestion, and the others. The simplest animals that show this type of organization are the nemertean worms, which have a complete digestive system distinct from the circulatory system. Most animal phyla demonstrate this type of organization.

The unicellular protozoa (Protista) are the simplest animal-like organisms. These unicellular forms are nonetheless complete organisms that perform all of the basic functions of life as seen in the more complex animals. Within the confines of their cell, they show remarkable organization and division of labor, possessing distinct supportive structures, locomotor devices, fibrils, and simple sensory structures. The diversity observed among unicellular organisms is achieved by varying the architectural patterns of subcellular structures, organelles, and the cell as a whole (Chapter 4).

The **metazoa,** or multicellular animals, evolved greater structural complexity by combining cells into larger units. The metazoan cell is a specialized part of the whole organism and, unlike the protozoan cell, it is not capable of independent existence. The cells of a multicellular organism are specialized for performing the various tasks accomplished by subcellular elements in the protozoa. The simplest metazoans show the **cellular** grade of organization in which cells demonstrate division of labor but are not strongly associated to perform a specific collective function (Table 2-1). In the more complex **tissue** grade, similar cells are grouped together and perform their common functions as a highly coordinated unit. In animals of the tissue-organ grade of organization, tissues are assembled into still larger functional units called **organs.** Usually one type of tissue carries the burden of an organ's chief function, as muscle tissue does in the heart; other tissues—epithelial, connective, and nervous—perform supportive roles. The chief functional cells of an organ are called the parenchyma (pa-ren'ka-ma; Gr. *para,* beside, + *enchyma,* infusion). The supportive tissues are its stroma (Gr. bedding). For instance, in the vertebrate pancreas the secreting cells are the parenchyma; the capsule and connective tissue framework represent the stroma.

Most metazoa (flatworms and all more structurally complex phyla) have an additional level of complexity in which different organs operate together as **organ systems.** Eleven different kinds of organ systems are observed in metazoans: skeletal, muscular, integumentary, digestive, respiratory, circulatory, excretory, nervous, endocrine, immune, and reproductive. The great evolutionary diversity of these organ systems is covered in Chapters 4–19.

Complexity and Body Size

The most complex grades of metazoan organization permit and to some extent even promote the evolution of large body size (Figure 2-1). Large size confers several important physical and ecological consequences for the organism. As animals

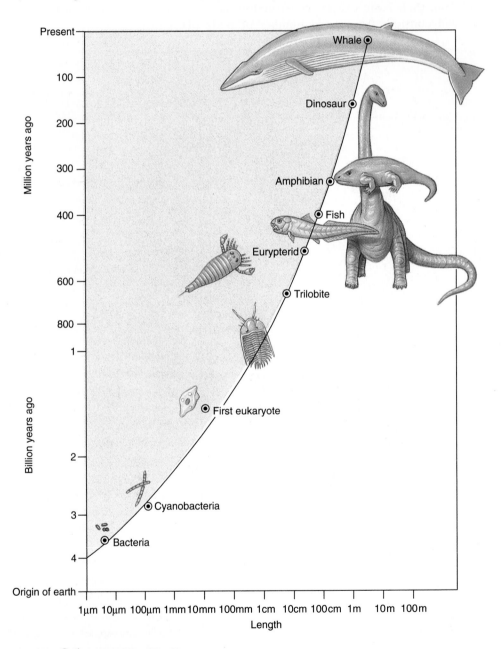

FIGURE 2-1

Graph showing the evolution of length increase in the largest organisms present at different periods of life on earth. Note that both scales are logarithmic.

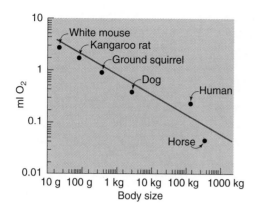

FIGURE *2-2*

Net cost of running for mammals of various sizes. Each point represents the cost (measured in rate of oxygen consumption) of moving 1 g of body over 1 km. The cost decreases with increasing body size.

become larger, the body surface increases much more slowly than body volume. This happens because surface area increases as the square of body length (length^2), whereas volume (and therefore mass) increases as the cube of body length (length^3). In other words, a large animal will have less surface area relative to its volume than will a small animal of the same shape. The surface area of a large animal may be inadequate for respiration and nutrition by cells located deep within the body. There are two possible solutions to this problem. One solution is to fold or invaginate the body surface to increase the surface area or, as exploited by the flatworms, flatten the body into a ribbon or disc so that no internal space is far from the surface. This solution allows the body to become large without internal complexity. However, most large animals adopted a second solution; they developed internal transport systems to shuttle nutrients, gases, and waste products between the cells and the external environment.

Larger size buffers the animal against environmental fluctuations; it provides greater protection against predation and enhances offensive tactics, and it permits a more efficient use of metabolic energy. A large mammal uses more oxygen than a small mammal, but the cost of maintaining its body temperature is less per gram of weight for the large mammal than for a small one. Large animals also can move themselves about at less energy cost than can small animals. A large mammal uses more oxygen in running than a small mammal, but the energy cost of moving 1 g of its body over a given distance is much less for a large mammal than for a small one (Figure 2-2). For all of these reasons, ecological opportunities of larger animals are very different from those of small ones. In subsequent chapters we will describe the extensive adaptive radiations observed in the taxa of large animals, covered in Chapters 7–19.

Extracellular Components of the Metazoan Body

In addition to the hierarchically arranged cellular structures discussed, the metazoan animal contains two important noncellular components: the body fluids and the extracellular structural elements. In all eumetazoans, the body fluids are subdivided into two fluid "compartments": those that occupy the **intracellular space,** within the body's cells, and those that occupy the **extracellular space,** outside the cells. In animals with closed vascular systems (such as segmented worms and vertebrates), the extracellular fluids are subdivided further into the **blood plasma** (the fluid portion of the blood outside the cells; blood cells are really part of the intracellular compartment) and **interstitial fluid.** The interstitial fluid, also called tissue fluid, occupies the space surrounding the cells. Many invertebrates have open blood systems, however, with no true separation of blood plasma from interstitial fluid.

The term intercellular, meaning "between cells," should not be confused with the term intracellular, meaning "within cells."

If we were to remove all the specialized cells and body fluids from the interior of the body, we would be left with the third element of the animal body: extracellular structural elements. This is the supportive material of the organism, including loose connective tissue (especially well developed in vertebrates but present in all metazoa), cartilage (molluscs and chordates), bone (vertebrates), and cuticle (arthropods, nematodes, annelids, and others). These elements provide mechanical stability and protection. In some instances, they act also as a depot of materials for exchange and serve as a medium for extracellular reactions. We will describe the diversity of extracellular skeletal elements characteristic of the different groups of animals in Chapters 4–6 and 9–19.

Types of Tissues

A **tissue** is a group of similar cells (together with associated cell products) specialized for the performance of a common function. The study of tissues is called **histology** (Gr. *histos,* tissue, + *logos,* discourse). All cells in metazoan animals take part in the formation of tissues. Sometimes the cells of a tissue may be of several kinds, and some tissues have a great many intercellular materials.

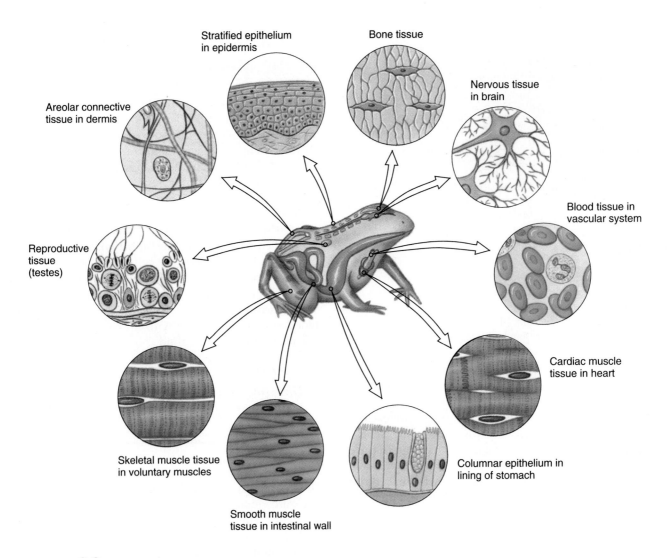

Stratified epithelium
in epidermis

Bone tissue

Nervous tissue
in brain

Areolar connective
tissue in dermis

Blood tissue in
vascular system

Reproductive
tissue
(testes)

Cardiac muscle
tissue in heart

Skeletal muscle tissue
in voluntary muscles

Columnar epithelium in
lining of stomach

Smooth muscle
tissue in intestinal wall

FIGURE 2-3

Types of tissues in a vertebrate, showing examples of where different tissues are located in a frog.

During embryonic development, the germ layers become differentiated into four kinds of tissues. These are epithelial, connective (including vascular), muscular, and nervous tissues (Figure 2-3). This is a surprisingly short list of basic tissue types that are able to meet the diverse requirements of animal life.

Epithelial Tissue

An **epithelium** (p1., epithelia) is a sheet of cells that covers an external or internal surface. On the outside of the body, the epithelium forms a protective covering. Inside, the epithelium lines all the organs of the body cavity, as well as ducts and passageways through which various materials and secretions move. On many surfaces the epithelial cells are often modified into glands that produce lubricating mucus or specialized products such as hormones or enzymes.

Epithelia are classified on the basis of cell form and number of cell layers. Simple epithelia (Figure 2-4) are found in all metazoan animals, while stratified epithelia (Figure 2-5) are mostly restricted to the vertebrates. All types of epithelia are supported by an underlying basement membrane, which is a condensation of the ground substance of connective tissue. Blood vessels never penetrate into epithelial tissues, so they are dependent on the diffusion of oxygen and nutrients from the underlying tissues.

Simple
squamous
epithelial   Basement   Free
cell   membrane   Nucleus   surface

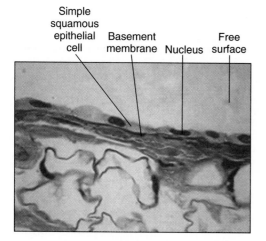

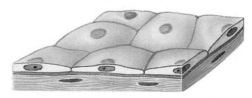

Simple squamous epithelium

Simple squamous epithelium, composed of flattened cells that form a continuous delicate lining of blood capillaries, lungs, and other surfaces where it permits the passive diffusion of gases and tissue fluids into and out of cavities.

Simple cuboidal   Basement   Lumen
epithelial cell   membrane   (free space)

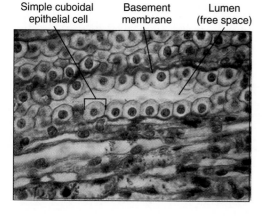

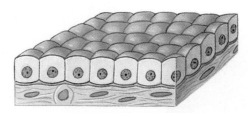

Simple cuboidal epithelium

Simple cuboidal epithelium is composed of short, boxlike cells. Cuboidal epithelium usually lines small ducts and tubules, such as those of the kidney and salivary glands, and may have active secretory or absorptive functions.

Basement   Epithelial   Microvilli on
membrane   cells   cell surface   Nuclei

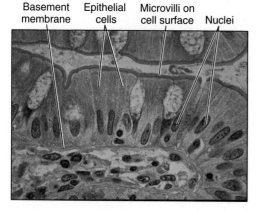

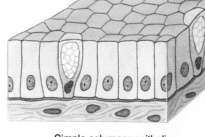

Simple columnar epithelium

Simple columnar epithelium resembles cuboidal epithelium but the cells are taller and usually have elongate nuclei. This type of epithelium is found on highly absorptive surfaces such as the intestinal tract of most animals. The cells often bear minute, fingerlike projections called microvilli that greatly increase the absorptive surface. In some organs, such as the female reproductive tract, the cells are ciliated.

FIGURE 2-4

Types of simple epithelium

Connective Tissue

Connective tissues are a diverse group of tissues that serve various binding and supportive functions. They are so widespread in the body that the removal of other tissues would still leave the complete form of the body clearly apparent.

Connective tissue is made up of relatively few **cells,** a great many extracellular **fibers,** and a fluid, known as ground substance (also called matrix), in which the fibers are embedded. We recognize several different types of connective tissue. **Connective tissue proper** in vertebrates includes both **loose connective tissue,** composed of fibers and fixed and

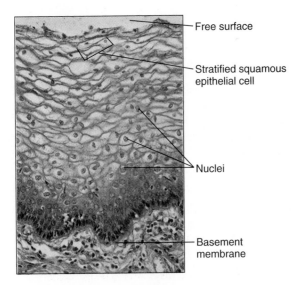

Free surface

Stratified squamous epithelial cell

Nuclei

Basement membrane

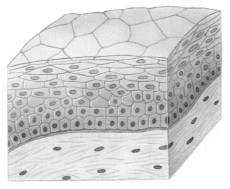

Stratified squamous epithelium

Stratified squamous epithelium consists of two to many layers of cells adapted to withstand mild mechanical abrasion. The basal layer of cells undergoes continuous mitotic divisions, producing cells that are pushed toward the surface where they are sloughed off and replaced by new cells beneath them. This type of epithelium lines the oral cavity, esophagus, and anal canal of many vertebrates, and the vagina of mammals.

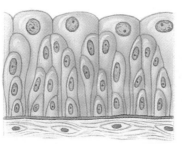

Transitional epithelium—unstretched

Transitional epithelium is a type of stratified epithelium specialized to accommodate great stretching. This type of epithelium is found in the urinary tract and bladder of vertebrates. In the relaxed state it appears to be four or five cell layers thick, but when stretched out it appears to have only two or three layers of extremely flattened cells.

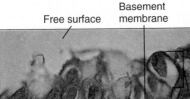

Basement membrane

Free surface

Connective tissue   Nucleus   Transitional epithelial cell

FIGURE 2-5

Types of stratified epithelium

Transitional epithelium—stretched

wandering cells suspended in a syrupy ground substance, and **dense connective tissue,** such as tendons and ligaments, composed largely of densely packed fibers (Figure 2-6). Much of the fibrous tissue of connective tissue is composed of **collagen** (Gr. *kolla,* glue, + *genos,* descent), a protein material of great tensile strength. Collagen is the most abundant protein in the animal kingdom, found in animal bodies wherever both flexibility and resistance to stretching are required. The connective tissue of invertebrates, as in vertebrates, consists of cells, fibers, and ground substance, but usually it is not as elaborately developed.

Other types of connective tissue include **blood, lymph,** and **tissue fluid** (collectively considered vascular tissue), composed of distinctive cells in a watery ground substance, the plasma. Vascular tissue lacks fibers under normal conditions. **Cartilage** is a semirigid form of connective tissue with closely packed fibers embedded in a gel-like ground substance (matrix). **Bone** is a calcified connective tissue containing calcium salts organized around collagen fibers.

Muscular Tissue

Muscle is the most common tissue in the body of most animals. It is made up of elongated cells often called fibers, specialized for contraction. It originates (with few exceptions) from the meso-

Nucleus   Collagen fiber   Elastic fiber

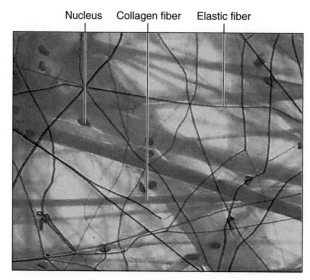

Nucleus   Fibers

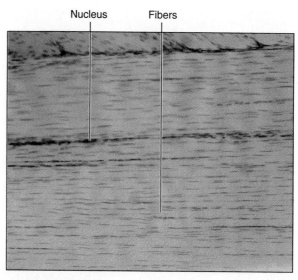

Loose connective tissue, also called areolar connective tissue, is the "packing material" of the body that anchors blood vessels, nerves, and body organs. It contains fibroblasts that synthesize the fibers and ground substance of connective tissue and wandering macrophages that phagocytize pathogens or damaged cells. The different fiber types include strong collagen fibers (thick and violet in micrograph) and elastic fibers (black and branching in micrograph) formed of the protein elastin. Adipose (fat) tissue is considered a type of loose connective tissue.

Dense connective tissue forms tendon, ligaments, and fasciae (fa'sha), the latter arranged as sheets or bands of tissue surrounding skeletal muscle. In tendon (shown here) the collagenous fibers are extremely long and tightly packed together.

Chondrocyte   Lacuna   Matrix

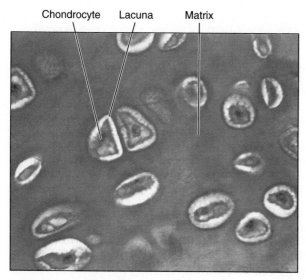

Central canal   Osteocytes in lacunae   Mineralized matrix

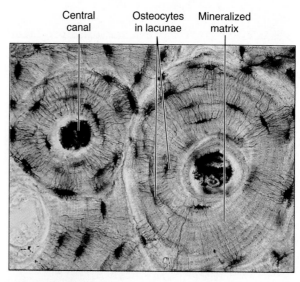

Cartilage is a vertebrate connective tissue composed of a firm gel ground substance (matrix) containing cells (chondrocytes) living in small pockets called lacunae, and collagen or elastic fibers (depending on the type of cartilage). In hyaline cartilage shown here, both collagen fibers and ground substance are stained uniformly purple, and cannot be distinguished one from the other. Because cartilage lacks a blood supply, all nutrients and waste materials must diffuse through the ground substance from surrounding tissues.

Bone, the strongest of vertebrate connective tissues, contains mineralized collagen fibers. Small pockets (lacunae) within the matrix contain bone cells, called osteocytes. The osteocytes communicate with blood vessels that penetrate into bone by means of a tiny network of channels called canaliculi. Unlike cartilage, bone undergoes extensive remodeling during an animal's life, and can repair itself following even extensive damage.

FIGURE 2-6

Types of connective tissue

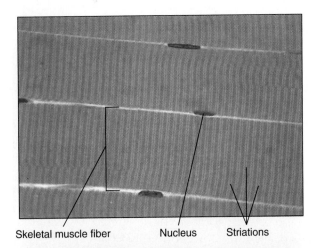

Skeletal muscle fiber   Nucleus   Striations

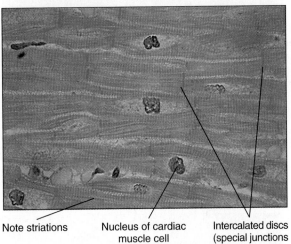

Note striations   Nucleus of cardiac muscle cell   Intercalated discs (special junctions between cells)

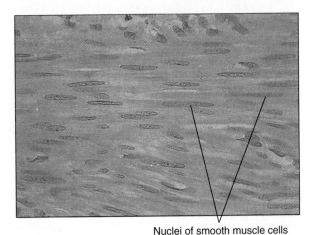

Nuclei of smooth muscle cells

FIGURE 2-7

Types of muscle tissue

Skeletal muscle is a type of striated muscle found in both invertebrates and vertebrates. It is composed of extremely long, cylindrical fibers, which are multinucleate cells that may reach from one end of the muscle to the other. Viewed through the light microscope, the cells appear to have a series of stripes, called striations, running across them. Skeletal muscle is called voluntary muscle (in vertebrates) because it contracts when stimulated by nerves under conscious cerebral control.

Cardiac muscle is another type of striated muscle found only in the vertebrate heart. The cells are much shorter than those of skeletal muscle and have only one nucleus per cell (uninucleate). Cardiac muscle tissue is a branching network of fibers with individual cells interconnected by junctional complexes called intercalated discs. Cardiac muscle is called involuntary muscle because it does not require nerve activity to stimulate contraction. Instead, heart rate is controlled by specialized pacemaker cells located in the heart itself. However, autonomic nerves from the brain may alter pacemaker activity.

Smooth muscle is nonstriated muscle found in both invertebrates and vertebrates. Smooth muscle cells are long, tapering strands, each containing a single nucleus. Smooth muscle is the most common type of muscle in invertebrates in which it serves as body wall musculature and lines ducts and sphincters. In vertebrates, smooth muscle cells are organized into sheets of muscle circling the walls of the alimentary canal, blood vessels, respiratory passages, and urinary and genital ducts. Smooth muscle is typically slow acting and can maintain prolonged contractions with very little energy expenditure. Its contractions are involuntary and unconscious. The principal functions of smooth muscles are to push the material in a tube, such as the intestine, along its way by active contractions or to regulate the diameter of a tube, such as a blood vessel, by sustained contraction.

derm, and its unit is the cell or **muscle fiber.** When viewed with a light microscope, **striated muscle** appears transversely striped (striated), with alternating dark and light bands (Figure 2-7). In vertebrates we can recognize two types of striated muscle: **skeletal** and **cardiac muscle.** A third kind of muscle is smooth (or visceral) muscle, which lacks the characteristic alternating bands of the striated type (Figure 2-7).The unspecialized cyto-

plasm of muscles is called sarcoplasm, and the contractile elements within the fiber are the **myofibrils.**

Nervous Tissue

Nervous tissue is specialized for the reception of stimuli and the conduction of impulses from one region to another. The

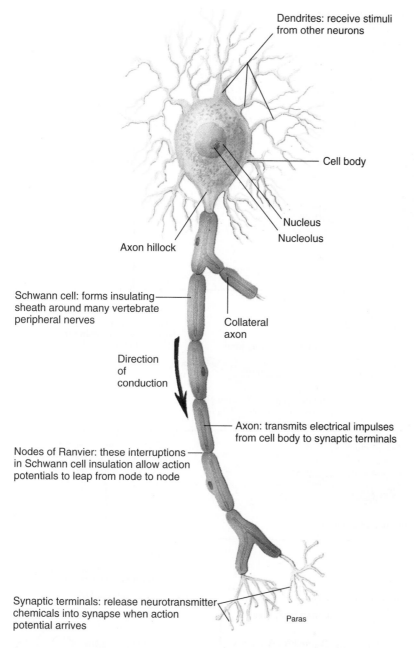

Dendrites: receive stimuli from other neurons

Cell body

Nucleus
Nucleolus

Axon hillock

Schwann cell: forms insulating sheath around many vertebrate peripheral nerves

Collateral axon

Direction of conduction

Axon: transmits electrical impulses from cell body to synaptic terminals

Nodes of Ranvier: these interruptions in Schwann cell insulation allow action potentials to leap from node to node

Synaptic terminals: release neurotransmitter chemicals into synapse when action potential arrives

Paras

FIGURE 2-8

Functional anatomy of a neuron. From the nucleated body, or **soma,** extend one or more **dendrites** (Gr. *dendron,* tree), which receive electrical impulses from receptors or other nerve cells, and a single **axon** that carries impulses away from the cell body to other nerve cells or to an effector organ. The axon is often called a **nerve fiber.** Nerves are separated from other nerves or from effector organs by specialized junctions called synapses.

two basic types of cells in nervous tissue are **neurons** (Gr. nerve), the basic functional unit of the nervous system, and **neuroglia** (nu-rog′le-a; Gr. nerve, + *glia,* glue), a variety of nonnervous cells that insulate neuron membranes and serve various supportive functions. Figure 2-8 shows the functional anatomy of a typical nerve cell.

Animal Body Plans

As pointed out in the prologue to this chapter, while the diversity of animal body form is enormous, the possibilities for designs for living are constrained by ancestral history. Nevertheless, animals are shaped by their particular habitat and way of life. A worm that adopts a parasitic life in a vertebrate's intestine will look and function very differently from a free-living member of the same group. Yet both will share the distinguishing hallmarks of the phylum.

Major evolutionary innovations in the forms of animals include multicellularity, bilateral symmetry, the "tube-within-a-tube" plan, and the eucoelomate (true coelom) body plan. These advancements with their principal alternatives can be arranged in a branching pattern, as shown in Figure 2-9.

Animal Symmetry

Symmetry refers to balanced proportions, or the correspondence in size and shape of parts on opposite sides of a median plane. **Spherical symmetry** means that any plane passing through the center divides the body into equivalent, or mirrored, halves (Figure 2-10). This type of symmetry is found chiefly among some of the protozoa and is rare in other groups of animals. Spherical forms are best suited for floating and rolling.

Radial symmetry (Figure 2-10) applies to forms that can be divided into similar halves by more than two planes passing through the longitudinal axis. These are the tubular, vase, or bowl shapes in which one end of the longitudinal axis is usually the mouth, as found, for example, in some sponges, the hydras, jellyfish, and sea urchins. A variant form is biradial symmetry, in which, because of some part that is single or paired rather than radial, only one plane passing through the longitudinal axis produces mirrored halves. Sea walnuts, which are more or less globular in form but have a pair of tentacles, are an example. Radial and biradial animals are usually sessile, freely floating, or weakly swimming. The two phyla that are primarily radial, Cnidaria and Ctenophora, are called the **Radiata.** The echinoderms (sea stars and their kin) are primarily bilateral animals (their larvae are bilateral) that have become secondarily radial as adults.

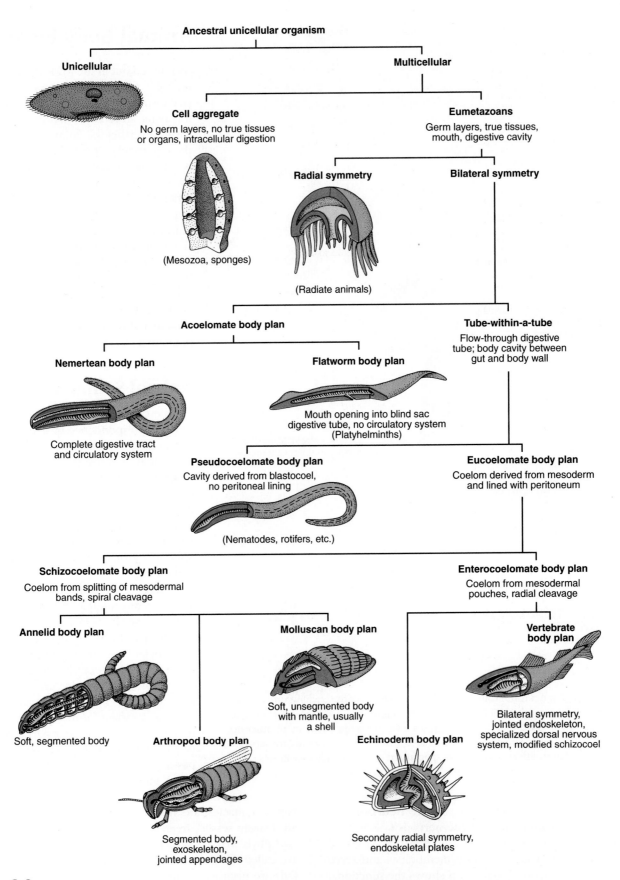

FIGURE 2-9

Architectural patterns of animals. These basic body plans have been variously modified during evolutionary descent to fit animals to a great variety of habitats. Ectoderm is shown in gray, mesoderm in red, and endoderm in yellow.

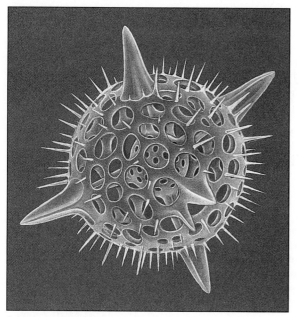

Spherical symmetry

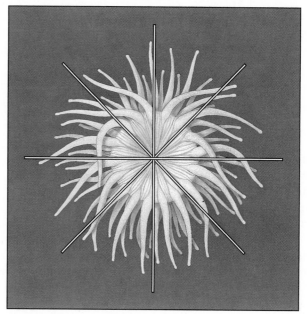

Radial symmetry

Bilateral symmetry applies to animals that can be divided along a sagittal plane into two mirrored portions, right and left halves (Figure 2-10). The appearance of bilateral symmetry in animal evolution was a major advancement because bilateral animals are much better fitted for directional (forward) movement than are radially symmetrical animals. Bilateral animals are collectively called the **Bilateria.** Bilateral symmetry is strongly associated with cephalization, discussed on page 5.

Bilateral symmetry

Some convenient terms used for locating regions of animal bodies (Figure 2-11) are **anterior,** used to designate the head end; **posterior,** the opposite or tail end; **dorsal,** the back side; and **ventral,** the front or belly side. **Medial** refers to the midline of the body; **lateral** refers to the sides. **Distal** parts are farther from the middle of the body than some point of reference; **proximal** parts are nearer. A **frontal plane** (also sometimes called coronal plane) divides a bilateral body into dorsal and ventral halves by running through the anteroposterior axis and the right-left axis at right angles to the **sagittal plane,** the plane dividing an animal into right and left halves. A **transverse plane** (also called a

FIGURE 2-10

The planes of symmetry as illustrated by spherically, radially, and bilaterally symmetrical animals.

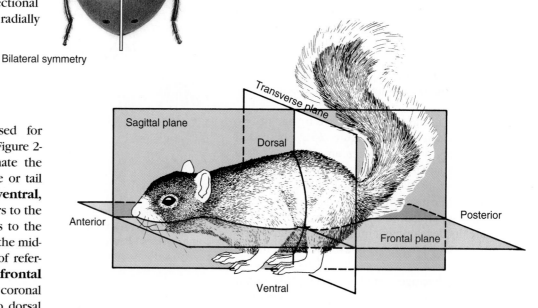

FIGURE 2-11

Descriptive terms used to identify positions on the body of a bilaterally symmetrical animal.

cross section) would cut through a dorsoventral and a right-left axis at right angles to both the sagittal and frontal planes and would result in anterior and posterior portions.

Developmental Patterns in Bilateral Animals

The expression of an animal's body plan is based on an inherited pattern of development. In sexual multicellular organisms, the extremely complex process of embryonic development of an animal from egg to fully differentiated adult is highly predictable and virtually flawless. When developmental alterations do appear during speciation events, the changes tend to be confined to the end stages of development. The organization of eggs and early cleavage stages remain stubbornly resistant to change because any deviation introduced at an early stage would be ruinous to the entire course of development. Nevertheless, several times in the history of life just such radical transformations did occur to herald the appearance of completely new designs for living, such as new classes, phyla, or even major divisions within the animal kingdom.

Patterns of Cleavage

One of the most fundamental aspects of animal development indicating evolutionary relationships is the symmetry of cleavage. Cleavage is the initial process of development following fertilization of the egg by a sperm: dividing up a fertilized egg, now called a **zygote**, into a large number of cells, called **blastomeres**.

Zygotes of most animals cleave by one of two patterns: radial or spiral (Figure 2-12). In **radial cleavage**, the cleavage planes are symmetrical to the polar axis and produce tiers, or layers, of cells on top of each other. Radial cleavage is also said to be **regulative** because each blastomere of the early embryo, if separated from the others, can adjust or "regulate" its development into a complete and well-proportioned (though possibly smaller) embryo (Figure 2-13).

Spiral cleavage, found in several phyla, differs from radial in several ways. Rather than the eggs dividing parallel or perpendicular to the animal-vegetal axis, they cleave oblique to this axis and typically produce quartets of cells that come to lie not on top of each other but in the furrows between the cells (Figure 2-12). In addition, spirally cleaving eggs tend to pack themselves tightly together much like a cluster

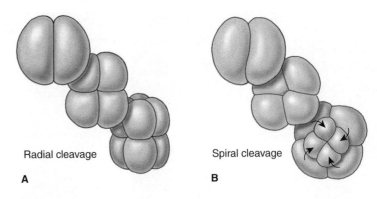

Radial cleavage   **A**   **Spiral cleavage**   **B**

FIGURE 2-12

Radial and spiral cleavage patterns shown at two-, four-, and eight-cell stages. **A,** Radial cleavage, typical of echinoderms and chordates. **B,** Spiral cleavage, typical of molluscs, annelids, and other protostomes. Arrows indicate clockwise movements of small cells (micromeres) following division of large cells (macromeres).

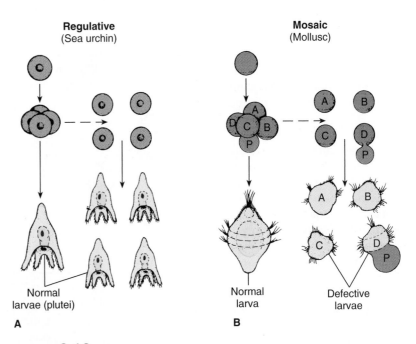

Normal larvae (plutei)   **A**   Normal larva   Defective larvae   **B**

FIGURE 2-13

Regulative and mosaic development. **A,** Regulative development. If the early blastomeres of a sea urchin embryo are separated, each will develop into a complete larva. **B,** Mosaic development. When the blastomeres of a mollusc embryo are separated, each gives rise to a partial, defective larva.

of soap bubbles, rather than just lightly contacting each other as do those in many radially cleaving embryos. Spirally cleaving embryos also differ from radial embryos in having a **mosaic** form of development. This means that the organ-forming determinants in the egg cytoplasm become strictly local-

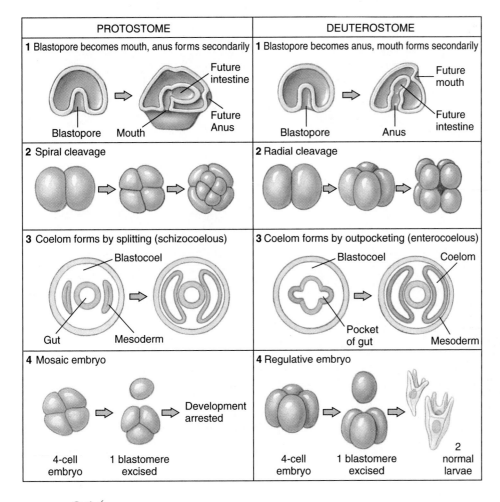

PROTOSTOME	DEUTEROSTOME
1 Blastopore becomes mouth, anus forms secondarily	**1** Blastopore becomes anus, mouth forms secondarily
Future intestine / Blastopore / Mouth / Future Anus	Future mouth / Blastopore / Anus / Future intestine
2 Spiral cleavage	**2** Radial cleavage
3 Coelom forms by splitting (schizocoelous) — Blastocoel / Gut / Mesoderm	**3** Coelom forms by outpocketing (enterocoelous) — Blastocoel / Coelom / Pocket of gut / Mesoderm
4 Mosaic embryo — 4-cell embryo / 1 blastomere excised / Development arrested	**4** Regulative embryo — 4-cell embryo / 1 blastomere excised / 2 normal larvae

FIGURE 2-14

Developmental tendencies of protostomes and deuterostomes. These tendencies are much modified in some groups, for example, the vertebrates. Cleavage in mammals is rotational rather than radial; in reptiles, birds, and many fishes cleavage is discoidal. Vertebrates have also evolved a derived form of coelom formation that is basically schizocoelous.

ized in the egg, even before the first cleavage division. The result is that if the early blastomeres are separated, each will continue to develop for a time as though it were still part of the whole. Each forms a defective, partial embryo (Figure 2-13). A curious feature of most spirally cleaving embryos is that at about the 29-cell stage a blastomere called the 4d cell is formed that will give rise to all the mesoderm of the embryo.

The importance of these two cleavage patterns extends well beyond the differences we have described. They signal a fundamental dichotomy, the early evolutionary divergence of bilateral metazoan animals into two separate lineages. Spiral cleavage is found in the annelids, molluscs, and several other invertebrate phyla; all are included in the **Protostomia** ("mouth first") division of the animal kingdom (see the illustration inside the front cover of this book). The name Protostomia refers to the formation of the mouth from the first embryologi-

cal opening, the blastopore. Radial cleavage is characteristic of the **Deuterostomia** ("mouth second") division of the animal kingdom, a grouping that includes the echinoderms (sea stars and their kin), chordates, and several minor phyla. In the Deuterostomia, the blastopore usually becomes the anus, while the mouth forms secondarily. Other distinguishing developmental hallmarks of these two divisions are summarized in Figure 2-14.

Body Cavities

Another major developmental event affecting the body plan of bilateral animals was the evolution of a fluid-filled cavity between the outer body wall and the gut (see Figure 2-9). The presence of a coelom provides a tube-within-a-tube arrangement that allows much greater body flexibility than is possible in animals lacking an internal body cavity. The coelom also provides space for visceral organs and permits greater size and complexity by exposing more cells to surface exchange. The fluid-filled coelom additionally serves as a hydrostatic skeleton in some forms, especially many worms, aiding in such activities as movement and burrowing.

As shown in Figure 2-9, the presence or absence of a coelom is a key determinant in the evolutionary success of the bilateral metazoa.

Acoelomate Bilateria

The flatworms (phylum Platyhelminthes) and ribbon worms (phylum Nemertea) have no body cavity surrounding the gut (Figure 2-15, *top*); they are "acoelomate" (Gr. *a*, without, + *koiloma*, cavity). The region between the ectodermal epidermis and the endodermal digestive tract is completely filled with mesoderm in the form of a spongy mass of space-filling cells called parenchyma.

Pseudocoelomate Bilateria

Nematodes and several other phyla have a body cavity surrounding the gut called a **pseudocoel** ("false cavity"), and its possessors have a tube-within-a-tube arrangement (Figure 2-15, *center*). The pseudocoel is derived from the blastocoel of the embryo and represents a persistent blastocoel. It lacks a

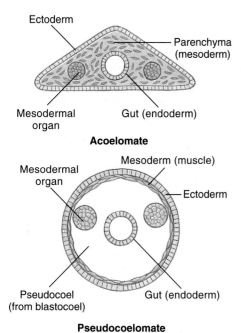

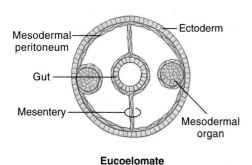

FIGURE 2-15

Acoelomate, pseudocoelomate, and
eucoelomate body plans.

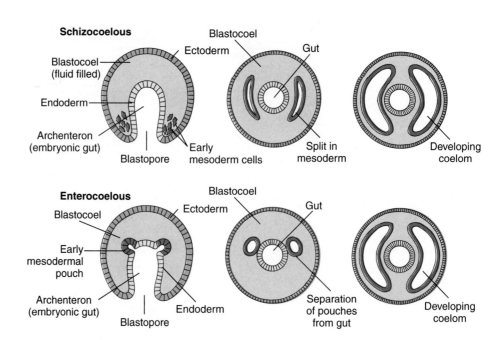

FIGURE 2-16

Types of mesoderm and coelomic formation. In schizocoelous formation, the mesoderm
originates from the wall of the archenteron near the blastopore and proliferates into a
band of tissue which splits to form the coelom. In enterocoelous formation, most
mesoderm originates as a series of pouches from the archenteron; these pinch off and
enlarge to form the coelom. In both formations the coeloms expand to obliterate the
blastocoel.

peritoneum, a thin cellular membrane derived from mesoderm that, in animals with a true coelom, lines the body cavity.

Eucoelomate Bilateria

The remainder of the bilateral animals possess a **true coelom**
lined with mesodermal peritoneum (Figure 2-15, *bottom*). The
true coelom arises within the mesoderm itself and may be
formed by one of two methods, **schizocoelous** or **enterocoelous** (Figure 2-16), or by modifying these methods. The
two terms are descriptive, for *schizo* comes from the Greek
schizein, meaning to split; *entero* is derived from the Greek
enteron, meaning gut; and *coelous* comes from the Greek *koilos,* meaning hollow or cavity. In schizocoelous formation the
coelom arises, as the word implies, from the splitting of mesodermal bands that originate from cells in the blastopore
region. (Mesoderm is one of the three primary germ layers

that appear very early in the development of all bilateral animals, lying between the innermost endoderm and outermost
ectoderm.) In enterocoelous formation, the coelom comes
from pouches of the archenteron, or primitive gut.

Once development is complete, the results of schizocoelous and enterocoelous formations are indistinguishable.
Both give rise to a true coelom lined with a mesodermal peritoneum (Gr. *peritonaios,* stretched around) and having
mesenteries in which the visceral organs are suspended.

Metamerism (Segmentation)

Metamerism is the serial repetition of similar body segments
along the longitudinal axis of the body. Each segment is called a
metamere, or **somite.** In forms such as the earthworm and
other annelids, in which metamerism is most clearly represented, the segmental arrangement includes both external and
internal structures of several systems. There is repetition of
muscles, blood vessels, nerves, and the setae of locomotion.
Some other organs, such as those of sex, may be repeated in
only a few somites. In higher animals much of the segmental
arrangement has become obscure.

True metamerism is found in only three phyla: Annelida,
Arthropoda, and Chordata (Figure 2-17), although superficial
segmentation of the ectoderm and the body wall may be found
among many diverse groups of animals.

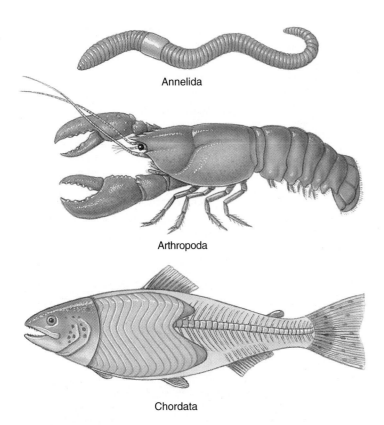

Annelida

Arthropoda

Chordata

Cephalization

The differentiation of a head end is called **cephalization** and is found only in bilaterally symmetrical animals. The concentration of nervous tissue and sense organs in the head bestows obvious advantages to an animal moving through its environment head first. This is the most efficient positioning of instruments for sensing the environment and responding to it. Usually the mouth of the animal is located on the head as well, since so much of an animal's activity is concerned with procuring food. Cephalization is always accompanied by differentiation along an anteroposterior axis (polarity). Polarity usually involves gradients of activities between limits, such as between the anterior and the posterior ends.

FIGURE 2-17

Segmented phyla. These three phyla illustrate an important principle in nature—metamerism, or repetition of structural units. Metamerism is homologous in annelids and arthropods, but chordates have derived their segmentation independently. Segmentation brings more varied specialization because segments, especially in arthropods, have become modified for different functions.

Summary

From the relatively simple organisms that make up the beginnings of life on earth, animal evolution has progressed through a history of ever more intricately organized forms. Adaptive modifications for different niche requirements within a phyletic line of ancestry, however, have been constrained by the ancestral pattern of body architecture.

Cells became integrated into tissues, tissues into organs, and organs into systems. Whereas a unicellular animal carries out all life functions within the confines of a single cell, an advanced multicellular animal is an organization of subordinate units that are united at successive levels. One correlate of increased body complexity is an increase in body size, which offers certain advantages, such as more effective predation, reduced energy cost of locomotion, and improved homeostasis.

The metazoan animal body consists of cells, most of which are functionally specialized; body fluids, divided into the intracellular and extracellular fluid compartments; and the extracellular structural elements, which are fibrous or formless elements that serve various structural functions in the extracellular space. The cells of metazoans develop into various tissues made up of similar cells performing common functions. The basic tissue types are nervous, connective, epithelial, and muscular. Tissues are organized into larger functional units called organs, and organs are associated to form systems.

Every organism has an inherited body plan that may be described in terms of broadly inclusive characteristics, such as symmetry, presence or absence of body cavities, partitioning of body fluids, presence or absence of segmentation, degree of cephalization, and type of nervous system.

Based on several developmental characteristics, the bilateral metazoan animals are divided into two great lineages. The Protostomia are characterized by spiral cleavage, mosaic development, and the mouth forming at or near the embryonic blastopore. The Deuterostomia are characterized by radial cleavage, regulative development, and the mouth forming secondarily to the anus and not from the blastopore.

Review Questions

1. Name the five levels of organization in animal complexity and explain how each successive level is more complex than the one preceding it.

2. Can you suggest why, during the evolution of separate animal lineages, there has been a tendency for complexity to increase when body size increases?

3. What are the meanings of the terms "parenchyma" and "stroma" as they relate to body organs?

4. Body fluids of eumetazoan animals are separated into fluid "compartments." Name these compartments and explain how compartmentalization may differ in animals with open and closed circulatory systems.

5. What are the four major types of tissues in the body of a metazoan?

6. How would you distinguish between simple and stratified epithelium?

7. What are the three elements present in all connective tissue? Give some examples of the different types of connective tissue.

8. What are three different kinds of muscle found among animals? Explain how each is specialized for particular functions.

9. Describe the principal structural and functional features of a neuron.

10. Match the animal group with its body plan:

_____ Unicellular   a. Nematode
_____ Cell aggregate   b. Vertebrate
_____ Blind sac,   c. Protozoan
       acoelomate   d. Flatworm
_____ Tube-within-a-tube,   e. Sponge
       pseudocoelomate   f. Arthropod
_____ Tube-within-a-tube,   g. Nemertean
       eucoelomate

11. Distinguish among spherical, radial, biradial, and bilateral symmetry.

12. Use the following terms to identify regions on your body and on the body of a frog: anterior, posterior, dorsal, ventral, lateral, distal, proximal.

13. How would frontal, sagittal, and transverse planes divide your body?

14. What is the difference between radial and spiral cleavage?

15. What are the distinguishing developmental hallmarks of the two great lineages of bilateral metazoans, the Protostomia and the Deuterostomia?

16. What is meant by metamerism? Name three phyla showing metamerism.

Selected References

See also general references on p. 395.

Arthur, W. 1997. The origin of animal body plans. Cambridge, United Kingdom, Cambridge University Press. *Explores the genetic, developmental, and population-level processes involved in the evolution of the 35 or so body plans that arose in the geological past.*

Bonner, J. T. 1988. The evolution of complexity by means of natural selection. Princeton, New Jersey, Princeton University Press. *Levels of complexity in organisms and how size affects complexity.*

Grene, M. 1987. Hierarchies in biology. Amer. Sci. **75**:504–510 (Sept.–Oct.). *The term "hierarchy" is used in many different senses in biology. The author points out that current evolutionary theory carries the hierarchical concept beyond the Darwinian restriction to the two levels of gene and organism.*

Kessel, R. G., and R. H. Kardon. 1979. Tissues and organs: a text-atlas of scanning electron microscopy. San Francisco, W. H. Freeman & Company. *Collection of excellent scanning electron micrographs with text.*

Radinsky, L. B. 1987. The evolution of vertebrate design. Chicago, University of Chicago Press. *A lucid functional analysis of vertebrate body plans and their evolutionary transformations over time.*

Welsch, U., and V. Storch. 1976. Comparative animal cytology and histology. London, Sidgwick & Jackson. *Comparative histology with good treatment of invertebrates.*

Willmer, P. 1990. Invertebrate relationships: patterns in animal evolution. Cambridge, Cambridge University Press. *Chapter 2 is an excellent discussion of animal symmetry, developmental patterns, origin of body cavities, and segmentation.*

Links to the Internet

Visit this textbook's web site at http://www.mhhe.com/zoology to find live Internet links for each of the references listed below.

1. Relationship between Cells, Tissues, Organs, and Organ Systems. This basic site provides information on how cells, tissues, and organs interact.

Classification and Phylogeny of Animals

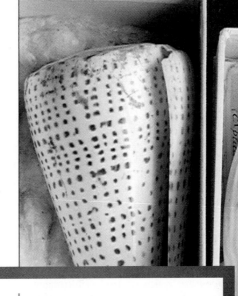

CHAPTER	three

Order in Diversity

Evolution of the animal kingdom has produced a great diversity of species. Zoologists have named more than 1.5 million species of animals, and thousands more are described each year. Some zoologists estimate that the species named so far constitute less than 20% of all living animals and less than 1% of all those that have existed in the past.

Despite its magnitude, the diversity of animals is not without limits. There are many conceivable forms that do not exist in nature, as our myths of minotaurs and winged horses show. Animal diversity is not random but has a definite order. The characteristic features of humans and cattle never occur together in the same organism as they do in the mythical minotaur. Nor do the characteristic wings of birds and bodies of horses occur together naturally as they do in the mythical horse Pegasus. Humans, cattle, birds, and horses are distinct groups of animals, yet they do share some important features, including vertebrae and homeothermy, that separate them from even more dissimilar forms such as insects and flatworms.

All human cultures classify their familiar animals according to patterns in animal diversity. These classifications have many purposes. Animals may be classified in some societies according to their usefulness or destructiveness to human endeavors. Others may group animals according to their roles in mythology. Biologists group animals according to their evolutionary relationships as revealed by ordered patterns in their sharing of homologous features. This classification is called a "natural system" because it reflects relationships that exist among animals in nature, outside the context of human activity. The systematic zoologist has three major goals: to discover all species of animals, to reconstruct their evolutionary relationships, and then to classify them accordingly.

Darwin's theory of common descent (Chapter 1) is the underlying principle that guides our search for order in the diversity of animal life. Our science of taxonomy ("arrangement law") produces a formal system for naming and classifying species that reflects this order. Animals that have very recent common ancestry share many features in common and are grouped most closely in our taxonomic classification; dissimilar animals that share only very ancient common ancestry are placed in different taxonomic groups except at the "highest" or most inclusive levels of taxonomy. Taxonomy is part of the broader science of **systematics,** which uses everything that is known about animals to understand their evolutionary relationships. Taxonomy predates evolutionary biology, however, and many taxonomic practices are relics of pre-evolutionary world views. Adjusting our taxonomic system to accommodate evolution has produced many problems and controversies. Taxonomy has reached an unusually active and controversial point in its development with several alternative taxonomic systems competing for use. To understand this controversy, we need to review the history of animal taxonomy.

FIGURE 3-1

Carolus Linnaeus (1707–1778). This portrait was made of Linnaeus at age 68, three years before his death.

Linnaeus and the Development of Classification

The Greek philosopher and biologist Aristotle was the first to classify organisms on the basis of their structural similarities. Following the Renaissance in Europe, the English naturalist John Ray (1627–1705) introduced a more comprehensive system of classification and a new concept of species. The flowering of systematics in the eighteenth century culminated in the work of Carolus Linnaeus (1707–1778; Figure 3-1), who gave us our current scheme of classification.

Linnaeus was a Swedish botanist at the University of Uppsala. He had a great talent for collecting and classifying objects, especially flowers. Linnaeus produced an extensive system of classification for both plants and animals. This scheme, published in his great work, *Systema Naturae,* used morphology (the comparative study of organismal form) for arranging specimens in collections. He divided the animal kingdom into species and gave each one a distinctive name. He grouped species into genera, genera into orders, and orders into classes. Because his knowledge of animals was limited, his lower categories, such as the genera, were very broad and included animals that are only distantly related. Much of his classification has been drastically altered, but the basic principle of his scheme is still followed.

Linnaeus's scheme of arranging organisms into an ascending series of groups of increasing inclusiveness is the **hierarchical system** of classification. The major categories, or **taxa** (sing., **taxon**), into which organisms are grouped were given one of several standard taxonomic ranks to indicate the general degree of inclusiveness of the group. The hierarchy of taxonomic ranks has been expanded considerably since Linnaeus's time (Table 3-1). It now includes seven mandatory ranks for the animal kingdom, in descending series: kingdom, phylum, class, order, family, genus, and species. All organisms being classified must be placed into at least seven taxa, one at each of the mandatory ranks. Taxonomists have the option of subdividing these seven ranks even further to recognize more than seven taxa (superclass, subclass, infraclass, superorder, suborder, and others) for any particular group of organisms. More than 30 taxonomic ranks now are recognized. For very large and complex groups, such as the fishes and insects, these additional ranks are needed to express different degrees of evolutionary divergence. Unfortunately, they also make the system more complex.

Linnaeus's system for naming species is known as **binomial nomenclature.** Each species has a Latinized name composed of two words (hence binomial) written in italics (underlined if handwritten or typed). The first word is the name of the **genus,** written with a capital initial letter; the second word is the **specific epithet** which is peculiar to the species within the genus and is written with a small initial letter (see Table 3-1). The name of the genus is always a noun, and the specific epithet is usually an adjective that must agree in gender with the genus. For instance, the scientific name of the common robin is *Turdus migratorius* (L. *turdus,* thrush; *migratorius,* of the migratory habit). The specific epithet never stands alone; the complete binomial must be used to name a species. Names of genera must refer only to single groups of organisms; the same name cannot be given to two different genera of animals. The same specific epithet may be used in different genera, however, to denote different and unrelated species. For example, the scientific name of the white-breasted nuthatch is *Sitta carolinensis.* The specific

TABLE 3-1

Examples of Taxonomic Categories to Which Representative Animals Belong

	Human	Gorilla	Southern Leopard Frog	Bush Katydid
Kingdom	Animalia	Animalia	Animalia	Animalia
Phylum	Chordata	Chordata	Chordata	Arthropoda
Subphylum	Vertebrata	Vertebrata	Vertebrata	Uniramia
Class	Mammalia	Mammalia	Amphibia	Insecta
Subclass	Eutheria	Eutheria	—	Pterygota
Order	Primates	Primates	Anura	Orthoptera
Suborder	Anthropoidea	Anthropoidea	—	Ensifera
Family	Hominidae	Pongidae	Ranidae	Tettigoniidae
Subfamily	—	—	Raninae	Phaneropterinae
Genus	*Homo*	*Gorilla*	*Rana*	*Scudderia*
Species	*Homo sapiens*	*Gorilla gorilla*	*Rana sphenocephala*	*Scudderia furcata*
Subspecies	—	—	—	*Scudderia furcata furcata*

The hierarchical system of classification applied to four species (human, gorilla, Southern leopard frog, and bush katydid). Higher taxa generally are more inclusive than lower-level taxa, although taxa at two different levels may be equivalent in content (for example, the family Hominidae contains only the genus Homo, making the content of these taxa equivalent, whereas the family Pongidae contains the genera Gorilla, Pan, and Pongo, making it more inclusive than any of these genera). Closely related species are united at a lower point in the hierarchy than are distantly related species. For example, humans and gorillas are united at the level of the suborder (Anthropoidea) and above; they are united with the Southern leopard frog at the subphylum level (Vertebrata) and with the bush katydid at the level of the kingdom (Animalia).

epithet *"carolinensis"* is used in other genera, including *Parus carolinensis* (Carolina chickadee) and *Anolis carolinensis* (green anole, a lizard) to mean "of Carolina." All ranks above the species are designated using uninomial nouns, written with a capital initial letter.

Some species are divided into subspecies, in which case a trinomial nomenclature is employed (see katydid example, Table 3-1). Thus to distinguish the southern form of the robin from the eastern robin, the scientific term Turdus migratorius achrustera *(duller color) is employed for the southern type. The generic, specific, and subspecific names are printed in italics (underlined if handwritten or typed). The subspecies name may be a repetition of the specific epithet. Formal recognition of subspecies has lost popularity among taxonomists because the boundaries between subspecies are rarely distinct. Recognition of subspecies usually is based on one or a few superficial characters and does not denote an evolutionarily distinct unit. Subspecies, therefore, should not be taken too seriously.*

The person who first describes a type specimen and publishes the name of a species is called the authority. This person's name and date of publication are often written after the species name. Thus, Didelphis marsupialis *Linnaeus, 1758, tells us that Linnaeus was the first person to publish the species name of the opossum. The authority citation is not part of the scientific name but rather is an abbreviated bibliographical reference. Sometimes, the generic status of a species is revised following its initial description. In this case, the name of the authority is presented in parentheses.*

Taxonomic Characters and Reconstruction of Phylogeny

A major goal of systematics is to reconstruct the evolutionary tree or **phylogeny** that relates all extant and extinct species. The tree is constructed by studying organismal features, formally called **characters,** that vary among species. A character is any feature that the taxonomist uses to study variation within or among species. We find potentially useful taxonomic characters in morphological, chromosomal, and molecular features. Taxonomists find characters by observing patterns of similarity among organisms. If two organisms possess similar features, they may have inherited these features from an equivalent one in a common ancestor. Character similarity that results from common ancestry is called **homology** (see p. 17). Similarity does not always reflect common ancestry, however. Independent evolutionary origins of similar features on different lineages produce patterns of similarity among organisms that do not reflect common descent; this occurrence complicates the work of taxonomists. Character similarity that misrepresents common descent is called nonhomologous similarity or **homoplasy.**

Using Character Variation to Reconstruct Phylogeny

To reconstruct the phylogeny of a group using characters that vary among its members, the first step is to determine which variant form of each character was present in the common ancestor of the entire group. This form is called the **ancestral character state** for the group as a whole. We presume that all other variant forms of the character arose later within the group, and these forms are called evolutionarily **derived character states.** The **polarity** of a character refers to the ancestral/descendant relationships among its different states. For example, if we consider as a character the dentition of amniotic vertebrates (reptiles, birds, and mammals), presence versus absence of teeth in the jaws constitute two different character states. Teeth are absent from birds but present in the other amniotes. To evaluate the polarity of this character, we must determine which character state, presence or absence of teeth, characterized the most recent common ancestor of amniotes and which state was derived subsequently within the amniotes.

The method that we use to examine the polarity of a variable character is called **outgroup comparison.** We begin by selecting an additional group of organisms, called the **outgroup,** that is phylogenetically close but not within the group being studied. The amphibians and different groups of bony fishes constitute appropriate outgroups to the amniotes for polarizing variation in the dentition of amniotes. Next, we infer that any character state found both within the group being studied and in the outgroup is ancestral for the study group. Teeth are usually present in amphibians and bony fishes; therefore, we infer that presence of teeth is ancestral for amniotes and absence of teeth is derived. The polarity of this character indicates that teeth were lost in the ancestral lineage of all modern birds. Polarity of characters is evaluated most effectively when several different outgroups are used. All character states found in the study group that are absent from appropriate outgroups are considered derived (see Figure 3-2 for additional examples).

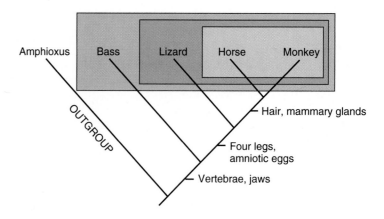

FIGURE 3-2

The cladogram as a nested hierarchy of taxa. *Amphioxus* (p. 268) is the outgroup, and the study group comprises four vertebrates (bass, lizard, horse, and monkey). Four characters that vary among vertebrates are used to generate a simple cladogram: presence versus absence of four legs, amniotic eggs, hair, and mammary glands. For all four characters, absence is the ancestral state in vertebrates because this is the condition found in the outgroup, *Amphioxus;* for each character, presence is the derived state in vertebrates. Because they share presence of four legs and amniotic eggs as synapomorphies, the lizard, horse, and monkey form a clade relative to the bass. This clade is subdivided further by two synapomorphies (presence of hair and mammary glands) that unite the horse and monkey relative to the lizard. We know from comparisons involving even more distantly related animals that presence of vertebrae and jaws constitute synapomorphies of the vertebrates and that *Amphioxus,* which lacks these features, falls outside the vertebrate clade.

The organisms or species that share derived character states form subsets within the study group; these subsets are called **clades** (Gr. *klados,* branch). A derived character shared by the members of a clade is formally called a **synapomorphy** (Gr. *synapsis,* joining together, + *morphē,* form) of that clade. Taxonomists use synapomorphies as evidence of homology to infer that a particular group of organisms forms a clade. Within the amniotes, absence of teeth and presence of feathers are synapomorphies that identify the birds as a clade. A clade corresponds to a unit of evolutionary common descent; it includes all descendants of a particular ancestral lineage. The pattern formed by the derived states of all characters within our study group will take the form of a **nested hierarchy** of clades within clades. The goal is to identify all of the different clades nested within the study group, which would reveal the patterns of common descent among all species in the group.

The nested hierarchy of clades is presented as a branching diagram called a **cladogram** (Figure 3-2; see also Figure 1-17). Taxonomists often make a technical distinction between a cladogram and a **phylogenetic tree.** The branches of a cladogram are only a formal device for indicating the nested hierarchy of clades within clades. The cladogram is not strictly equivalent to a phylogenetic tree, whose branches represent real lineages that occurred in the evolutionary past. To obtain a phylogenetic tree, we must add to the cladogram information concerning ancestors, the durations of evolutionary lineages, or the amounts of evolutionary change that occurred on the lineages. Because the branching order of a cladogram matches that of the corresponding phylogenetic tree, however, the cladogram often is used as a first approximation of the phylogenetic tree.

Sources of Phylogenetic Information

We find the characters used to construct cladograms in comparative morphology (including embryology), comparative cytology, and comparative biochemistry. **Comparative morphology** examines the varying shapes and sizes of organismal structures, including their developmental origins. As we will see in later chapters, the variable structures of skull bones, limb bones, and integument (scales, hair, feathers) are particularly important for reconstructing the phylogeny of vertebrates. Comparative morphology uses specimens obtained from both living organisms and fossilized remains. **Comparative biochemistry** uses the sequences of amino acids in proteins and the sequences of nucleotides in nucleic acids to identify variable characters for constructing a cladogram (Figure 3-3). Recent work has shown that some fossils retain enough DNA for use in comparative biochemical studies. **Comparative cytology** uses variation in the numbers, shapes, and sizes of chromosomes and their parts to obtain variable characters for constructing cladograms. Comparative cytology is used almost exclusively on living rather than fossilized organisms because chromosomal structure is not well preserved in fossils.

To add the evolutionary timescale necessary for producing a phylogenetic tree, we must consult the fossil record. We can look for the earliest appearance of derived morphological characters in fossils to estimate the ages of clades distinguished by those characters. The age of a fossil showing the derived characters of a particular clade is determined by radioactive dating (p. 15) to estimate the age of the clade. A lineage representing the most recent common ancestor of all species in the clade is then added to the phylogenetic tree.

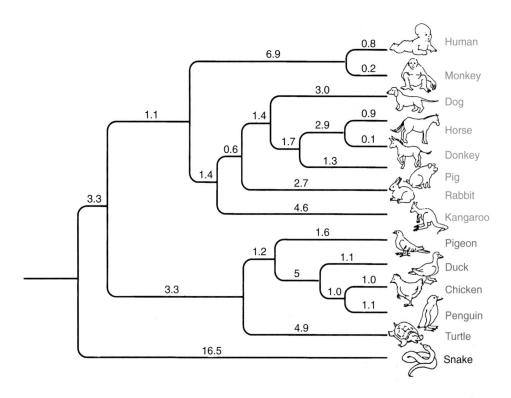

FIGURE 3-3

A phylogenetic tree of representative amniotes based on inferred base substitutions in the gene that encodes the respiratory protein, cytochrome *c*. Numbers on the branches are the estimated numbers of mutational changes that occurred in this gene along the different evolutionary lineages.

Theories of Taxonomy

A theory of taxonomy establishes the principles that we use to recognize and to rank taxonomic groups. There are two currently popular theories of taxonomy: (1) traditional evolutionary taxonomy and (2) phylogenetic systematics (cladistics). Both are based on evolutionary principles. We will see, however, that these two theories differ on how evolutionary principles are used. These differences have important implications for how we use a taxonomy to study the evolutionary process.

The relationship between a taxonomic group and a phylogenetic tree or cladogram is important for both of these theories. This relationship can take one of three forms: **monophyly, paraphyly,** or **polyphyly** (Figure 3-4). A taxon is monophyletic if it includes the most recent common ancestor of all members of the group and all descendants of that ancestor (see Figure 3-4A). A taxon is paraphyletic if it includes the most recent common ancestor of all members of a group and some but not all of the descendants of that ancestor (see Figure 3-4B). A taxon is polyphyletic if it does not include the most recent common ancestor of all members of a group; this situation requires that the group has had at least two separate evolutionary origins, usually requiring independent

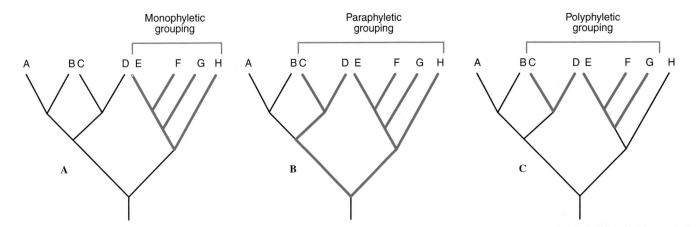

FIGURE 3-4

Relationships between phylogeny and taxonomic groups illustrated for a hypothetical phylogeny of eight species (A through H). **A,** *Monophyly*—a monophyletic group contains the most recent common ancestor of all members of the group and all of its descendants. **B,** *Paraphyly*—a paraphyletic group contains the most recent common ancestor of all members of the group and some but not all of its descendants. **C,** *Polyphyly*—a polyphyletic group does not contain the most recent common ancestor of all members of the group, thereby requiring that the group has had at least two separate phylogenetic origins.

evolutionary acquisition of a diagnostic feature (see Figure 3-4C). For example, if birds and mammals were grouped in a taxon called Homeothermia, we would have a polyphyletic taxon because birds and mammals descend from two quite separate amniotic lineages that have evolved homeothermy independently. The most recent common ancestor of birds and mammals is not homeothermic and does not occur in the polyphyletic Homeothermia just described. Both evolutionary and cladistic taxonomy accept monophyletic groups and reject polyphyletic groups in their classifications. They differ on the acceptance of paraphyletic groups.

Traditional Evolutionary Taxonomy

Traditional **evolutionary taxonomy** incorporates two different evolutionary principles for recognizing and ranking higher taxa: (1) common descent and (2) amount of adaptive evolutionary change, as shown on a phylogenetic tree. Evolutionary taxa must have a single evolutionary origin, and must show unique adaptive features.

The mammalian paleontologist George Gaylord Simpson (Figure 3-5) was highly influential in developing and formalizing the principles of evolutionary taxonomy. According to Simpson, a particular branch on the evolutionary tree is given the status of a higher taxon if it represents a distinct **adaptive zone.** Simpson describes an adaptive zone as "a characteristic reaction and mutual relationship between environment and organism, a way of life and not a place where life is led." By entering a new adaptive zone through a fundamental change in organismal structure and behavior, an evolving population can use environmental resources in a completely new way.

FIGURE 3-5

George Gaylord Simpson (1902–1984) formulated the principles of evolutionary taxonomy.

A taxon that comprises a distinct adaptive zone is termed a **grade.** Simpson gives the example of penguins as a distinct adaptive zone within birds. The lineage immediately ancestral to all penguins underwent fundamental changes in the form of the body and wings to permit a switch from aerial to aquatic locomotion (Figure 3-6). Aquatic birds that can fly both in the air and underwater are somewhat intermediate in habitat, morphology, and behavior between aerial and aquatic adaptive zones. Nonetheless, the obvious modifications of the wings and body of penguins for swimming represent a new grade of organization. The penguins are therefore recognized as a distinct taxon within the birds, the family Spheniscidae. The broader the adaptive zone when fully occupied by a group of organisms, the higher the rank that the corresponding taxon is given.

Evolutionary taxa may be either monophyletic or paraphyletic. Recognition of paraphyletic taxa requires, however,

FIGURE 3-6

A, Penguin. **B,** Diving petrel. Penguins (avian family Spheniscidae) were considered by George G. Simpson a distinct adaptive zone within birds because of their adaptations for submarine flight. Simpson believed that the adaptive zone ancestral to the penguins resembled that of the diving petrels, which display adaptations for combined aerial and aquatic flight. The adaptive zones of penguins and diving petrels are distinct enough to be recognized taxonomically as different families within a common order (Ciconiiformes).

that our taxonomies distort patterns of common descent. An evolutionary taxonomy of the anthropoid primates provides a good example (Figure 3-7). This taxonomy places humans (genus *Homo*) and their immediate fossil ancestors in the family Hominidae, and it places the chimpanzees (genus *Pan*), gorillas (genus *Gorilla*), and orangutans (genus *Pongo*) in the family Pongidae. However, the pongid genera *Pan* and *Gorilla* share more recent common ancestry with the *Hominidae* than they do with the remaining pongid genus, *Pongo*. The family Pongidae is therefore paraphyletic because it does not include humans, who also descend from the most recent common ancestor of all pongids (see Figure 3-7). Evolutionary taxonomists nonetheless recognize the pongid genera as a single, family-level grade of arboreal, herbivorous primates having limited mental capacity; in other words, they show the same family-level adaptive zone. Humans are terrestrial, omnivorous primates who

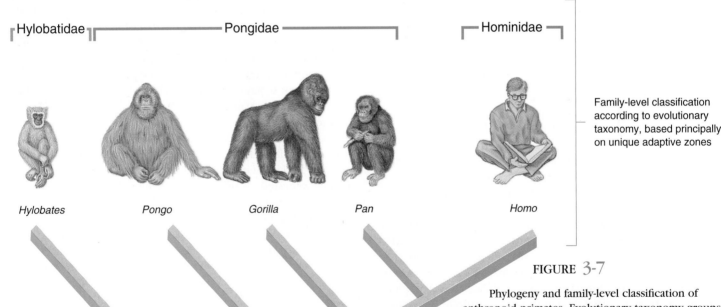

Family-level classification according to evolutionary taxonomy, based principally on unique adaptive zones

Hylobatidae — Pongidae — Hominidae

Hylobates — *Pongo* — *Gorilla* — *Pan* — *Homo*

FIGURE 3-7

Phylogeny and family-level classification of anthropoid primates. Evolutionary taxonomy groups the genera *Gorilla, Pan,* and *Pongo* into the paraphyletic family Pongidae because they share the same adaptive zone or grade of organization. Humans (genus *Homo*) are phylogenetically closer to *Gorilla* and *Pan* than any of these genera are to *Pongo,* but humans are placed in a separate family (Hominidae) because they represent a different grade of organization. Cladistic taxonomy requires either that the family Pongidae be split into three monophyletic family-level taxa or that *Homo* be included in the same taxonomic family as some or all of the apes. For example, *Pongo, Gorilla, Pan,* and *Homo* could be combined into a single monophyletic family Hominidae. The gibbons (genus *Hylobates*) form the monophyletic family Hylobatidae, which is compatible with both evolutionary and cladistic classifications.

possess greatly expanded mental and cultural attributes, thereby comprising a distinct adaptive zone at the taxonomic level of the family. Unfortunately, if we want our taxa to constitute adaptive zones, we compromise our ability to present common descent in the most straightforward taxonomic manner.

Traditional evolutionary taxonomy has been challenged from two opposite directions. One challenge states that because phylogenetic trees can be very difficult to obtain, it is impractical to base our taxonomic system on common descent and adaptive evolution. We are told that our taxonomy should represent a more easily measured feature, the overall similarity of organisms evaluated without regard to phylogeny. This principle is known as **phenetic taxonomy.** Phenetic taxonomy did not have a strong impact on animal classification, and scientific interest in this approach is in decline.

Despite the difficulties of reconstructing phylogeny, zoologists still consider this endeavor a central goal of their systematic work, and they are unwilling to compromise this goal for purposes of methodological simplicity.

Phylogenetic Systematics/Cladistics

A second and stronger challenge to evolutionary taxonomy is one known as **phylogenetic systematics** or **cladistics.** As the first name implies, this approach emphasizes the criterion of common descent and, as the second name implies, it is based on the cladogram of the group being classified. This approach to taxonomy was first proposed in 1950 by the German entomologist Willi Hennig (Figure 3-8) and therefore is sometimes called "Hennigian systematics." All taxa recognized by Hennig's cladistic system must be monophyletic. We saw previously how the evolutionary taxonomists' recognition of the primate families Hominidae and Pongidae distorts genealogical relationships to emphasize the adaptive uniqueness of the Hominidae. Because the most recent common ancestor of the paraphyletic family Pongidae is also an ancestor of the Hominidae, recognition of the Pongidae is incompatible with cladistic taxonomy.

The disagreement on the validity of paraphyletic groups may seem trivial at first, but its important consequences become clear when we discuss evolution. For example, the claims that amphibians evolved from bony fish, that birds evolved from reptiles, or that humans evolved from apes may be made by an evolutionary taxonomist but are meaningless to a cladist. We imply by these statements that a descendant group (amphibians, birds, or humans) evolved from part of an ancestral group (bony fish, reptiles, and apes, respectively) to which the descendant does not belong. This usage automatically makes the ancestral group paraphyletic, and indeed bony fish, reptiles, and apes as traditionally recognized are paraphyletic groups. How are such paraphyletic groups recognized? Do they share distinguishing features that are not shared by the descendant group?

Paraphyletic groups are usually defined in a negative manner. They are distinguished only by absence of features found in a particular descendant group, because any traits that they share from their common ancestry are present also in the excluded

FIGURE 3-8

Willi Hennig (1913–1976), German entomologist who formulated the principles of phylogenetic systematics/cladistics.

descendants (unless secondarily lost). For example, apes are those "higher" primates that are not humans. Likewise, fish are those vertebrates that lack the distinguishing characteristics of tetrapods (amphibians and amniotes). What does it mean then to say that humans evolved from apes? To the evolutionary taxonomist, apes and humans are different adaptive zones or grades of organization; to say that humans evolved from apes states that bipedal, tailless organisms of large brain capacity evolved from arboreal, tailed organisms of smaller brain capacity. To the cladist, however, the statement that humans evolved from apes says essentially that humans evolved from something that they are not, a trivial statement that contains no useful information. Extinct ancestral groups are always paraphyletic because they exclude a descendant that shares their most recent common ancestor. Although many such groups have been recognized by evolutionary taxonomists, none are recognized by cladists.

Cladists denote the common descent of different taxa by identifying **sister taxa.** Sister taxa share more recent common ancestry with each other than either one does with any other taxon. The sister group of humans appears to be the chimpanzees, with the gorillas forming a sister group to the humans and chimpanzees combined. The orangutans are the sister group of the clade that includes humans, chimpanzees, and gorillas, and the gibbons form the sister group of the clade that includes the orangutans, chimpanzees, gorillas, and humans (see Figure 3-7).

Current State of Animal Taxonomy

The formal taxonomy of animals that we use today was established using the principles of evolutionary systematics and has been revised recently in part using the principles of cladistics.

Introduction of cladistic principles initially has the effect of replacing paraphyletic groups with monophyletic subgroups while leaving the remaining taxonomy mostly unchanged. A thorough revision of taxonomy along cladistic principles, however, will require profound changes, one of which almost certainly will be abandonment of the Linnean ranks. In our coverage of animal taxonomy, we will try as much as possible to use taxa that are monophyletic and therefore consistent with the criteria of both evolutionary and cladistic taxonomy. We will continue, however, to use Linnean ranks. In some cases in which familiar taxa are clearly paraphyletic grades, we will note this fact and suggest alternative taxonomic schemes that contain only monophyletic taxa.

When discussing patterns of descent, we will avoid statements such as "mammals evolved from reptiles" that imply paraphyly and will instead specify appropriate sister-group relationships. We will avoid referring to groups of organisms as being primitive, advanced, specialized, or generalized because all groups of animals contain combinations of primitive, advanced, specialized, and generalized features; these terms are best restricted to describing specific characteristics and not an entire group.

Revision of taxonomy according to cladistic principles can cause confusion. In addition to new taxonomic names, we see old ones used in unfamiliar ways. For example, the cladistic use of "bony fishes" includes amphibians and amniotes (including reptilian groups, birds, and mammals) in addition to the finned, aquatic animals that we normally group under the term "fish." The cladistic use of "reptiles" includes birds in addition to the snakes, lizards, turtles, and crocodilians; however, it excludes some fossil forms, such as the synapsids, that traditionally were placed in the Reptilia (see Chapter 17). Taxonomists must be very careful to specify when using these seemingly familiar terms whether the traditional evolutionary taxa or newer cladistic taxa are being discussed.

Species

While discussing Darwin's book, *On the Origin of Species,* in 1859, Thomas Henry Huxley (p. 10) asked, "In the first place, what is a species? The question is a simple one, but the right answer to it is hard to find, even if we appeal to those who should know most about it." We have used the term "species" so far as if it had a simple and unambiguous meaning. Actually, Huxley's commentary is as valid today as it was 140 years ago. Our concepts of species have become more sophisticated, but the diversity of different concepts and the disagreements surrounding their use are as evident now as they were in Darwin's time.

Criteria for Recognition of Species

Despite widespread disagreement about the nature of species, biologists have repeatedly designated certain criteria as being important to their identification of species. First, the criterion of **common descent** is central to nearly all modern concepts of species. The members of a species must trace their ancestry to a common ancestral population although not necessarily to a single pair of parents. Species are thus historical entities. A second criterion is that species must be the smallest distinct groupings of organisms sharing patterns of ancestry and descent; otherwise, it would be difficult to separate species from higher taxa whose members also share common descent. Morphological characters traditionally have been important in identifying such groupings, but chromosomal and molecular characters increasingly are being used for this purpose. A third important criterion is that of reproductive community, which applies only to sexually reproducing organisms; members of a species must form a reproductive community that excludes members of other species. This criterion is very important to many modern concepts of species.

Concepts of Species

Before Darwin, a species was considered a distinct and immutable entity. The concept that species were defined by fixed, essential features (usually morphological) is called the **typological species concept.** This concept was discarded following the establishment of Darwinian evolutionary theory.

The most influential concept of species inspired by Darwinian evolutionary theory is the biological species concept formulated by Theodosius Dobzhansky and Ernst Mayr. In 1983, Mayr stated the biological species concept as follows: *"A species is a reproductive community of populations (reproductively isolated from others) that occupies a specific niche in nature."* Note that the species is identified here according to the reproductive properties of populations, not according to morphology. The species is an *interbreeding* population of individuals having *common descent.* By adding the criterion of the **niche,** an ecological concept denoting an organism's role in its ecological community, we recognize that members of a reproductive community constitute an ecological entity in nature. Because reproductive community should maintain genetic cohesiveness, organismal variation should be relatively smooth and continuous within species and discontinuous between them. Although the biological species is based on reproductive properties of populations rather than organismal morphology, morphology nonetheless can help us to diagnose biological species.

The biological species concept has been strongly criticized. To understand why, we must keep in mind several important facts about species. First, a species has dimensions in space and time, which often creates problems for locating discrete boundaries between species. Second, we view the species both as a unit of evolution and as a rank in the taxonomic hierarchy. These roles sometimes conflict, as we will see in the following text. A third problem is that according to the biological species concept, species do not exist in groups of organisms that reproduce only asexually. It is common systematic practice, however, to describe species in all groups of organisms.

The **evolutionary species concept** was proposed by Simpson (see Figure 3-5) in the 1940s to add an evolutionary time dimension to the biological species concept. This concept persists in a modified form today. A current definition of the evolutionary species is *a single lineage of ancestor-descendant populations that maintains its identity from other such lineages and that has its own evolutionary tendencies and historical fate.* Note that the criterion of common descent is retained here in the need for a species to have a distinct historical identity. Unlike the biological species concept, the evolutionary species concept applies both to sexually and asexually reproducing forms. As long as continuity of diagnostic features is maintained by the evolving lineage, it will be recognized as a single species. Abrupt changes in diagnostic features will mark the boundaries of different species in evolutionary time.

The last concept that we present is the **phylogenetic species concept.** The phylogenetic species concept is defined as an *irreducible (basal) grouping of organisms diagnosably distinct from other such groupings and within which there is a parental pattern of ancestry and descent.* This concept also emphasizes common descent, and both asexual and sexual groups are covered. The phylogenetic species is a strictly monophyletic unit, making it ideal for cladistic systematics. Any population that has become separated from others and has undergone character evolution that distinguishes it will be recognized as a species. The criterion of irreducibility requires that no more than one such population can be placed in a single species. The main difference in practice between the evolutionary and phylogenetic species concepts is that the latter emphasizes recognizing as species the smallest groupings of organisms that have undergone independent evolutionary change. The evolutionary species concept would group into a single species geographically disjunct populations that demonstrate some genetic divergence but are judged similar in their major "evolutionary tendencies," whereas the phylogenetic species concept would treat them as separate species. In general, a larger number of species would be described using the phylogenetic species concept than any other concept. The phylogenetic species concept is intended to encourage us to reconstruct patterns of evolutionary common descent on the finest scale possible.

Current disagreements concerning concepts of species should not be discouraging. Whenever a field of scientific investigation enters a phase of dynamic growth, old concepts will be reevaluated and either refined or replaced with newer, more progressive ones. The active debate occurring within systematics shows that this field has acquired unprecedented activity and importance in biology. Just as Thomas Henry Huxley's time was one of enormous advances in biology, so is the present time. Both times are marked by fundamental reconsiderations of the meaning of species. We cannot predict yet which, if any, of these concepts of species will prevail. Understanding the conflicting perspectives, rather than learning a single concept of species, is therefore of greatest importance for people now entering the study of zoology.

Major Divisions of Life

From Aristotle's time, people have tried to assign every living organism to one of two kingdoms: plant or animal. Unicellular forms were arbitrarily assigned to one of these kingdoms, whose recognition was based primarily on the properties of multicellular organisms. This system has outlived its usefulness. It does not represent common descent among organisms accurately. Under the traditional, two-kingdom system, neither the animals nor the plants constitute monophyletic groups.

Several alternative systems have been proposed to solve the problem of classifying unicellular forms. In 1866 Haeckel proposed the new kingdom Protista to include all single-celled organisms. At first the bacteria and cyanobacteria (blue-green algae), forms that lack nuclei bounded by a membrane, were included with nucleated unicellular organisms. Finally, the important differences between the anucleate bacteria and cyanobacteria (prokaryotes) and all other organisms that have membrane-bound nuclei (eukaryotes) were recognized. In 1969 R. H. Whittaker proposed a five-kingdom system that incorporated the basic prokaryote-eukaryote distinction. The kingdom Monera contained the prokaryotes. The kingdom Protista contained the unicellular eukaryotic organisms (protozoa and unicellular eukaryotic algae). The multicellular organisms were split into three kingdoms on the basis of mode of nutrition and other fundamental differences in organization. The kingdom Plantae included multicellular photosynthesizing organisms (higher plants and multicellular algae). Kingdom Fungi contained the molds, yeasts, and fungi that obtain their food by absorption. The invertebrates (except the protozoa) and the vertebrates form the kingdom Animalia. Most of these forms ingest their food and digest it internally, although some parasitic forms are absorptive.

All of these different systems were proposed without regard to the phylogenetic relationships that are needed to construct evolutionary or cladistic taxonomies. The oldest phylogenetic events in the history of life have been obscure, because the different forms of life share very few characters that can be compared among these higher taxa to reconstruct their phylogeny. Recently, however, a cladistic classification of all life forms has been proposed based on phylogenetic information obtained from molecular data (the nucleotide base sequence of ribosomal RNA, Figure 3-9). According to this tree, Woese, Kandler, and Wheelis (1990) recognized three monophyletic domains above the kingdom level: Eucarya (all eukaryotes), Bacteria (the true bacteria), and Archaea (prokaryotes differing from bacteria in structure of the membrane and in ribosomal RNA sequences). They did not divide the Eucarya into kingdoms, although if we retain Whittaker's kingdoms Plantae, Animalia, and Fungi, the Protista is paraphyletic because this group does not contain all descendants of its most recent common ancestor (Figure 3-10). To maintain a cladistic classification, the Protista must be discontinued by recognizing as separate kingdoms the Ciliata, Flagellata, and Microsporidia as shown in Figure 3-9, and phylogenetic

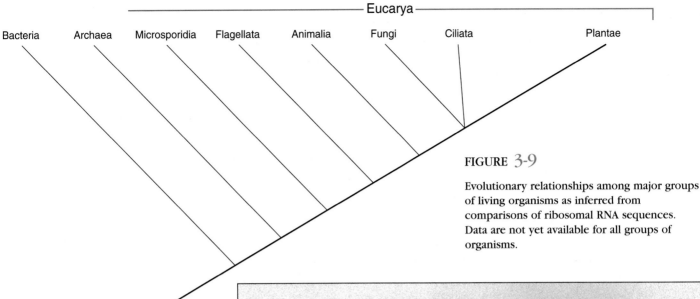

Eucarya

Bacteria   Archaea   Microsporidia   Flagellata   Animalia   Fungi   Ciliata   Plantae

FIGURE 3-9

Evolutionary relationships among major groups of living organisms as inferred from comparisons of ribosomal RNA sequences. Data are not yet available for all groups of organisms.

information must be gathered for additional protistan groups, including the amebas. This taxonomic revision has not been made; however, if the phylogenetic tree in Figure 3-9 is supported by further evidence, revision of the taxonomic kingdoms will be necessary.

Until recently, the animal-like protistans were traditionally studied in zoology courses as the animal phylum Protozoa. Given current knowledge and the principles of phylogenetic systematics, this practice commits two taxonomic errors; "protozoa" are neither animals nor are they a valid monophyletic taxon at any level. The kingdom Protista is likewise invalid because it is not monophyletic. The animal-like protistans, now divided into seven separate phyla, are nonetheless of interest to students of zoology because of their animal-like properties. We therefore include coverage of them in this book.

Major Subdivisions of the Animal Kingdom

The phylum is the largest formal taxonomic category in the Linnean classification of the animal kingdom. Animal phyla are often grouped together to produce

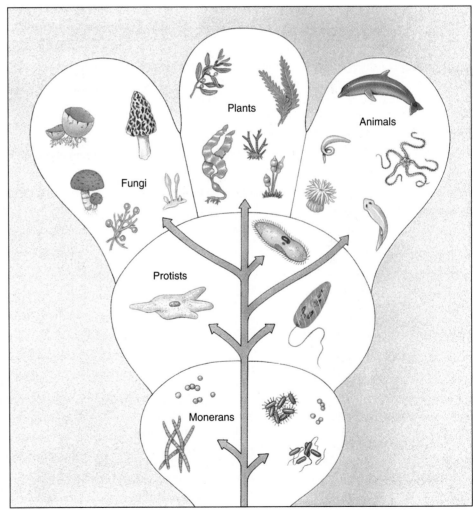

FIGURE 3-10

Whittaker's five-kingdom classification superimposed on a phylogenetic tree showing living representatives of these kingdoms. Note that the kingdoms Monera and Protista constitute paraphyletic groups (because they do not include all of their descendants) and are therefore unacceptable to cladistic systematics.

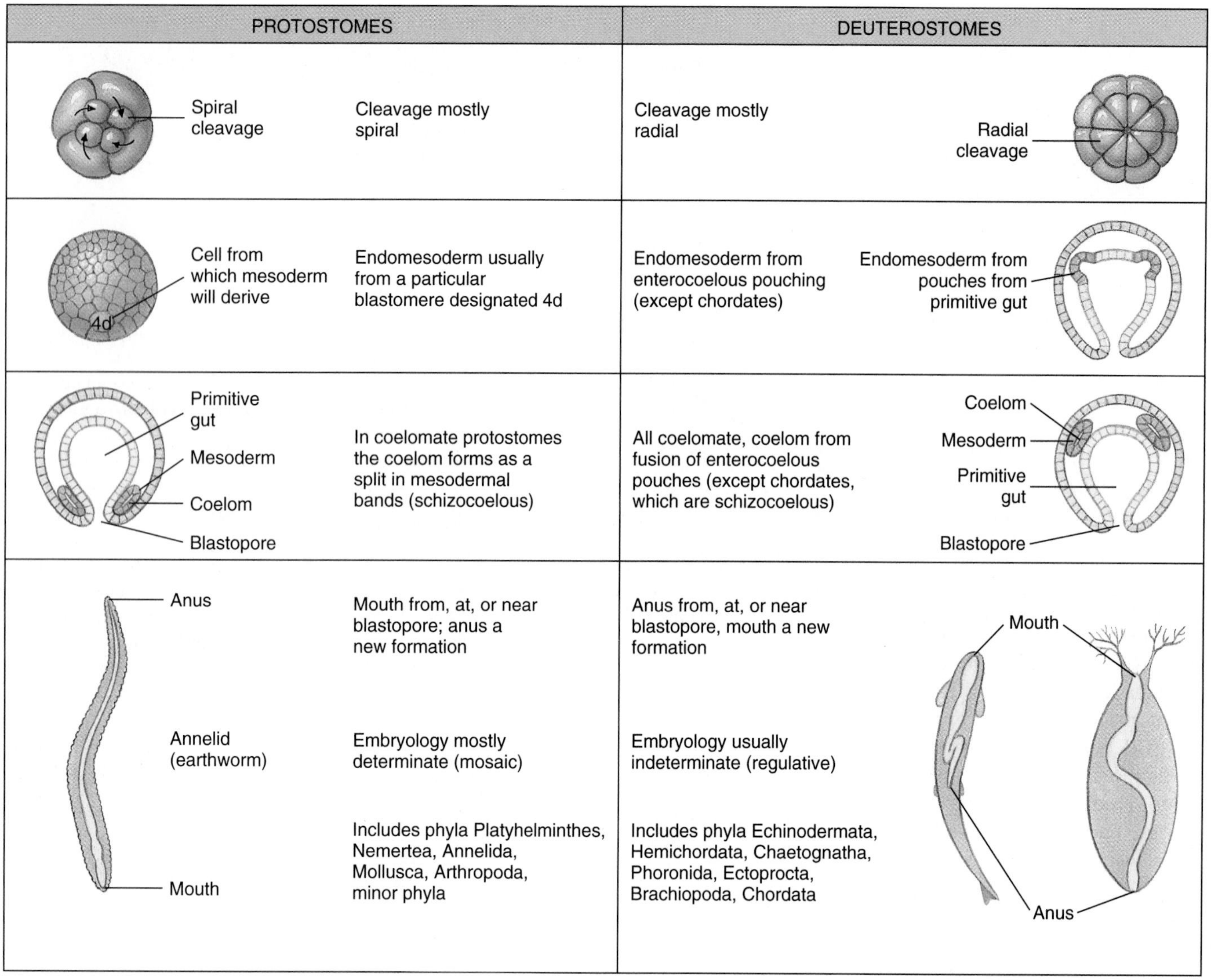

PROTOSTOMES		DEUTEROSTOMES	
Spiral cleavage	Cleavage mostly spiral	Cleavage mostly radial	Radial cleavage
Cell from which mesoderm will derive (4d)	Endomesoderm usually from a particular blastomere designated 4d	Endomesoderm from enterocoelous pouching (except chordates)	Endomesoderm from pouches from primitive gut
Primitive gut / Mesoderm / Coelom / Blastopore	In coelomate protostomes the coelom forms as a split in mesodermal bands (schizocoelous)	All coelomate, coelom from fusion of enterocoelous pouches (except chordates, which are schizocoelous)	Coelom / Mesoderm / Primitive gut / Blastopore
Anus / Annelid (earthworm) / Mouth	Mouth from, at, or near blastopore; anus a new formation	Anus from, at, or near blastopore, mouth a new formation	Mouth / Anus
	Embryology mostly determinate (mosaic)	Embryology usually indeterminate (regulative)	
	Includes phyla Platyhelminthes, Nemertea, Annelida, Mollusca, Arthropoda, minor phyla	Includes phyla Echinodermata, Hemichordata, Chaetognatha, Phoronida, Ectoprocta, Brachiopoda, Chordata	

FIGURE 3-11

Basis for the distinction between divisions of bilateral animals.

additional, informal taxa intermediate between the phylum and the animal kingdom. These taxa are based on embryological and anatomical characters that reveal the phylogenetic affinities of the different animal phyla. Zoologists in the past have recognized subkingdom Protozoa, which contains the primarily unicellular phyla, and the subkingdom Metazoa, which contains the multicellular phyla. As noted previously, however, the Protozoa are not a valid taxonomic group and do not belong within the animal kingdom, which is synonymous with the Metazoa. The higher-level groupings of the true animal phyla are as follows:

Branch A (Mesozoa): phylum Mesozoa, the mesozoa

Branch B (Parazoa): phylum Porifera, the sponges, and phylum Placozoa

Branch C (Eumetazoa): all other phyla

Grade I (Radiata): phyla Cnidaria, Ctenophora

Grade II (Bilateria): all other phyla

Division A (Protostomia): characteristics in Figure 3-11

Acoelomates: phyla Platyhelminthes, Gnathostomulida, Nemertea

Pseudocoelomates: phyla Rotifera, Gastrotricha, Kinorhyncha, Nematoda, Nematomorpha, Acanthocephala, Entoprocta, Priapulida, Loricifera

Eucoelomates: phyla Mollusca, Annelida,
Arthropoda, Echiurida, Sipunculida, Tardigrada,
Pentastomida, Onychophora, Pogonophora

Division B (Deuterostomia): characteristics in
Figure 3-11; phyla Phoronida, Ectoprocta,
Chaetognatha, Brachiopoda, Echinodermata,
Hemichordata, Chordata

As in the outline, the bilateral animals are customarily divided into **Protostomia** and **Deuterostomia** on the basis of their embryological development (Figure 3-11). Note, however, that the individual characters listed in Figure 3-11 are not completely diagnostic in separating protostomes from deuterostomes. Some of the phyla are difficult to place into one of these two categories because they possess characteristics of each group.

Summary

Animal systematics has three major goals: (1) to identify all species of animals, (2) to evaluate the evolutionary relationships among animal species, and (3) to group animal species hierarchically in taxonomic groups (taxa) that convey evolutionary relationships. The taxa are ranked to denote increasing inclusiveness as follows: species, genus, family, order, class, phylum, and kingdom. All of these ranks can be subdivided to signify taxa that are intermediate between them. The names of species are binomial, with the first name designating the genus to which the species belongs (first letter capitalized) followed by a species epithet (lowercase), both written in italics. Taxa at all other ranks are given single nonitalicized names.

Two major schools of taxonomy are currently active. Traditional evolutionary taxonomy groups species into higher taxa according to the joint criteria of common descent and adaptive evolution; such taxa have a single evolutionary origin and occupy a distinctive adaptive zone. A second approach, known as phylogenetic systematics or cladistics, emphasizes common descent exclusively in grouping species into higher taxa. Only monophyletic taxa (those having a single evolutionary origin and containing all descendants of the group's most recent common ancestor) are used in cladistics. In addition to monophyletic taxa, evolutionary taxonomy recognizes some taxa that are paraphyletic (having a single evolutionary origin but excluding some descendants of the most recent common ancestor of the group). Both schools of taxonomy exclude polyphyletic taxa (those having more than one evolutionary origin).

Both evolutionary taxonomy and cladistics require that patterns of common descent among species be assessed before higher taxa are recognized. Comparative morphology (including development), cytology, and biochemistry are used to reconstruct the nested hierarchical relationships among taxa that reflect the branching of evolutionary lineages through time. The fossil record provides estimates of the ages of evolutionary lineages. Comparative studies and the fossil record jointly permit us to reconstruct a phylogenetic tree representing the evolutionary history of the animal kingdom.

The biological species concept has guided recognition of most animal species. A biological species is defined as a reproductive community of populations (reproductively isolated from others) that occupies a specific niche in nature. A biological species is not immutable through time but changes during the course of evolution. Because the biological species concept may be difficult to apply in spatial and temporal dimensions, and because it excludes asexually reproducing forms, alternative concepts have been proposed. These alternatives include the evolutionary species concept and the phylogenetic species concept. No single concept of species is universally accepted by all zoologists.

Traditionally, all living forms were placed into two kingdoms (animal and plant) but more recently, a five-kingdom system (animals, plants, fungi, protistans, and monerans) has been followed. Neither of these systems conforms to the principles of evolutionary or cladistic taxonomy because they place single-celled organisms into either paraphyletic or polyphyletic groups. Based on our current knowledge of the phylogenetic tree of life, "protozoa" do not form a monophyletic group and they do not belong within the animal kingdom, which comprises multicellular forms (metazoa).

Review Questions

1. List in order, from most inclusive to least inclusive, the principal categories (ranks of taxa) in Carolus Linnaeus's system of classification.

2. Explain why the system for naming species that originated with Linnaeus is "binomial."

3. How do monophyletic, paraphyletic, and polyphyletic taxa differ? How do these differences affect the validity of such taxa for both evolutionary and cladistic taxonomies?

4. How are taxonomic characters recognized? How are such characters used to construct a cladogram?

5. What is the difference between a cladogram and a phylogenetic tree? Given a cladogram for a group of species, what additional information is needed to obtain a phylogenetic tree?

6. How would cladists and evolutionary taxonomists differ in their interpretations of the statement that humans evolved from apes, which evolved from monkeys?

7. How does the biological species concept differ from earlier typological concepts of a species? Why do evolutionary biologists prefer it to typological species concepts?

8. What problems have been identified with the biological species concept? How do other concepts of species attempt to overcome these problems?

9. What are the five kingdoms distinguished by Whittaker? How does their recognition conflict with the principles of cladistic taxonomy?

Selected References

See also general references on page 395.

Ereshefsky, M. (ed.). 1992. The units of evolution. Cambridge, Massachusetts, MIT Press. *A thorough coverage of concepts of species, including reprints of important papers on the subject.*

Hall, B. K. 1994. Homology: the hierarchical basis of comparative biology. San Diego, Academic Press. *A collection of papers discussing the many dimensions of homology, the central concept of comparative biology and systematics.*

Hillis, D. M., C. Moritz, and B. K. Mable (eds.). 1996. Molecular systematics, ed. 2. Sunderland, Massachusetts, Sinauer Associates, Inc. *A detailed coverage of the biochemical and analytical procedures of comparative biochemistry.*

Hull, D. L. 1988. Science as a process. Chicago, University of Chicago Press. *A study of the working methods and interactions of systematists, containing a thorough review of the principles of evolutionary, phenetic, and cladistic taxonomy.*

Maddison, W. P., and D. R. Maddison. 1992. MacClade version 3.01. Sunderland, Massachusetts, Sinauer Associates, Inc. *A program for MacIntosh computers that conducts phylogenetic analyses of systematic characters. The instruction manual stands alone as an excellent introduction to phylogenetic procedures. The computer program is user-friendly and excellent for instruction in addition to serving as a tool for analyzing real data.*

Mayr, E., and P. D. Ashlock. 1991. Principles of systematic zoology. New York, McGraw-Hill. *A detailed survey of systematic principles as applied to animals.*

Panchen, A. L. 1992. Classification, evolution, and the nature of biology. New York, Cambridge University Press. *Excellent explanations of the methods and philosophical foundations of biological classification.*

Links to the Internet

Visit this textbook's web site at http://www.mhhe.com/zoology to find live Internet links for each of the references listed below.

1. The Tree of Life. This site is a must-see for anyone interested in information on the classification and phylogeny of animals. Its navigator provides the ability to search for phylogenetic information on a wide range of animal groups. It provides links to much biological information on the web.

2. BIOSIS. BIOSIS is an Internet resource guide to animal taxa, animal diversity, various taxa, and mailing lists. Search for information by animal taxa or subject categories. This resource has an enormous number of links applicable for many zoological topics.

3. Taxonomy Resources. This site, maintained by the National Center of Biotechnology Information, provides information on systematics and molecular genetics.

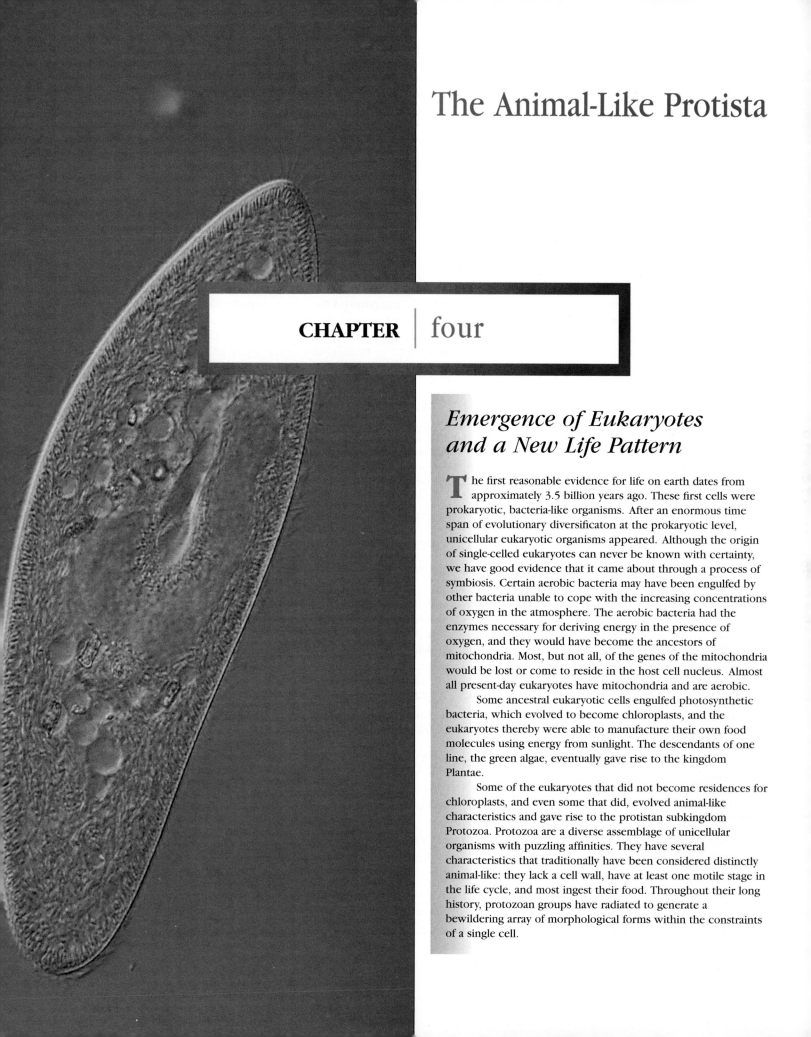

The Animal-Like Protista

CHAPTER four

Emergence of Eukaryotes and a New Life Pattern

The first reasonable evidence for life on earth dates from approximately 3.5 billion years ago. These first cells were prokaryotic, bacteria-like organisms. After an enormous time span of evolutionary diversificaton at the prokaryotic level, unicellular eukaryotic organisms appeared. Although the origin of single-celled eukaryotes can never be known with certainty, we have good evidence that it came about through a process of symbiosis. Certain aerobic bacteria may have been engulfed by other bacteria unable to cope with the increasing concentrations of oxygen in the atmosphere. The aerobic bacteria had the enzymes necessary for deriving energy in the presence of oxygen, and they would have become the ancestors of mitochondria. Most, but not all, of the genes of the mitochondria would be lost or come to reside in the host cell nucleus. Almost all present-day eukaryotes have mitochondria and are aerobic.

Some ancestral eukaryotic cells engulfed photosynthetic bacteria, which evolved to become chloroplasts, and the eukaryotes thereby were able to manufacture their own food molecules using energy from sunlight. The descendants of one line, the green algae, eventually gave rise to the kingdom Plantae.

Some of the eukaryotes that did not become residences for chloroplasts, and even some that did, evolved animal-like characteristics and gave rise to the protistan subkingdom Protozoa. Protozoa are a diverse assemblage of unicellular organisms with puzzling affinities. They have several characteristics that traditionally have been considered distinctly animal-like: they lack a cell wall, have at least one motile stage in the life cycle, and most ingest their food. Throughout their long history, protozoan groups have radiated to generate a bewildering array of morphological forms within the constraints of a single cell.

The animal-like Protista are those organisms known traditionally as protozoa. The protozoan is a complete organism in which all life activities are carried on within the limits of a single plasma membrane. For many years, "Protozoa" was considered a single phylum, but the several phyla of protozoa now recognized are included in this single chapter only for convenience.

The protozoan phyla do demonstrate a basic body plan or grade—the single eukaryotic cell (Figure 4-1)—and they amply demonstrate the enormous adaptive potential of that grade. Over 64,000 species have been named, and over half of these are fossil. Although they are unicellular, protozoa are not simple. They are functionally complete organisms with many complicated microanatomical structures. Their various organelles tend to be more specialized than those of the average cell in a multicellular organism. Particular organelles may perform as skeletons, sensory structures, conducting mechanisms, and feeding structures.

In its 1980 classification the Society of Protozoologists recognized seven phyla (see classification on pp. 84 to 85). The most important of these are the Sarcomastigophora (containing the flagellates and amebas), the Apicomplexa (important intracellular parasites, including the malarial organisms), and the Ciliophora (ciliates).

Protozoa are found wherever life exists. They are highly adaptable and easily distributed from place to place. They require moisture, whether they live in marine or freshwater habitats, soil, decaying organic matter, or plants and animals. They may be sessile or free swimming, and they form a large part of the floating plankton. Some species may have spanned geological eras of more than 100 million years.

Protozoa play an enormous role in the economy of nature. Their fantastic numbers are attested by the gigantic ocean and soil deposits formed by their skeletons. About 10,000 species of protozoa are symbiotic in or on animals or plants, or sometimes even other protozoa. The relationship may be **mutualistic** (both partners benefit), **commensalistic** (one partner benefits without affecting the other), or **parasitic** (one partner benefits at the expense of the other). Some of the most important diseases of humans and domestic animals are caused by parasitic protozoa.

A number of species are colonial and some have multicellular stages in their life cycles, which may lead one to wonder why such protozoa are not considered metazoa. The reasons are that they usually have clearly recognizable, noncolonial relatives and, more arbitrarily, that they do not have more than one kind of nonreproductive cell and they do not undergo embryonic development. By definition, metazoa have more than one kind of nonreproductive cell in their bodies and undergo embryogenesis.

Position Relative to the Animal Kingdom

The protozoan is a complete organism in which all life activities are carried on within the limits of a single cell membrane. Because their protoplasmic mass is not subdivided into cells, protozoa sometimes have been termed "acellular," but most people prefer "unicellular" to emphasize the many structural similarities to the cells of multicellular animals.

In the five-kingdom classification of living organisms, the protozoan phyla are considered members of the Protista. This group contains all the unicellular eukaryotes, as well as the multicellular algae, which are included because of their simple structure and clear relationship to unicellular algae. Because the word "protist" has been associated with small, unicellular organisms, some biologists object to using it for a group that contains multicellular algae, some of which are many meters long. They prefer **Protoctista** (Gr., *protos,* very first, + *ktistos,* to establish). Whether Protista or Protoctista, it is an extremely heterogeneous, paraphyletic group, some of whose members are more closely related to the multicellular kingdoms Plantae, Fungi, and Animalia than they are to each other. There can be little doubt, however, that the Animalia share common ancestry with one or more groups of animal-like protista, or protozoa.

Biological Contributions

1. **Intracellular specialization** (division of labor within the cell) involves the organization of functional organelles in the cell.
2. The simplest example of **division of labor between cells** is seen in certain colonial protozoa that have both somatic and reproductive zooids (individuals) in the colony.
3. **Asexual reproduction** by mitotic division appears in the protists.
4. **True sexual reproduction** with zygote formation is found in some protozoa.
5. The responses (taxes) of protozoa to stimuli represent the **simplest reflexes and instincts** as we know them in metazoans.
6. The simplest animal-like organisms with **exoskeletons** are certain shelled protozoa.
7. **All types of nutrition** are developed in the protozoa: autotrophic, saprozoic, and holozoic. **Basic enzyme systems** to accomplish these types of nutrition are developed.
8. Means of **locomotion** in aqueous media are developed.

Characteristics of Protozoan Phyla

1. **Unicellular;** some colonial, and some with multicellular stages in their life cycles
2. **Mostly microscopic,** although some are large enough to be seen with the unaided eye
3. All symmetries represented in the group; shape variable or constant (oval, spherical, or other)
4. **No germ layer present**
5. No organs or tissues, but **specialized organelles** are found; nucleus single or multiple
6. Free living, mutualism, commensalism, parasitism all represented in the group
7. Locomotion by **pseudopodia, flagella, cilia,** and direct cell movements; some sessile
8. Some provided with a **simple endoskeleton** or **exoskeleton,** but mostly naked
9. **Nutrition of all types:** autotrophic (manufacturing own nutrients by photosynthesis), heterotrophic (depending on other plants or animals for food), saprozoic (using nutrients dissolved in the surrounding medium)
10. Aquatic or terrestrial habitat; free-living or symbiotic mode of life
11. Reproduction **asexually** by fission, budding, and cysts, and **sexually** by conjugation or by syngamy (union of male and female gametes to form a zygote)

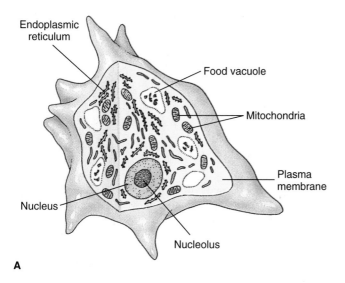

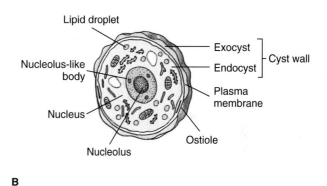

FIGURE 4-1

Structure of *Acanthamoeba palestinensis.* **A,** Active, feeding form. **B,** Cyst.

Form and Function

Inasmuch as protozoa are cells, in many aspects their structure and physiology are the same as those of cells of multicellular organisms. However, because they must carry on all the functions of life as individual organisms, and because they show such enormous diversity in form, habitat, feeding, and organization, many features are unique to various protozoan cells.

Locomotor Organelles

Cilia and Flagella

Cilia and **flagella** are major locomotory structures of protozoa, and they are no less important than pseudopodia for mul-

ticellular animals. Not only do many small metazoa use cilia for locomotion, the cilia of many create water currents for their feeding and respiration. Ciliary movement is vital to many species in such functions as handling food, reproduction, excretion, and osmoregulation. Spermatozoa of most animals, including humans, use flagella for locomotion.

Although we can observe that cilia and flagella differ in their beating patterns, their internal structure is the same. Each contains nine pairs of longitudinal microtubules arranged in a circle around a central pair (Figure 4-2), and this is true for all flagella and cilia in the animal kingdom, with a few notable exceptions. This "9 + 2" tube of microtubules in the flagellum or cilium is its **axoneme;** the axoneme is covered by a membrane continuous with the cell membrane covering the rest of the organism. At about the point where the axoneme enters

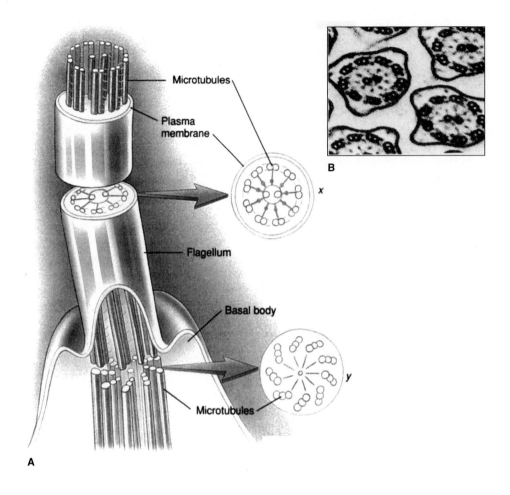

Microtubules

Plasma membrane

Flagellum

Basal body

Microtubules

A

B

FIGURE 4-2

A, The axoneme is composed of nine pairs of microtubules plus a central pair, and it is enclosed within the cell membrane. The central pair ends at about the level of the cell surface in a basal plate (axosome). The peripheral microtubules continue inward for a short distance to make up two of each of the triplets in the kinetosome (at level *y* in **A**). **B,** Electron micrograph of section through several flagella, corresponding to section at *x* in **A.** (×133,000)

the cell proper, the central pair of fibrils ends at a small plate within the circle of nine pairs (Figure 4-2). Also at about that point, another microtubule joins each of the nine pairs, so that these form a short tube extending from the base of the flagellum into the cell and consisting of nine *triplets* of microtubules. The short tube of nine triplets is the **kinetosome** and is exactly the same in structure as the **centriole** (see a general biology or a cell biology text). The centrioles of some flagellates may give rise to the kinetosomes, or the kinetosomes may function as centrioles. All typical flagella and cilia have a kinetosome at their base, regardless of whether they are borne by a protozoan or metazoan cell. The kinetosomes of protozoa have older, traditional names **(blepharoplast, basal body, basal granule)** that are still in common usage.

Description of the axoneme as "9 + 2" is traditional, but it is also misleading because there is only a single pair of microtubules in the center. To be consistent we would have to describe the axoneme as "9 + 1."

The current explanation for ciliary and flagellar movement is the **sliding microtubule hypothesis.** The movement is powered by the release of chemical bond energy in ATP. Two little arms are visible in electron micrographs on each of the pairs of peripheral tubules in the axoneme (Figure 4-2), and these bear the enzyme adenosine triphosphatase (ATPase), which cleaves the ATP. When the bond energy in ATP is released, the arms "walk along" one of the microtubules in the adjacent pair, causing it to slide relative to the other tubule. Shear resistance, causing the axoneme to bend when the microtubules slide past each other, is provided by "spokes" from one of the tubules in each doublet projecting toward the central pair of microtubules. These spokes are also visible in electron micrographs.

Pseudopodia

Although **pseudopodia** are the chief means of locomotion of members of the Sarcodina (see Classification, p. 84), they can be formed by a variety of flagellate protozoa, as well as by ameboid cells of many invertebrates. In fact, much of the

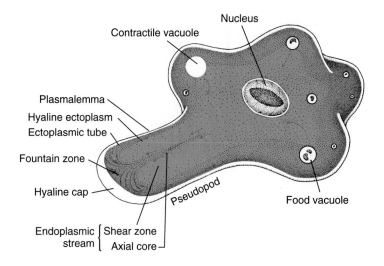

FIGURE 4-3

Ameba in active locomotion. Arrows indicate direction of streaming protoplasm. The first sign of a new pseudopodium is thickening of the ectoplasm to form a clear hyaline cap, into which the fluid endoplasm flows. As endoplasm reaches the forward tip, it fountains out and is converted into ectoplasm, forming a stiff outer tube that lengthens as the forward flow continues. Posteriorly the ectoplasm is converted into fluid endoplasm, replenishing the flow. Substratum is necessary for ameboid movement.

defense against disease in the human body depends on ameboid white blood cells, and ameboid cells in many other animals, vertebrate and invertebrate, play similar roles.

In the protozoa, pseudopodia exist in several forms. The most familiar are the lobopodia (Figures 4-3 and 4-4), which are rather large, blunt extensions of the cell body containing both central, granular **endoplasm** and **ectoplasm,** a peripheral, nongranular layer. The ectoplasm is in the gel state (jellylike semisolid), and the endoplasm is in the sol state (fluid). **Filipodia** are thin extensions, usually branching, and containing only ectoplasm. They are found in members of the sarcodine class Filosea, such as *Euglypha* (see Figure 4-8). **Reticulopodia** (see Figure 4-14) are distinguished from filipodia in that reticulopodia repeatedly rejoin to form a netlike mesh, although some protozoologists believe that the distinction between filipodia and reticulopodia is artificial. Members of the superclass Actinopoda have **axopodia** (see Figure 4-14), which are long, thin pseudopodia supported by axial rods of microtubules (Figure 4-5). The microtubules are arranged in a definite spiral or geometrical array, depending on the species, and constitute the axoneme of the axopod. Axopodia can be extended or retracted, apparently by addition or removal of microtubular material. Cytoplasm can flow along the axonemes, toward the body on one side and in the reverse direction on the other.

How pseudopodia work has long attracted the interest of zoologists, but we have only recently gained some insight into the phenomenon. When a typical lobopodium begins to form, an extension of ectoplasm called the **hyaline cap** appears, and endoplasm begins to flow toward and into the hyaline cap (Figure 4-3). As the endoplasmic material flows into the hyaline

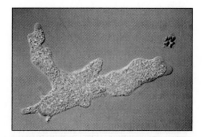

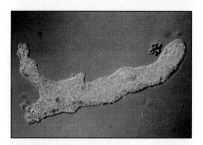

FIGURE 4-4

Ameboid movement leading to phagocytosis. *Top* and *center,* the ameba extends a pseudopodium toward a *Pandorina* colony. *Bottom,* the ameba surrounds the *Pandorina* before engulfing it by phagocytosis.

cap, it fountains out to the periphery and changes from the sol (fluid) to the gel (semisolid) state; that is, it becomes ectoplasm. Thus the ectoplasm is a tube through which the endoplasm flows as the pseudopodium extends. On the trailing side of the organism, ectoplasm becomes endoplasm. At some point the pseudopodium becomes anchored to the substrate, and the cell is drawn forward. The current hypothesis of pseudopodial movement involves the participation of actin, myosin, and other components. As endoplasm fountains out in the hyaline cap, actin subunits become polymerized into microfilaments, which in turn become cross-linked to each other by actin-binding protein (ABP) to form a gel; that is, the endoplasm becomes ectoplasm. At the "posterior" the ABP releases the microfilaments, which return to the sol state of endoplasm. Before the microfilaments are disassembled into actin subunits, they interact with myosin, contracting and causing the endoplasm to flow in the direction of the pseudopodium by hydrostatic pressure.

Nutrition and Digestion

Protozoa can be categorized broadly into autotrophs, which synthesize their own organic constituents from inorganic

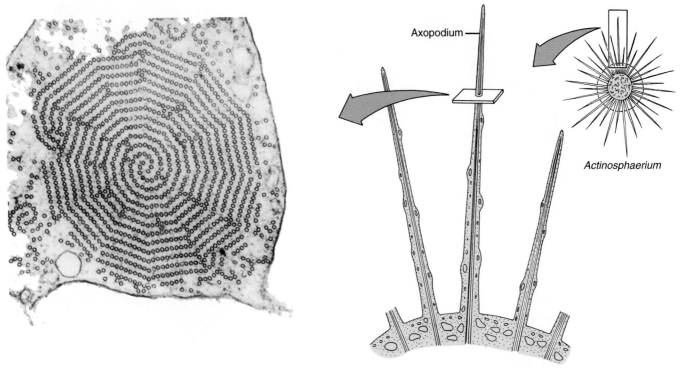

FIGURE 4-5

Diagram of axopodium *(right)* to show orientation of electron micrograph of axopodium (from *Actinosphaerium nucleofilum*) in cross section *(left)*. Protozoa with axopodia are shown in Figure 4-14 (*Actinophrys* and *Clathrulina*). The axoneme of the axopodium is composed of an array of microtubules, which may vary from three to many in number depending on the species. Some species can extend or retract their axopodia quite rapidly (×99,000).

substrates, and heterotrophs, which must have organic molecules synthesized by other organisms. Another kind of classification, usually applied to heterotrophs, involves those that ingest visible particles of food (**phagotrophs,** or **holozoic** feeders) as contrasted with those ingesting food in a soluble form (**osmotrophs,** or **saprozoic** feeders).

Holozoic nutrition implies phagocytosis (Figures 4-4 and 4-6), in which there is an infolding or invagination of the cell membrane around the food particle. As the invagination extends farther into the cell, it is pinched off at the surface. The food particle is thus contained in an intracellular, membrane-bound vesicle, the **food vacuole** or **phagosome.** Lysosomes, small vesicles containing digestive enzymes, fuse with the phagosome and pour their contents into it, where digestion begins. As the digested products are absorbed across the vacuole membrane, the phagosome becomes smaller. Any undigestible material may be released to the outside by exocytosis, the vacuole again fusing with the cell surface membrane. In most ciliates, many flagellates, and many apicomplexans, the site of phagocytosis is a definite mouth structure, the **cytostome** (Figures 4-6, 4-17, and 4-21). In amebas, phagocytosis can occur at almost any point by envelopment of the particle with pseudopodia. Many ciliates have a chracteristic structure for expulsion of waste matter,

the **cytopyge** or **cytoproct** (Figure 4-21), found in a characteristic location.

Excretion and Osmoregulation

Water balance, or osmoregulation, is a function of the one or more **contractile vacuoles** (see Figures 4-3, 4-13, and 4-21) possessed by most protozoa, particularly freshwater forms, which live in a hypoosmotic environment. These vacuoles are often absent in marine or parasitic protozoa, which live in a nearly isosmotic medium.

The contractile vacuoles usually are located in the ectoplasm and act as pumps to remove excess water from the cytoplasm. They are filled by droplets fed by a system of collecting canals in some species and from smaller vesicles in others (Figure 4-7). When full, they empty through a canal to the outside. The rate of pulsation varies; in *Paramecium* the posterior vacuole may pulsate faster than the anterior one because of water delivered there along with ingested food. Vacuoles in marine forms pulsate more slowly than in similar freshwater forms.

Nitrogenous wastes from metabolism apparently diffuse through the cell membrane, but some may also be emptied by way of the contractile vacuoles.

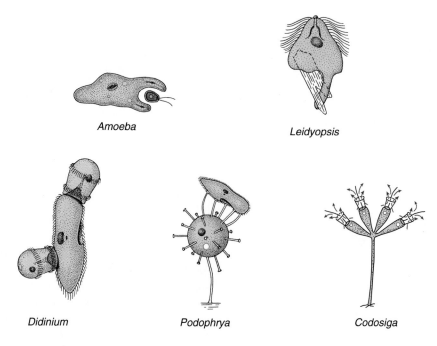

Amoeba

Leidyopsis

Didinium

Podophrya

Codosiga

FIGURE 4-6

Some feeding methods among protozoa. *Amoeba* surrounds a small flagellate with pseudopodia. *Leidyopsis,* a flagellate living in the intestine of termites, forms pseudopodia and ingests wood chips. *Didinium,* a ciliate, feeds only on *Paramecium,* which it swallows through a temporary cytostome in its anterior end. Sometimes more than one *Didinium* feed on the same *Paramecium. Podophrya* is a suctorian ciliophoran. Its tentacles attach to its prey and suck prey cytoplasm into its body, where the cytoplasm is pinched off to form food vacuoles. *Codosiga,* a sessile flagellate with a collar of microvilli, feeds on particles suspended in the water drawn through its collar by the beat of its flagellum. The particles are moved to the cell body and ingested (surrounded by small pseudopods). Technically all of these methods are types of phagocytosis.

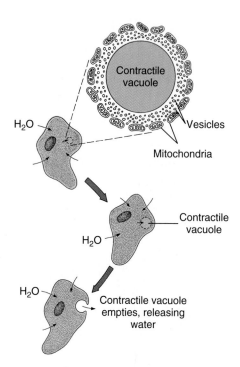

FIGURE 4-7

How an ameba pumps out water. Water enters the ameba's body by osmosis and is removed by the rhythmic filling and emptying of the contractile vacuole. The contractile vacuole of *Amoeba proteus* is surrounded by tiny vesicles that fill with fluid, which then empties into the vacuole. Note the numerous mitochondria that apparently provide energy needed to adjust the salt content of the tiny vesicles.

Reproduction

Sexual phenomena occur widely among the protozoa, and sexual processes may precede certain phases of asexual reproduction, but embryonic development does not occur; protozoa do not have embryos. The essential features of sexual processes include a reduction division of the chromosome number to half (diploid number to haploid number), the development of sex cells (gametes) or at least gamete nuclei, and usually a fusion of the gamete nuclei (p. 82).

Asexual Reproduction: Fission

The cell multiplication process that results in more individuals in the protozoa is called **fission.** The most common type of fission is **binary,** in which two essentially identical individuals result (Figures 4-8 and 4-9). When the progeny cell is considerably smaller than the parent and then grows to adult size, the process is called **budding.** This process occurs in some ciliates. In **multiple fission,** division of the cytoplasm (cytokinesis) is preceded by several nuclear divisions, so that a number of individuals is produced almost simultaneously (Figure 4-18).

Multiple fission, or **schizogony,** is common among the Apicomplexa and some amebas. When multiple fission leads to spore or sporozoite formation, it is called **sporogony.**

Protozoan Colonies Protozoan colonies are formed when the daughter zooids (individuals in the colony), derived asexually, remain associated instead of moving apart and living a separate existence (Figures 4-10 and 4-12). Protozoan colonies vary from individuals embedded together in a gelatinous substance to those having protoplasmic connections among them. The arrangement of the individuals results in a variety of colony types, each characteristic of the protozoan that forms it. Usually the individuals of a colony are structurally and physiologically the same, although there may be some division of labor, such as differentiation of reproductive and somatic zooids.

Sexual Processes

Although all protozoa reproduce asexually, and some are apparently exclusively asexual, the widespread occurrence of

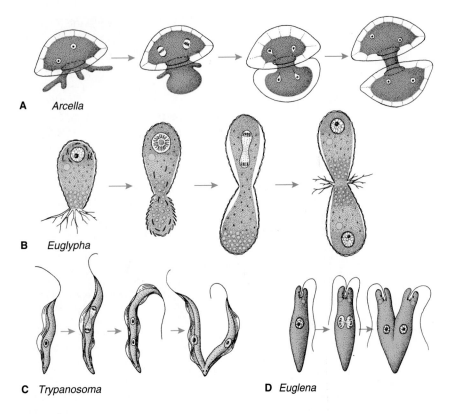

A *Arcella*

B *Euglypha*

C *Trypanosoma*

D *Euglena*

FIGURE 4-8

Binary fission in some sarcodines and flagellates. **A,** The two nuclei of *Arcella* divide as some of its cytoplasm is extruded and begins to secrete a new shell for the daughter cell. **B,** The shell of another sarcodine, *Euglypha,* is constructed of secreted platelets. Secretion of the platelets for the daughter cell begins before the cytoplasm begins to move out the aperture. As these form the shell of the daughter cell, the nucleus divides. **C,** *Trypanosoma* has a kinetoplast (part of the mitochondrion) near the kinetosome of its flagellum close to its posterior end in the stage shown. All of these parts must replicate before the cell divides. **D,** Division of *Euglena.*

sex among the protozoa testifies to the value of genetic recombination. The gamete nuclei, or pronuclei, which fuse in fertilization to restore the diploid number of chromosomes, are usually borne in special gamete cells. The gametes may be similar in appearance or unlike.

We are accustomed to thinking of the timing of meiosis as it occurs in metazoa, during or just before gamete formation, which is called **gametic meiosis** (Figure 4-11). Such is indeed the case in the Ciliophora and some flagellates and amebas. However, in other flagellates and in the Apicomplexa, the first divisions *after* fertilization (zygote formation) are meiotic **(zygotic meiosis)** (Figures 4-11 and 4-18), and all the individuals produced asexually (mitotically) in the life cycle up to the next zygote are haploid. In some sarcodines (foraminiferans) haploid and diploid generations alternate **(intermediary meiosis),** a phenomenon widespread among plants.

The fertilization of an individual gamete by another is called **syngamy,** but some sexual phenomena in protozoa do not involve that process. Examples are **autogamy,** in which gametic nuclei arise by meiosis and fuse to form a zygote within

the same organism that produced them, and **conjugation,** in which there is an exchange of gametic nuclei between paired organisms (conjugants). Conjugation will be described further in the discussion of the paramecium.

Life Cycles

Many protozoa have very complex life cycles; others have simple ones. A simple life cycle may consist of an active phase and a **cyst** (Figure 4-1). A cyst is a resistant, quiescent stage within a cyst wall. In some cases the cyst may be lacking. *Amoeba* has a relatively simple life cycle. The more complex life cycles include two or more stages in the active phase and a reproductive phase that may include sexual as well as asexual phenomena. For example, some protozoa have both a ciliated and a nonciliated stage; others have ameboid and flagellate stages; and still others have free-swimming and sessile stages. The most complex protozoan life cycles are found in the parasitic phylum Apicomplexa, a good example of which is *Plasmodium* (Gr. *plasma,* something molded), the malarial parasite (see Figure 4-18).

Encystment is common among protozoa, helping them withstand unfavorable conditions. Usually a complex series of events occurs when the organism encysts. It becomes quiescent, and many organelles, such as cilia, flagella, and the contractile vacuole, may dedifferentiate and disappear. A cyst wall is secreted over the surface so that the animal can withstand desiccation, temperature changes, and other harsh conditions. Reproductive cycles, such as budding, fission, and syngamy, may also occur in the encysted condition. The cysts of some species may be viable for many years.

Cysts of some soil-inhabiting and freshwater protozoa have amazing durability. The cysts of the soil ciliate Colpoda *can survive 7 days in liquid air and 3 hours at 100° C.* Colpoda *can survive in dried soil for up to 38 years, and cysts of a certain small flagellate (Podo) can survive up to 49 years! Not all cysts are so sturdy, however. Those of* Entamoeba histolytica *will tolerate gastric acidity but not desiccation, temperature above 50° C, or sunlight.*

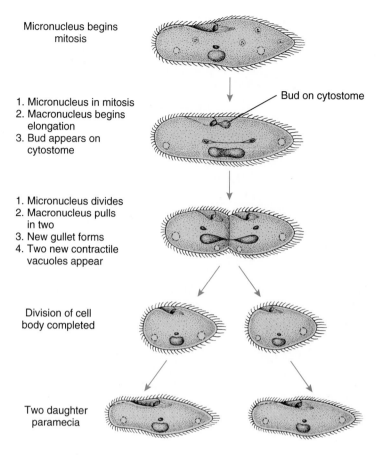

Micronucleus begins mitosis

1. Micronucleus in mitosis
2. Macronucleus begins elongation
3. Bud appears on cytostome

Bud on cytostome

1. Micronucleus divides
2. Macronucleus pulls in two
3. New gullet forms
4. Two new contractile vacuoles appear

Division of cell body completed

Two daughter paramecia

FIGURE 4-9

Binary fission in a ciliophoran (*Paramecium*). Division is transverse, across rows of cilia.

Phylum Sarcomastigophora

The Sarcomastigophora includes both protozoa that move by flagella (Mastigophora) and those that move by pseudopodia (Sarcodina). These characteristics are not mutually exclusive; some mastigophorans (flagellates) form and use pseudopodia, and a number of sarcodines (amebas) have flagellated stages in their life cycles.

Subphylum Mastigophora

The name "Mastigophora" means "whip bearing," which refers to the protozoans' flagella. These organisms are common in both fresh and marine waters. The group is subdivided into the **phytoflagellates** (Phytomastigophorea), most of which contain chlorophyll and are thus plantlike, and the **zooflagellates** (Zoomastigophorea), which, lacking chlorophyll, are holozoic or saprozoic and are thus animal-like (Figure 4-12).

The phytoflagellates are commonly called the "green flagellates," although they may appear green, yellow, brown,

or even colorless, according to the pigments present. Some are important producers, making up much of the base of the food chains of pelagic communities. As in *Euglena* (Figure 4-13), these forms have one or more characteristic **chloroplasts,** colored bodies that contain chlorophylls and other pigments. Some of the green flagellates have a light-sensitive **stigma,** or eyespot, a shallow pigment cup that allows light from only one direction to strike a light-sensitive receptor.

> *Among the most important of the phytoflagellate producers are the dinoflagellates. However, they can also damage other organisms, such as when they produce a "red tide." The name "red tide" was originally applied to situations in which the organisms reproduced in such profusion (producing a "bloom") that the water turned red from their color, but any instance of a dinoflagellate or algal bloom producing detectable levels of toxic substances is now called a red tide. The water may be red, brown, yellow, or not remarkably colored at all, and the phenomenon has nothing to do with tides! Red tides have resulted in considerable economic losses to the shellfish industry.*

Although phytoflagellates are primarily autotrophic, as just described, some are also saprozoic, some holozoic, and some use a combination of methods.

The zooflagellates, colorless because they lack chloroplasts, are holozoic or saprozoic. Many are parasitic, such as the various species of the leaf-shaped *Trypanosoma* (Gr. *trypanon,* auger, + *soma,* body) (Figure 4-12), a blood parasite that causes African sleeping sickness. Some have pseudopodia as well as flagella.

> *Two subspecies of* Trypanosoma brucei (T. b. rhodesiense *and* T. b. gambiense) *cause clinically distinct forms of African trypanosomiasis (sleeping sickness) in humans. Another subspecies, T. b. brucei, along with several other trypanosomes, causes a similar disease in domestic animals. This makes agriculture very difficult in large areas of Africa.* Trypanosoma cruzi *causes American trypanosomiasis, or Chagas' disease, a very serious disease in South and Central America.*

The movement of the flagellum is varied, making it a versatile propulsive organ. The flagellar beat is undulating, proceeding from one end to the other as succeeding waveforms or as a corkscrew or spiral form. In various groups the

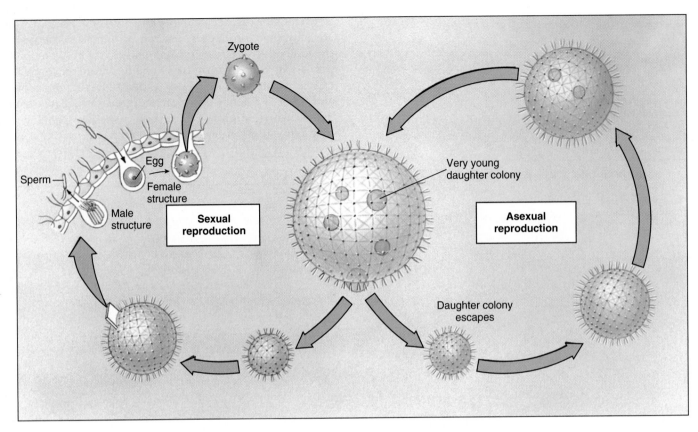

FIGURE 4-10

Life cycle of *Volvox.* Asexual reproduction occurs in spring and summer when specialized diploid reproductive cells divide to form young colonies that remain in the mother colony until large enough to escape. Sexual reproduction occurs largely in autumn when haploid sex cells develop. The zygote may encyst and so survive the winter, developing into a mature asexual colony in the spring. In some species the colonies have separate sexes; in others both eggs and sperm are produced in the same colony.

FIGURE 4-11

Timing of fertilization and meiosis (reduction division) demonstrated among various protozoa. *1,* Asexual reproduction only, all individuals presumably haploid. *2,* Zygotic meiosis, in which reduction division occurs immediately after fertilization (zygote formation). *3,* Intermediary meiosis, in which asexually reproducing, diploid individuals undergo meiosis to produce asexually reproducing, haploid individuals. Fertilization restores the diploid condition. *4,* Gametic meiosis, in which meiosis occurs during gamete formation, and all cells except gametes are diploid.

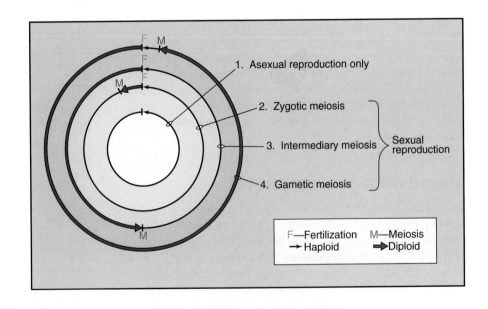

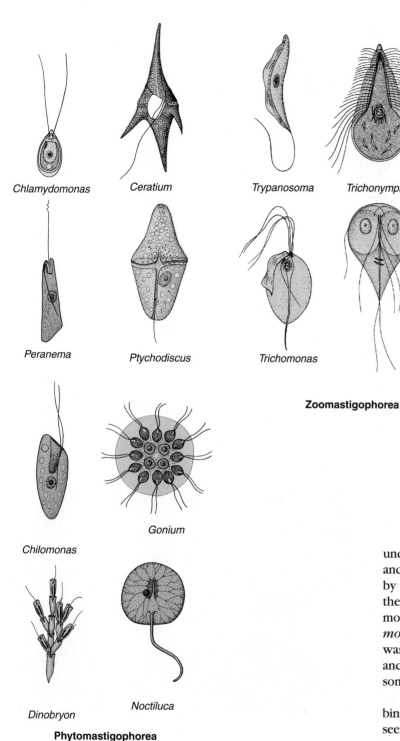

Zoomastigophorea

Chlamydomonas Ceratium Trypanosoma Trichonympha

Peranema Ptychodiscus Trichomonas Giardia

Chilomonas Gonium

Dinobryon Noctiluca

Phytomastigophorea

FIGURE 4-12

Some flagellate protozoa. *Gonium* and *Dinobryon* are colonial. *Ptychodiscus, Ceratium,* and *Noctiluca* are dinoflagellates.

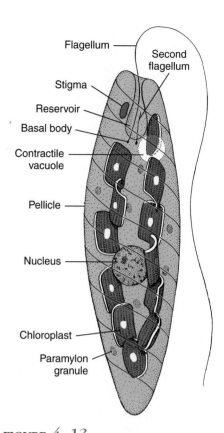

FIGURE 4-13

Euglena. Features shown are a combination of those visible in living and stained preparations.

undulation may begin at the tip or the base of the flagellum, and the protozoan may be pushed or pulled through the water by this beat. Many species have tiny, hairlike projections from the flagella to increase the efficiency of the beat. There may be more than one flagellum; *Trichomonas* (Gr. *thrix,* hair, + *monas,* single), for example, has four of them extending forward and one trailing backward, which apparently helps anchor and steer the animal. Flagella may also form rows somewhat like ciliary rows.

Most flagellates reproduce asexually by longitudinal binary fission (Figure 4-8C and D). However, multiple fission is seen in some stages of trypanosomes and others. The sexual process is rare except in some of the colonial forms such as *Volvox* (Figure 4-10). In *Volvox* certain reproductive zooids develop into eggs or bundles of sperm. Each fertilized egg (zygote) undergoes repeated divisions to produce a small colony, which is released in the spring. A number of asexual generations may follow before sexual reproduction occurs again.

Subphylum Sarcodina

The sarcodines characteristically move and feed by means of pseudopodia (Figures 4-3 and 4-14). Classification of sarcodines

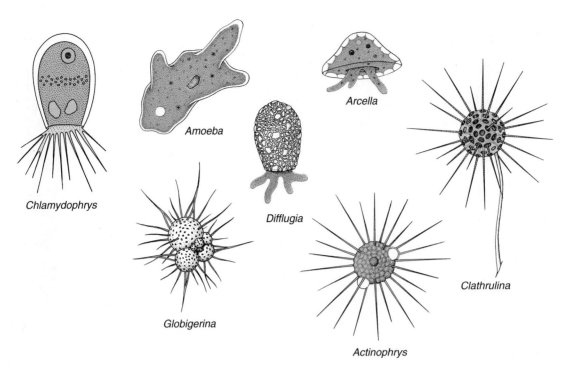

FIGURE 4-14

Diversity among the Sarcodina. *Difflugia*, *Arcella*, and *Amoeba* have lobopodia. *Chlamydophrys* has filopodia. The foraminiferan *Globigerina* shows reticulopodia. *Actinophrys* and *Clathrulina* have axopodia.

is based in part on the characteristics of their pseudopodia and and on characteristics of their protective **tests** (skeletons), if any. Sarcodines are found in both fresh and salt water and in moist soils. Some are planktonic; some prefer a substratum. A few are parasitic.

Nutrition in amebas and other sarcodines is holozoic; that is, they ingest and digest liquid or solid foods. Most amebas are omnivorous, living on algae, bacteria, protozoa, rotifers, and other microscopic organisms. An ameba may take in food at any part of its body surface merely by putting out a pseudopodium to enclose the food **(phagocytosis)**. The enclosed food particle, along with some of the environmental water, becomes a food vacuole, which is carried about by the streaming movements of endoplasm. As digestion occurs within the vacuole by enzymatic action, water and digested materials pass into the cytoplasm. Undigested particles are eliminated through the cell membrane.

Most sarcodines reproduce by binary fission. Sporulation and budding occur in some of the sarcodines.

Some sarcodines are covered with a protective test, or shell, with openings for the pseudopodia; in others the test is internal. Some tests are constructed of secreted siliceous material reinforced with sand grains (Figure 4-14). The **foraminiferans** (Foraminiferida [L. *foramen,* hole, + *fero,* bear]) are mostly marine forms that secrete complex many-chambered tests of calcium carbonate (Figure 4-15). They

A   **B**

FIGURE 4-15

A, Air-dried foraminiferan, showing spines extending from the test. **B,** A test of the foraminiferan *Vertebralima striata.* Foraminiferans (subclass Sarcodina) are ameboid marine protozoans that secrete a calcareous, many-chambered test in which to live, and then extrude protoplasm through pores to form a layer over the outside. The animal begins with one chamber and, as it grows, it secretes a succession of new and larger chambers, continuing this process throughout life. Many foraminiferans are planktonic, and when they die their shells are added to the ooze on the ocean's bottom.

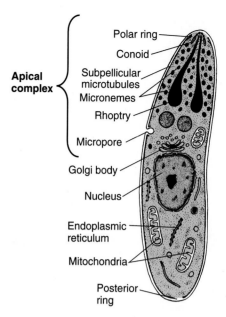

FIGURE 4-17

Diagram of an apicomplexan sporozoite or merozoite at the electron microscopic level, illustrating the apical complex. The polar ring, conoid, micronemes, rhoptries, subpellicular microtubules, and micropore (cytostome) are all considered components of the apical complex.

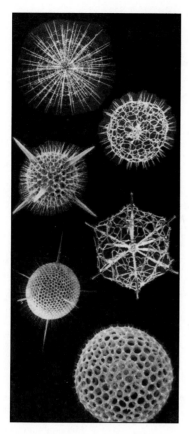

FIGURE 4-16

Types of radiolarian tests (subclass Sarcodina). In his study of these beautiful forms collected on the famous *Challenger* expedition of 1872 to 1876, Ernst Haeckel worked out our present concepts of symmetry.

usually add sand grains to the secreted material, using great selectivity in choosing colors. Slender pseudopodia extend through openings in the test and then run together to form a protoplasmic net, in which they ensnare their prey. They are beautiful little creatures with many pseudopodia radiating out from a central test. The **radiolarians** are marine forms, mostly living in plankton, that have intricate and beautiful siliceous skeletons (Figure 4-16). (Radiolaria [L. *radiolus,* small ray], used here as a common name, are now separated into two classes.) A central capsule that separates the inner and outer cytoplasm is perforated to allow cytoplasmic continuity.

The radiolarians are among the oldest known protists, and they and the foraminiferans have left excellent fossil records. For millions of years the tests of dead protozoa have been dropping to the seafloor, forming deep-sea sediments that are estimated to be from 600 to 3600 m deep (approximately 2000 to 12,000 feet) and containing as many as 50,000 foraminiferans per gram of sediment. Many limestone and chalk deposits on land were laid down by these small creatures when the land was covered by the sea.

Some of the amebas are endoparasitic, mostly in the intestine of humans or other animals. Most are harmless, but *Entamoeba histolytica* (Gr. *entos,* within, + *amoibē,* change; *histos,* tissue, + *lysis,* a loosing), the most common amebic parasite harmful to humans, causes amebic dysentery. In rare cases, certain free-living amebas cause a brain disease that is almost always fatal. They apparently gain access through the nose while a person is swimming in a pond or lake.

Phylum Apicomplexa

Class Sporozoea

All apicomplexans are endoparasites, and their hosts are found in many animal phyla. The presence of a certain combination of organelles, the **apical complex,** distinguishes this phylum (Figure 4-17). The apical complex is usually present only in certain developmental stages of the organisms; for example, **merozoites** and **sporozoites** (Figure 4-18). Some of the structures, especially the **rhoptries** and **micronemes,** apparently aid in penetrating the host's cells or tissues.

Locomotor organelles are not as obvious in this group as they are in other protozoa. Pseudopodia occur in some intracellular stages, and gametes of some species are flagellated. Tiny

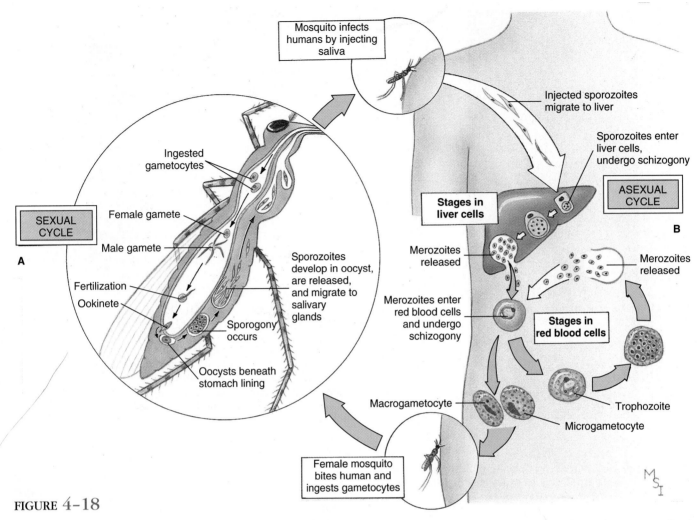

FIGURE 4-18

Life cycle of *Plasmodium vivax,* one of the protozoa (class Sporozoea) that causes malaria in humans. **A,** Sexual cycle produces sporozoites in the body of the mosquito. **B,** Sporozoites infect humans and reproduce asexually, first in liver cells, then in red blood cells. Malaria is spread by the *Anopheles* mosquito, which sucks up gametocytes along with human blood, then, when biting another victim, leaves sporozoites in the new wound.

contractile fibrils can form waves of contraction across the body surfaces to propel the organism through a liquid medium.

The life cycle usually includes both asexual and sexual reproduction, and there is sometimes an invertebrate intermediate host. At some point in the life cycle most apicomplexans develop a **spore (oocyst),** which is infective for the next host and is often protected by a resistant coat.

Plasmodium is the sporozoan parasite that causes **malaria.** The vectors (carriers) of the parasites are female *Anopheles* mosquitos, and the life cycle is depicted in Figure 4-18. Malaria is one of the most important diseases in the world, with 100 million cases occurring each year. It causes 1 million deaths per year worldwide. Though this represents a great decline from the situation 45 years ago, recent years have seen a resurgence caused by increasing resistance of mosquitos to insecticides, increasing numbers of *Plasmodium* strains that are resistant to drugs, and socioeconomic conditions and civil strife that interfere with malaria control efforts in tropical countries.

Although malaria has been recognized as a disease and a scourge of humanity since antiquity, the discovery that it was caused by a protozoan was made only 100 years ago by Charles Louis Alphonse Laveran, a French army physician. At that time, the mode of transmission was still mysterious, and "bad air" (hence the name malaria) was a popular candidate. Ronald Ross, an English physician in the Indian Medical Service, determined some years later that the malarial organism was carried by Anopheles *mosquitos. Because Ross knew nothing about mosquitos or their normal parasites, his efforts were long frustrated by trying to use the wrong kinds of mosquitos and confusion caused by the other parasites he found in them. That he persisted is a tribute to his determination. The discovery was nonetheless momentous, and it earned Ross the Nobel Prize in 1902 and knighthood in 1911.*

Toxoplasma (Gr. *toxo,* a bow, + *plasma,* molded) is a common parasite in the intestinal tissues of cats, but this parasite can produce extraintestinal stages as well. The extraintestinal stages can develop in a wide variety of animals other than cats—for example, rodents, cattle, and humans. Gametes and oocysts are not produced by the extraintestinal forms, but they can initiate the intestinal cycle in a cat, if the cat eats infected prey. In humans *Toxoplasma* causes few or no ill effects except in a woman infected during pregnancy, particularly in the first trimester. Such infection greatly increases the chances of a birth defect in the baby; about 2% of all mental retardation in the United States is a result of congenital toxoplasmosis. Humans can become infected from eating insufficiently cooked beef, pork, or lamb or from accidentally ingesting oocysts from the feces of cats. Pregnant women should not eat raw meat or empty cats' litterboxes.

Other common sporozoan parasites are the **coccidians** (Coccidia [Gr. *kokkos,* kernel or berry]), which infect epithelial tissues in both vertebrates and invertebrates, and the **gregarines,** which live mainly in the digestive tract and body cavity of certain invertebrates.

Some 20% or more of adults in the United States are infected with Toxoplasma gondii; *we have no symptoms because the parasite is held in check by our immune systems. However,* T. gondii *is one of the most important opportunistic infections in AIDS patients. In between 5 and 15% of AIDS patients the latent infection is activated, often in the brain, with serious consequences. Another coccidian,* Cryptosporidium parvum, *first was reported in humans in 1976. We now recognize it is a major cause of diarrheal disease worldwide, especially in children in tropical countries. Waterborne outbreaks have occurred in the United States, and the diarrhea can be life-threatening in immunocompromised patients (such as those with AIDS). The latest coccidian pathogen to emerge has been* Cyclospora cayetanensis. *About 850 cases of diarrhea due to* Cyclospora *in the United States and Canada were reported in May and June, 1996. We do not yet know how the parasite is transmitted nor even the normal host.*

Phylum Ciliophora

The ciliates are a large and interesting group, with a great variety of forms living in all types of freshwater and marine habitats. They are the most structurally complex and diversely specialized of all the protozoa. The majority are free living, although some are commensal or parasitic. They are usually solitary and motile, but some are colonial and some are sessile. There is a great diversity of shape and size. In general they are larger than most other protozoa, ranging from 10 μm to 3 mm long. At some stage all have cilia that beat in a coor-

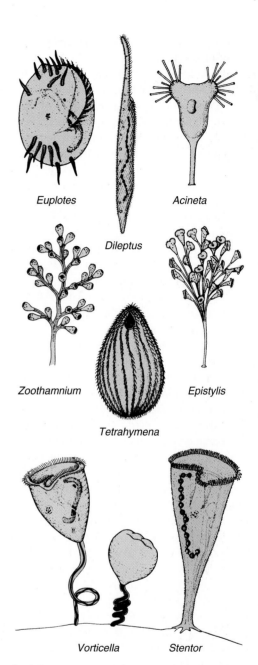

FIGURE 4-19

Some representative ciliates. *Euplotes* have stiff cirri used for crawling about. Contractile myonemes in the ectoplasm of *Stentor* and in the stalks of *Vorticella* allow great expansion and contraction. Note the macronuclei, long and curved in *Euplotes* and *Vorticella,* shaped like a string of beads in *Stentor.*

dinated rhythmical manner, although the arrangement of the cilia varies.

Ciliates are always multinucleate, possessing at least one **macronucleus** and one **micronucleus,** but varying from one to many of either type. The macronuclei are apparently responsible for metabolic, synthetic, and developmental functions. Macronuclei are varied in shape among the different species (Figures 4-19 and 4-21). The micronuclei participate in sexual reproduction and give rise to macronuclei after exchange of micronuclear material between individuals. The micronuclei divide mitotically, and the macronuclei divide amitotically.

The **pellicle** of ciliates may consist only of the cell membrane or in some species may form a thickened armor. The cilia are short and usually arranged in longitudinal or diagonal rows. Cilia may cover the surface of the animal or may be restricted to the oral region or to certain bands. In some forms the cilia are fused into a sheet called an **undulating membrane** or into smaller **membranelles,** both used to propel food into the **cytopharynx** (gullet). In other forms there may be fused cilia forming stiffened tufts called **cirri,** often used in locomotion by the creeping ciliates (Figure 4-19).

The kinetosomes and a structural system of fibrils just beneath the pellicle make up the **infraciliature** (Figure 4-20A). The infraciliature apparently does not coordinate the ciliary beat, as formerly believed. Coordination of the ciliary movement seems to be by waves of depolarization of the cell membrane moving down the animal, similar to the conduction of a nerve impulse.

Most ciliates are holozoic, possessing a cytostome (mouth) that in some forms is a simple opening and in others is connected to a gullet or ciliated groove. The mouth in some is strengthened with stiff, rodlike structures for swallowing larger prey; in others, such as the paramecia, ciliary water currents carry microscopic food particles toward the mouth. *Didinium* has a proboscis for engulfing the paramecia on which it feeds (Figure 4-6). Suctorians paralyze their prey and then ingest their contents through tube-like tentacles by a complex feeding mechanism that apparently combines phagocytosis with a sliding filament action of microtubules in the tentacles (Figure 4-6).

Some ciliates have curious small bodies in their ectoplasm between the bases of the cilia. Examples are **trichocysts** (Figures 4-20 and 4-21) and **toxicysts.** Upon mechanical or chemical stimulation, these bodies explosively expel a long, threadlike structure. The mechanism of expulsion is unknown. The function of trichocysts is probably defensive. When a paramecium is attacked by a *Didinium,* it expels its trichocysts but to no avail. Toxicysts, however, release a poison that paralyzes the prey of carnivorous ciliates. Toxicysts are structurally quite distinct from trichocysts. Many dinoflagellates have structures very similar to trichocysts.

Contractile vacuoles are found in all freshwater and some marine ciliates; the number varies from one to many among the different species. In most ciliates the vacuole is fed by one or more collecting canals (Figure 4-21). The vacuoles occupy a fixed position, and each discharges through a more-or-less permanent pore.

Reproduction and Life Cycles

The life cycles of ciliates usually involve both **asexual binary fission** and a type of sexual reproduction called **conjugation** (Figure 4-22). Conjugation is the temporary union of two individuals for the purpose of exchanging chromosomal material. During union the micronucleus of each individual undergoes meiosis, giving rise to four haploid micronuclei, three of which degener-

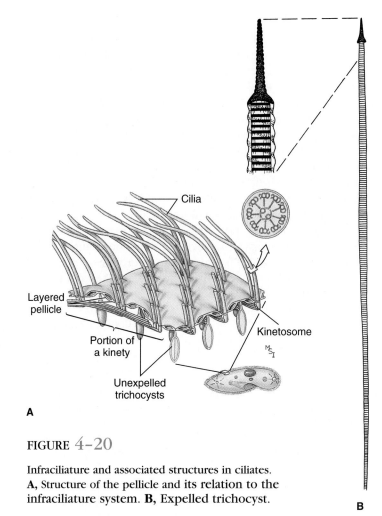

A

FIGURE 4–20

Infraciliature and associated structures in ciliates. **A,** Structure of the pellicle and its relation to the infraciliature system. **B,** Expelled trichocyst.

B

ate. The remaining micronucleus then divides into two haploid pronuclei, one of which is exchanged for a pronucleus of the conjugant partner. When the exchanged pronucleus unites with the pronucleus of the partner, the diploid number of chromosomes is restored. The two partners, each with fused pronuclei (now comparable to a zygote), separate, and each divides twice by mitosis, thereby giving rise to four daughter paramecia each.

Conjugation always involves two individuals of different **mating types.** This prevents inbreeding. Most ciliate species are divided into several varieties, each variety made up of two mating types. Mating occurs only between the differing mating types within each variety.

Phylogeny and Adaptive Radiation

Phylogeny

Protozoa represent an early phylogenetic split from all multicellular (metazoan) animals. The common ancestor of protozoan and metazoan animals was almost certainly unicellular, and the evolutionary derivation of multicellular forms probably has occurred more than once. Sponges, for example, may well have been derived separately from other metazoa. Some protozoa,

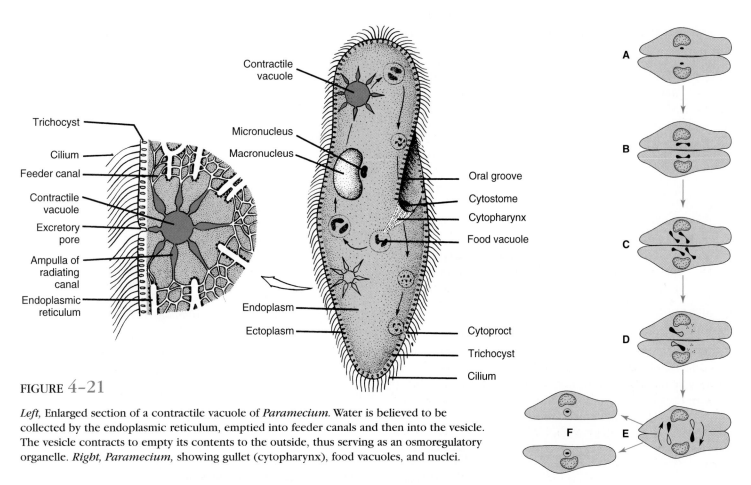

FIGURE 4-21

Left, Enlarged section of a contractile vacuole of *Paramecium.* Water is believed to be collected by the endoplasmic reticulum, emptied into feeder canals and then into the vesicle. The vesicle contracts to empty its contents to the outside, thus serving as an osmoregulatory organelle. *Right, Paramecium,* showing gullet (cytopharynx), food vacuoles, and nuclei.

FIGURE 4-22

Scheme of conjugation in *Paramecium.* **A,** Two individuals come in contact on oral surface. **B** and **C,** Micronuclei divide twice (meiosis), resulting in four haploid nuclei in each partner. **D,** Three micronuclei degenerate; remaining one divides to form "male" and "female" pronuclei. **E,** Male pronuclei are exchanged between conjugants. **F,** Male and female pronuclei fuse, and individuals separate. Subsequently old macronuclei are absorbed and replaced by new macronuclei.

particularly among the colonial and multicellular flagellates (Figure 4-10), show various degrees of cell aggregation and some differentiation that may parallel the body plans of early metazoa.

With the exception of certain shell-bearing Sarcodina, such as foraminiferans and radiolarians, protozoa have left no fossil records. Mastigophorans may be the oldest of all protozoa, perhaps having arisen from a combination of bacteria and spirochaetes, but the group is probably polyphyletic. Some of the phytoflagellates are more closely related to the multicellular algae than they are to zooflagellates. Some colorless phytoflagellates have chlorophyll-bearing relatives, and some autotrophic forms are facultatively saprophytic in darkness. Hence the common origin of animals and plants may lie in the Phytomastigophorea. That the amebas and flagellates probably share a common ancestor is indicated by the ameboid stages in some flagellates and the flagellated stages of some amebas.

Apicomplexa, which are all specialized parasites, probably come from flagellated ancestors; they often have ameboid feeding stages and flagellated gametes. The origin of the ciliates is somewhat obscure, but the basic structural similarity of the flagellum to the cilium seems to indicate that the ciliates and flagellates share common ancestry.

Adaptive Radiation

We have described some of the wide range of adaptations of protozoa in the previous pages. The Sarcodina range from bottom-dwelling, naked species to planktonic forms with beautiful, intricate tests such as the foraminiferans and radiolarians. There are many symbiotic species of amebas. Flagellates likewise show adaptations for a similarly wide range of habitats, with the added variation of photosynthetic ability in many species of Phytomastigophorea. In fact, botanists usually refer to these organisms as unicellular algae and divide them into some six to nine divisions (botanical equivalents of phyla).

Within a single-cell body plan, the division of labor and specialization of organelles are carried furthest by the ciliates, the most complex of all protozoa. Specializations for intracellular parasitism have been adopted by the Apicomplexa, Microspora, and Myxozoa.

Classification of Protozoan Phyla

The four main groups of protozoa traditionally recognized were flagellates, amebas, spore formers, and ciliates. The system that follows reflects a more nearly correct arrangement, including the recognition that amebas and flagellates are more closely related to each other than they are to other groups, and that the "spore formers" represent several completely unrelated forms. This taxonomy continues to follow the principles of evolutionary taxonomy rather than cladistic taxonomy because some clearly paraphyletic groups (marked by *) are included (see Chapter 3).

Kingdom Protista* (pro-tees´ta) (Gr. *protistos*, first of all). Single-celled eukaryotes and multicellular organisms with no more than one type of somatic cell (for example, multicellular algae).

Subkingdom Protozoa* (pro-to-zo´a) (Gr. *protos*, first, primary, + *zoōn*, animal). Animal-like protistans.

Phylum Sarcomastigophora* (sar´ko-mas-ti-gof´o-ra) (Gr. *sarkos*, flesh, + *mastix*, whip, + *phora*, bearing). Flagella, pseudopodia, or both types of locomotory organelle; usually with only one type of nucleus; typically no spore formation; sexuality, when present, essentially syngamy.

Subphylum Mastigophora* (mas-ti-gof´o-ra) (Gr. *mastix*, whip, + *phora*, bearing). One or more flagella typically present in adult stages; autotrophic or heterotrophic or both; reproduction usually asexual by fission.

Class Phytomastigophorea* (fi´to-mas-ti-go-for´e-a) (Gr. *phyton*, plant, + *mastix*, whip, + *phora*, bearing). Plantlike flagellates, usually bearing chloroplasts, which contain chlorophylls and other pigments. Examples: *Chilomonas, Euglena, Volvox, Ceratium, Peranema.*

Class Zoomastigophorea (zo´o-mas-ti-go-for´e-a) (Gr. *zoōn*, animal, + *mastix*, whip, + *phora*, bearing). Flagellates without chloroplasts; one to many flagella; ameboid forms with or without flagella in some groups; species predominantly symbiotic. Examples: *Trichomonas, Trichonympha, Trypanosoma, Leishmania, Dientamoeba.*

Subphylum Opalinata (o´pa-lin-a´ta) (NF *opaline*, like opal in appearance, + *-ata*, group suffix). Body covered with longitudinal rows of cilium-like organelles; parasitic; cytostome lacking; two to many nuclei of one type. Examples: *Opalina, Protoopalina.*

Subphylum Sarcodina (sar-ko-di´na) (Gr. *sarkos*, flesh, + *ina*, belonging to). Pseudopodia typically present; flagella present in developmental stages of some; cortical zone of cytoplasm relatively undifferentiated compared with other major taxa; body naked or with external or internal skeleton; free living or parasitic.

Superclass Actinopoda (ak´ti-nop´o-da) (Gr. *aktis, aktinos*, ray, + *pous, podos*, foot). Often spherical; usually planktonic; pseudopodia in form of axopodia, with microtubular supporting structure. Examples: *Actinosphaerium, Actinophrys, Thalassicolla.*

Superclass Rhizopoda (ri-zop´o-da) (Gr. *rhiza*, root, + *pous, podos*, foot). Locomotion by one of several pseudopodial types or by protoplasmic flow without production of discrete pseudopodia. Examples: *Amoeba, Entamoeba, Arcella.*

Phylum Labyrinthomorpha (la´bi-rinth-o-morf´a) (Gr. *labyrinth*, maze, labyrinth, + *morph*, form + *a*, suffix). Small group living on algae; mostly marine or estuarine.

Phylum Apicomplexa (a´pi-com-plex´a) (L. *apex*, tip or summit, + *complex*, twisted around, + *a*, suffix). Characteristic set of organelles (apical complex) associated with anterior end present in some developmental stages; cilia and flagella absent except for flagellated microgametes in some groups; cysts often present; all species parasitic.

Class Perkinsea (per-kin´se-a). Small group parasitic in oysters.

Class Sporozoea (spor´o-zo´e-a) (Gr. *sporos*, seed, + *zoōn*, animal). Spores or oocysts typically present that contain infective sporozoites; locomotion of mature organisms by body flexion, gliding, or undulation of longitudinal ridges; flagella present only in microgametes of some groups; pseudopodia ordinarily absent, if present they are used for feeding, not locomotion; one or two host life cycles. Examples: *Monocystis, Gregarina, Eimeria, Plasmodium, Toxoplasma, Babesia.*

Phylum Myxozoa (mix-o-zo´a) (Gr. *myxa*, slime, mucus, + *zoōn*, animal). Parasites of ectothermic vertebrates, mainly fishes, and invertebrates.

Phylum Microspora (mi-cros´por-a) (Gr. *micro*, small, + *sporos*, seed). Parasites of invertebrates, especially arthropods, and some vertebrates.

Phylum Ascetospora (as-e-tos´por-a) (Gr. *asketos*, curiously wrought, + *sporos*, seed). Small group that is parasitic in invertebrates and a few vertebrates.

Phylum Ciliophora (sil-i-of'or-a) (L. *cilium*, eyelash, + Gr. *phora*, bearing). Cilia or ciliary organelles in at least one stage of life cycle; two types of nuclei, with rare exceptions; binary fission across rows of cilia; budding and multiple fission; sexuality involving conjugation and autogamy; nutrition heterotrophic; contractile vacuole typically present; most species free living, but many commensal, some parasitic. (This is a very large group, now divided by the Society of Protozoologists into three classes and numerous orders and suborders. The classes are separated on technical characteristics of the ciliary patterns, expecially around the cytostome, the development of the cystostome, and other characteristics.) Examples: *Paramecium, Colpoda, Tetrahymena, Balantidium, Stentor, Blepharisma, Epidinium, Euplotes, Vorticella, Carchesium, Trichodina, Podophrya, Ephelota.*

Summary

The assemblage of protists known as protozoa is a large, heterogeneous group now recognized as being composed of seven phyla. The largest and most important of the phyla are the Sarcomastigophora (flagellates and amebas), the Apicomplexa (coccidians, malaria-causing organisms, and others), and the Ciliophora (ciliates). They demonstrate the great adaptive potential of the basic body plan, the single eukaryotic cell, and occupy a vast array of niches and habitats. Many species have complex and specialized organelles.

Ciliary movement is important in both protozoa and metazoa. The most widely accepted mechanism to account for ciliary movement is the sliding microtubule hypothesis. Pseudopodial or ameboid movement is a locomotory and food-gathering mechanism in protozoa and plays a vital role as a defense mechanism in metazoa. It is accomplished by microfilaments moving past each other, and it requires expenditure of energy from ATP.

Various protozoa feed by holophytic, holozoic, or saprozoic means. The excess water that enters their bodies is expelled by contractile vacuoles. Respiration and waste elimination are through the body surface. Protozoa reproduce asexually by binary fission, multiple fission, and budding; sexual processes are common.

Most phytoflagellates are photosynthetic, and many zooflagellates are important parasites. They move by beating one or more flagella. Sarcodines move by pseudopodia; many are important members of planktonic communities, and some are parasites. Many have a test, or shell. All apicomplexans are parasitic, including *Plasmodium*, which causes malaria. The Ciliophora move by means of cilia or ciliary organelles. They are a large and diverse group, and many are complex in structure.

Review Questions

1. Explain why a protozoan may be very complex, even though it is composed of only one cell.

2. Distinguish among the following protozoan phyla: Sarcomastigophora, Apicomplexa, Ciliophora.

3. Explain the transitions of endoplasm and ectoplasm in ameboid movement. What is a current hypothesis regarding the role of actin in ameboid movement?

4. Distinguish lobopodia, filipodia, reticulopodia, and axopodia.

5. Contrast the structure of an axoneme to that of a kinetosome.

6. What is the sliding microtubule hypothesis?

7. Explain how protozoa eat, digest their food, osmoregulate, and respire.

8. Distinguish the following: sexual and asexual reproduction; binary fission, budding, and multiple fission.

9. What is the survival value of encystment?

10. Contrast and give examples of phytoflagellates and zooflagellates.

11. Name three kinds of sarcodines and tell where they are found (their habitats).

12. Outline the general life cycle of malaria organisms. How do you account for the resurgence of malaria in recent years?

13. What is the public health importance of *Toxoplasma,* and how do humans become infected with it? What is the public health importance of *Cryptosporidium* and *Cyclospora?*

14. Define the following with reference to ciliates: macronucleus, micronucleus, pellicle, undulating membrane, cirri, infraciliature, trichocysts, toxicysts, conjugation.

15. What are indications that the Sarcodina, Apicomplexa, and Ciliophora may share common ancestry at some level with Phytomastigophorea?

Selected References

See also general references on page 395.

Anderson, O. R. 1988. Comparative protozoology: ecology, physiology, life history. New York, Springer-Verlag. *Good treatment of the aspects mentioned in the subtitle.*

Bonner, J. T. 1983. Chemical signals of social amoebae. Sci. Am. **248:**114–120 (April). *Social amebas of two different species secrete different chemical compounds that act as aggregation signals. The evolution of the aggregation signals in these protozoa may provide clues to the origin of the diverse chemical signals (neurotransmitters and hormones) in more complex organisms.*

Fenchel, T. 1987. Ecology of protozoa: the biology of free-living phagotrophic protists. Madison, Wisconsin, Science Tech Publishers.

Harrison, G. 1978. Mosquitoes, malaria and man: a history of the hostilities since 1880. New York, E. P. Dutton. *A fascinating story, well told.*

Kreier, J. P., and J. R. Baker. 1987. Parasitic protozoa. Boston, Allen and Unwin. *Concise summary of parasitic forms.*

Lee, J. 1993. On a piece of chalk—Updated. J. Eukaryotic Microbiol. **40:**395–410. *Excellent summary of current knowledge of foraminiferans.*

Lee, J. J., S. H. Hutner, and E.C. Bovee (eds.). 1985. An illustrated guide to the protozoa. Lawrence, Kansas, Society of Protozoologists. *A comprehensive guide and essential reference for students of the protozoa.*

Lee, J. J., and A. T. Soldo. 1992. Protocols in protozoology. Lawrence, Kansas, Society of Protozoologists. *A treasure trove of experimental and laboratory techniques for protozoa.*

Sleigh, M. A. 1989. Protozoa and other protists. London, Edward Arnold. *Extensively updated version of the author's The biology of protozoa.*

Stossel, T. P. 1994. The machinery of cell crawling. Sci. Am. **271:**54–63 (Sept.). *Ameboid movement—how cells crawl—is important throughout the animal kingdom, as well as in Protista. We now understand quite a bit about its mechanism.*

Links to the Internet

Visit this textbook's web site at *http://www.mhhe.com/zoology* to find live Internet links for each of the references listed below.

1. Society of Protozoologists. This site contains links to images of protozoans, databases, culture collections and discussion groups.

2. Protozoa. Various lectures on different protozoan groups are located at this site.

3. *Pfiesteria.* North Carolina Aquatic Botany Laboratory *P. piscicida* page with links and information on the toxic dinoflagellate *Pfiesteria.*

4. National Center for Infectious Diseases. This CDC site has many links to information on bacterial, viral, protozoan, and worm-related diseases (primarily affecting humans).

5. Trypanosomiasis. This CDC site contains information on both East African and West African trypanosomiasis.

6. Malaria. This CDC site has an enormous amount of information and links on malaria.

7. *Giardia lamblia.* A FDA-supported page with information on *G. lamblia,* a flagellated protozoan that causes diarrhea in humans.

8. *Entamoeba histolytica.* A FDA-supported page with information on *E. histolytica,* a protozoan that affects human and other mammals, causing gastrointestinal distress and diarrhea.

Sponges:
Phylum Porifera

CHAPTER | five

The Advent of Multicellularity

Sponges are the simplest multicellular animals. Because the cell is the elementary unit of life, the evolution of organisms larger than unicellular protozoa arose as an aggregate of such building units. Nature has experimented with producing larger organisms without cellular differentiation—certain large, single-celled marine algae, for example—but such examples are rarities. Typically nature has held stubbornly to multicellular construction in the progress toward higher organization. There are many advantages to multicellularity as opposed to simply increasing the mass of a single cell. Since it is at cell surfaces that exchange takes place, dividing a mass into smaller units greatly increases the surface area available for metabolic activities. It is impossible to maintain a workable surface-to-mass ratio by simply increasing the size of a single-celled organism. Thus multicellularity is a highly adaptive path toward increasing body size.

Strangely, although sponges are multicellular, they are phylogenetically distinct from other metazoans, forming a sister group to the Eumetazoa. They apparently split off from the metazoan line before the origin of the radiates (Chapter 6). The sponge body is an assemblage of cells embedded in a gelatinous matrix and supported by a skeleton of minute needlelike spicules and protein. Because sponges neither look nor behave like other animals, it is understandable that they were not completely accepted as animals by zoologists until well into the nineteenth century.

Sponges, in contrast to protozoa, are many-celled animals, or metazoa. However, they have the simplest type of metazoan organization—a **cellular level** of organization (p. 38). Since the sponge lacks true tissues or organs, the division of labor is confined to a few types of cells that have become specialized for certain functions.

Sponges belong to the phylum Porifera (po-rif′er-a) (L. *porus,* pore, + *fera,* bearing). They are sessile and their bodies bear myriads of tiny pores and canals that make up a filter-feeding system adequate for their inactive life-style, depending on the water currents carried through their unique canal systems to bring them food and oxygen and to carry away body wastes. Their bodies are little more than masses of cells embedded in a gelatinous matrix and stiffened by a skeleton of minute **spicules** of calcium or silica or by fibers of a collagenous substance called **spongin.** They have no organs or true tissues, and even their cells show a certain degree of independence. As sessile animals with only negligible body movement, they have not evolved a nervous system or sense organs and have only the simplest of contractile elements.

Most sponges are colonial, and they vary in size from a few millimeters to the great loggerhead sponges, which may reach 2 m or more across. Many sponge species are brightly colored because of pigments in the surface cells. Red, yellow, orange, green, and purple sponges are not uncommon. However, the color fades quickly when they are removed from water. Some sponges, including those with the simplest body plan, are radially symmetrical, but many are quite irregular in shape. Some stand erect, some are branched or lobed, and others are low, even encrusting, in form (Figure 5-1). Some bore holes into shells or rocks.

The sponges are an ancient group, with an abundant fossil record extending back to the Cambrian period and even, according to some claims, the Precambrian. Living poriferans have been traditionally assigned to three classes: Calcarea (with calcareous spicules), Hexactinellida (six-rayed siliceous spicules), and Demospongiae (with a skeleton of siliceous spicules or spongin or both). A fourth class (Sclerospongiae) was proposed to contain sponges with a massive calcareous skeleton and siliceous spicules. Some zoologists* believe that the class Sclerospongiae is not needed, but we will retain it for the present.

Ecological Relationships

Most of the 5000 or more sponge species are marine, although some 150 species of the Demospongiae live in fresh water. Marine sponges are much more abundant than most people realize; they are present in all seas and at all depths, and a few even exist in brackish water. Although the larvae are free swimming, the adults are always attached, usually to rocks, shells, corals, or other submerged objects (Figure 5-2). Some benthic forms even grow on sand and mud bottoms.

---

*Wood, R. 1990. Reef-building sponges. Am. Sci. **78**:224–235.

Position in Animal Kingdom

The multicellular organisms of the Kingdom Animalia (the metazoa) are typically divided into three grades: (1) Mesozoa (a single phylum), (2) Parazoa (phylum Porifera, the sponges; and phylum Placozoa), and (3) Eumetazoa (all other phyla). Space limitations preclude us from covering the very small phyla Mesozoa and Placozoa in this text.

Although Mesozoa and Parazoa are multicellular, their plan of organization is distinct from that in the eumetazoan phyla. Such cellular layers as they possess are not homologous to the germ layers of the Eumetazoa, and neither group has developmental patterns in line with the other metazoa. The name Parazoa means the "beside-animals."

Biological Contributions

1. Although the simplest in organization of all the metazoa, these groups do compose a higher level of morphological and physiological integration than that found in protozoan colonies. The Mesozoa and Parazoa may be said to belong to a **cellular level of organization.**
2. The sponges (poriferans) have several types of cells differentiated for various functions, some of which are organized into **incipient tissues** of a low level of integration.
3. The developmental patterns of the poriferans are different from those of other phyla, and their embryonic layers are not homologous to the germ layers of Eumetazoa.
4. The sponges have developed a unique system of **water currents** on which they depend for food and oxygen.

Many animals (crabs, nudibranchs, mites, bryozoans) live as commensals or parasites in or on sponges. The larger sponges particularly tend to harbor a large variety of invertebrate commensals. On the other hand, sponges grow on many other living animals, such as molluscs, barnacles, brachiopods, corals, or hydroids. Some crabs attach pieces of sponge to their carapaces for camouflage and for protection, since most predators seem to find sponges distasteful. A number of toxic or noxious substances have been found in sponges. Some reef fishes, however, are known to graze on shallow-water sponges. One sponge has been described that preys on shrimp.

Form and Function

The only body openings of these unusual animals are pores, usually many tiny ones called **ostia** for incoming water, and one to a few large ones called **oscula** (sing., **osculum**) for water outlet. These openings are connected by a system of canals, some of which are lined with peculiar flagellated collar cells called **choanocytes,** whose flagella maintain a current of environmental water through the canals. Water enters the canals through a multitude of tiny incurrent pores (**dermal ostia**) and leaves by way of one or more large oscula. The cells lining the passageways are very loosely organized. The skeleton prevents collapse of the canals. Depending on the species, the skeleton may be made up of needlelike calcareous or siliceous spicules, a meshwork of organic spongin fibers, or a combination of the two.

Sessile animals make few movements and therefore need little in the way of nervous, sensory, or locomotor parts. Sponges apparently have lived as sessile animals from their earliest appearance and have never acquired specialized nervous or sensory structures, and they have only the very simplest of contractile systems.

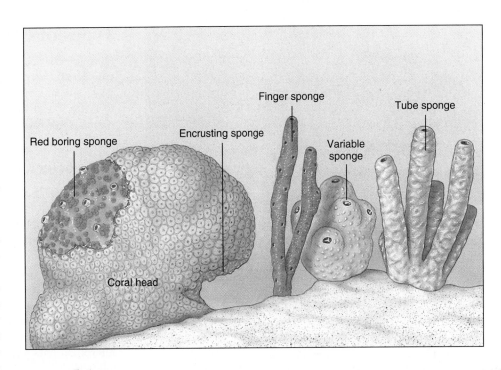

FIGURE 5-1

Some growth habits and forms of sponges.

Types of Canal Systems

Most sponges fall into one of three types, based on their canal systems—asconoid, syconoid, or leuconoid (Figure 5-3).

Asconoids—Flagellated Spongocoels

The asconoid sponges have the simplest type of organization. They are small and tube shaped. Water enters through microscopic dermal pores into a large cavity called the **spongocoel,** which is lined with choanocytes. The choanocyte flagella pull the water through the pores and expel it through the single osculum (Figure 5-3). *Leucosolenia* (Gr. *leukos,* white, + *solen,* pipe) is an asconoid type of sponge. Its slender, tubular individuals grow in groups attached by a common **stolon,** or stem, to objects in shallow seawater. *Clathrina* (L. *clathri,* lattice work) is an asconoid with bright yellow, intertwined tubes (Figure 5-4). Asconoids are found only in the Calcarea.

Syconoids—Flagellated Canals

Syconoid sponges look somewhat like larger editions of asconoids, from which they were derived. They have the tubular body and single osculum, but the body wall, which is thicker and more complex than that of asconoids, receives water through **incurrent canals** that deliver it to the choanocyte-lined **radial canals,** which empty into the spongocoel (Figures 5-3 and 5-5). The spongocoel in syconoids is

Characteristics of Phylum Porifera

1. Multicellular; body a loose aggregation of cells of mesenchymal origin
2. Body with **pores (ostia), canals,** and **chambers** that serve for passage of water
3. All aquatic; mostly marine
4. Symmetry radial or none
5. **Epidermis of flat pinacocytes;** most interior surfaces lined with **flagellated collar cells (choanocytes)** that create water currents; a gelatinous protein matrix called mesoglea contains amebocytes, collencytes, and skeletal elements
6. Skeletal structure of **fibrillar collagen** (a protein) and **calcareous** or **siliceous crystalline spicules,** often combined with variously modified **collagen (spongin) fibrils**
7. No organs or true tissues; digestion intracellular; excretion and respiration by diffusion
8. Reactions to stimuli apparently local and independent; nervous system probably absent
9. All adults sessile and attached to substratum
10. Asexual reproduction by buds or gemmules and sexual reproduction by eggs and sperm; free-swimming ciliated larvae

FIGURE 5-2

This orange demosponge, *Mycale laevis,* often grows beneath platelike colonies of the stony coral, *Montastrea annularis.* The large oscula of the sponge are seen at the edges of the plates. Unlike some other sponges, *Mycale* does not burrow into the coral skeleton and may actually protect the coral from invasion by more destructive species. Pinkish radioles of a Christmas tree worm, *Spirobranchus giganteus* (phylum Annelida, class Polychaeta), also project from the coral colony. An unidentified reddish sponge can also be seen to the right of the Christmas tree worm.

lined with epithelial-type cells, rather than the choanocytes as found in asconoids. Syconoids are found in Calcarea. *Sycon* (Gr. *sykon,* a fig) is a commonly studied example of the syconoid type of sponge (Figure 5-5).

Leuconoids—Flagellated Chambers

Leuconoid organization is the most complex of the sponge types and the best adapted for increase in sponge size. Most leuconoids form large colonial masses, each member of the mass having its own osculum, but individual members are poorly defined and often impossible to distinguish (Figure 5-2). Clusters of flagellated chambers are filled from **incurrent canals** and discharge water into **excurrent canals** that eventually lead to the osculum (Figure 5-3). Most sponges are of the leuconoid type, which occurs in most Calcarea and in all other classes.

These three types of canal systems—asconoid, syconoid, and leuconoid—demonstrate an increase in complexity and efficiency of the water-pumping system, but they do not imply an evolutionary or developmental sequence. Possession of the leuconoid plan is of clear adaptive value; it increases the proportion of flagellated surfaces compared with the volume, thus providing more collar cells to meet food demands. It makes possible a much larger body size than asconoid or syconoid grades.

Types of Cells

Sponge cells are loosely arranged in a gelatinous matrix called **mesohyl** (also called mesoglea, or mesenchyme) (Figure 5-6). The mesohyl is the "connective tissue" of the sponges; in it are found various ameboid cells, fibrils, and skeletal elements. There are several types of cells in sponges.

Pinacocytes

The nearest approach to a true tissue in sponges is found in the arrangement of the **pinacocyte** (Figure 5-6) cells of the external epithelium. These are thin, flat, epithelial-type cells that cover the exterior surface and some interior surfaces. Some are T-shaped, with their cell bodies extending into the mesohyl. Pinacocytes are somewhat contractile and help regulate the surface area of the sponge. Some of the pinacocytes are modified as contractile **myocytes,** which are usually arranged in circular bands around the oscula or pores, where they help regulate the rate of water flow. Myocytes contain microfilaments similar to those found in muscle cells of other animals.

Porocytes

The tubular cells that pierce the wall of asconoid sponges, through which water flows, are called **porocytes** (Figure 5-3).

Choanocytes

The **choanocytes,** which line the flagellated canals and chambers, are ovoid cells with one end embedded in the mesohyl and the other exposed. The exposed end bears a flagellum surrounded by a collar (Figures 5-6 and 5-7). The electron microscope shows that the collar is made up of adjacent microvilli, connected to each other by delicate microfibrils, so that the collar forms a fine filtering device for straining food particles from the water (Figure 5-7B and C). The beat of the flagellum pulls water through the sievelike collar and forces it out through the open top of the collar. Particles too large to enter the collar become trapped in secreted mucus and slide down the collar to the base where they are phagocytized by the cell body. Larger particles have already been screened out by the small size of the dermal pores and prosopyles. The food engulfed by the cells is passed on to a neighboring archaeocyte for digestion.

Archaeocytes

Archaeocytes are ameboid cells that move about in the mesohyl (Figure 5-6) and carry out a number of functions. They can phagocytize particles at the external epithelium and receive particles for digestion from the choanocytes. Archaeocytes apparently can differentiate into any of the other types of more specialized cells in the sponge. Some, called **sclerocytes,**

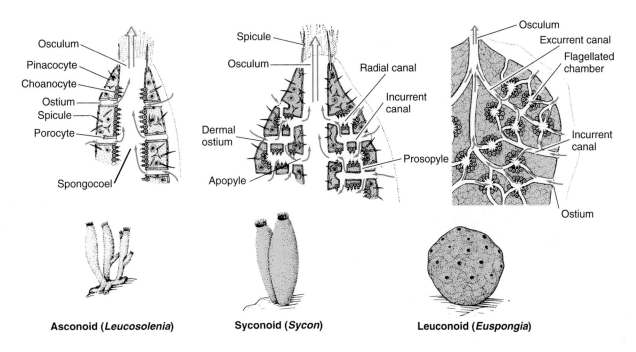

Asconoid (*Leucosolenia*) **Syconoid (*Sycon*)** **Leuconoid (*Euspongia*)**

FIGURE 5-3

Three types of canal systems. The degree of complexity from simple asconoid to complex leuconoid type has involved mainly the water canal and skeletal systems, accompanied by outfolding and branching of the collar cell layer. The leuconoid type is considered the major plan for sponges, because it permits greater size and more efficient water circulation.

secrete spicules. Others, called **spongocytes,** secrete the spongin fibers of the skeleton, and **collencytes** secrete fibrillar collagen.

Types of Skeletons

The skeleton gives support to the sponge, preventing collapse of the canals and chambers. The major structural protein in the animal kingdom is collagen, and fibrils of collagen are found throughout the intercellular matrix of all sponges. In addition, various Demospongiae secrete a form of collagen traditionally known as spongin. Demospongiae also secrete siliceous spicules, as do Sclerospongiae. Calcareous sponges secrete spicules composed mostly of crystalline calcium carbonate that have one, three, or four rays (Figure 5-8). Glass sponges have siliceous spicules with six rays arranged in three planes at right angles to each other. There are many variations in the shape of spicules, and these structural variations are of taxonomic importance.

FIGURE 5-4

Clathrina canariensis (class Calcarea) is common on Caribbean reefs in caves and under ledges.

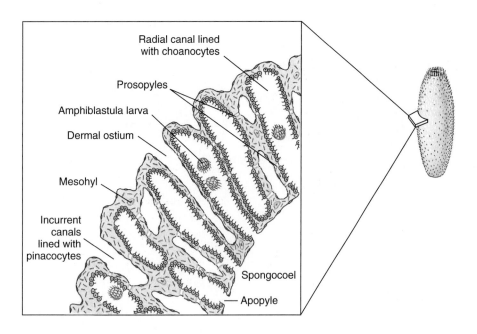

FIGURE 5-5

Cross section through wall of sponge *Sycon,* showing canal system.

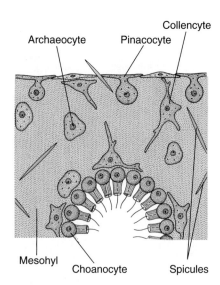

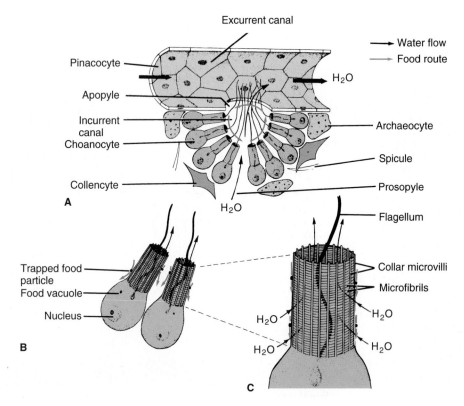

FIGURE 5-7

Food trapping by sponge cells. **A,** Cutaway section of canals showing cellular structure and direction of water flow. **B,** Two choanocytes. **C,** Structure of the collar. Small red arrows indicate movement of food particles.

FIGURE 5-6

Small section through sponge wall, showing four types of sponge cells. Pinacocytes are protective and contractile; choanocytes create water currents and engulf food particles; archaeocytes have a variety of functions, including phagocytosis of food particles and differentiation into other cell types; collencytes appear to have a contractile function.

Sponge Physiology

Sponges feed primarily on particles suspended in the water pumped through their canal systems. Detritus particles, planktonic organisms, and bacteria are consumed nonselectively in the size range from 50 μm (average diameter of ostia) to 0.1 μm (width of spaces between the microvilli of the choanocyte collar). Pinacocytes may phagocytize particles at the surface, but most of the larger particles are consumed in the canals by archaeocytes that move close to the lining of the canals. The smallest particles, accounting for about 80% of the particulate organic carbon, are phagocytized by the choanocytes. Digestion is entirely **intracellular** (occurs within cells), a chore performed by the archaeocytes.

Sponges consume a significant proportion of their nutrients in the form of organic matter dissolved in the water cir-

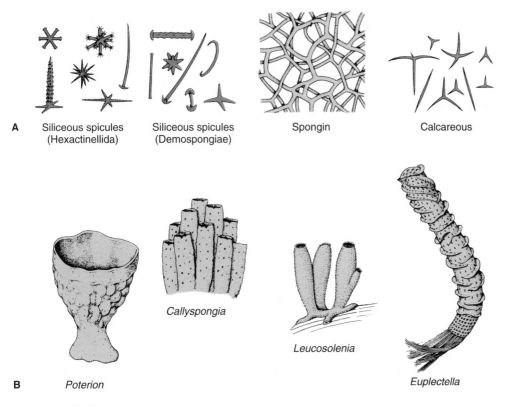

A Siliceous spicules (Hexactinellida)   Siliceous spicules (Demospongiae)   Spongin   Calcareous

Callyspongia

Leucosolenia

Euplectella

B *Poterion*

FIGURE 5-8

A, Types of spicules found in sponges. There is amazing diversity, beauty, and complexity of form among the many types of spicules. **B,** Some examples of sponge body forms.

Other sponges are oviparous, and both oocytes and sperm are expelled free into the water. The free-swimming larva of most sponges is a solid-bodied **parenchymula** (Figure 5-9). The outwardly directed, flagellated cells migrate to the interior after the larva settles and become the choanocytes in the flagellated chambers. In sexual reproduction ova are fertilized by motile sperm in the mesohyl; there the zygotes develop into flagellated larvae, which break loose and are carried away by water currents.

Certainly one reason for the success of sponges as a group is that they have few enemies. Because of a sponge's elaborate skeletal framework and often noxious odor or taste, most potential predators find sampling a sponge about as pleasant as eating a mouthful of glass splinters embedded in fibrous gelatin.

culating through the system. Such material is apparently taken up by pinocytosis (or potocytosis).

There are no respiratory or excretory organs; these functions are performed by diffusion. Contractile vacuoles have been found in archaeocytes and choanocytes of freshwater sponges.

All the life activities of the sponge depend on the current of water flowing through the body. A sponge pumps a remarkable amount of water. Some large sponges can filter 1500 liters of water a day.

Reproduction and Development

All sponges are capable of both sexual and asexual reproduction. In **sexual reproduction** most sponges are **monoecious** (have both male and female sex cells in one individual). Sperm arise from transformation of choanocytes. In Calcarea and at least some Demospongiae, oocytes also develop from choanocytes; in other demosponges oocytes apparently are derived from archaeocytes. Sperm are released into the water by one individual and are taken into the canal system of another. There choanocytes phagocytize them, then transform into carrier cells and carry the sperm through the mesohyl to the oocytes.

The loose organization of sponges is ideally suited for the regeneration of injured and lost parts, and for asexual reproduction. Sponges reproduce asexually by forming external buds that detach or remain to form colonies. In addition to external buds, which all sponges can form, freshwater sponges and some marine sponges reproduce asexually by the regular formation of internal buds called **gemmules** (Figure 5-10). These dormant masses of encapsulated archaeocytes are produced during unfavorable conditions. They can survive periods of drought and freezing; later, with the return of favorable conditions for growth, the archaeocytes in the gemmules escape and develop into new sponges.

Brief Survey of Sponges

Class Calcarea (Calcispongiae)

Calcarea are the calcareous sponges, so called because their spicules are composed of calcium carbonate. The spicules are straight monaxons or have three or four rays (Figure 5-8A). The sponges tend to be small—10 cm or less in height—and tubular or vase shaped. They may be asconoid, syconoid, or leuconoid in structure. Although many are drab, some are

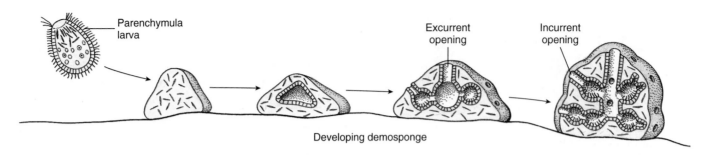

FIGURE 5-9

Development of demosponges.

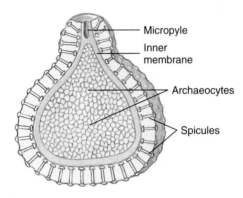

FIGURE 5-10

Section through a gemmule of a freshwater sponge (Spongillidae). Gemmules are a mechanism for survival of the harsh conditions of winter. On return of favorable conditions, the archaeocytes exit through the micropyle to form a new sponge. The archaeocytes of the gemmule give rise to all the cell types of the new sponge.

bright yellow, red, green, or lavender. *Leucosolenia, Clathrina* (Figure 5-4), and *Sycon* are common examples.

Class Hexactinellida (Hyalospongiae)

The glass sponges are nearly all deep-sea forms. Most of them are radially symmetrical and range from 7 to 10 cm to more than 1 m in length. They differ dramatically in structure from other sponges. One distinguishing feature, reflected in the class name, is the skeleton of six-rayed siliceous spicules bound together in an exquisite glasslike latticework (Figure 5-8A). The living tissue is mostly syncytical networks and sheets produced by the fusion of the pseudopodia of archaeocytes. Some free archaeocytes are present, but they have limited motility. The protoplasmic units that function as choanocytes have no nuclei. There is little mesohyl, and apparently they do not have myocytes. The structure of glass sponges is adapted to the slow, constant current of the sea bot-

tom, because the channels and pores are relatively large and permit an easy flow of water. The beautiful *Euplectella* (NL from Gr. *euplektos,* well-plaited), or the Venus' flower basket, is a classic example of this class.

Class Demospongiae

Demospongiae comprise approximately 80% of all sponge species, including most of the larger sponges. Their skeletons may be of siliceous spicules, spongin fibers, or both. All members of the class are leuconoid, and all are marine except one family, the freshwater Spongillidae. Freshwater sponges are widely distributed in well-oxygenated ponds and streams, where they are found encrusting plant stems and old pieces of submerged wood. They resemble a bit of wrinkled scum, pitted with pores, and are brownish or greenish in color. Freshwater sponges die and disintegrate in late autumn, leaving gemmules to survive the winter.

The marine Demospongiae are varied in both color and shape. Some are encrusting; some are tall and fingerlike; and some are shaped like fans, vases, cushions, or balls (Figure 5-11). Some sponges bore into and excavate molluscan shells and coral skeletons. Loggerhead sponges may grow several meters in diameter. The so-called bath sponges belong to the group called horny sponges, which have only spongin skeletons. They can be cultured by cutting out pieces of the individual sponges, fastening them to a weight, and dropping them into the proper water conditions. It takes many years for them to grow to market size. Most of the commercial "sponges" now on the market are synthetic, but the harvest and use of bath sponges persist.

Class Sclerospongiae

The Sclerospongiae are a small group of sponges that secrete a massive skeleton and are thus often called coralline sponges. Living tissue may extend as far as 3 cm into the skeleton but only 1 mm above it. With leuconoid organization and siliceous spicules and spongin in most, sclerosponges live

A   **B**   **C**

FIGURE 5-11

Marine Demospongiae on Caribbean coral reefs. **A,** *Pseudoceratina crassa* is a colorful sponge growing at moderate depths. **B,** *Ectyoplasia ferox* is irregular in shape and its oscula form small, volcano-like cones. It is toxic and may irritate the skin if touched. **C,** *Monanchora unguifera* with commensal brittle star, *Ophiothrix suensoni* (phylum Echinodermata, class Ophiuroidea).

in caves, crevices, and other such cryptic habitats on coral reefs or in deep water. They are thought to be relict representatives of ancient groups with a geological history extending from the Cambrian period.

Phylogeny and Adaptive Radiation

Phylogeny

Sponges originated before the Cambrian period. Two groups of calcareous spongelike organisms occupied early Paleozoic reefs. The Devonian period saw rapid development of many glass sponges. The possibility that sponges arose from choanoflagellates (protozoa that bear collars and flagella) earned support for a time. However, many zoologists object to that hypothesis because sponges do not acquire collars until late in their embryological development. The outer cells of the larvae are flagellated but not collared, and they do not become collar cells until they become internal. Also, collar cells are found in certain corals and echinoderms, so they are not unique to the sponges.

However, these objections are countered by new evidence based on the sequences of ribosomal RNA. This evidence supports the hypothesis of a common ancestor for choanoflagellates and metazoans. It suggests also that sponges and Eumetazoa are sister groups, with the Porifera having split off before the origin of the radiates and placozoans, but sharing a common ancestor.

Adaptive Radiation

The Porifera have been a highly successful group that has branched out into several thousand species and a variety of marine and freshwater habitats. Their diversification centers largely around their unique water-current system and its various degrees of complexity. The asconoid body plan is not efficient enough to move water through a spongocoel with a large cross-sectional area; therefore, asconoid sponges are all small. Development of the syconoid plan, with its flagellated canals, was an improvement, but the complex canals and flagellated chambers of the leuconoid sponges provided a more efficient means of moving large amounts of water through the colony. Their increased capacity for feeding and gaseous exchange have made larger size and adaptive radiation possible among the leuconoid sponges.

Classification of Phylum Porifera

Class Calcarea (cal-ca′re-a) (L. *calcis*, lime, + Gr. *spongos*, sponge) **(Calcispongiae).** Have spicules of calcium carbonate that often form a fringe around the osculum; spicules needle-shaped or three- or four-rayed; all three types of canal systems (asconoid, syconoid, leuconoid) represented; all marine. Examples: *Sycon, Leucosolenia, Clathrina.*

Class Hexactinellida (hex-ak-tin-el′i-da) (Gr. *hex*, six, + *aktis*, ray) **(Hyalospongiae).** Have six-rayed, siliceous spicules extending at right angles from a central point; spicules often united to form network; body often cylindrical or funnel shaped. Flagellated chambers in simple syconoid or leuconoid arrangement. Habitat mostly deep water; all marine. Examples: Venus' flower basket (*Euplectella*), *Hyalonema.*

Class Demospongiae (de-mo-spun′je-e) (tolerated misspelling of Gr. *desmos*, chain, tie, bond, + *spongos*, sponge). Have skeleton of siliceous spicules that are not six-rayed, or spongin, or both. Leuconoid-type canal systems. One family found in fresh water; all others marine. Examples: *Thenea, Cliona, Spongilla, Myenia*, and all bath sponges.

Class Sclerospongiae (skler′o-spun′je-e) (Gr. *skleros*, hard, + *spongos*, sponge). Secrete massive basal skeleton of calcium carbonate, with living tissue extending into skeleton from 1 mm to 3 cm or more, extending above skeleton less than 1 mm; have siliceous spicules similar to Demospongiae (sometimes absent), and spongin fibers; leuconoid organization; inhabit caves, crevices, tunnels, and deep water on coral reefs. Examples: *Astrosclera, Calcifibrospongia.*

Summary

The sponges (phylum Porifera) are an abundant marine group with some freshwater representatives. They have various specialized cells, but these cells are not organized into tissues or organs. They depend on the flagellar beat of their choanocytes to circulate water through their bodies for gathering food and exchange of respiratory gases. They are supported by secreted skeletons of fibrillar collagen, collagen in the form of large fibers or filaments (spongin), calcareous or siliceous spicules, or a combination of spicules and spongin in most species.

Sponges reproduce asexually by budding, fragmentation, and gemmules (internal buds). Most sponges are monoecious but produce sperm and oocytes at different times. Embryogenesis is unusual, with a migration of flagellated cells at the surface to the interior (parenchymella). Sponges have great regenerative abilities.

Sponges are an ancient group, remote phylogenetically from other metazoa, but some evidence suggests that they are a sister group to the Eumetazoa. Their adaptive radiation is centered on elaboration of the water circulation and filter feeding system.

Review Questions

1. Give six characteristics of sponges.
2. Briefly describe asconoid, syconoid, and leuconoid body types in sponges.
3. What sponge body type is most efficient and makes possible the largest body size?
4. Define the following: ostia, osculum, spongocel, mesohyl.
5. Define the following: pinacocytes, choanocytes, archaeocytes, sclerocytes, collencytes.
6. What material is found in the skeleton of all sponges?
7. Describe the skeletons of each of the classes of sponges.
8. Describe how sponges feed, respire, and excrete.
9. What is a gemmule?
10. Describe how gametes are produced and the process of fertilization in most sponges.
11. What is the largest class of sponges, and what is its body type?
12. What are the closest relatives of sponges? Justify your answer.
13. It has been suggested that despite being large, multicellular animals, sponges function more like protozoa. What aspects of sponge biology support this statement and how? Consider, for example, nutrition, reproduction, gas exchange, and cellular organization.

Selected References

See also general references on page 395.

Bergquist, P. R. 1978. Sponges. Berkeley, California, University of California Press. *Excellent monograph on sponge structure, classification, evolution, and general biology.*

Gould, S. J. 1995. Reversing established orders. Nat. Hist. **104**(9):12–16. *Describes several anomalous animal relationships, including the sponge that preys on shrimp.*

Hartman, W. D. 1982. Porifera. In S. P. Parker (ed.). Synopsis and classification of living organisms, vol. 1. New York, McGraw-Hill Book Company. *Review of sponge classification.*

Simpson, T. L. 1984. The cell biology of sponges. New York, Springer-Verlag. *A review and synthesis, points out many problems yet to be solved.*

Wainright, P. O., G. Hinkle, M. L. Sogin, S. K. Stickel. 1993. Monophyletic origins of the Metazoa: an evolutionary link with Fungi. Science **260**:340–342. *Reports molecular evidence that the sister group of metazoans is Fungi and that multicellular animals are monophyletic.*

Wood, R. 1990. Reef-building sponges. Am. Sci. **78**:224–235. *The author presents evidence that the known sclerosponges belong to either the Calcarea or the Demospongiae and that a separate class Sclerospongiae is not needed.*

Links to the Internet

Visit this textbook's website at *http://www.mhhe.com/zoology* to find live Internet links for each of the references listed below.

1. Porifera: Fossil Record. The fossil history, life history, ecology, systematics, and morphology of sponges are described at this site.

2. Phylum Porifera. University of Michigan site on phylum Porifera. Pictures, information on the phylum. Links to the four (some identify three) classes of sponges.

3. Introduction to Porifera. This site describes the evolutionary history, life history and ecology of the sponges. Many links to other sites are available.

4. Porifera. Arizona's Tree of Life Web Page. Pictures, references, and information on sponges.

The Radiate Animals:
Cnidarians and Ctenophores

CHAPTER | six

A Fearsome Tiny Weapon

Although members of the phylum Cnidaria are more highly organized than sponges, they are still relatively simple animals. Most are sessile; those that are unattached, such as jellyfish, can swim only feebly. None can chase their prey. Indeed, we might easily get the false impression that the cnidarians were placed on earth to provide easy meals for other animals. The truth is, however, many cnidarians are very effective predators that are able to kill and eat prey that are much more highly organized, swift, and intelligent. They manage these feats because they possess tentacles that bristle with tiny, remarkably sophisticated weapons called nematocysts.

As it is secreted within the cell that contains it, a nematocyst is endowed with potential energy to power its discharge. It is as though a factory manufactured a gun, cocked and ready with a bullet in its chamber, as it rolls off the assembly line. Like the cocked gun, the completed nematocyst requires only a small stimulus to make it fire. Rather than a bullet, a tiny thread bursts from the nematocyst. Achieving a velocity of 2 m/sec and an acceleration of $40,000 \times$ gravity, it instantly penetrates its prey and injects a paralyzing toxin. A small animal unlucky enough to brush against one of the tentacles is suddenly speared with hundreds or even thousands of nematocysts and quickly immobilized. Some nematocyst threads can penetrate human skin, resulting in sensations ranging from minor irritation to great pain, even death, depending on the species. A fearsome, but wondrous, tiny weapon.

Phylum Cnidaria

The phylum Cnidaria (ny-dar′e-a) (Gr. *knidē*, nettle, + L. *aria* [pl. suffix]; like or connected with) is an interesting group of more than 9000 species. It takes its name from cells called **cnidocytes,** which contain the stinging organelles **(nematocysts)** characteristic of the phylum. Nematocysts are *formed and used* only by cnidarians. Another name for the phylum, Coelenterata (se-len′te-ra′ta) (Gr. *koilos,* hollow, + *enteron,* gut, + L. *ata* [pl. suffix], characterized by), is used less commonly than formerly, and it sometimes now refers to both radiate phyla, since its meaning is equally applicable to both.

The cnidarians are generally regarded as originating close to the basal stock of the metazoan line. They are an ancient group with the longest fossil history of any metazoan, reaching back more than 700 million years. Although their organization has a structural and functional simplicity not found in other metazoans, they form a significant proportion of the biomass in some locations. They are widespread in marine habitats, and there are a few in fresh water. Although they are mostly sessile or, at best, fairly slow moving or slow swimming, they are quite efficient predators of organisms that are much swifter and more complex. The phylum includes

Position in Animal Kingdom

The two phyla Cnidaria and Ctenophora make up the radiate animals, which are characterized by **primary radial** or **biradial symmetry,** which we believe is ancestral for the eumetazoans. Radial symmetry, in which the body parts are arranged concentrically around the oral-aboral axis, is particularly suitable for sessile or sedentary animals and for free-floating animals because they approach their environment (or it approaches them) from all sides equally. Biradial symmetry is basically a type of radial symmetry in which only two planes through the oral-aboral axis divide the animal into mirror images because of the presence of some part that is paired. All other eumetazoans have a primary bilateral symmetry; that is, they are bilateral or were derived from an ancestor that was bilateral.

Neither phylum has advanced generally beyond the **tissue level of organization,** although a few organs occur. In general, the ctenophores' structure is more complex than that of the cnidarians.

Biological Contributions

1. Both phyla have developed two well-defined **germ layers,** ectoderm and endoderm; a third, or mesodermal, layer, which is derived embryologically from the ectoderm, is present in some. The body plan is saclike, and the body wall is composed of two distinct layers, epidermis and gastrodermis, derived from the ectoderm and endoderm, respectively. The gelatinous matrix, mesoglea, between these layers may be structureless, may contain a few cells and fibers, or may be composed largely of mesodermal connective tissue and muscle fibers.

2. An internal body cavity, the **gastrovascular cavity,** is lined by the gastrodermis and has a single opening, the mouth, which also serves as the anus.

3. **Extracellular digestion** occurs in the gastrovascular cavity, and intracellular digestion takes place in the gastrodermal cells. Extracellular digestion allows ingestion of larger food particles.

4. Most radiates have **tentacles,** or extensible projections around the oral end, that aid in capturing food.

5. Radiates are the simplest animals to possess true **nerve cells** (protoneurons), but the nerves are arranged as a nerve net, with no central nervous system.

6. Radiates are the simplest animals to possess sense organs, which include well-developed statocysts (organs of equilibrium) and ocelli (photosensitive organs).

7. Locomotion in the free-moving forms is achieved by either **muscular contractions** (cnidarians) or **ciliary comb plates** (ctenophores). However, both groups are still better adapted to floating or being carried by currents than to strong swimming.

8. **Polymorphism**[1] in the cnidarians has widened their ecological possibilities. In many species the presence of both a polyp (sessile and attached) stage and a medusa (free-swimming) stage permits occupation of a benthic (bottom) and a pelagic (open-water) habitat by the same species. Polymorphism also widens the possibilities of structural complexity.

9. Some unique features are found in these phyla, such as **nematocysts** (stinging organelles) in cnidarians and **colloblasts** (adhesive organelles) and **ciliary comb plates** in ctenophores.

---

[1]Note that polymorphism here refers to more than one structural form of individual within a species, as contrasted with the use of the word in genetics, in which it refers to different allelic forms of a gene in a population.

Characteristics of Phylum Cnidaria

1. Entirely aquatic, some in fresh water but mostly marine
2. **Radial symmetry** or biradial symmetry around a longitudinal axis with **oral** and **aboral** ends; no definite head
3. Two basic types of individuals: **polyps** and **medusae**
4. Exoskeleton or endoskeleton of chitinous, calcareous, or protein components in some
5. Body with two layers, epidermis and gastrodermis, with mesoglea **(diploblastic)**; mesoglea with cells and connective tissue (ectomesoderm) in some **(triploblastic)**
6. **Gastrovascular cavity** (often branched or divided with septa) with a single opening that serves as both mouth and anus; extensible tentacles usually encircling the mouth or oral region
7. Special stinging cell organelles called **nematocysts** in either epidermis or gastrodermis or in both; nematocysts abundant on tentacles, where they may form batteries or rings
8. **Nerve net** with symmetrical and asymmetrical synapses; with some sensory organs; diffuse conduction
9. Muscular system (epitheliomuscular type) of an outer layer of longitudinal fibers at base of epidermis and an inner one of circular fibers at base of gastrodermis; modifications of this plan in some cnidarians, such as separate bundles of independent fibers in the mesoglea
10. Asexual reproduction by budding (in polyps) or sexual reproduction by gametes (in all medusae and some polyps); sexual forms monoecious or dioecious; **planula larva;** holoblastic indeterminate cleavage
11. No excretory or respiratory system
12. No coelomic cavity

some of nature's strangest and loveliest creatures: the branching, plantlike hydroids; the flowerlike sea anemones; the jellyfishes; and those architects of the ocean floor, the horny corals (sea whips, sea fans, and others), and the stony corals whose thousands of years of calcareous house-building have produced great reefs and coral islands (p. 114).

We recognize four classes of Cnidaria: Hydrozoa (the most variable class, including hydroids, fire corals, Portuguese man-of-war, and others), Scyphozoa (the "true" jellyfishes), Cubozoa (cube jellyfishes), and Anthozoa (the largest class, including sea anemones, stony corals, soft corals, and others).

Ecological Relationships

Cnidarians are found most abundantly in shallow marine habitats, especially in warm temperatures and tropical regions. There are no terrestrial species. Colonial hydroids are usually found attached to mollusc shells, rocks, wharves, and other animals in shallow coastal water, but some species are found at great depths. Floating and free-swimming medusae are found in open seas and lakes, often far from the shore. Floating colonies such as the Portuguese man-of-war and *Velella* (L. *velum*, veil, + *ellus*, dim. suffix) have floats or sails by which the wind carries them.

Some ctenophores, molluscs, and flatworms eat hydroids bearing nematocysts and use these stinging structures for their own defense. Some other animals, such as some molluscs and fishes, feed on cnidarians, but cnidarians rarely serve as food for humans.

Cnidarians sometimes live symbiotically with other animals, often as commensals on the shell or other surface of their host. Certain hydroids (Figure 6-1) and sea anemones commonly live on snail shells inhabited by hermit crabs, providing the crabs some protection from predators. Algae frequently live as mutuals in the tissues of cnidarians, notably in some freshwater hydras and in reef-building corals. The presence of the algae in reef-building corals limits the occurrence of coral reefs to relatively shallow, clear water where sunlight is sufficient for the photosynthetic requirements of the algae. These corals are an essential component of coral reefs, and reefs are extremely important habitats in tropical waters. Coral reefs are discussed further later in the chapter.

Although many cnidarians have little economic importance, reef-building corals are an important exception. Fish and other animals associated with reefs provide substantial amounts of food for humans, and reefs are of economic value as tourist attractions. Precious coral is used for jewelry and ornaments, and coral rock serves for building purposes.

Planktonic medusae may be of some importance as food for fish that are of commercial value; the reverse is also true—the young fish fall prey to cnidarians.

Form and Function

Dimorphism and Polymorphism in Cnidarians

One of the most interesting—and sometimes puzzling—aspects of this phylum is the dimorphism and often polymorphism displayed by many of its members. All cnidarian forms fit into one of two morphological types (dimorphism): the

for a certain function, such as feeding, reproduction, or defense (Figure 6-1).

> *The name "medusa" was suggested by a fancied resemblance to the Gorgon Medusa, a mythological lass with snaky tresses that turned to stone any who gazed upon them.*

Medusae are usually free swimming and have bell-shaped or umbrella-shaped bodies and tetramerous symmetry (body parts arranged in fours). The mouth is usually centered on the concave side, and tentacles extend from the rim of the umbrella.

The sea anemones and corals (class Anthozoa) are all polyps: hence, they are not dimorphic. The true jellyfishes (class Scyphozoa) have a conspicuous medusoid form, but many have a polypoid larval stage. The colonial hydroids of class Hydrozoa, however, sometimes have life histories that feature both the polyp, or hydroid, stage and the free-swimming medusa stage—rather like a Jekyll-and-Hyde existence. A species that has both the attached polyp and the floating medusa within its life history can take advantage of the feeding and distribution possibilities of both pelagic (open-water) and benthic (bottom) environments. Many hydrozoans are also polymorphic, with several distinct types of polyps in a colony.

Superficially the polyp and medusa seem very different. But actually each has retained the saclike body plan that is basic to the phylum (Figure 6-2). The medusa is essentially an unattached polyp with the tubular portion widened and flattened into the bell shape.

Both the polyp and the medusa possess the three body wall layers typical of the cnidarians, but the jellylike layer of mesoglea is much thicker in the medusa, constituting the bulk of the animal and making it more buoyant. Because of this mass of mesoglea ("jelly"), the medusae are commonly called jellyfishes.

Nematocysts: The Stinging Organelles

One of the most characteristic structures in the entire cnidarian group is the stinging organelle called the **nematocyst**

A

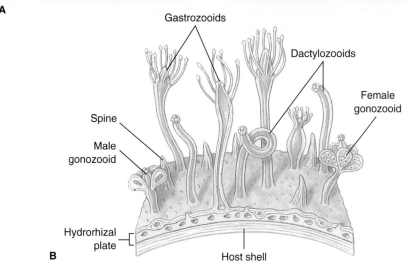

B

FIGURE 6-1

A, A hermit crab with its cnidarian mutuals. The shell is blanketed with polyps of the hydrozoan *Hydractinia milleri.* The crab gets some protection from predation by the cnidarians, and the cnidarians get a free ride and bits of food from their host's meals. **B,** Portion of a colony of *Hydractinia,* showing the types of zooids and the stolon (hydrorhiza) from which they grow.

polyp, or hydroid form, which is adapted to a sedentary or sessile life, and the **medusa,** or jellyfish form, which is adapted for a floating or free-swimming existence (Figure 6-2).

Most polyps have tubular bodies with a mouth at one end surrounded by tentacles. The aboral end is usually attached to a substratum by a pedal disc or other device. Polyps may live singly or in colonies. Colonies of some species include morphologically differing individuals (polymorphism), each specialized

(Figure 6-3). Over 20 different types of nemato-cysts (Figure 6-4) have been described in the cnidarians so far; they are important in taxonomic determinations. The nematocyst is a tiny capsule composed of material similar to chitin and containing a coiled tubular "thread" or filament, which is a continuation of the narrowed end of the capsule. This end of the capsule is covered by a little lid, or **operculum.** The inside of the undischarged thread may bear tiny barbs, or spines.

The nematocyst is enclosed in the cell that has produced it, the cnidocyte (during its development, the cnidocyte is properly called the **cnidoblast**). Except in Anthozoa, cnidocytes are equipped with a triggerlike **cnidocil,** which is a modified cilium. Anthozoan cnidocytes have a somewhat different ciliary mechanoreceptor. In some sea anemones, and perhaps other cnidarians, small organic molecules from the prey "tune" the mechanoreceptors, sensitizing them to the frequency of vibration caused by the prey swimming. Tactile stimulation causes the nematocyst to discharge. Cnidocytes are borne in invaginations of ectodermal cells and, in some forms, in gastrodermal cells, and they are especially abundant on the tentacles. After a nematocyst discharges, its cnidocyte is absorbed and a new one replaces it. Not all nematocysts have barbs or inject poison. Some, for example, do not penetrate the prey but rapidly recoil like a spring after discharge, grasping and holding any part of the prey caught in the coil (Figure 6-4). Adhesive nematocysts generally are not used to capture food.

The mechanism of nematocyst discharge is remarkable. Present evidence indicates that discharge is due to a combination of tensional forces generated during nematocyst formation and also to an astonishingly high osmotic pressure within the nematocyst: 140 atmospheres. When stimulated to discharge, the high internal osmotic pressure causes water to rush into the capsule. The operculum opens, and the rapidly increasing *hydrostatic pressure* within the capsule forces the thread out with great force, the thread turning inside out as it goes. At the everting end of the thread, the barbs flick to the outside like tiny switchblades. This minute but awesome weapon then injects poison when it penetrates the prey.

Note the distinction between osmotic and hydrostatic pressure. The nematocyst is never required actually to contain 140 atmospheres of hydrostatic pressure within itself; such a hydrostatic pressure would doubtless cause it to explode. As the water rushes in during discharge, the osmotic pressure falls rapidly, while the hydrostatic pressure rapidly increases.

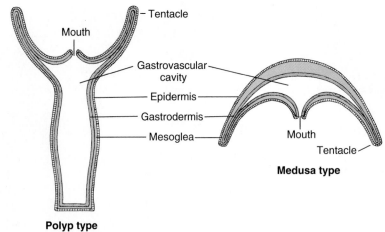

FIGURE 6-2

Comparison between the polyp and medusa types of individuals.

The nematocysts of most cnidarians are not harmful to humans and are a nuisance at worst. However, the stings of the Portuguese man-of-war (see Figure 6-12) and certain jellyfish are quite painful and sometimes dangerous.

Nerve Net

The nerve net of the cnidarians is one of the best examples of a diffuse nervous system in the animal kingdom. This plexus of nerve cells is found both at the base of the epidermis and at the base of the gastrodermis, forming two interconnected nerve nets. Nerve processes (axons) end on other nerve cells at synapses or at junctions with sensory cells or effector organs (nematocysts or epitheliomuscular cells). Nerve impulses are transmitted from one cell to another by release of a neurotransmitter from small vesicles on one side of the synapse or junction. One-way transmission between nerve cells in higher animals is ensured because the vesicles are located on only one side of the synapse. However, cnidarian nerve nets are peculiar in that many of the synapses have vesicles of neurotransmitters on both sides, allowing transmission across the synapse in either direction. Another peculiarity of cnidarian nerves is the absence of any sheathing material (myelin) on the axons.

There is no concentrated grouping of nerve cells to suggest a "central nervous system." Nerves are grouped, however, in the "ring nerves" of hydrozoan medusae and in the marginal sense organs of scyphozoan medusae. In some cnidarians the nerve nets form two or more systems: in Scyphozoa there is a fast conducting system to coordinate swimming movements and a slower one to coordinate movements of tentacles.

The nerve cells of the net have synapses with slender sensory cells that receive external stimuli, and the nerve cells have junctions with epitheliomuscular cells and nematocysts.

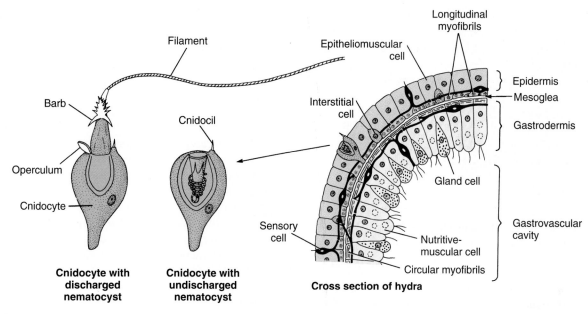

Cnidocyte with discharged nematocyst

Cnidocyte with undischarged nematocyst

Cross section of hydra

FIGURE 6-3

At left, structure of a stinging cell. At right, portion of the body wall of a hydra. Cnidocytes, which contain the nematocysts, arise in the epidermis from interstitial cells.

FIGURE 6-4

Several types of nematocysts shown after discharge. At bottom are two views of a type that does not impale the prey; it recoils like a spring, catching any small part of the prey in the path of the recoiling thread.

Together with the contractile fibers of the epitheliomuscular cells, the sensory cell and nerve net combination is often referred to as a **neuromuscular system,** an important landmark in the evolution of the nervous system. The nerve net arose early in metazoan evolution, and it has never been completely lost phylogenetically. Annelids have it in their digestive systems. In the human digestive system it is represented by nerve plexuses in the musculature. The rhythmical peristaltic movements of the stomach and intestine are coordinated by this counterpart of the cnidarian nerve net.

Body Structure

The mouth opens into the **gastrovascular cavity** (coelenteron), which communicates with the cavities in the tentacles. The mouth may be surrounded by an elevated **manubrium** or by elongated **oral lobes.**

Body Wall   The body wall surrounding the gastrovascular cavity consists of an outer **epidermis** (ectodermal) and an inner **gastrodermis** (endodermal) with **mesoglea** between them (Figure 6-3).

Epidermis   The epidermal layer contains epitheliomuscular, interstitial, gland, cnidocyte, and sensory and nerve cells.

Epitheliomuscular cells (Figure 6-5) make up most of the epidermis and serve both for covering and for muscular contraction. The bases of most of these cells are extended parallel to the tentacle or body axis and contain myofibrils, thus forming a layer of longitudinal muscle next to the mesoglea.

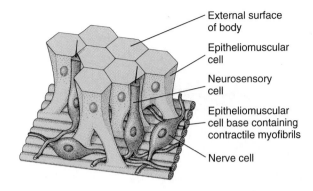

External surface of body

Epitheliomuscular cell

Neurosensory cell

Epitheliomuscular cell base containing contractile myofibrils

Nerve cell

FIGURE 6-5

Epitheliomuscular and nerve cells in hydra.

Contraction of these fibrils shortens the body or tentacles. Modifications of these cells in various cnidarians include emphasis on the epithelial or muscular portions of the cells and flattened rather than columnar shape.

Interstitial cells are the undifferentiated stem cells found among the bases of the epitheliomuscular cells. Differentiation of the interstitial cells gives rise to cnidoblasts, sex cells, buds, nerve cells, and others, but generally not to epitheliomuscular cells (which reproduce themselves).

Gland cells are tall cells particularly abundant around the mouth and in the pedal disc of hydra. They secrete mucus or adhesive material.

Cnidocytes containing nematocysts are found throughout the epidermis. They may be between the epitheliomuscular cells or housed in invaginations of these cells, and they are most abundant on the tentacles. There are three functional types of nematocysts in the hydra: those that penetrate the prey and inject poison (penetrants, Figure 6-3); those that recoil and entangle the prey (volvents, Figure 6-4); and those that secrete an adhesive substance used in locomotion and attachment (glutinants).

Sensory cells are scattered among the other epidermal cells, especially around the mouth and the tentacles. The free end of each sensory cell bears a flagellum, which is the sensory receptor for chemical and tactile stimuli. The other end branches into fine processes, which synapse with the nerve cells.

Nerve cells of the epidermis are often multipolar (have many processes), and in more highly organized cnidarians the cells may be bipolar (with two processes). Their processes (axons) form synapses with sensory cells and other nerve cells, and junctions with epitheliomuscular cells and cnidocytes. Both one-way and two-way synapses with other nerve cells are present.

Gastrodermis The gastrodermis, a layer of cells lining the gastrovascular cavity, is made up chiefly of large, ciliated, columnar epithelial cells with irregular flat bases. The cells of the gastrodermis include nutritive-muscular, interstitial, and gland cells and, in classes other than Hydrozoa, cnidocytes.

Nutritive-muscular cells are usually tall columnar cells that have laterally extended bases containing myofibrils. In hydrozoans the myofibrils run at right angles to the body or tentacle axis and so form a circular muscle layer. However, this muscle layer is very weak, and longitudinal extension of the body and tentacles is brought about mostly by increasing the volume of water in the gastrovascular cavity. The water is brought in through the mouth by the beating of the cilia on the nutritive-muscular cells in hydrozoans or by ciliated cells in the pharynx of anthozoans. Thus, the water in the gastrovascular cavity serves as a **hydrostatic skeleton.** The two cilia on the free end of each cell also serve to circulate food and fluids in the digestive cavity. The cells often contain large numbers of food vacuoles. Gastrodermal cells of many cnidarians contain algal cells. In the green hydra (*Chlorohydra* [Gr. *chlōros,* green, + *hydra,* a mythical nine-headed monster slain by Hercules]), these are green algae **(zoochlorellae),** but in marine cnidarians they are a type of dinoflagellate (p. 75) **(zooxanthellae).** Both are cases of mutualism, with the algae furnishing organic compounds they have synthesized to their cnidarian hosts.

Interstitial cells scattered among the bases of the nutritive cells can transform into other cell types. **Gland cells** secrete digestive enzymes; feeding and digestion are discussed later.

Mesoglea The mesoglea lies between the epidermis and gastrodermis and adheres to both layers. It is gelatinous, or jellylike, and has no fibers or cellular elements in hydrozoan polyps. It is thicker in medusae and has elastic fibers, and in scyphozoan medusae it has ameboid cells. The mesoglea of anthozoans is a mesenchyme containing ameboid cells.

Locomotion

Colonial polyps are permanently attached, but hydras can move about freely by gliding on the basal disc, aided by mucus secretions. Sea anemones can move similarly on their basal discs. Hydras can also use a "measuring worm" movement, looping along by bending over and attaching their tentacles to the substratum. They may even turn handsprings or detach and, by forming a gas bubble on the basal disc, float to the surface.

Most medusae can move freely, and they swim by contracting the bell, expelling water from the concave, oral side. The muscular contractions are antagonized by the compressed mesoglea and the elastic fibers within it. Usually, they contract several times and move generally upward, then sink slowly. Cubozoan medusae, however, can swim strongly.

Feeding and Digestion

Cnidarians prey on a variety of organisms of appropriate size; larger species are usually capable of killing and eating larger prey. Normally the prey is drawn into the gastrovascular cavity where gland cells discharge enzymes onto the food. Digestion is started in the gastrovascular cavity (**extracellular digestion**), but nutritive-muscular cells phagocytize many food particles for **intracellular digestion**. Ameboid cells may carry undigested particles to the gastrovascular cavity, where they are eventually expelled with other indigestible matter.

> *The effectiveness of cnidarians in snaring prey was amply demonstrated recently in a fjord in the western Baltic Sea where a heavy bloom of jellyfish reduced the larval herring population to less than half its former density in just a few weeks. One jellyfish only 4.2 cm in diameter contained 68 larval herring!*

Reproduction

Most cnidarians are dioecious, and many shed their gametes directly into the water. Zygotes may be retained by the female and brooded for some period. Gonads are epidermal in hydrozoans and gastrodermal in the other groups. The embryo characteristically develops into a free-swimming **planula** larva (see Figure 6-9).

Cnidarians are capable of asexual reproduction, usually by budding, but sea anemones commonly practice a peculiar form of fission known as **pedal laceration** (p. 113).

Class Hydrozoa

The majority of Hydrozoa are marine and colonial in form, and the typical life cycle includes both the asexual polyp and the sexual medusa stages. Some, however, such as the freshwater hydras, have no medusa stage. Some marine hydroids do not have free medusae (Figure 6-6), whereas some hydrozoans occur only as medusae and have no polyp.

Hydras, although not typical hydrozoans, have become favorites as an introduction to Cnidaria because of their size and ready availability. Combining study of a hydra with that of a representative colonial marine hydroid such as *Obelia* (Gr. *obelias*, round cake) gives an excellent idea of the class Hydrozoa.

Hydra: A Freshwater Hydrozoan

The common freshwater hydra (Figure 6-7) is a solitary polyp and one of the few cnidarians found in fresh water. Its normal habitat is the underside of aquatic leaves and lily pads in cool, clean fresh water of pools and streams. The hydra family is found throughout the world, with 16 species occurring in North America.

Reduced medusae (gonophores)

FIGURE 6-6

In some hydroids, such as this *Tubularia crocea,* medusae are reduced to gonadal tissue and do not detach. These reduced medusae are known as gonophores.

> *Over 230 years ago, Abraham Trembley was astonished to discover that isolated sections of the stalk of hydra could regenerate and each become a complete animal. Since then, over 2000 investigations of hydra have been published, and the organism has become a classic model for the study of morphological differentiation. The mechanisms governing morphogenesis have great practical importance, and the simplicity of hydra lends itself to these investigations. Substances controlling development (morphogens), such as those determining which end of a cut stalk will develop a mouth and tentacles, have been discovered, and they may be present in the cells in extremely low concentrations (10^{-10}M).*

The body of the hydra can extend to a length of 25 to 30 mm or can contract to a tiny, jellylike mass. It is a cylindrical tube with the lower (aboral) end drawn out into a slender stalk, ending in a basal or **pedal disc** for attachment. This pedal disc has gland cells that enable the hydra to adhere to a substratum and also to secrete a gas bubble for floating. In the center of the disc there may be an excretory pore. The **mouth,** located on a conical elevation called the **hypostome,**

FIGURE 6-7

Hydra with developing bud and ovary.

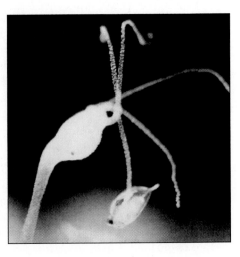

FIGURE 6-8

Hydra catches an unwary water flea with the nematocysts of its tentacles. This hydra already contains one water flea eaten previously.

is encircled by six to ten hollow tentacles that, like the body, are greatly extended when the animal is hungry.

The mouth opens into the gastrovascular cavity, which communicates with the cavities in tentacles. In some individuals **buds** may project from the sides, each with a mouth and tentacles like the parent. Testes or ovaries, when present, appear as rounded projections on the surface of the body (Figure 6-7).

Hydras feed on a variety of small crustaceans, insect larvae, and annelid worms. The hydra awaits its prey with tentacles extended (Figure 6-8). The food organism that brushes against its tentacles may find itself harpooned by scores of nematocysts that render it helpless, even though it may be larger than the hydra. The tentacles move the prey toward the mouth, which slowly widens. Well moistened with mucous secretions, the mouth glides over and around the prey, totally engulfing it.

The activator that actually causes the mouth to open is the reduced form of **glutathione,** which is found to some extent in all living cells. Glutathione is released from the prey through the wounds made by the nematocysts, but only those animals releasing enough of the chemical to activate the feeding response are eaten by the hydra. This explains how a hydra distinguishes between *Daphnia,* which it relishes, and some other forms that it refuses. When glutathione is added to water containing hydras, each hydra will go through the motions of feeding, even though no prey is present.

In asexual reproduction, buds appear as outpocketings of the body wall and develop into young hydras that eventually detach from the parent. In sexual reproduction, temporary gonads (Figure 6-7) usually appear in the autumn, stimulated by the lower temperatures and perhaps also by the reduced aeration of stagnant waters. Eggs in the ovary usually mature one at a time and are fertilized by sperm shed into the water.

A cyst forms around the embryo before it breaks loose from the parent, enabling it to survive the winter. Young hydras hatch out in spring when the weather is favorable.

Hydroid Colonies

Far more representative of class Hydrozoa than the hydras are those hydroids that have a medusa stage in their life cycle. *Obelia* is often used in laboratory exercises for beginning students to illustrate the hydroid type (Figure 6-9).

A typical hydroid has a base, a stalk, and one or more terminal polyps (zooids). The base by which the colonial hydroids are attached to the substratum is a rootlike stolon, which gives rise to one or more stalks. The living cellular part of the stalks secretes a nonliving chitinous sheath. Attached to the ends of the branches of the stalks are the individual zooids. Most of the zooids are feeding polyps called **hydranths,** or **gastrozooids.** They may be tubular, bottle shaped, or vaselike, but all have a terminal mouth and a circlet of tentacles. In some forms, such as *Obelia,* the chitinous sheath continues as a protective cup around the polyp into which it can withdraw for protection (Figure 6-9). In others the polyp is naked. **Dactylozooids** are polyps specialized for defense.

The hydranths, much like hydras, capture and ingest prey, such as tiny crustaceans, worms, and larvae, thus providing nutrition for the entire colony. After partial digestion in the hydranth, the digestive broth passes into the common gastrovascular cavity where intracellular digestion occurs.

Circulation within the gastrovascular cavity is a function of the ciliated gastrodermis, but rhythmical contractions and pulsations of the body, which occur in many hydroids, also aid circulation.

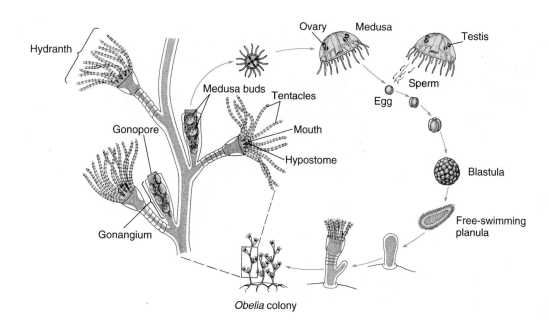

FIGURE 6-9

Life cycle of *Obelia*, showing alternation of polyp (asexual) and medusa (sexual) stages. *Obelia* is a calyptoblastic hydroid; that is, its polyps as well as its stems are protected by continuations of the nonliving covering.

In contrast to hydras, new individuals that bud do not detach from the parent; thus the size of the colony increases. The new polyps may be hydranths or reproductive polyps known as **gonangia.** Medusae are produced by budding within the gonangia. The young medusae leave the colony as free-swimming individuals that mature and produce gametes (eggs and sperm) (Figure 6-9). In some species the medusae remain attached to the colony and shed their gametes there. In other species the medusae never develop, the gametes being shed by male and female gonophores. Development of the zygote results in a ciliated planula larva that swims about for a time. Then it settles down to a substratum to develop into a minute polyp that gives rise, by asexual budding, to the hydroid colony, thus completing the life cycle.

Hydroid medusae are usually smaller than their scyphozoan counterparts, ranging from 2 or 3 mm to several centimeters in diameter (Figure 6-10). The margin of the bell projects inward as a shelflike **velum,** which partly closes the open side of the bell and is used in swimming (Figure 6-11). Muscular pulsations that alternately fill and empty the bell propel the animal forward, aboral side first, with a sort of "jet propulsion." The tentacles attached to the bell margin are richly supplied with nematocysts.

The mouth opening at the end of a suspended **manubrium** leads to a stomach and four radial canals that connect with a ring canal around the margin. This in turn connects with the hollow tentacles. Thus the coelenteron is continuous from mouth to tentacles, and the entire system is lined with gastrodermis. Nutrition is similar to that of the hydranths.

FIGURE 6-10

Bell medusa, *Polyorchis penicillatus,* medusa stage of an unknown attached polyp.

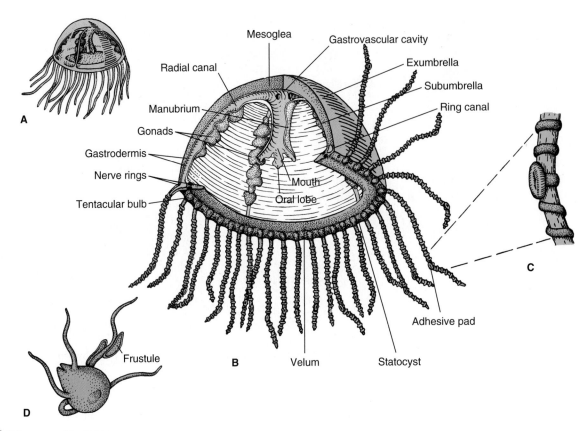

FIGURE 6-11

Structure of *Gonionemus*. **A,** Medusa with typical tetramerous arrangement. **B,** Cutaway view showing morphology. **C,** Portion of a tentacle with its adhesive pad and ridges of nematocysts. **D,** Tiny polyp, or hydroid stage, that develops from the planula larva. It can produce more polyps by budding (frustules) or produce medusa buds.

The nerve net is usually concentrated into two nerve rings at the base of the velum. The bell margin is liberally supplied with sensory cells. It usually also bears two kinds of specialized sense organs: **statocysts,** which are small organs of equilibrium (Figure 6-11), and **ocelli,** which are light-sensitive organs.

Other Hydrozoans

Some hydrozoans form floating colonies, such as *Physalia* (Gr. *physallis,* bladder), the Portuguese man-of-war (Figure 6-12). These colonies include several types of modified medusae and polyps. *Physalia* has a rainbow-hued float, probably a modified polyp, which carries it along at the mercy of the winds and currents. It contains an air sac filled with secreted gas and acts as a carrier for the generations of individuals that bud from it and hang suspended in the water. There are several types of individuals, including the feeding polyps, reproductive polyps, long stinging tentacles, and the so-called jelly polyps. Many swimmers have experienced the uncomfortable sting that these colonial floaters can inflict. The pain, along with the panic of the swimmer, can increase the danger of drowning.

Other hydrozoans secrete massive calcareous skeletons that resemble true corals (Figure 6-13). They are sometimes called **hydrocorals.**

Class Scyphozoa

Class Scyphozoa (si-fo-zo′a) (Gr. *skyphos,* cup) includes most of the larger jellyfishes, or "cup animals." A few, such as *Cyanea* (Gr. *kyanos,* dark-blue substance), may attain a bell diameter exceeding 2 m and tentacles 60 to 70 m long (Figure 6-14). Most scyphozoans, however, range from 2 to 40 cm in diameter. Most are found floating in the open sea, some even at depths of 3000 m, but one unusual order is sessile and attaches by a stalk to seaweeds and other objects on the sea bottom (Figure 6-15). Their coloring may range from colorless to striking orange and pink hues.

Scyphomedusae, unlike hydromedusae, have no velum. The bells of different species vary in depth from a shallow saucer shape to a deep helmet or goblet shape, and in many the margin is scalloped, each notch bearing a sense organ called a **rhopalium** and a pair of lobelike projections called lappets. *Aurelia* (L. *aurum,* gold) has eight such notches (Figure 6-16); others may have four or sixteen. Each rhopalium bears a statocyst for balance, two sensory pits containing concentrations of sensory cells, and sometimes an ocellus (simple eye) for photoreception. The mesoglea is thick and contains cells as well as fibers. The stomach is usually divided into pouches containing small tentacles with nematocysts.

The mouth is centered on the subumbrellar side. The manubrium is usually drawn out into four frilly **oral lobes** that are used in food capture and ingestion. The marginal tentacles may be many or few and may be short, as in *Aurelia*, or long, as in *Cyanea*. The tentacles, manubrium, and often the entire body surface of scyphozoans are well supplied with nematocysts. Scyphozoans feed on all sorts of small organisms, from protozoa to fishes. Capture of prey involves stinging and manipulation with tentacles and oral arms, but the methods vary. *Aurelia* feeds on small planktonic animals. These are caught in the mucus of the umbrella surface, carried to "food pockets" on the umbrella margin by cilia, and picked up from the pockets by the oral lobes whose cilia carry the food to the gastrovascular cavity. Cilia on the gastrodermis keep a

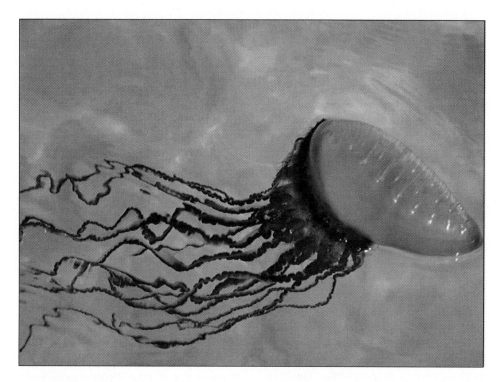

FIGURE 6-12

A Portuguese man-of-war colony, *Physalia physalis* (order Siphonophora, class Hydrozoa). Colonies often drift onto southern ocean beaches, where they are a hazard to bathers. Each colony of medusa and polyp types is integrated to act as one individual. As many as a thousand zooids may be found in one colony. The nematocysts secrete a powerful neurotoxin.

A

B

FIGURE 6-13

These hydrozoans form calcareous skeletons that resemble true coral. **A,** *Stylaster roseus* (order Stylasterina) occurs commonly in caves and crevices in coral reefs. These fragile colonies branch in only a single plane and may be white, pink, purple, red, or red with white tips. **B,** Species of *Millepora* (order Milleporina) form branching or platelike colonies and often grow over the horny skeleton of gorgonians (see Figure 6-26). They have a generous supply of powerful nematocysts that produce a burning sensation on human skin, justly earning the common name fire coral.

current of water moving to bring food and oxygen into the stomach and carry out wastes.

Internally four **gastric pouches** containing nematocysts connect with the stomach in scyphozoans, and a complex system of **radial canals** that branch from the pouches to the **ring canal** (Figure 6-16) completes the gastrovascular cavity, through which nutrients circulate.

The sexes are separate, and fertilization and early development occur in the gastric pouches or on the frilled oral arms. In most scyphozoans, a ciliated planula larva develops into a little polypoid larva called a **scyphistoma** (Figure 6-16), which looks somewhat like a hydra. During the summer the scyphistoma produces more scyphistomas by budding. In winter and spring the scyphistoma buds off a series of juvenile, saucer-shaped medusae, and the series of buds is known as a **strobila.** The juvenile medusae, called **ephyrae,** break loose, swim away, and grow into mature, sexual medusae (Figure 6-16).

FIGURE 6-14

Giant jellyfish, *Cyanea capillata* (order Semeaeostomeae, class Scyphozoa). A North Atlantic species of *Cyanea* reaches a bell diameter exceeding 2 m. It is known as the "sea blubber" by fishermen.

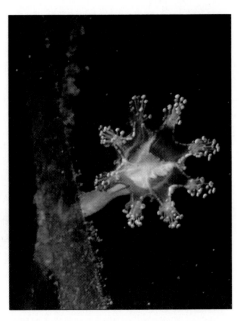

FIGURE 6-15

Thaumatoscyphus hexaradiatus (order Stauromedusae, class Scyphozoa). Members of this order are unusual scyphozoans in that the medusae are sessile and attached to seaweed or other objects.

Class Cubozoa

The Cubozoa until recently were considered an order (Cubomedusae) of Scyphozoa. The medusa is the predominant form (Figure 6-17); the polyp is inconspicuous and in most cases unknown. In transverse section the bells are almost square. A tentacle or group of tentacles is found at each corner of the square at the umbrella margin. The base of each tentacle is differentiated into a flattened, tough blade called a **pedalium** (Figure 6-17). Rhopalia are present. The umbrella margin is not scalloped, and the subumbrella edge turns inward to form a **velarium.** The velarium functions as a velum does in hydrozoan medusae, increasing swimming efficiency, but it differs structurally. Cubomedusae are strong swimmers and voracious predators, feeding mostly on fish.

Chironex fleckeri *(Gr. cheir, hand, + nexis, swimming) is a large cubomedusa known as the sea wasp. Its stings are quite dangerous and sometimes fatal. Most of the fatal stings have been reported from tropical Australian waters, usually following quite massive stings. Witnesses have described victims as being covered with "yards and yards of sticky wet string." The stings are very painful, and death, if it is to occur, ensues within a matter of minutes from sudden cardiac arrest. If death does not occur within 20 minutes after stinging, complete recovery is likely.*

Class Anthozoa

The anthozoans, or "flower animals," are polyps with a flowerlike appearance (Figure 6-18). There is no medusa stage. Anthozoa are all marine and are found in both deep and shallow water and in polar seas as well as tropical seas. They vary greatly in size and may be solitary or colonial. Many are supported by skeletons.

The class has three subclasses: **Zoantharia** (or **Hexacorallia**), made up of the sea anemones, hard corals, and others; the **Ceriantipatharia,** which includes only the tube anemones and thorny corals; and the **Alcyonaria** (or **Octocorallia**), containing the soft and horny corals, such as sea fans, sea pens, sea pansies, and others. The zoantharians and ceriantipatharians have a **hexamerous** plan (of six or multiples of six) or polymerous symmetry and have simple tubular tentacles arranged in one or more circlets on the oral disc. The alcyonarians are **octomerous** (built on a plan of eight) and always have eight pinnate (featherlike) tentacles arranged around the margin of the oral disc (Figure 6-19).

The gastrovascular cavity is large and partitioned by septa, or mesenteries, that are inward extensions of the body wall. Where one septum extends into the gastrovascular cavity from the body wall, another extends from the diametrically opposite side; thus, they are said to be **coupled.** In the Zoantharia, the septa are not only coupled, they are also **paired** (Figure 6-20). The muscular arrangement varies among the different groups, but there are usually circular muscles in the body wall and longitudinal and transverse muscles in the septa.

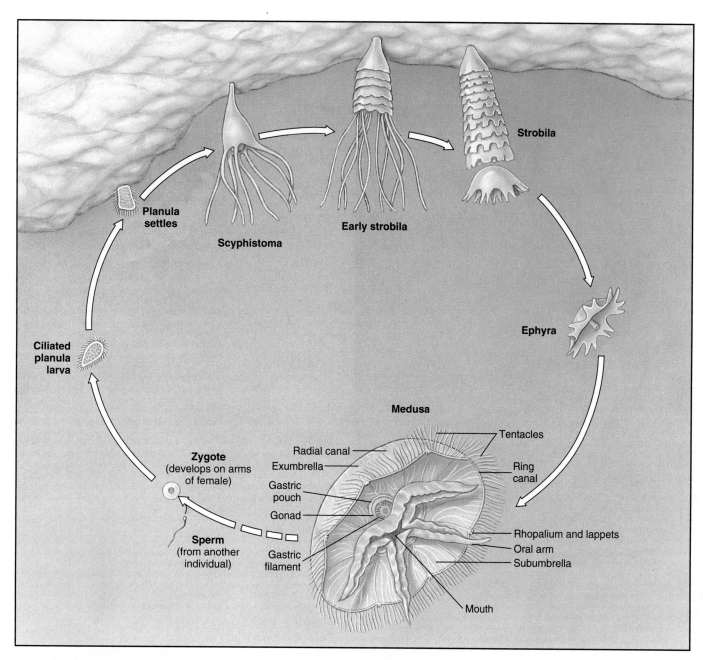

FIGURE 6-16

Life cycle of *Aurelia,* a marine scyphozoan medusa.

The subclass *Ceriantipatharia* has been created from the *Ceriantharia* and *Antipatharia,* formerly considered orders of *Zoantharia.* Ceriantharians are tube anemones and live in soft bottom sediments, buried to the level of the oral disc. Antipatharians are the thorny or black corals. They are colonial and have a skeleton of a horny material. Both of these groups are small in numbers of species and are limited to the warmer waters of the sea.

The mesoglea is a mesenchyme containing ameboid cells. There is a general tendency towards biradial symmetry in the septal arrangement and in the shape of the mouth and pharynx. There are no special organs for respiration or excretion.

Sea Anemones

Sea anemone polyps (order Actiniaria) are larger and heavier than hydrozoan polyps (Figures 6-18 and 6-20). Most of them range from 5 mm or less to 100 mm in diameter, and from 5 mm to 200 mm long, but some grow much larger. Some of

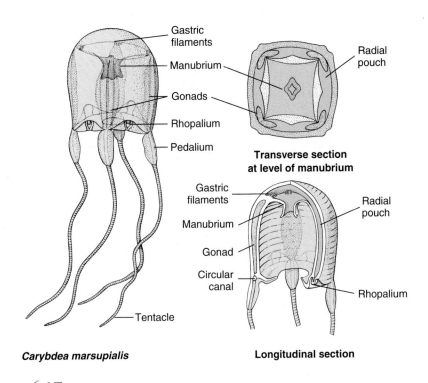

Transverse section at level of manubrium

Longitudinal section

Carybdea marsupialis

FIGURE 6-17

Carybdea, a cubozoan medusa.

FIGURE 6-18

Sea anemones are the familiar and colorful "flower animals" of tide pools, rocks, and pilings of the intertidal zone. Most, however, are subtidal, their beauty seldom revealed to human eyes. These are rose anemones, *Tealia piscivora.*

Sea anemones are cylindrical in form with a crown of tentacles arranged in one or more circles around the mouth on the flat **oral disc** (Figure 6-20). The slit-shaped mouth leads into a **pharynx.** At one or both ends of the mouth is a ciliated groove called a **siphonoglyph,** which extends into the pharynx. The siphonoglyphs create water currents directed into the pharynx. The cilia elsewhere on the pharynx direct water outward. The currents thus created carry in oxygen and remove wastes. They also help maintain an internal fluid pressure or a hydrostatic skeleton that serves as a support for opposing muscles.

The pharynx leads into a large gastrovascular cavity that is divided into radial chambers by means of pairs of septa that extend vertically from the body wall toward the pharynx (Figure 6-20). These chambers communicate with each other and are open below the pharynx. In many anemones the lower ends of the septal edges are prolonged into **acontia threads,** also provided with nematocysts and gland cells, that can be protruded through the mouth or through pores in the body wall to help overcome prey or provide defense. The pores also aid in the rapid discharge of water from the body when the animal is endangered and contracts to a small size.

Anemones form some interesting mutualistic relationships with other organisms. Many anemones house unicellular algae in their tissues (as do reef-building corals), from which they undoubtedly derive some nutrients. Some hermit crabs place anemones on the snail shells in which the crabs live, gaining some protection by the presence of the anemone, while the anemone dines on particles of food dropped by the crab. The anemonefishes (Figure 6-21) of the tropical Indo-Pacific form associations with large anemones. An unknown property of the skin mucus of the fish causes the anemone's nematocysts not to discharge, but if some other fish is so unfortunate as to brush the anemone's tentacles, it is likely to become a meal.

them are quite colorful. Anemones are found in coastal areas all over the world, especially in the warmer waters, and they attach by means of their pedal discs to shells, rocks, timber, or whatever submerged substrata they can find. Some burrow in the bottom mud or sand.

Sea anemones are carnivorous, feeding on fish or almost any live animals of suitable size. Some species live on minute forms caught by ciliary currents.

The sexes are separate in some sea anemones, and some are hermaphroditic. The gonads are arranged on the

A

B

FIGURE 6-19

A, Orange sea pen *Ptilosarcus gurneyi* (order Pennatulacea, class Anthozoa). Sea pens are colonial forms that inhabit soft bottoms. The base of the fleshy body of the primary polyp is buried in the bottom. It gives rise to numerous secondary, branching polyps. **B,** Close-up of a gorgonian. The pinnate tentacles characteristic of the subclass Alcyonaria are apparent.

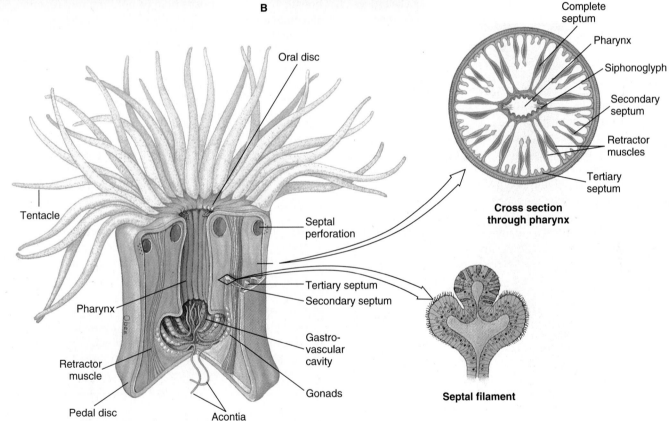

FIGURE 6-20

Structure of the sea anemone. The free edges of the septa and the acontia threads are equipped with nematocysts to complete the paralyzation of prey begun by the tentacles.

margins of the septa. Fertilization is external in some species, whereas in others the sperm enter the gastrovascular cavity to fertilize the eggs. The zygote develops into a ciliated larva. Asexual reproduction commonly occurs by pedal laceration. Small pieces of the pedal disc break off as the animal moves, and each of these regenerates a small anemone.

Zoantharian Corals

The zoantharian corals belong to the order Scleractinia, sometimes known as the true or stony corals. The stony corals might be described as miniature sea anemones that live in calcareous cups they themselves have secreted (Figures 6-23 and 6-24). Like that of the anemones, the coral polyp's gastrovascular

FIGURE 6-21

Orangefin anemonefish *(Amphiprion chrysopterus)* nestles in the tentacles of its sea anemone host. Anemonefishes do not elicit stings from their hosts but may lure unsuspecting other fish to become meals for the anemone.

cavity is subdivided by septa arranged in multiples of six (hexamerous) and its hollow tentacles surround the mouth, but there is no siphonoglyph.

Instead of a pedal disc, the epidermis at the base of the column secretes the limy skeletal cup, including the sclerosepta, which project up into the polyp between the true septa (Figure 6-24). The living polyp can retract into the safety of the cup when not feeding. Since the skeleton is secreted below the living tissue rather than within it, the calcareous material is an exoskeleton. In many colonial corals, the skeleton may become massive, building up over many years, with the living coral forming a sheet of tissue over the surface. The gastrovascular cavities of the polyps are all connected through this sheet of tissue.

Alcyonarian Corals

Alcyonarians are often referred to as octocorals because of their strict octomerous symmetry, with eight pinnate tentacles and eight unpaired, complete septa (Figure 6-25). They are all colonial, and the gastrovascular cavities of the polyps communicate through a system of gastrodermal tubes called **solenia.** The tubes run through an extensive mesoglea in most alcyonarians, and the surface of the colony is covered by epidermis. The skeleton is secreted within the mesoglea and consists of limy spicules, fused spicules, or a horny protein, often in combination. Thus the skeletal support of most alcyonarians is an endoskeleton. The variation in pattern among the species of alcyonarians lends great variety to the form of the colonies.

The graceful beauty of the alcyonarians—in hues of yellow, red, orange, and purple—helps create the "submarine gardens" of the coral reefs (Figure 6-26).

Coral Reefs

Coral reefs are among the most productive of all ecosystems, and they have a diversity of life forms rivaled only by the tropical rain forest. They are large formations of calcium carbonate (limestone) in shallow tropical seas laid down by living organisms over thousands of years; living plants and animals are confined to the top layer of reefs where they add more calcium carbonate to that deposited by their predecessors. The most important organisms that take dissolved calcium and carbonate ions from seawater and precipitate it as limestone to form reefs are the **reef-building corals** and **coralline algae.** Reef-building corals have mutualistic algae (zooxanthellae) living in their tissues. Coralline algae are several types of red algae, and they may be encrusting or form upright, branching growths. Not only do they contribute to the total mass of calcium carbonate, but their deposits help to hold the reef together. Some alcyonarians and hydrozoans (especially *Millepora* [L. *mille,* a thousand, + *porus,* pore] spp., the "fire coral," Figure 6-13) contribute in some measure to the calcareous material, and an enormous variety of other organisms contributes small amounts. However, reef-building corals seem essential to the formation of large reefs, since such reefs do not occur where these corals cannot live.

Because zooxanthellae are vital to reef-building corals, and water absorbs light, reef-building corals rarely live below a depth of 30 m (100 feet). Interestingly, some deposits of coral reef limestone, particularly around Pacific islands and atolls, reach great thickness—even thousands of feet. Clearly the corals and other organisms could not have grown from the bottom in the abyssal blackness of the deep sea and reached shallow water where light could penetrate. Charles Darwin was the first to realize that such reefs began their growth in shallow water around volcanic islands; then, as the islands slowly sank beneath the sea, the growth of the reefs kept up with the rate of sinking, thus accounting for the depth of the deposits.

Despite their great intrinsic and economic value, coral reefs in many areas are threatened today by a variety of factors, mostly of human origin. These include nutrients from sewage and agricultural fertilizer that wash into the water from land. Agricultural pesticides, as well as sediment from tilled fields, also contribute to reef degradation. Corals in the Persian Gulf have withstood a surprising amount of pollution, high salinity, and temperature swings—much more than reefs in other parts of the world have been able to endure. They apparently have survived the greatest oil slick ever created by humans (in the Gulf War of 1991).

Classification of Phylum Cnidaria

Class Hydrozoa (hi-dro-zo′a) (Gr. *hydra,* water serpent, + *zōon,* animal). Solitary or colonial; asexual polyps and sexual medusae, although one type may be suppressed; hydranths with no mesenteries; medusae (when present) with a velum; both fresh water and marine. Examples: *Hydra, Obelia, Physalia, Tubularia.*

Class Scyphozoa (si-fo-zo′a) (Gr. *skyphos,* cup, + *zōon,* animal). Solitary; polyp stage reduced or absent; bell-shaped medusae without velum; gelatinous mesoglea much enlarged; margin of bell or umbrella typically with eight notches that are provided with sense organs; all marine. Examples: *Aurelia, Cassiopeia, Rhizostoma.*

Class Cubozoa (ku′bo-zo′a) (Gr. *kybos,* a cube, + *zōon,* animal). Solitary; polyp stage reduced; bell-shaped medusae square in cross section, with tentacle or group of tentacles hanging from a bladelike pedalium at each corner of the umbrella; margin of umbrella entire, without velum but with velarium; all marine. Examples: *Tripedalia, Carybdea, Chironex, Chiropsalmus.*

Class Anthozoa (an-tho-zo′a) (Gr. *anthos,* flower, + *zōon,* animal). All polyps; no medusae; solitary or colonial; enteron subdivided by mesenteries or septa bearing nematocysts; gonads endodermal; all marine.

Subclass Zoantharia (zo′an-tha′re-a) (N.L. from Gr. *zōon,* animal, + *anthos,* flower, + L. *aria,* like or connected with) **(Hexacorallia).** With simple unbranched tentacles; mesenteries in pairs, in multiples of six; sea anemones, hard corals, and others. Examples: *Metridium, Anthopleura, Tealia, Astrangia, Acropora.*

Subclass Ceriantipatharia (se′re-ant-ip′a-tha′re-a) (N.L. combination of Cerianthraria and Antipatharia). With simple unbranched tentacles; mesenteries unpaired, initially six; tube anemones and black or thorny corals. Examples: *Cerianthus, Antipathes, Stichopathes.*

Subclass Alcyonaria (al′ce-o-na′re-a) (Gr. *alkonion,* kind of sponge resembling nest of kingfisher [*alkyon,* kingfisher], + L. *aria,* like or connected with) **(Octocorallia).** With eight pinnate tentacles; eight complete, unpaired mesenteries; soft and horny corals. Examples: *Tubipora, Alcyonium, Gorgonia, Plexaura, Renilla.*

FIGURE 6-22

A sea anemone that swims. When attacked by a predatory sea star *Dermasterias,* the anemone *Stomphia didemon* detaches from the bottom and rolls or swims spasmodically to a safer location.

The distribution of coral reefs in the world is limited to locations that offer optimal conditions for their zooxanthellae. They require warmth, light, and the salinity of undiluted sea-water, thus limiting coral reefs to shallow waters between 30° N and 30° S latitude and excluding them from areas with upwelling of cold water or areas near major river outflows with attendant low salinity and high turbidity. Photosynthesis and fixation of carbon dioxide by the zooxanthellae furnish food molecules for their hosts, they recycle phosphorus and nitrogenous waste compounds that otherwise would be lost, and they enhance the ability of the coral to deposit calcium carbonate.

A

B

C

FIGURE 6-23

A, Cup coral *Tubastrea* sp. The polyps form clumps resembling groups of sea anemones. Although often found on coral reefs, *Tubastrea* is not a reef-building coral and has no symbiotic zooxanthellae in its tissues. **B,** The polyps of *Montastrea cavernosa* are tightly withdrawn in the daytime but open to feed at night, as in **C.**

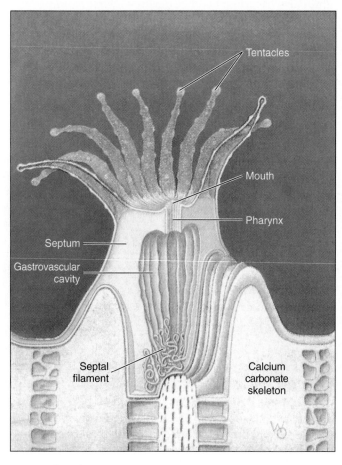

FIGURE 6-24

Polyp of a zoantharian coral (order Scleractinia) showing calcareous cup (exoskeleton), gastrovascular cavity, sclerosepta, septa, and septal filaments.

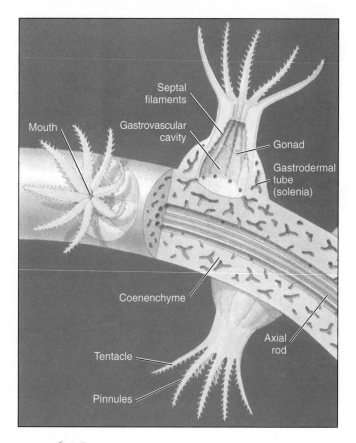

FIGURE 6-25

Polyps of an alcyonarian coral (octocoral). Note the eight pinnate tentacles, coenenchyme, and solenia. They have an endoskeleton of limy spicules often with a horny protein, which may be in the form of an axial rod.

FIGURE 6-26

FIGURE 6-26

Sea fan, *Subergorgia mollis* (order Gorgonacea, subclass Alcyonaria), on a Pacific coral reef. The colonial gorgonian, or horny, corals are conspicuous components of reef faunas.

Phylum Ctenophora

Ctenophora (te-nof'o-ra) (Gr. *kteis, ktenos,* comb, + *phora,* pl. of bearing) is composed of fewer than 100 species. All are marine forms occurring in all seas but especially in warm waters. They take their name from the eight rows of comblike plates they bear for locomotion. Common names for ctenophores are "sea walnuts" and "comb jellies." Ctenophores, along with cnidarians, represent the only two phyla having primary radial symmetry, in contrast to other metazoans, which have primary bilateral symmetry.

Ctenophores do not have nematocysts, except in one species (*Haeckelia rubra,* after Ernst Haeckel, nineteenth-century German zoologist) that carries nematocysts on certain regions of its tentacles but lacks colloblasts. These nematocysts are apparently appropriated from cnidarians on which it feeds.

In common with cnidarians, ctenophores have not advanced beyond the tissue grade of organization. There are no definite organ systems in the strict meaning of the term.

Except for a few creeping and sessile forms, ctenophores are free-swimming. Although they are feeble swimmers and are more common in surface waters, ctenophores are sometimes found at considerable depths. Highly modified forms such as *Cestum* (L. *cestus,* girdle) use sinuous body movements as well as their comb plates in locomotion (Figure 6-27).

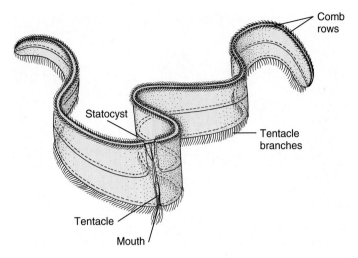

FIGURE 6-27

Venus' girdle (*Cestum* sp.), a highly modified ctenophore. It may reach a length of 5 feet but is usually much smaller.

The fragile, transparent bodies of ctenophores are easily seen at night when they emit light (luminesce).

Form and Function

Pleurobrachia (Gr. *pleuron,* side, + L. *brachia,* arms), a pretty little sea walnut, is often used as a representative example of the ctenophores (Figure 6-28). Its surface bears eight longitudinal rows of transverse plates bearing long fused cilia and called **comb plates.** The beating of the cilia in each row starts at the aboral end and proceeds along the rows to the oral end, thus propelling the animal forward. All rows beat in unison. A reversal of the wave direction drives the animal backward. Ctenophores may be the largest animals that swim exclusively by cilia.

Two long tentacles are carried in a pair of tentacle sheaths (Figure 6-29) from which they can stretch to a length of perhaps 15 cm. The surface of the tentacles bears specialized glue cells called **colloblasts,** which secrete a sticky substance that facilitates the catching of small prey. When covered with food, the tentacles contract and the food is wiped off on the mouth. The gastrovascular cavity consists of a pharynx, stomach, and a system of gastrovascular canals. Rapid digestion occurs in the pharynx, then partly digested food circulates through the rest of the system where digestion is completed intracellularly. Residues are regurgitated or expelled through small pores in the aboral end.

A nerve net system similar to that of the cnidarians includes a subepidermal plexus concentrated under each comb plate.

The sense organ at the aboral pole is a **statocyst,** or organ of equilibrium, and is also concerned with the beating of the comb rows but does not trigger their beat. Other sensory cells are abundant in the epidermis.

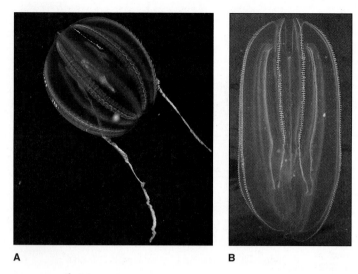

A   **B**

FIGURE 6-28

A, Comb jelly *Pleurobrachia* sp. (order Cydippida, class Tentaculata). Its fragile beauty is especially evident at night when it luminesces from its comb rows. **B,** *Mnemiopsis* sp. (order Lobata, class Tentaculata).

> *Since the 1980s population explosions of* **Mnemiopsis leidyi** *in the Black and Azov Seas have led to catastrophic declines in fisheries there. Inadvertently introduced from the coast of the Americas with ballast water of ships, the ctenophores feed on zooplankton, including small crustaceans and the eggs and larvae of fish. The normally inoffensive* M. leidyi *is kept in check in the Atlantic by certain specialized predators, but introduction of such predators into the Black Sea carries its own dangers.*

All ctenophores are monoecious, bearing both an ovary and a testis. Gametes are shed into the water, except in a few species that brood their eggs, and there is a free-swimming larva.

Phylogeny and Adaptive Radiation

Phylogeny

Although the origin of the cnidarians and ctenophores is obscure, the most widely supported hypothesis today is that the radiate phyla arose from a radially symmetrical, planula-like ancestor. Such an ancestor could have been common to the

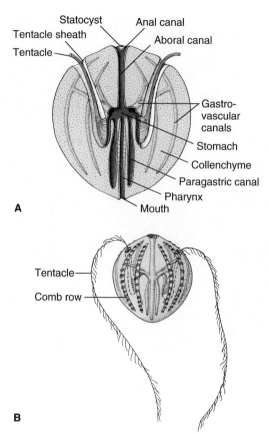

A

B

FIGURE 6-29

The comb jelly *Pleurobrachia*, a ctenophore. **A,** Hemisection. **B,** External view.

radiates and to the higher metazoans, the latter having been derived from a branch whose members habitually crept about on the sea bottom. Such a habit would select for bilateral symmetry. Others became sessile or free floating, conditions for which radial symmetry is a selective advantage. A planula larva in which an invagination formed to become the gastrovascular cavity would correspond roughly to a cnidarian with an ectoderm and an endoderm.

Some researchers believe the trachyline medusae (an order of class Hydrozoa) resemble the ancestral cnidarian because of their direct development from the planula and actinula larvae to the medusa (Figure 6-30). The trachyline-like ancestor would have given rise to other cnidarian lines after the evolution of the polyp stage and alternation of sexual (medusa) and asexual (polyp) generations. Subsequently, the medusa was completely lost in the anthozoan line. If the order Trachylina is retained within the class Hydrozoa, however, then the Hydrozoa becomes paraphyletic. Future investigators may resolve this problem.

In the past it was assumed that the ctenophores arose from a medusoid cnidarian, but this assumption has been questioned recently. The similarities between the groups are

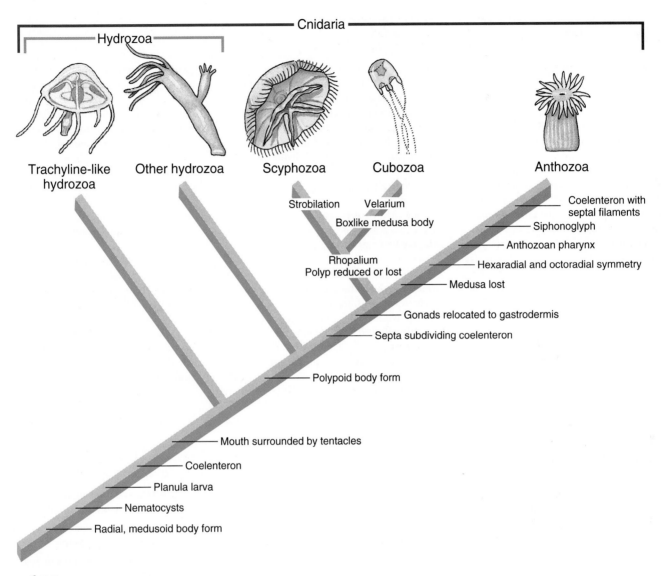

FIGURE 6-30

Cladogram showing hypothetical relationships of cnidarian classes with some shared derived characters indicated. This hypothesis suggests that the hydrozoan order Trachylina retains the ancestral cnidarian life cycle, having branched off before the evolution of the polyp stage. Note that this arrangement makes the Hydrozoa paraphyletic; the trachyline-like Hydrozoa is a sister group to all the other Cnidaria.

(Source: *R. C. Brusca and G. J. Brusca,* Invertebrates, *1990, Sinauer Associates, Sunderland, MA.*)

mostly of a general nature and do not seem to indicate a close relationship. Some molecular evidence suggests that the ctenophores branched off the metazoan line after the sponges but before the cnidarians.

Adaptive Radiation

In their evolution neither phylum has deviated far from its basic plan of structure. In the Cnidaria, both the polyp and medusa are constructed on the same scheme. Likewise, the ctenophores have adhered to the arrangement of the comb plates and their biradial symmetry.

Nonetheless, the cnidarians have achieved large numbers of individuals and species, demonstrating a surprising degree of diversity considering the simplicity of their basic body plan. They are efficient predators, many feeding on prey quite large in relation to themselves. Some are adapted for feeding on small particles. The colonial form of life is well explored, with some colonies growing to great size among the corals, and others, such as the siphonophores, showing astonishing polymorphism and specialization of individuals within the colony.

Summary

The phyla Cnidaria and Ctenophora have a primary radial symmetry; radial symmetry is an advantage for sessile or free-floating organisms because environmental stimuli come from all directions equally. The Cnidaria are surprisingly efficient predators because they possess stinging organelles called nematocysts. Both phyla are essentially diploblastic (some triploblastic, according to definition of mesoderm), with a body wall composed of epidermis and gastrodermis and a mesoglea between. The digestive-respiratory (gastrovascular) cavity has a mouth and no anus. Cnidarians are at the tissue level of organization. They have two basic body types (polyp and medusa), and in many hydrozoans and scyphozoans the life cycle involves both the asexually reproducing polyp and the sexually reproducing medusa.

An organelle unique to cnidarians, the nematocyst, is produced by a cnidoblast (which becomes the cnidocyte) and is coiled within a capsule. When discharged, some types of nematocysts penetrate the prey and inject poison. Discharge is effected by a change in permeability of the capsule and an increase in internal hydrostatic pressure because of the high osmotic pressure within the capsule.

Most hydrozoans are colonial and marine, but the freshwater hydras are commonly demonstrated in class laboratories. They have a typical polypoid form but are not colonial and have no medusoid stage. Most marine hydrozoans are in the form of a branching colony of many polyps (hydranths). The medusa may be free-swimming or remain attached to the colony.

The scyphozoans are typical jellyfishes, in which the medusa is the dominant body form, and many have an inconspicuous polypoid stage. Cubozoans are predominantly medusoid. They include the dangerous sea wasps.

Anthozoans are all marine and are polypoid; there is no medusoid stage. The most important subclasses are the Zoantharia (with hexamerous or polymerous symmetry) and Alcyonaria (with octomerous symmetry). The largest zoantharian orders contain the sea anemones, which are solitary and do not have a skeleton, and the stony corals, which are mostly colonial and secrete a calcareous exoskeleton. Stony corals are the critical component in coral reefs, which are habitats of great beauty, productivity, and ecological and economic value. The Alcyonaria contain the soft and horny corals, many of which are important and beautiful components of coral reefs.

The Ctenophora are biradial and swim by means of eight comb rows. Colloblasts, with which they capture small prey, are characteristic of the phylum.

Cnidaria and Ctenophora are probably derived from an ancestor that resembled the planula larva of the cnidarians. Despite their relatively simple level of organization, the cnidarians are an important phylum.

Review Questions

1. Explain the selective advantage of radial symmetry for sessile and free-floating animals.

2. What characteristics of the phylum Cnidaria do you think are most important in distinguishing it from other phyla?

3. Name and distinguish the classes in the phylum Cnidaria.

4. Many cnidarians are dimorphic, assuming two completely different body forms during their life histories. Explain what these two different forms are and why the life history of some cnidarians is described as "alternation of generations." How do these different forms differ ecologically? How has dimorphism diverged and developed in the different cnidarian classes?

5. Explain the mechanism of nematocyst discharge. How can a hydrostatic pressure of one atmosphere be maintained within the nematocyst until it receives an expulsion stimulus?

6. What is an unusual feature of the nervous system of cnidarians?

7. Diagram a hydra and label the main body parts.

8. Name and give the functions of the main cell types in the epidermis and in the gastrodermis of hydra.

9. What stimulates feeding behavior in hydras?

10. Give an example of a highly polymorphic, floating, colonial hydrozoan.

11. Distinguish the following from each other: statocyst and rhopalium; scyphomedusae and hydromedusae; scyphistoma, strobila, and ephyrae; velum, velarium, and pedalium; Zoantharia and Alcyonaria.

12. Define the following with regard to sea anemones: siphonoglyph; primary septa or mesenteries; incomplete septa; septal filaments; acontia threads; pedal laceration.

13. Describe three specific interactions of anemones with nonprey organisms.

14. Contrast the skeletons of zoantharian and alcyonarian corals.

15. Coral reefs generally are limited in geographic distribution to shallow marine waters. How do you account for this?

16. Specifically, what kinds of organisms are most important in deposition of calcium carbonate on coral reefs?

17. How do zooxanthellae contribute to the welfare of hermatypic corals?

18. What characteristics of Ctenophora do you think are most important in distinguishing it from other phyla?

19. How do ctenophores swim, and how do they obtain food?

20. What is a widely held hypothesis on the origin of the radiate phyla?

Selected References

See also general references on page 395.

Brown, B. E., and J. C. Ogden. 1993. Coral bleaching. Sci. Am. **268**:64-70 (Jan.). *Abnormally warm water is apparently the cause of reef corals losing their zooxanthellae.*

Crossland, C. J., B. G. Hatcher, and S. V. Smith. 1991. Role of coral reefs in global ocean production. Coral Reefs **10**:55-64. *Because of extensive recycling of nutrients within reefs, their net energy production for export is relatively minor. However, they play a major role in inorganic carbon precipitation by biologically-mediated processes.*

Fishman, D. J. 1991. Corals in a troubled sea of crude: the hardy marine ecosystem of the Persian Gulf. Ocean Realm (Spring):9-11. *The corals in the Persian Gulf are tougher than corals in other areas, but there was doubt that they could withstand the disastrous oil spill from the Persian Gulf War. (See Vogt, below.)*

Goreau, T. F., N. I. Goreau, and T. J. Goreau. 1979. Corals and coral reefs. Sci. Am. **241**:124-135 (Aug.). *A good summary of the biology, ecology, and physiology of reef corals.*

Hamner, W. M. 1994. Deadly jellyfish of Australia. National Geographic **186**(2):116-130. *The delicate box jellyfish, also called the sea wasp, is considered the earth's most venomous creature.*

Humann, P. 1992. Reef creature identification. Florida, Caribbean, Bahamas. Jacksonville, Florida, New World Publications, Inc. *This is the best field guide available for identification of "non-coral" cnidarians of the Caribbean.*

Humann, P. 1993. Reef coral identification. Florida, Caribbean, Bahamas. Jacksonville, Florida, New World Publications, Inc. *Superb color photographs and accurate identifications make this by far the best field guide to corals now available; includes sea grasses and some algae.*

Kenchington, R., and G. Kelleher. 1992. Crown-of-thorns starfish management conundrums. Coral Reefs **11**:53-56. *The first article of an entire issue on the starfish:* Acanthaster planci, *a predator of corals. Another entire issue was devoted to this predator in 1990.*

Lenhoff, H. M., and S. G. Lenhoff. 1988. Trembley's polyps. Sci. Am. **258**:108-113 (Apr.) *Trembley's elegant experiments on hydras in the 1740s marked the dawn of experimental zoology.*

Vogt, H. P. 1995. Coral reefs in Saudi Arabia: 3.5 years after the Gulf War oil spill. Coral Reefs **14**:271-273. *Reefs in the Persian Gulf have shown a surprising resiliency after a very serious oil spill.*

Ward, F. 1990. Florida's coral reefs are imperiled. National Geographic **178**(1):115-132. *Describes the degradation suffered by reefs in the Florida Keys, along with the probable causes.*

Links to the Internet

Visit this textbook's web site at http://www.mhhe.com/zoology to find live Internet links for each of the references listed below.

Cnidaria

1. Cnidaria WWW Server. Many aspects of cnidarian biology are described here, from morphology to molecular evolution. This site, supported by the University of California-Irvine, has many links to other sites.

2. Introduction to the Cnidaria. University of California at Berkeley Museum of Paleontology. Links to the fossil record, life history and ecology, systematics, and more on morphology. Many links to the different classes, and each site typically includes at least one picture of a member of the taxon.

3. Introduction to the Scleractinia. University of California at Berkeley Museum of Paleontology. Pictures, much information, and links to information on corals.

4. Cnidaria. Arizona's Tree of Life Web Page. Pictures, characteristics, phylogenetic relationships, references on cnidarians.

5. Phylum Cnidaria. University of Michigan site on phylum Cnidaria. Pictures, links to the four classes of cnidarians.

6. Introduction to the Hydrozoa. University of California at Berkeley, Museum of Paleontology. Images, photographs, systematics, more information, and links.

7. Corals and Coral Reefs. Information from the education department of Sea World on coral reefs, destruction of coral, classification, many links.

8. Corals and Coral Reefs. Destruction of coral ecosystems.

9. Fact Sheet—Coral Reefs. A fact sheet from the Ecological Society of America on coral reef losses and causes.

Ctenophores

10. Ctenophora. Arizona's Tree of Life Web Page. Pictures, references on ctenophores.

11. Introduction to the Ctenophora. University of California at Berkeley, Museum of Paleontology. Images, photographs, systematics, more information, and links.

Acoelomate Animals:
Flatworms, Ribbon Worms, and Jaw Worms

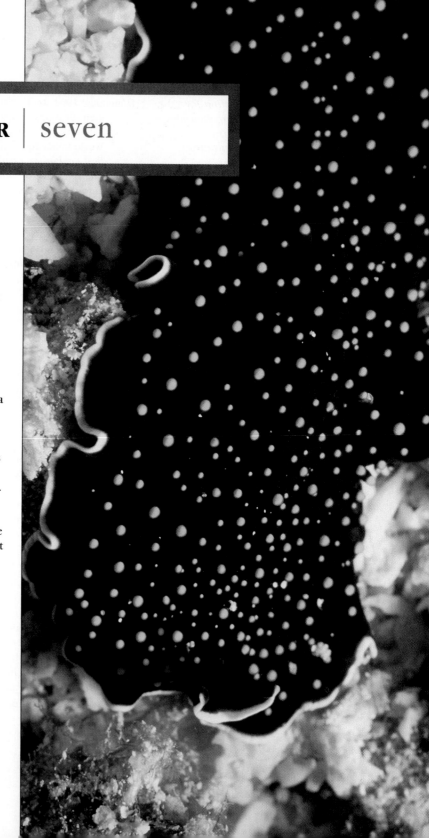

CHAPTER | seven

Getting Ahead

For animals that spend their lives sitting and waiting, as do most members of the two radiate phyla we considered in the preceding chapter, radial symmetry is ideal. One side of the animal is just as important as any other for snaring prey coming from any direction. But if an animal is active in seeking food, shelter, home sites, and reproductive mates, it requires a different set of strategies and a new body organization. Active, directed movement requires an elongated body form with head (anterior) and tail (posterior) ends. In addition, one side of the body faces up (dorsal) and the other side, specialized for locomotion, faces down (ventral). These conditions resulted in a bilaterally symmetrical animal in which the body could be divided along only one plane of symmetry to yield two halves which were mirror images of each other. Furthermore, because it is better to determine where one is going than where one has been, sense organs and centers for nervous control came to be located at the anterior end. This process is called cephalization. Thus cephalization and primary bilateral symmetry evolved together.

The three acoelomate phyla considered in this chapter are not much more complex in organization than the Radiata except for symmetry. The evolutionary consequence of that difference alone was enormous, however, for bilaterality is the type of symmetry assumed by all more complex animals.

The term "worm" has been applied loosely to elongated, bilateral invertebrate animals without appendages. At one time zoologists considered worms (Vermes) a group in their own right. Such a group included a highly diverse assortment of forms. This unnatural assemblage has been reclassified into various phyla. By tradition, however, zoologists still refer to the various groups of these animals as flatworms, ribbon worms, roundworms, segmented worms, and the like. In this chapter we will consider the Platyhelminthes (Gr. *platys,* flat, + *helmins,* worm), or flatworms, the Nemertea (Gr. *Nemertes,* one of the nereids, unerring one), or ribbon worms, and the Gnathostomulida (Gr. *gnathos,* jaw, + *stoma,* mouth, + L. *ulus,* diminutive), or jaw worms. The Platyhelminthes is by far the most widely prevalent and abundant of the three phyla, and some flatworms are very important pathogens of humans and domestic animals.

Phylum Platyhelminthes

The Platyhelminthes were derived from an ancestor that probably had many cnidarian-like characteristics, including a gelatinous mesoglea. Nonetheless, replacement of the gelatinous mesoglea with a cellular, mesodermal **parenchyma** laid the basis for a more complex organization. Parenchyma is a form of tissue containing more cells and fibers than the mesoglea of the cnidarians.

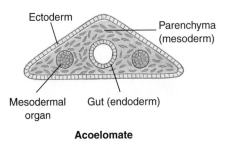

Acoelomate

FIGURE 7-1

Acoelomate body plan.

Flatworms range in size from a millimeter or less to some of the tapeworms that are many meters in length. Their typically flattened bodies may be slender, broadly leaflike, or long and ribbonlike.

Ecological Relationships

The flatworms include both free-living and parasitic forms, but the free-living members are found exclusively in the class Turbellaria. A few turbellarians are symbiotic (commensals or parasites), but the majority are adapted as bottom dwellers in marine or fresh water or live in moist places on land. Many,

Position in Animal Kingdom

1. The Platyhelminthes, or flatworms, the Nemertea, or ribbon worms, and the Gnathostomulida, or jaw worms, are the simplest animals to have **primary bilateral symmetry.**

2. These phyla have only one internal space, the digestive cavity, with the region between the ectoderm and endoderm filled with mesoderm in the form of muscle fibers and mesenchyme (parenchyma). Since they lack a coelom or a pseudocoelom, they are termed **acoelomate animals** (Figure 7-1), and because they have three well-defined germ layers, they are termed triploblastic.

3. Acoelomates show more specialization and division of labor among their organs than do the radiate animals because having mesoderm makes more elaborate organs possible. Thus the acoelomates are said to have reached the **organ-system level of organization.**

4. They belong to the protostome division of the Bilateria and have spiral cleavage, and at least the platyhelminths and nemerteans have mosaic (determinate) cleavage.

Biological Contributions

1. The acoelomates developed the basic **bilateral** plan of organization that has been widely exploited in the animal kingdom.

2. The **mesoderm** developed into a well-defined embryonic germ layer **(triploblastic),** making available a great source of tissues, organs, and systems.

3. Along with bilateral symmetry, **cephalization** was established. There is some centralization of the nervous system evident in the **ladder type of system** found in flatworms.

4. Along with the subepidermal musculature, there is also a mesenchymal system of muscle fibers.

5. They are the simplest animals with an **excretory system.**

6. The nemerteans are the simplest animals to have a **circulatory system** with blood and a **one-way alimentary canal.** Although not stressed by zoologists, the rhynchocoel cavity in ribbon worms is technically a true coelom, but because it is merely a part of the proboscis mechanism, it probably is not homologous to the coelom of eucoelomate phyla.

7. Unique and specialized structures occur in all three phyla. The parasitic habit of many flatworms has led to many specialized adaptations, such as organs of adhesion.

Characteristics of Phylum Platyhelminthes

1. Three germ layers (**triploblastic**)
2. **Bilateral symmetry;** definite polarity of anterior and posterior ends
3. **Body flattened dorsoventrally** in most; oral and genital apertures mostly on ventral surface
4. Body with multiple reproductive units in one class (Cestoda)
5. Epidermis may be cellular or syncytial (ciliated in some); **rhabdites** in epidermis of most Turbellaria; epidermis a syncytial **tegument** in Monogenea, Trematoda, and Cestoda
6. Muscular system of mesodermal origin, in the form of a sheath of circular, longitudinal, and oblique layers beneath the epidermis or tegument
7. No internal body space (acoelomate) other than digestive tube; spaces between organs filled with parenchyma
8. Digestive system incomplete (gastrovascular type); absent in some
9. Nervous system consisting of a **pair of anterior ganglia** with **longitudinal nerve cords** connected by transverse nerves and located in the parenchyma in most forms; in forms with more primitive characters the nervous system is similar to that of cnidarians
10. Simple sense organs; eyespots in some
11. Excretory system of two lateral canals with branches bearing **flame cells (protonephridia);** lacking in some forms
12. Respiratory, circulatory, and skeletal systems lacking; lymph channels with free cells in some trematodes
13. Most forms monoecious; reproductive system complex, usually with well-developed gonads, ducts, and accessory organs; internal fertilization; life cycle simple in free-swimming forms and those with single hosts; complicated life cycle often involving several hosts in many internal parasites.
14. Class Turbellaria mostly free-living; classes Monogenea, Trematoda, and Cestoda entirely parasitic.

especially of the larger species, are found on the underside of stones and other hard objects in freshwater streams or in the littoral zones of the ocean.

Relatively few turbellarians live in fresh water. Planarians (Figure 7-2) and some others frequent streams and spring pools; others prefer flowing water of mountain streams. Some species occur in moderately hot springs. Terrestrial turbellarians are found in fairly moist places under stones and logs (Figure 7-3).

All members of the classes Monogenea and Trematoda (the flukes) and the class Cestoda (the tapeworms) are parasitic. Most of the Monogenea are ectoparasites, but all the trematodes and cestodes are endoparasitic. Many species have indirect life cycles with more than one host; the first host is often an invertebrate, and the final host is usually a vertebrate. Humans serve as hosts for a number of species. Certain larval stages may be free-living.

Form and Function

Body Form

The body of turbellarians is covered by a ciliated epidermis resting on a basement membrane. It contains rod-shaped **rhabdites** (Figure 7-4) that, when discharged into water, swell and form a protective mucous sheath around the body. Single-cell mucous glands open on the surface of the epidermis. The body covering in all the other classes is a **tegument** (Figure 7-5), which does not bear cilia in the adult. The cell bodies are sunk beneath the outer layer and superficial muscle layers and communicate with the outer layer (**distal cytoplasm**) by processes extending between the muscles. Because the distal cytoplasm is continuous, with no intervening cell membranes, the tegument is **syncytial.** This peculiar epidermal arrangement probably is related to adaptations for parasitism in ways that are still unclear.

In the body wall beneath the basement membrane are layers of **muscle fibers** that run circularly, longitudinally, and diagonally. Other muscle fibers may cross through the body from one side of the outer muscle sheath to the other (Figure 7-4). A meshwork of parenchyma cells, developed from mesoderm, fills the spaces between the muscles and

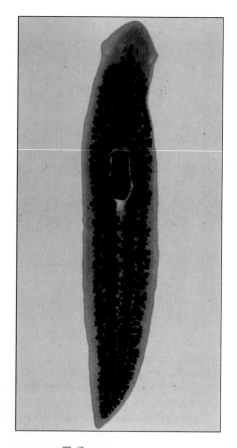

FIGURE 7-2

Stained planarian.

FIGURE 7-3

Terrestrial turbellarian (order Tricladida) from the Amazon Basin, Peru.

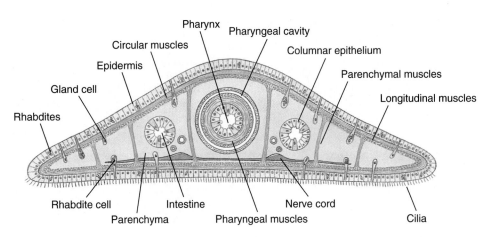

FIGURE 7-4

Cross section of planarian through pharyngeal region, showing relationships of body structures.

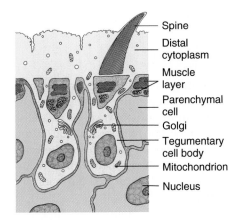

FIGURE 7-5

Diagram of the structure of the tegument of a trematode *Fasciola hepatica*.

visceral organs. Parenchyma cells in some, perhaps all, flatworms are not a separate cell type, but are the noncontractile portions (cell bodies) of muscle cells.

Earlier investigators had called the distal cytoplasm of parasitic flatworms a cuticle *(L. cuticula, dim. of cutis, pertaining to the skin). In invertebrate zoology, the term cuticle means an outer, protective layer secreted by an epidermis, which was what the tegument looked like at the light microscope level. When these worms were studied at the electron microscope level, however, biologists found abundant vesicles and mitochondria, that is, it was a living tissue, not a dead, secreted layer. They adopted the word* tegument *(L. tegmen, a cover) as being more noncommital. Zoologists often use the word* integument *(also from L. tegmen) to refer generally to the body coverings of animals.*

Nutrition and Digestion

The digestive system includes a mouth, pharynx, and intestine. In the turbellarians the muscular **pharynx** opens posteriorly just inside the mouth, through which it can extend (Figure 7-6). The mouth is usually at the anterior end in flukes, and the pharynx is not protrusible. The intestine may be simple or branched.

Intestinal secretions contain proteolytic enzymes for some **extracellular digestion.** Food is sucked into the intestine, where cells of the gastrodermis often phagocytize it and complete the digestion **(intracellular).** Undigested food is egested through the pharynx. The entire digestive system is lacking in the tapeworms. They must absorb all of their nutrients as small molecules (predigested by the host) directly through their tegument.

Excretion and Osmoregulation

Except in some turbellarians, the osmoregulatory system consists of canals with tubules that end in **flame cells (protonephridia)** (Figure 7-6A). The flame cell surrounds a small space into which a tuft of flagella projects. In some turbellarians and in the other classes of flatworms, the protonephridia form a **weir** (Old English *wer*, a fence placed in a stream to catch fish). In a weir the rim of the cup formed by the flame cell bears fingerlike projections that interdigitate with similar projections of a tubule cell. The beat of the flagella (resembling a flickering flame) provides a negative pressure to draw fluid through the weir into the space (lumen) enclosed by the tubule cell. The lumen continues into collecting ducts that finally open to the outside by pores. The wall of the duct beyond the flame cell commonly bears folds or microvilli that probably function in reabsorption of certain ions or molecules. It is likely that this system is osmoregulatory in most forms; it is reduced or absent in marine turbellarians, which do not have to expel excess water.

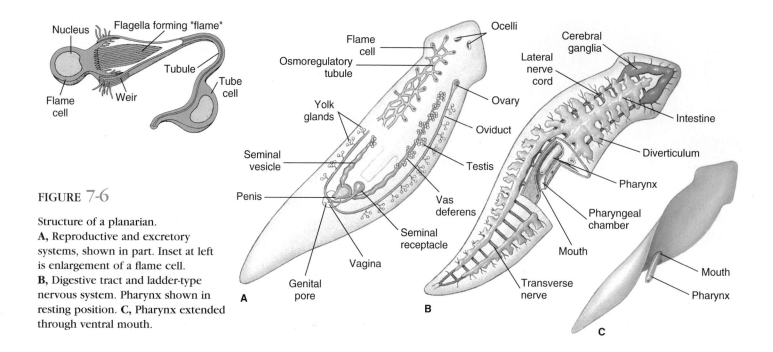

FIGURE 7-6

Structure of a planarian.
A, Reproductive and excretory systems, shown in part. Inset at left is enlargement of a flame cell.
B, Digestive tract and ladder-type nervous system. Pharynx shown in resting position. **C,** Pharynx extended through ventral mouth.

Metabolic wastes are largely removed by diffusion through the body wall.

Nervous System

The most primitive flatworm nervous system, found in some turbellarians, is a **subepidermal nerve plexus** resembling the nerve net of the cnidarians. Other flatworms have, in addition to a nerve plexus, one to five pairs of **longitudinal nerve cords** lying under the muscle layer (Figure 7-6B). Connecting nerves form a "ladder-type" pattern. The brain is a mass of ganglion cells arising anteriorly from the nerve cords. Except in simpler turbellarians, which have a diffuse system, the neurons are organized into sensory, motor, and association types—an important advance in the evolution of the nervous system.

Sense Organs

Active locomotion in flatworms has favored not only cephalization in the nervous system but also advancements in the development of sense organs. **Ocelli,** or light-sensitive eyespots, are found in the turbellarians (Figure 7-6A) and some flukes.

Tactile cells and chemoreceptive cells are abundant over the body, and in planarians they form definitive organs on the **auricles** (the earlike lobes on the sides of the head). Some also have **statocysts** for equilibrium and **rheoreceptors** for sensing water current direction.

Reproduction

Many flatworms reproduce both asexually and sexually. Many freshwater turbellarians can reproduce by fission, merely con-

stricting behind the pharynx and separating into two animals, each of which regenerates the missing parts. In some forms such as *Stenostomum* and *Microstomum,* the individuals do not separate at once but remain attached, forming chains of zooids (Figure 7-7B and C). Flukes reproduce asexually in the snail intermediate host (described further below), and some tapeworms, such as *Echinococcus,* can bud off thousands of juveniles in the intermediate host.

Most flatworms are monoecious (hermaphroditic) but practice cross-fertilization. In some turbellarians the yolk for nutrition of the developing embryo is contained within the egg cell itself **(endolecithal),** just as it is normally in other phyla of animals. The endolecithal egg is considered ancestral for flatworms. The other turbellarians plus all trematodes, monogeneans, and cestodes share a derived condition in which the egg cell contains little or no yolk, and the yolk is contributed by cells released from separate organs **(yolk glands).** Usually a number of yolk cells surrounds the zygote within the eggshell **(ectolecithal).** The ectolecithal turbellarians therefore appear to form a clade with the Trematoda, Monogenea, and Cestoda to the exclusion of endolecithal turbellarians.

In some freshwater planarians the capsules are attached by little stalks to the underside of stones or plants, and embryos emerge as juveniles that resemble miniature adults. In other flatworms, the embryo becomes a larva, which may be ciliated or not, according to the group.

Class Turbellaria

Turbellarians are mostly free-living worms that range in length from 5 mm or less to 50 cm. Usually covered with ciliated epi-

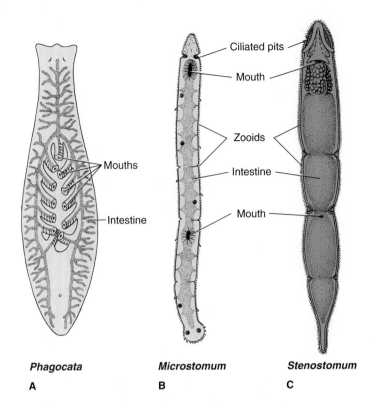

Phagocata
A

Microstomum
B

Stenostomum
C

FIGURE 7-7

Some small freshwater turbellarians. **A,** *Phagocata* has numerous mouths and pharynges. **B** and **C,** Incomplete fission results for a time in a series of attached zooids.

dermis, they are typically creeping worms that combine muscular with ciliary movements to achieve locomotion. The mouth is on the ventral side. Unlike the flukes (trematodes) and tapeworms (cestodes), they have simple life cycles.

Members of the order Acoela (Gr. *a,* without, + *koilos,* hollow) are often regarded as having changed least from the ancestral form. Its members are small and have a mouth but no gastrovascular cavity or excretory system. Food is merely passed through the mouth or pharynx into temporary spaces that are surrounded by a syncytial mesenchyme where gastrodermal phagocytic cells digest the food intracellularly. The order has a syncytial epidermis and a diffuse nervous system.

The freshwater planarians, such as *Dugesia* (formerly called *Euplanaria* but changed by priority to *Dugesia* after A. L. Dugès, who first described the form in 1830), belong to one of the more complex orders and are used extensively in introductory laboratories.

Planarians move by gliding, head slightly raised, over a slime track secreted by the marginal adhesive glands. The beating of the epidermal cilia in the slime track drives the animal along. Rhythmical muscular waves can be seen passing backward from the head as the worm glides.

Planarians are mainly carnivorous, feeding largely on small crustaceans, nematodes, rotifers, and insects. They can detect food from some distance by means of chemoreceptors.

They entangle their prey in mucous secretions from the mucous glands and rhabdites. Planarians then grip their prey, encircle it with their bodies, and suck nutrients from it with the proboscis. They also feed on carrion.

Class Trematoda

Trematodes are all parasitic flukes, and as adults they are almost all found as endoparasites of vertebrates. They are chiefly leaflike in form and are structurally similar in many respects to the more complex Turbellaria. A major difference is found in the tegument (described earlier, Figure 7-5).

Other structural adaptations for parasitism are apparent: various penetration glands or glands to produce cyst material; organs for adhesion such as suckers and hooks; and increased reproductive capacity. Otherwise, trematodes share several characteristics with turbellarians, such as a well-developed alimentary canal (but with the mouth at the anterior, or cephalic, end) and similar reproductive, excretory, and nervous systems, as well as a musculature and parenchyma that differ only slightly from those of the Turbellaria. Sense organs are poorly developed.

Of the three subclasses of Trematoda, two are small and poorly known groups, but Digenea (Gr. *dis,* double, + *genos,* descent) is a large group with many species of medical and economic importance.

Digenea

With rare exceptions, digenetic trematodes have a complex life cycle, the first (**intermediate**) host being a mollusc and the final (**definitive**) host being a vertebrate. In some species a second, and sometimes even a third, intermediate host intervenes. The group has many species, and they can inhabit diverse sites in their hosts: all parts of the digestive tract, respiratory tract, circulatory system, urinary tract, and reproductive tract.

One of the world's most amazing biological phenomena is the digenean life cycle. Although the cycles of different species vary widely in detail, a typical example would include the adult, shelled zygote, miracidium, sporocyst, redia, cercaria, and metacercaria stage (Figure 7-8). The shelled embryo or larva usually passes from the definitive host in the excreta and must reach water to develop further. There, it hatches to a free-swimming, ciliated larva, the **miracidium.** The miracidium penetrates the tissues of a snail, where it transforms into a **sporocyst.** The sporocyst reproduces asexually to yield either more sporocysts or a number of **rediae.** The rediae, in turn, reproduce asexually to produce more rediae or to produce **cercariae.** In this way a single zygote can give rise to an enormous number of progeny. The cercariae emerge from the snail and penetrate a second intermediate host or encyst on vegetation or other objects to become **metacercariae,** which are juvenile flukes. The adult grows from the metacercaria when that stage is eaten by the definitive host.

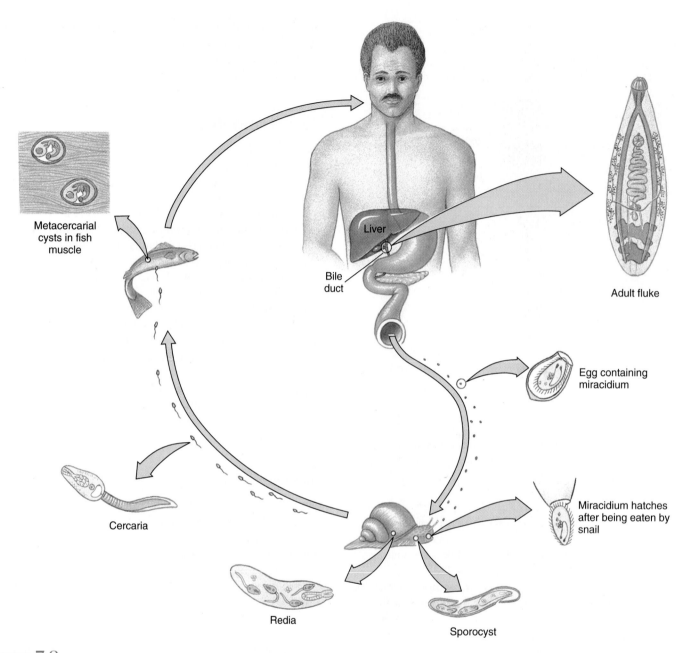

Metacercarial
cysts in fish
muscle

Liver

Bile
duct

Adult fluke

Egg containing
miracidium

Miracidium hatches
after being eaten by
snail

Cercaria

Redia

Sporocyst

FIGURE 7-8

Life cycle of human liver fluke, *Clonorchis sinensis.*

Some of the most serious parasites of humans and domestic animals belong to the Digenea (Table 7-1).

Clonorchis (Gr. *clon,* branch, + *orchis,* testis) (Figure 7-8) is the most important liver fluke of humans and is common in many regions of the Orient, especially in China, southern Asia, and Japan. Cats, dogs, and pigs are also often infected. The adult lives in the bile passages, and shelled miracidia are shed in the feces. If ingested by certain freshwater snails, the sporocyst and redia stages develop, and free-swimming cercariae emerge. Cercariae that manage to find a suitable fish encyst in the skin or muscles as metacercariae. When the fish is eaten raw, the juveniles migrate up the bile duct to mature and may survive

there for 15 to 30 years. The effect of the flukes on humans depends mainly on the extent of the infection. A heavy infection may cause a pronounced cirrhosis of the liver and death.

Schistosomiasis, an infection with blood flukes of the genus *Schistosoma* (Gr. *schistos,* divided, + *soma,* body) (Figure 7-9), ranks as one of the major infectious diseases in the world, with 200 million people infected. The disease is widely prevalent over much of Africa and parts of South America, the West Indies, the Middle East, and the Far East. It is spread when shelled miracidia shed in human feces and urine get into water containing host snails (Figure 7-9B). Cercariae that contact human skin penetrate through the skin to enter blood vessels,

TABLE 7-1

Examples of Flukes Infecting Humans

Common and scientific names	Means of infection; distribution and prevalence in humans
Blood flukes (*Schistosoma* spp.); three widely prevalent species, others reported	Cercariae in water penetrate skin; 200 million people infected with one or more species
S. mansoni	Africa, South and Central America
S. haematobium	Africa
S. japonicum	Eastern Asia
Chinese liver flukes (*Clonorchis sinensis*)	Eating metacercariae in raw fish; about 30 million cases in Eastern Asia
Lung flukes (*Paragonimus* spp.), seven species, most prevalent is *P. westermani*	Eating metacercariae in raw freshwater crabs, crayfish; Asia and Oceania, sub-Saharan Africa, South and Central America; several million cases in Asia
Intestinal fluke (*Fasciolopsis buski*)	Eating metacercariae on aquatic vegetation; 10 million cases in Eastern Asia
Sheep liver fluke (*Fasciola hepatica*)	Eating metacercariae on aquatic vegetation; widely prevalent in sheep and cattle, occasional in humans

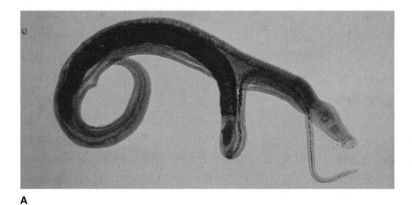

A

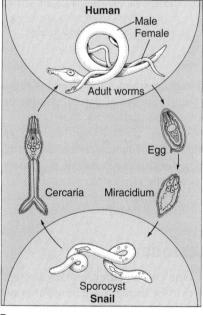

B

FIGURE 7-9

A, Adult male and female *Schistosoma mansoni* in copulation. The blood flukes differ from most other flukes in being dioecious. The male is broader and heavier and has a large, ventral gynecophoric canal, posterior to the ventral sucker. The gynecophoric canal embraces the long, slender female (darkly stained individual) during insemination and oviposition. **B,** Life cycle of *Schistosoma mansoni.*

which they follow to certain favorite regions depending on the type of fluke. *Schistosoma mansoni* lives in the venules draining the large intestine; *S. japonicum* localizes more in the venules draining the small intestine; and *S. haematobium* lives in the venules draining the urinary bladder. In each case many of the eggs released by female worms do not find their way out of the body but lodge in the liver or other organs. There they are sources of chronic inflammation. The inch-long adults may live for years in the human host, and their eggs cause such disturbances as severe dysentery, anemia, liver enlargement, bladder inflammation, and brain damage.

Cercariae of several genera whose normal hosts are birds often enter the skin of human bathers in their search for a suitable bird host, causing a skin irritation known as "swimmer's itch" (Figure 7-10). In this case the human is a dead end in the fluke's life cycle because the fluke cannot develop further in the human.

Class Monogenea

The monogenetic flukes traditionally have been placed as an order of the Trematoda, but they are sufficiently different to deserve a separate class. Cladistic analysis places them closer to the Cestoda. Monogeneans are mostly external parasites that clamp onto the gills and external surfaces of fish using a hooked attachment organ called an **opisthaptor** (Figure 7-11). A few are found in the urinary bladders of frogs and turtles, and one has been reported from the eye of a hippopotamus. Although

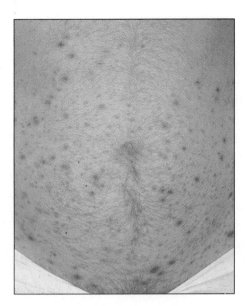

FIGURE 7-10

Human abdomen, showing schistosome dermatitis caused by penetration of schistosome cercariae that are unable to complete development in humans. Sensitization to allergenic substances released by cercariae results in rash and itching.

widespread and common, monogeneans seem to cause little damage to their hosts under natural conditions. However, like numerous other fish pathogens, they become a serious threat when their hosts are crowded together, as in fish farming.

The life cycles of monogeneans are simple, with a single host, as suggested by the name of the group, which means "single descent." The egg hatches a ciliated larva that attaches to the host or swims around awhile before attachment.

Class Cestoda

Cestoda, or tapeworms, differ in many respects from the preceding classes: they usually have long flat bodies composed of many reproductive units, or **proglottids** (Figure 7-12), and they completely lack a digestive system. As in Monogenea and Trematoda, there are no external, motile cilia in the adult, and the tegument is of a distal cytoplasm with sunken cell bodies beneath the superficial layer of muscle (Figure 7-5). In contrast to the monogeneans and trematodes, however, their entire surface is covered with minute projections that are similar in certain respects to the microvilli of the vertebrate small intestine (Figure 7-13). The microvilli greatly amplify the surface area of the tegument, which is a vital adaptation of the tapeworm, since it must absorb all its nutrients across the tegument.

Tapeworms are nearly all monoecious. They have well-developed muscles, and their excretory system and nervous system are somewhat similar to those of other flatworms. They have no special sense organs but do have sensory endings in the tegument that are modified cilia (Figure 7-13). One

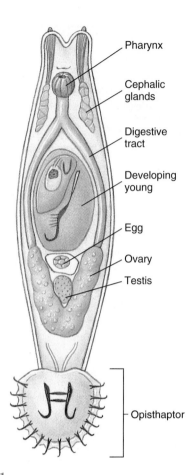

FIGURE 7-11

Gyrodactylus sp. (class Monogenea), ventral view.

of their most specialized structures is the **scolex,** or holdfast, which is the organ of attachment. It is usually provided with suckers or suckerlike organs and often with hooks or spiny tentacles (Figure 7-14).

With rare exceptions, all cestodes require at least two hosts, and the adult is a parasite in the digestive tract of vertebrates. Often one of the intermediate hosts is an invertebrate.

The main body of the worms, the chain of proglottids, is called the **strobila** (Figure 7-14). Typically, there is a **germinative zone** just behind the scolex where new proglottids form. As new proglottids differentiate in front of it, each individual unit moves posteriorly in the strobila, and its gonads mature. The proglottid usually is fertilized by another proglottid in the same or a different strobila. The shelled embryos form in the uterus of the proglottid, and they are either expelled through a uterine pore, or the entire proglottid detaches from the worm as it reaches the posterior end of the strobila.

About 4000 species of tapeworms are known to parasitologists. Almost all vertebrate species are infected. Normally, adult tapeworms do little harm to their hosts. Table 7-2 lists the most common tapeworms in humans.

In the beef tapeworm, *Taeniarhynchus saginatus* (Gr. *tainia,* band, ribbon, + *rhynchos,* snout, beak), shelled lar-

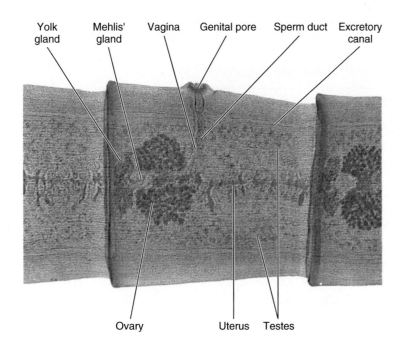

FIGURE **7-12**

Mature proglottid of *Taenia pisiformis,* a dog tapeworm. Portions of two other proglottids also shown.

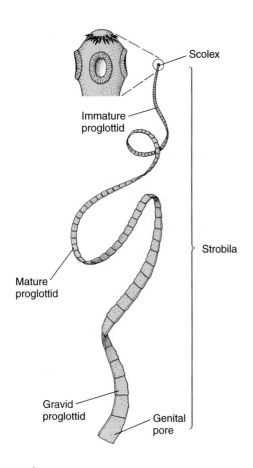

FIGURE **7-14**

A tapeworm, showing strobila and scolex. The scolex is the organ of attachment.

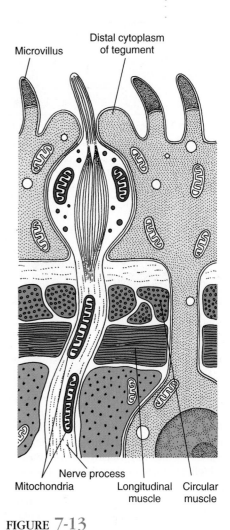

FIGURE **7-13**

Schematic drawing of a longitudinal section through a sensory ending in the tegument of *Echinococcus granulosus.*

vae shed from the human host are ingested by cattle (Figure 7-15). The six-hooked larvae (oncospheres) hatch, burrow into blood or lymph vessels, and migrate to skeletal muscle where they encyst to become "bladder worms" (cysticerci). Each of these juveniles develops an invaginated scolex and remains quiescent until the uncooked muscle is eaten by humans. In the new host the scolex evaginates, attaches to the intestine, and matures in 2 or 3 weeks; then ripe proglottids may be expelled daily for many years. Humans become infected by eating raw or rare infested ("measly") beef. The adult worm may attain a length of 7 m or more, folded back and forth in the host intestine.

The pork tapeworm, *Taenia solium,* uses humans as definitive hosts and pigs as intermediate hosts. Humans can also serve as intermediate hosts by ingesting the shelled larvae from contaminated hands or food or, in persons harboring an adult worm, by regurgitating segments into the stomach. The juveniles may encyst in the central nervous system, where great damage may result (Figure 7-16).

TABLE 7-2

Common Cestodes of Humans

Common and scientific names	Means of infection; prevalence in humans
Beef tapeworm *(Taeniarhynchus saginatus)*	Eating rare beef; most common of large tapeworms in humans
Pork tapeworm *(Taenia solium)*	Eating rare pork; less common than *T. saginatus*
Fish tapeworm *(Diphyllobothrium latum)*	Eating rare or poorly cooked fish; fairly common in Great Lakes region of United States, and other areas of world where raw fish is eaten
Dog tapeworm *(Dipylidium caninum)*	Unhygienic habits of children (juveniles in flea and louse); moderate frequency
Dwarf tapeworm *(Vampirolepis nana)*	Juveniles in flour beetles; common
Unilocular hydatid *(Echinococcus granulosus)*	Cysts of juveniles in humans; infection by contact with dogs; common wherever humans are in close relationship with dogs and ruminants
Multilocular hydatid *(Echinococcus multilocularis)*	Cysts of juveniles in humans; infection by contact with foxes; less common than unilocular hydatid

FIGURE 7-15

Life cycle of the beef tapeworm, *Taeniarhynchus.* Ripe proglottids break off in the human intestine, pass out in the feces, crawl out of the feces onto grass, and are ingested by cattle. The larvae hatch in the cow's intestine, freeing oncospheres, which penetrate into muscles and encyst, developing into "bladder worms." A human eats infected rare beef, and the cysticercus is freed in the intestine, where it attaches to the intestine wall, forms a strobila, and matures.

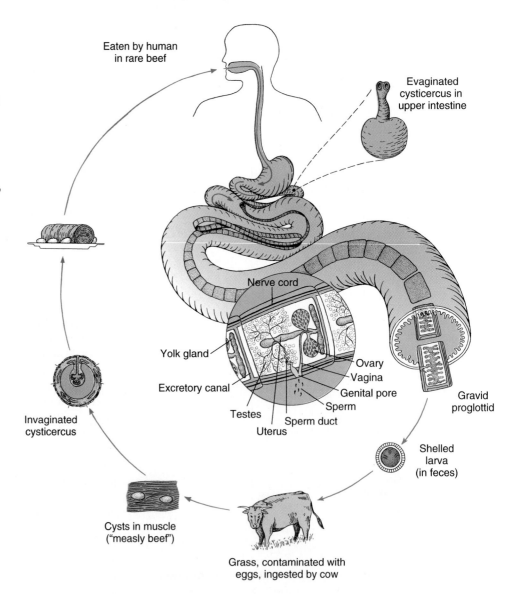

Classification of Phylum Platyhelminthes

Class Turbellaria (tur′bel-lar′e-a) (L. *turbellae* [pl.], stir, bustle, + *-aria,* like or connected with). The turbellarians. Usually free-living forms with soft flattened bodies; covered with ciliated epidermis containing secreting cells and rodlike bodies (rhabdites); mouth usually on ventral surface sometimes near center of body; no body cavity except intercellular spaces in parenchyma; mostly hermaphroditic; some have asexual fission. Examples: *Dugesia* (planaria), *Microstomum, Planocera.*

Class Trematoda (trem′a-to′da) (Gr. *trēma,* a hole, + *eidos,* form). The digenetic flukes. Body covered with a syncytial tegument without cilia; leaflike or cylindrical in shape; usually with oral and ventral suckers, no hooks; alimentary canal usually with two main branches; mostly monoecious; life cycle complex, with first host a mollusc, final host usually a vertebrate; parasitic in all classes of vertebrates. Examples: *Fasciola, Clonorchis, Schistosoma.*

Class Monogenea (mon′o-gen′e-a) (Gr. *mono,* single, + *genos,* descent). The monogenetic flukes. Body covered with a syncytial tegument without cilia; body usually leaflike to cylindrical in shape; posterior attachment organ with hooks, suckers, or clamps, usually in combination; monoecious; life cycle simple, with single host and usually with free-swimming, ciliated larva; all parasitic, mostly on skin or gills of fish. Examples: *Dactylogyrus, Polystoma, Gyrodactylus.*

Class Cestoda (ses-to′da) (Gr. *kestos,* girdle, + *eidos,* form). The tapeworms. Body covered with nonciliated, syncytial tegument; general form of body tapelike; scolex with suckers or hooks, sometimes both, for attachment; body usually divided into series of proglottids; no digestive organs; usually monoecious; parasitic in digestive tract of all classes of vertebrates; life cycle complex, with two or more hosts; first host may be vertebrate or invertebrate. Examples: *Diphyllobothrium, Hymenolepis, Taeniarhynchus, Taenia.*

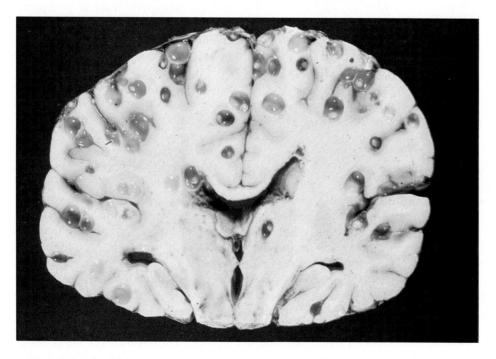

FIGURE **7-16**

Section through the brain of a person who died of cerebral cysticercosis, an infection with the cysticerci of *Taenia solium.*

Phylum Nemertea (Rhynchocoela)

Nemerteans are often called the ribbon worms. Their name (Gr. *Nemertes,* one of the nereids, unerring one) refers to the unerring aim of the proboscis, a long muscular tube (Figures 7-17 and 7-18) that can be thrust out swiftly to grasp the prey. The phylum is also called Rhynchocoela (Gr. *rhynchos,* beak, + *koilos,* hollow), which also refers to the proboscis. They are thread-shaped or ribbon-shaped worms. A few of the nemerteans are found in moist soil and fresh water, but by far the larger number are marine. At low tide they are often coiled up under stones. It seems probable that they are active at high tide and quiescent at low tide. Some live in secreted gelatinous tubes. There are about 650 species in the group.

Nemertean worms are usually less than 20 cm long, although a few are several meters in length (Figure 7-19). Some are brightly colored, although most are dull or pallid.

With few exceptions, the general body plan of the nemerteans is similar to that of Turbellaria. Like the latter,

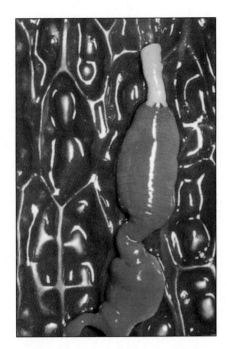

FIGURE **7-17**

Ribbon worm *Amphiporus bimaculatus* (phylum Nemertea) is 6 to 10 cm long, but other species range up to several meters. The proboscis of this specimen is partially extended at the top; the head is marked by two brown spots.

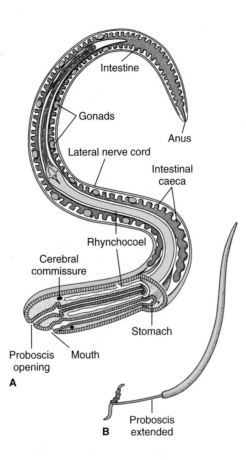

A

B Proboscis extended

FIGURE **7-18**

A, Internal structure of female ribbon worm (diagrammatic). Dorsal view to show proboscis. **B,** *Amphiporus* with proboscis extended to catch prey.

FIGURE **7-19**

Baseodiscus is a genus of nemerteans whose members typically measure several meters in length. This *B. mexicanus* from the Galápagos Islands was over 10 m long.

their epidermis is ciliated and has many gland cells. Another striking similarity is the presence of flame cells in the excretory system. Rhabdites occur in several nemerteans. In the marine forms there is a ciliated larva that has some resemblance to the trochophore larva found in annelids and molluscs. Other flatworm characteristics are the presence of bilateral symmetry and a mesoderm and the lack of coelom. All in all, the present evidence seems to indicate that the nemerteans share a close common ancestor with the Platyhelminthes. Nevertheless, they differ from flatworms in several important aspects.

Form and Function

Many nemerteans are difficult to examine because they are so long and fragile. *Amphiporus* (Gr. *amphi,* on both sides, + *poros,* pore), one of the smaller genera that ranges from 2 to 10 cm in length, is fairly typical of the nemertean structure (Figure 7-18). Its body wall consists of ciliated epidermis and layers of circular and longitudinal muscles. Locomotion con-

sists largely of gliding over a slime track, although larger species move by muscular contractions.

The mouth is anterior and ventral, and the **digestive tract** is **complete,** extending the full length of the body and ending at the anus. The development of an anus marks a significant advancement over the gastrovascular systems of the flatworms and radiates. Regurgitation of wastes is no longer necessary; ingestion and egestion can occur simultaneously. Cilia move food through the intestine. Digestion is largely extracellular.

Nemerteans are carnivorous, feeding primarily on annelids and other small invertebrates. They seize their prey with a **proboscis** that lies in an interior cavity of its own, the **rhynchocoel,** above the digestive tract (but not connected with it). The proboscis itself is a long, blind muscular tube that opens at the anterior end at a proboscis pore above the mouth. (In a few nemerteans the esophagus opens through the proboscis pore rather than through a separate mouth.) Muscular pressure on the fluid in the rhynchocoel causes the long tubular proboscis to be everted rapidly through the proboscis pore. Eversion of the proboscis exposes a sharp barb, called a stylet (absent in some nemerteans). The sticky, slime-covered proboscis coils around the prey and stabs it repeatedly with the stylet, while pouring a toxic secretion on the prey. Then, retracting the proboscis, the nemertean draws the prey near the mouth, where it is engulfed by the esophagus which is thrust out to meet it.

Unlike other acoelomates, the nemerteans have a true circulatory system, and the irregular flow is maintained by the contractile walls of the vessels. Many flame-bulb protonephridia are closely associated with the circulatory system, so that their function appears to be truly excretory (for disposal of metabolic wastes), in contrast to their apparently osmoregulatory role in Platyhelminthes.

Nemerteans have a pair of nerve ganglia, and one or more pairs of longitudinal nerve cords are connected by transverse nerves.

Some species reproduce asexually by fragmentation and regeneration. In contrast to flatworms, most nemerteans are dioecious.

Phylum Gnathostomulida

The first species of the Gnathostomulida (nath′o-sto-myu′lid-a) (Gr. *gnatho*, jaw, + *stoma*, mouth, + L. *-ulus*, diminutive) was observed in 1928 in the Baltic, but its description was not published until 1956. Since then these animals have been found in many parts of the world, including the Atlantic coast of the United States, and over 80 species in 18 genera have been described.

Gnathostomulids are delicate wormlike animals 0.5 to 1 mm long (Figure 7-20). They live in the interstitial spaces of very fine sandy coastal sediments and silt and can endure conditions of very low oxygen.

Lacking a pseudocoel, a circulatory system, and an anus, the gnathostomulids show some similarities to the turbellarians and were at first included in that group. However, their parenchyma is poorly developed, and their pharynx is reminiscent of the rotifer mastax (p. 140). The pharynx is armed with a pair of lateral jaws used to scrape fungi and bacteria off the substratum. Although the epidermis is ciliated, each epidermal cell has but one cilium, which is a condition not found in other bilateral animals except in some gastrotrichs (p. 141).

Phylogeny and Adaptive Radiation

Phylogeny

There can be little doubt that the bilaterally symmetrical flatworms were derived from a radial ancestor, perhaps one very similar to the planula larva of the cnidarians. Some investigators believe that this **planuloid ancestor** may have given rise to one branch of descendants that were sessile or free floating and radial, which became the Cnidaria, and another that acquired a creeping habit and bilateral symmetry. Bilateral symmetry is a selective advantage for creeping or swimming animals because sensory structures are concentrated anteriorly (cephalization), which is the end that first encounters environmental stimuli.

According to Ax,[1] an early branch from the bilateral line would have given rise to the Platyhelminthes and Gnathostomulida, and these two would form a sister group to all the

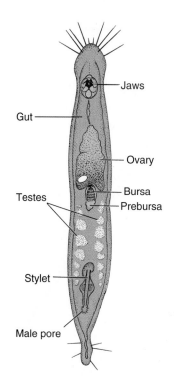

FIGURE 7-20

Gnathostomula jenneri (phylum Gnathostomulida) is a tiny member of the interstitial fauna between grains of sand or mud. Species in this family are among the most commonly encountered jaw worms, found in shallow water and down to depths of several hundred meters.

other bilateral animals (Eubilateria). This scheme would separate the Platyhelminthes and Nemertea, traditionally considered closely related, because the Nemertea have a flow-through gut. If this view is correct, the Nemertea must have branched off the Eubilateria line soon after the one-way intestine with anus was established.

Among the Platyhelminthes, it seems clear that the Turbellaria is paraphyletic,[2] but we are retaining the taxon for the present because presentation based on thorough cladistic analysis would require introduction of many more taxa and characteristics beyond the scope of this book. For example, if the possession of separate yolk glands is a derived characteristic, the ectolecithal Turbellaria should be placed with the Trematoda and Cestoda, to compose a sister group with the endolecithal Turbellaria. The unique architecture of the

[1]Ax, P. 1985. The position of the Gnathostomulida and Platyhelminthes in the phylogenetic system of the Bilateria. In Conway Morris, S., J. D. George, R. Gibson, and H. M. Platt (eds.). The origins and relationships of lower invertebrates. Oxford, Clarendon Press.

[2]Ehlers, U. 1985. Phylogenetic relationships within the Platyhelminthes. In Conway Morris, S., J. D. George, R. Gibson, and H. M. Platt (eds.). The origins and relationships of lower invertebrates. Oxford, Clarendon Press.

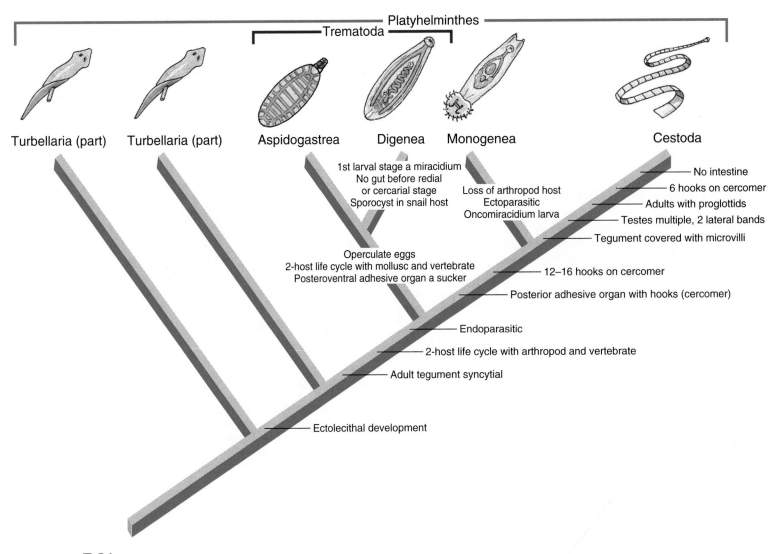

Platyhelminthes

Trematoda

Turbellaria (part) Turbellaria (part) Aspidogastrea Digenea Monogenea Cestoda

1st larval stage a miracidium
No gut before redial
or cercarial stage
Sporocyst in snail host

Loss of arthropod host
Ectoparasitic
Oncomiracidium larva

No intestine
6 hooks on cercomer
Adults with proglottids
Testes multiple, 2 lateral bands
Tegument covered with microvilli

Operculate eggs
2-host life cycle with mollusc and vertebrate
Posteroventral adhesive organ a sucker

12–16 hooks on cercomer

Posterior adhesive organ with hooks (cercomer)

Endoparasitic

2-host life cycle with arthropod and vertebrate

Adult tegument syncytial

Ectolecithal development

FIGURE 7-21

Phylogenetic relationships among parasitic Platyhelminthes. The traditionally accepted class Turbellaria is paraphyletic. Some turbellarians have ectolecithal development and, together with the Trematoda, Monogenea, and Cestoda, form a monophyletic clade and a sister group of the endolecithal turbellarians. For the sake of simplicity, the synapomorphies of those turbellarians and of the Aspidogastrea, as well as many others given by Brooks (1989) are omitted. All of these organisms comprise a clade (called Cercomeria) with a posterior adhesive organ.

(Source: *D. R. Brooks, "The Phylogeny of the* Cercomeria *(Platyhelminthes: Rhabdocoela) and General Evolutionary Principles," in* Journal of Parasitology, **75**:*606–616, 1989.*)

tegument in cestodes, monogeneans, and trematodes indicates with high probability that these groups share a common ancestor (Figure 7-21).

Adaptive Radiation

The flatworm body plan, with its creeping adaptation, placed a selective advantage on bilateral symmetry and further development of cephalization, ventral and dorsal regions, and caudal differentiation. Because of their body shape and metabolic requirements, early flatworms must have been well preadapted for parasitism and gave rise to symbiotic lines on numerous occasions. These lines produced descendants that radiated abundantly as parasites, and many flatworms became highly specialized for that mode of existence.

The ribbon worms have stressed the proboscis apparatus in their evolutionary diversity. Its use in capturing prey may have been secondarily evolved from its original function as a highly sensitive organ for exploring the environment. Although the ribbon worms have evolved beyond the flatworms in their complexity of organization, they have been dramatically less abundant as a group.

Likewise, the jaw worms have neither radiated nor become nearly as numerous as the flatworms. However, they have exploited the marine interstitial environment, particularly zones of very low oxygen concentration.

Summary

The Platyhelminthes, the Nemertea, and the Gnathostomulida are the simplest phyla that are bilaterally symmetrical, a condition of adaptive value for actively crawling or swimming animals. They have neither a coelom nor a pseudocoel and are thus acoelomate. They are triploblastic and at the organ-system level of organization.

The body surface of turbellarians is a cellular epithelium, at least in part ciliated, containing mucous cells and rod-shaped rhabdites that function together in locomotion. Members of all other classes of flatworms are covered by a nonciliated, syncytial tegument with a vesicular distal cytoplasm and cell bodies beneath superficial muscle layers. Digestion is extracellular and intracellular in most; cestodes must absorb predigested nutrients across their tegument because they have no digestive tract. Osmoregulation is by flame-cell protonephridia, and removal of metabolic wastes and respiration occur across the body wall. Except for some turbellarians, flatworms have a ladder-type nervous system with motor, sensory, and association neurons. Most flatworms are hermaphroditic, and asexual reproduction occurs in some groups.

The class Turbellaria is a paraphyletic group with mostly free-living and carnivorous members. The digenetic trematodes have mollusc intermediate hosts and almost always a vertebrate definitive host. The great amount of asexual reproduction that occurs in the intermediate host helps to increase the chance that some of the offspring will reach a definitive host. Aside from the tegument, digeneans share many basic structural characteristics with the Turbellaria. The Digenea includes a number of important parasites of humans and domestic animals. These contrast with the Monogenea, which are important ectoparasites of fishes and have a direct life cycle (without intermediate hosts).

Cestodes (tapeworms) generally have a scolex at their anterior end, followed by a long chain of proglottids, each of which contains a complete set of reproductive organs of both sexes. Cestodes live as adults in the digestive tract of vertebrates. Their tegument bears microvilli, which increase its surface area for absorption. The shelled larvae are passed in the feces of their host, and the juveniles develop in a vertebrate or invertebrate intermediate host.

Members of the Nemertea have a complete digestive system with an anus and a true circulatory system. They are free-living, mostly marine, and they ensnare prey with their long, eversible proboscis.

The Gnathostomulida are a curious phylum of little wormlike marine animals living among sand grains and silt. They have no anus, and they share certain characteristics with such widely diverse groups as turbellarians and rotifers.

The flatworms and the cnidarians both probably evolved from a common ancestor (planuloid), some of whose descendants became sessile or free floating and radial (cnidarians), while others became creeping and bilateral (flatworms).

Review Questions

1. Why is bilateral symmetry of adaptive value for actively motile animals?

2. Match the terms in the right column with the classes in the left column:
   ___ Turbellaria   a. Endoparasitic
   ___ Monogenea   b. Free-living and
   ___ Trematoda   commensal
   ___ Cestoda   c. Ectoparasitic

3. Give several major characteristics that distinguish the Platyhelminthes.

4. Distinguish two mechanisms by which flatworms supply yolk for their embryos. Which system is evolutionarily ancestral for flatworms and which one is derived?

5. How do flatworms digest their food?

6. Briefly describe the osmoregulatory system and the nervous system and sense organs of Platyhelminthes.

7. Contrast asexual reproduction in Turbellaria and Trematoda.

8. Contrast the typical life cycle of a monogenean with that of a digenetic trematode.

9. Describe and contrast the tegument of turbellarians and the other classes of platyhelminths. Could this be evidence that the trematodes, monogeneans, and cestodes form a clade within the Platyhelminthes? Why?

10. Answer the following questions with respect to both *Clonorchis* and *Schistosoma:* (a) How do humans become infected? (b) What is the general geographical distribution? (c) What are the main disease conditions produced?

11. Define each of the following with reference to cestodes: scolex, microvilli, proglottids, strobila.

12. Why is *Taenia solium* a more dangerous infection than *Taeniarhynchus saginatus?*

13. Give three differences between nemerteans and platyhelminths.

14. Where do gnathostomulids live?

15. Explain how a planuloid ancestor could have given rise to both the Cnidaria and the Bilateria.

16. What is an important character of the Nemertea that might suggest that the phylum is closer to other bilateral protostomes than to the Platyhelminthes? What are some characters of nemerteans suggesting that they share an ancestor with platyhelminths?

Selected References

See also general references on page 395.

Arme, C., and P. W. Pappas (eds.). 1983. Biology of the Eucestoda, 2 vols. New York, Academic Press, Inc. *The most up-to-date reference available on the biology of tapeworms. Advanced; not an identification manual.*

Brooks, D. R. 1989. The phylogeny of the Cercomeria (Platyhelminthes: Rhabdocoela) and general evolutionary principles. J. Parasitol.

75:606–616. *Cladistic analysis of parasitic flatworms.*

Desowitz, R. S. 1981. New Guinea tapeworms and Jewish grandmothers. New York, W.W. Norton & Company. *Accounts of parasites and parasitic diseases of humans. Entertaining and instructive. Recommended for all students.*

Schell, S. C. 1985. Handbook of trematodes of North America north of Mexico. Moscow, Idaho, University Press of

Idaho. *Good for trematode identification.*

Schmidt, G. D. 1985. Handbook of tapeworm identification. Boca Raton, Florida, CRC Press. *Good for cestode identification.*

Strickland, G. T. 1991. Hunter's tropical medicine, ed. 7. Philadelphia, W.B. Saunders Company. *A valuable source of information on parasites of medical importance.*

Links to the Internet

Visit this textbook's web site at http://www.mhhe.com/zoology to find live Internet links for each of the references listed below.

1. Platyhelminthes. Arizona's Tree of Life Web Page. Pictures, characteristics, phylogenetic relationships, references on flatworms. A picture of a flatworm, and a link to trematodes (aspidogastreans).

2. Phylum Platyhelminthes. University of Michigan site on phylum

Platyhelminthes. Pictures, information on the phylum. Links to the four classes of flatworms with a wealth of information on each group and many species in particular.

3. National Center for Infectious Diseases. This CDC site has many links to information on bacterial, viral, protozoan, and worm-related diseases (primarily affecting humans).

4. Schistosomiasis. This CDC site contains information on the effects of the schistosome worm on humans.

5. Turbellarians. Keys to Marine Invertebrates of the Woods Hole Region. Descriptive information, definition of terminology, and keys to the flatworms of the Woods Hole Region.

6. Diphyllobothrium. A FDA-supported page with information on *Diphyllobothrium*.

Pseudocoelomate Animals

CHAPTER | eight

A World of Nematodes

Without any doubt, nematodes are the most important pseudocoelomate animals, in terms of both numbers and their impact on humans. Nematodes are abundant over most of the world, yet most people are only occasionally aware of them as parasites of humans or of their pets. We are not aware of the billions of these worms in the soil, in ocean and freshwater habitats, in plants, and in all kinds of other animals. Their dramatic abundance moved N. A. Cobb to write in 1914:

> If all the matter in the universe except the nematodes were swept away, our world would still be dimly recognizable, and if, as disembodied spirits, we could then investigate it, we should find its mountains, hills, vales, rivers, lakes and oceans represented by a thin film of nematodes. The location of towns would be decipherable, since for every massing of human beings there would be a corresponding massing of certain nematodes. Trees would still stand in ghostly rows representing our streets and highways. The location of the various plants and animals would still be decipherable, and, had we sufficient knowledge, in many cases even their species could be determined by an examination of their erstwhile nematode parasites.

Source: N. A. Cobb. 1914. *Yearbook of the United States Department of Agriculture*, 1914.

Nine distinct phyla of animals belong to the pseudocoelomate category. These are Rotifera, Gastrotricha, Kinorhyncha, Loricifera, Priapulida, Nematoda, Nematomorpha, Acanthocephala, and Entoprocta. They are a heterogeneous assemblage of animals. Most of them are small; some are microscopic; some are fairly large. Some, such as the nematodes, are found in freshwater, marine, terrestrial, and parasitic habitats; others, such as the Acanthocephala, are strictly parasitic. Some have unique characteristics such as the lacunar system of the acanthocephalans and the ciliary corona of the rotifers.

Even in such a diversified grouping, however, a few characteristics are shared. All have a body wall of epidermis (often syncytial) and muscles surrounding the pseudocoel. The digestive tract is complete (except in Acanthocephala), and it, along with the gonads and excretory organs, is within the pseudocoel and bathed in perivisceral fluid. The epidermis in many secretes a nonliving cuticle with some specializations such as bristles and spines.

A constant number of cells or nuclei in the individuals of a species, a condition known as **eutely,** is present in several of the groups, and in most of them there is an emphasis on the longitudinal muscle layer.

Phylum Rotifera

Rotifera (ro-tif′e-ra) (L. *rota,* wheel, + *fero,* to bear) derive their name from their characteristic ciliated crown, or **corona,** which, when beating, often gives the impression of rotating wheels. Rotifers range in size from 40 μm to 3 mm in length, but most are between 100 and 500 μm long. Some have beautiful colors, although most are transparent, and some have odd and bizarre shapes. Their shapes are often correlated with their mode of life. The floaters are usually globular and saclike; the creepers and swimmers are somewhat elongated and wormlike; and the sessile types are commonly vaselike, with a cuticular envelope. Some are colonial.

Rotifers are a cosmopolitan group of about 1800 recognized species, some of which are found throughout the world. Most of the species are freshwater inhabitants, a few are marine, some are terrestrial, and some are epizoic (live on the surface of other animals) or parasitic. Most often they are benthic, occurring on the bottom or in vegetation of ponds or along the shores of large freshwater lakes where they swim or creep about on the vegetation.

The body is usually made up of a **head,** a **trunk,** and a **foot.** The head region bears the corona. The corona may form a ciliated funnel with its upper edges folded into lobes bearing bristles, or the corona may be made up of a pair of ciliated discs (Figure 8-2). The cilia create currents of water toward the mouth that draw in small planktonic forms for food. The corona may be retractile. The mouth, surrounded by some part of the corona, opens into a modified muscular pharynx called

Position in Animal Kingdom

In the nine phyla covered in this chapter, the original blastocoel of the embryo persists as a space, or body cavity, between the enteron and body wall. Because this cavity lacks the peritoneal lining found in the true coelomates, it is called a **pseudocoel,** and the animals possessing it are called **pseudocoelomates** (Figure 8-1). Pseudocoelomates belong to the Protostomia division of the bilateral animals, but they may be polyphyletic (not derived from a common ancestor).

Biological Contributions

1. The pseudocoel is a distinct gradation in body plan compared with the solid body structure of the acoelomates. The pseudocoel may be filled with fluid or may contain a gelatinous substance with some mesenchyme cells. In common with a true coelom, it presents certain adaptive potentials, although these are by no means realized in all members: (1) greater freedom of movement; (2) space for the development and differentiation of digestive, excretory, and reproductive systems; (3) a simple means of circulation or distribution of materials throughout the body; (4) a storage place for waste products to be discharged to the outside by excretory ducts; and (5) a hydrostatic organ. Since most pseudocoelomates are quite small, the most important functions of the pseudocoel are probably in circulation and as a means to maintain a high internal hydrostatic pressure.

2. A complete, mouth-to-anus digestive tract is found in these phyla and in all more complex phyla.

a **mastax,** which is a characteristic unique to rotifers. The mastax has a set of intricate jaws used for grasping and chewing. The trunk contains the visceral organs, and the terminal foot, when present, is segmented and, in some, is ringed with joints that can telescope to shorten. The one to four toes secrete a sticky substance from the **pedal glands** for attachment.

Rotifers have a pair of protonephridial tubules with flame bulbs, which eventually drain into a **cloacal bladder** that collects excretory and digestive wastes. They have a bilobed "brain" and sense organs that include eyespots, sensory pits, and papillae.

Female rotifers have one or two syncytial ovaries (**germovitellaria**) that produce yolk as well as oocytes. Although rotifers are dioecious, males are unknown in many species; in these reproduction is entirely parthenogenetic. In parthenogenetic species females produce only diploid eggs that have not undergone reduction division, cannot be fertilized, and

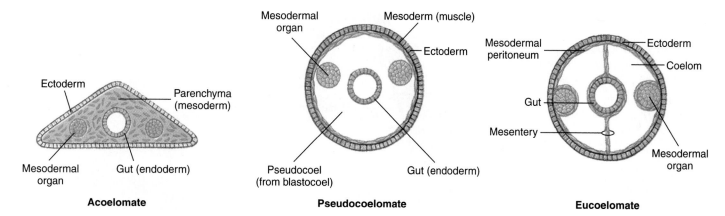

FIGURE 8-1

Acoelomate, pseudocoelomate, and eucoelomate body plans.

develop only into females. Such eggs are called **amictic eggs.** Other species can produce two kinds of eggs—amictic eggs, which develop parthenogenetically into females, and **mictic eggs,** which have undergone meiosis and are haploid. Mictic eggs, if unfertilized, develop quickly and parthenogenetically into males; if fertilized, they secrete a thick shell and become dormant for several months before developing into females. Such dormant eggs can withstand desiccation and other adverse conditions and permit rotifers to live in temporary ponds that dry up during certain seasons.

> *Mictic (Gr., miktos, mixed, blended) refers to the capacity of the haploid eggs to be fertilized (that is, "mixed") with the male's sperm nucleus to form a diploid embryo. Amictic ("without mixing") eggs are already diploid and can develop only parthenogenetically.*

Phylum Gastrotricha

The phylum Gastrotricha (gas-trot're-ka) (Gr. *gaster,* belly, + *thrix,* hair) is a small group (about 460 species) of microscopic animals, approximately 65 to 500 µm long (Figure 8-3). They are usually bristly or scaly in appearance, are flattened on the ventral side, and many move by gliding on ventral cilia. Others move in a leechlike fashion by briefly attaching the posterior end by means of adhesive glands. There are both marine and freshwater species, and they are common in lakes, ponds, and seashore sands. They feed on bacteria, diatoms, and small protozoa. Gastrotrichs are hermaphroditic. However, the male system of many freshwater species is nonfunctional, and the female system produces offspring parthenogenetically.

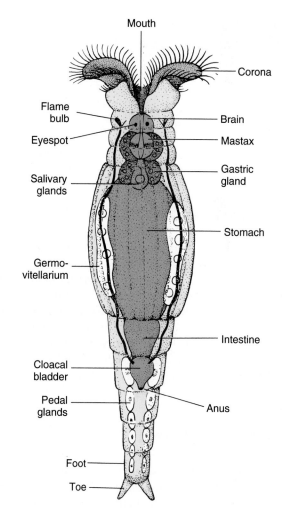

FIGURE 8-2

Structure of *Philodina*, a common rotifer.

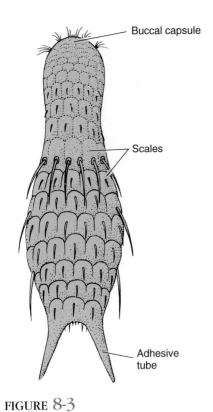

FIGURE 8-3

Chaetonotus, a common gastrotrich.

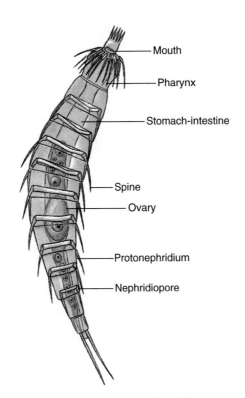

FIGURE 8-4

Echinoderes, a kinorhynch, is a minute marine worm. Segmentation is superficial. The head with its circle of spines is retractile.

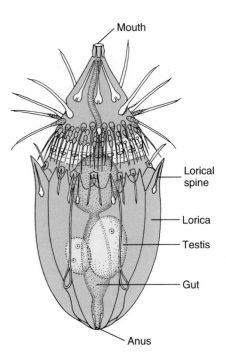

FIGURE 8-5

Dorsal view of an adult loriciferan, *Nanoloricus mysticus.*

Phylum Kinorhyncha

There are only about 100 species of Kinorhyncha (Gr. *kineo,* to move, + *rhynchos,* beak or snout). They are tiny marine worms, usually less than 1 mm long, that prefer mud bottoms. They have no external cilia, but their cuticle is divided into 13 segments (Figure 8-4). Kinorhynchs burrow into the mud by extending the head, anchoring it by its recurved spines **(scalids),** and drawing the body forward until the head is retracted. They feed on organic sediment in the mud, and some feed on diatoms. Kinorhynchs are dioecious.

Phylum Loricifera

The Loricifera (L. *lorica,* corselet, + *fero,* to bear) are a very recently described phylum of animals (1983). They live in the spaces between the grains of marine gravel, to which they cling tightly. Although they were described from specimens collected off the coast of France, they are apparently widely distributed in the world. They have oral styles and scalids rather similar to those of the kinorhynchs, and the entire forepart of the body can be retracted into the circular **lorica** (Figure 8-5). Loriciferans are dioecious.

Phylum Priapulida

The Priapulida (pri′a-pyu′li-da) (Gr. *priapos,* phallus, + *-ida,* pl. suffix) are a small group (only nine species) of marine worms found chiefly in the colder water of both hemispheres. Their cylindrical bodies are rarely more than 12 to 15 cm long. Most of them are burrowing predaceous animals that usually orient themselves upright in the mud with the mouth at the surface.

Long considered pseudocoelomate, they were judged coelomate when nuclei were found in the membranes lining the body cavity, the membranes thus representing a peritoneum. However, electron microscopy showed that the nuclei of their muscle cells were peripheral, and the muscles secreted an extracellular membrane. The muscle nuclei and extracellular membrane gave the appearance of an epithelial lining.

The body includes a proboscis, a trunk, and usually one or two caudal appendages (Figure 8-6). The eversible proboscis usually ends with rows of curved spines around the mouth; it is used in sampling the surroundings as well as for the capture of small soft-bodied prey. Priapulids are not metameric.

There is no circulatory system, but coelomocytes in the body fluids contain the respiratory pigment hemerythrin. There is a nerve ring and ventral cord with nerves

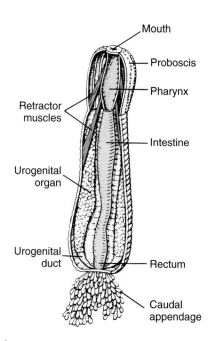

FIGURE 8-6

Major internal structures of *Priapulus*.

Characteristics of Phylum Nematoda

1. Body bilaterally symmetrical, cylindrical in shape
2. Body covered with a secreted, flexible, nonliving cuticle
3. Motile **cilia and flagella completely lacking;** some sensory endings derived from cilia present
4. **Muscles in body wall running in longitudinal direction only**
5. Excretory system of either one or more **gland cells** opening by an excretory pore, a **canal system** without gland cells, or both gland cells and canals together; **flame cell protonephridia lacking**
6. **Pharynx** usually muscular and **triradiate** in cross section
7. Male reproductive tract opening into rectum to form a cloaca; female reproductive tract opening a separate gonopore
8. Fluid in pseudocoel enclosed by cuticle forming a **hydrostatic skeleton**

and a protonephridial tubule that serves also as a gonoduct. The anus and urogenital pores open at the end of the trunk.

Sexes are separate, and embryogenesis is only poorly known. Their relationship to other coelomates is obscure. Some authorities believe that they are remnants of groups that were once more successful and widely distributed.

Phylum Nematoda: Roundworms

Nematodes, (Gr., *nematos,* thread) are present in nearly every conceivable kind of ecological niche. Approximately 12,000 species have been named, but Libbie Hyman estimated that, if all species were known, the number would be nearer 500,000.[1] They live in the sea, in fresh water, and in soil, from polar regions to the tropics, and from mountaintops to the depths of the sea. Good topsoil may contain billions of nematodes per acre. Nematodes also parasitize virtually every type of animal and many plants. The effects of nematode infestation on crops, domestic animals, and humans make this phylum one of the most important of all parasitic animal groups.

---

[1]Hyman, L. H. 1951. The invertebrates, vol. III. Acanthocephala, Aschelminthes, and Entoprocta, the pseudocoelomate Bilateria. New York, McGraw-Hill Book Company.

Libbie Henrietta Hyman (1888-1969), author of the monumental multi-volume treatise, The invertebrates, *was a distinguished American zoologist. She earned all her academic degrees from the University of Chicago, receiving her Ph.D. in 1915 and remaining there doing research until 1931. She published a laboratory manual for elementary zoology in 1919 and one for comparative vertebrate anatomy in 1922. The latter remained popular for the next 50 years and allowed her to live modestly on proceeds from her books while pursuing her first love, the study of invertebrates. She became a research associate at the American Museum of Natural History and published 135 research papers, chiefly on turbellarians, and the treatise for which she is best remembered,* The invertebrates. *She completed six volumes of this treatise, covering the protozoan groups through molluscs (to gastropods). Thousands of invertebrate zoologists owe her a debt of gratitude.*

Form and Function

Most nematodes are under 5 cm long, and many are microscopic, but some parasitic nematodes are over a meter in length.

Use of the pseudocoel as a hydrostatic skeleton is a conspicuous feature of nematodes, and we can best understand much of their functional morphology in the context of the high **hydrostatic pressure** in the pseudocoel.

The outer body covering is a relatively thick, noncellular **cuticle,** secreted by the underlying epidermis **(hypodermis).** The cuticle is of great functional importance to the worm, serving to contain the high hydrostatic pressure exerted by the fluid in the pseudocoel. The several layers of the cuticle are primarily of **collagen,** a structural protein also abundant in vertebrate connective tissue. Crisscrossing fibers make up three of the layers, conferring some longitudinal elasticity on the worm but severely limiting its capacity for lateral expansion.

Beneath the hypodermis is a layer of **longitudinal muscles.** There are no circular muscles in the body wall. The muscles are arranged in four bands, or quadrants, marked off by four epidermal cords that project inward to the pseudocoel (Figure 8-7). Another unusual feature of the body-wall muscles in nematodes is that the muscles extend processes to synapse with the nerve cords, rather than the nerve extending an axon to synapse with the muscle.

The fluid-filled pseudocoel, in which the internal organs lie, constitutes the hydrostatic skeleton. Hydrostatic skeletons are found in many invertebrates; they lend support by transmitting the force of muscle contraction to the enclosed, noncompressible fluid. Normally, muscles are arranged antagonistically, so that as movement is effected in one direction by contraction of one group of muscles, movement back in the opposite direction is effected by the antagonistic set of muscles. However, nematodes do not have circular body-wall muscles to antagonize the longitudinal muscles; therefore, the cuticle must serve that function. As the muscles on one side of the body contract, they compress the cuticle on that side, and the force of the contraction is transmitted (by the fluid in the pseudocoel) to the other side of the nematode, stretching the cuticle on that side. This compression and stretching of the cuticle serve to antagonize the muscle and are the forces that return the body to resting position when the muscles relax; this action produces the characteristic thrashing motion seen in nematode movement. An increase in efficiency of this system can be achieved only by an increase in hydrostatic pressure. Consequently, the hydrostatic pressure in the nematode pseudocoel is much higher than is usually found in other kinds of animals with hydrostatic skeletons but that also have antagonistic muscle groups.

In 1963 Sydney Brenner started studying a free-living nematode, Caenorhabditis elegans, *which was the beginning of some extremely fruitful research. Now this small worm has become one of the most important experimental models in biology. The origin and lineage of all the cells in its body (959) have been traced from the zygote to the adult, and the complete "wiring diagram" of its nervous system is known—all the neurons and all connections between them. The genome has been almost completely mapped, and scientists are working toward sequencing the entire genome of 3 million bases. Many basic discoveries have been made and will be made using* C. elegans.

The alimentary canal of the nematode consists of a **mouth** (Figure 8-7), a muscular **pharynx,** a long nonmuscular **intestine,** a short **rectum,** and a terminal **anus.** The cylindrical pharynx has radial muscles that insert on the cuticular lining of its lumen and on a basement membrane at its periphery. At rest the lumen is closed. When the muscles in the anterior of the pharynx contract, they open the lumen and suck in food material. Relaxation of the muscles anterior to the food mass closes the lumen of the pharynx and forces the food posteriorly toward the intestine. The intestine is only one cell layer thick and has no muscles. Food matter moves posteriorly in the intestine by body movements and by additional food being passed into the intestine from the pharynx. Defecation is accomplished by muscles that simply pull the anus open, and pseudocoelomic pressure surrounding the rectum provides expulsive force.

The adults of many parasitic nematodes have an anaerobic energy metabolism, and a Krebs cycle and cytochrome system characteristic of aerobic metabolism are absent. They derive energy through glycolysis and some additional electron transport sequences. Interestingly, some free-living nematodes and free-living stages of parasitic nematodes are obligate aerobes (require oxygen) and have a Krebs cycle and cytochrome system.

A **ring of nerve tissue** and ganglia around the pharynx gives rise to small nerves to the anterior end and to two **nerve cords,** one dorsal and one ventral. Some sensory organs around the lips and around the posterior end are rather elaborate.

Most nematodes are dioecious. The male is smaller than the female, and its posterior end usually bears a pair of copulatory spicules (Figure 8-8). Fertilization is internal, and the shelled zygotes or embryos are stored in the uterus until deposition. After embryonic development, a juvenile worm hatches from the egg. There are four juvenile stages, each separated by a molt, or shedding, of the cuticle. Many parasitic nematodes have free-living juvenile stages, and others require an intermediate host to complete their life cycles.

The copulatory spicules of male nematodes are not true intromittent organs, since they do not conduct the sperm, but are another adaptation to cope with the high internal hydrostatic pressure. The spicules must hold the vulva of the female open while the ejaculatory muscles overcome the hydrostatic pressure in the female and rapidly inject sperm into her reproductive tract. Furthermore, nematode spermatozoa are unique among those studied in the animal kingdom in that they lack a flagellum and acrosome. Within the female reproductive tract, the sperm become ameboid and move by pseudopodial movement.

Some Nematode Parasites

Nearly all vertebrates and many invertebrates are parasitized by nematodes. A number of these are very important pathogens of humans and domestic animals. A few nematodes are

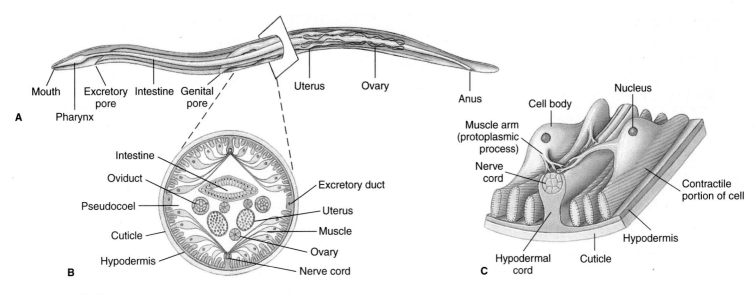

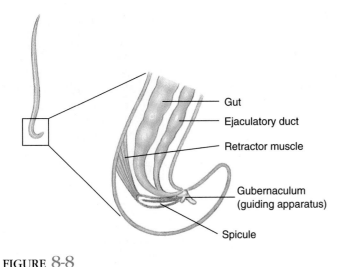

FIGURE 8-7

A, Structure of a nematode as illustrated by an *Ascaris* female. *Ascaris* has two ovaries and uteri, which open to the outside by a common genital pore. **B,** Cross section. **C,** Relationship of muscle cells to hypodermis and hypodermal cord.

FIGURE 8-8

Posterior end of a male nematode.

common in humans in North America (Table 8-1), but they and many others usually abound in tropical countries. We have space to mention only a few.

Ascaris lumbricoides: The Large Roundworm of Humans

The large human roundworm (Gr. *askaris,* intestinal worm) is one of the most common worm parasites of humans (Figure 8-9). Recent surveys have shown a prevalence of up to 64% in some areas of the southeastern United States, and it is estimated that 1.26 billion people in the world (20% of the world's population) have this worm. ***Ascaris suum*** (Figure 8-9), a parasite of pigs, is morphologically similar and was long

considered the same species. Females of both are up to 30 cm in length and can produce 200,000 eggs a day. The adults live in their host's small intestine, and the eggs pass out in the feces. They are extremely resistant to adverse conditions other than direct sunlight and high temperatures and can survive for months or years in the soil.

When the shelled juveniles are eaten with uncooked vegetables or when children put soiled fingers or toys in their mouths, the tiny juveniles are consumed and hatch in the intestine. They penetrate the intestinal wall and travel through the heart in the blood to the lungs, where they break out into the alveoli. They may cause a serious pneumonia at this stage. From the alveoli, the juveniles make their way up the bronchi, trachea, and pharynx, to be swallowed and finally reach the intestine again and grow to maturity. In the intestine the worms cause abdominal symptoms and allergic reactions, and in large numbers they may produce intestinal blockage.

Other ascarids are common in wild and domestic animals. Species of Toxocara, *for example, are found in dogs and cats. Their life cycle is generally similar to that of* Ascaris, *but the juveniles often do not complete their tissue migration in adult dogs, remaining in the host's body in a stage of arrested development. Pregnancy in the female dog, however, stimulates the juveniles to wander, and they infect the embryos in the uterus. The puppies are then born with worms. These ascarids also survive in humans but do not complete their development, leading to an occasionally serious condition in children known as* **visceral larva migrans.** *This condition is a good reason for pet owners to practice hygienic disposal of canine wastes!*

TABLE 8-1

Common Parasitic Nematodes of Humans in North America

Common and scientific names	Means of infection; prevalence in humans
Hookworm (*Ancylostoma duodenale* and *Necator americanus*)	Contact with juveniles in soil that burrow into skin; common in southern states
Pinworm (*Enterobius vermicularis*)	Inhalation of dust with ova and by contamination with fingers; most common worm parasite in United States
Intestinal roundworm (*Ascaris lumbricoides*)	Ingestion of embryonated ova in contaminated food; common in rural areas of Appalachia and southeastern states
Trichina worm (*Trichinella spiralis*)	Ingestion of infected muscle; occasional in humans throughout North America
Whipworm (*Trichuris trichiura*)	Ingestion of contaminated food or by unhygienic habits; usually common wherever *Ascaris* is found

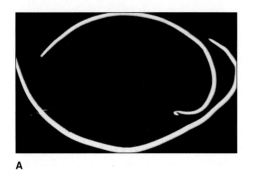

A

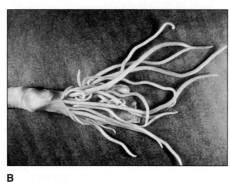

B

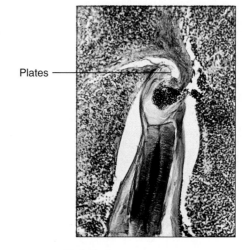

Plates

FIGURE 8-9

A, Intestinal roundworm *Ascaris lumbricoides,* male and female. Male *(top)* is smaller and has characteristic sharp kink in the end of the tail. The females of this large nematode may be over 30 cm long. **B,** Intestine of a pig, nearly completely blocked by *Ascaris suum.* Such heavy infections are also fairly common with *A. lumbricoides* in humans.

FIGURE 8-10

Section through anterior end of hookworm attached to dog intestine. Note cutting plates of mouth pinching off bit of mucosa from which the thick muscular pharynx sucks blood. Esophageal glands secrete an anticoagulant to prevent blood clotting.

Hookworms

Hookworms are so named because the anterior end curves dorsally, suggesting a hook. The most common species is *Necator americanus* (L. *necator,* killer), whose females are up to 11 mm long. The males can reach 9 mm in length. Large plates in their mouths (Figure 8-10) cut into the intestinal mucosa of the host where they suck blood and pump it through their intestines, partially digesting it and absorbing the nutrients. They suck much more blood than they need for food, and heavy infections cause anemia in the patient. Hookworm disease in children may result in retarded mental and physical growth and general loss of energy.

Shelled embryos are passed in the feces, and the juveniles hatch in the soil, where they live on bacteria. When human skin comes in contact with infested soil, the juveniles burrow through the skin to the blood, and reach the lungs and finally the intestine in a manner similar to that described for *Ascaris.*

Trichina Worm

Trichinella spiralis (Gr. *trichinos,* of hair, + *-ella,* diminutive) is one of the species of tiny nematodes responsible for the potentially lethal disease trichinosis. Adult worms burrow in the mucosa of the small intestine where the female produces living young. The juveniles penetrate into blood vessels and are carried throughout the body, where they may be found in almost any tissue or body space. Eventually, they penetrate skeletal muscle cells, becoming one of the largest known intracellular parasites. The juvenile causes astonishing redirection of gene expression in its host cell, which loses its striations and becomes a **nurse cell** that nourishes the worm

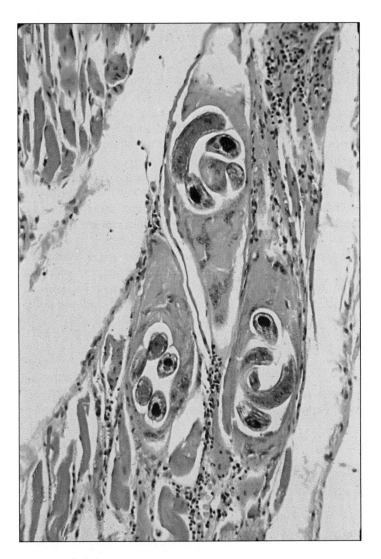

FIGURE 8-11

Section of human muscle infected with trichina worm *Trichinella spiralis.* The juveniles lie within muscle cells that the worms have induced to transform into nurse cells (commonly called cysts). An inflammatory reaction is evident around the nurse cells. The juveniles may live 10 to 20 years, and the nurse cells may eventually calcify.

(Figure 8-11). When meat containing live juveniles is swallowed, the worms are liberated into the intestine where they mature.

Trichinella spp. can infect a wide variety of mammals in addition to humans, including hogs, rats, cats, and dogs. Hogs become infected by eating garbage containing pork scraps with juveniles or by eating infected rats. In addition to *T. spiralis,* we now know there are four other sibling species in the genus. They differ in geographic distribution, infectivity to different host species, and freezing resistance. Heavy infections may cause death, but lighter infections are much more common— about 2.4% of the population of the United States is infected.

Pinworms

The pinworm, *Enterobius vermicularis* (Gr. *enteron,* intestine, + *bios,* life), causes relatively little disease, but it is the most common helminth parasite in the United States, estimated at 30% in children and 16% in adults. The adult parasites

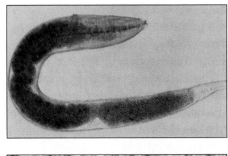

FIGURE 8-12

Pinworms, *Enterobius vermicularis.* **A,** Female worm from human large intestine (slightly flattened in preparation), magnified about 20 times. **B,** Group of shelled juveniles of pinworms, which are usually discharged at night around the anus of the host, who, by scratching during sleep, may get fingernails and clothing contaminated. This may be the most common and widespread of all human helminth parasites.

(Figure 8-12) live in the large intestine and cecum. The females, up to about 12 mm in length, migrate to the anal region at night to lay their eggs (Figure 8-12). Scratching the resultant itch effectively contaminates the hands and bedclothes. Eggs develop rapidly and become infective within 6 hours at body temperature. After they are swallowed, they hatch in the duodenum, and the worms mature in the large intestine.

Members of this order of nematodes have **haplodiploidy,** a characteristic shared with a few other animal groups, notably many hymenopteran insects (p. 225). Males are haploid and are produced parthenogenetically; females are diploid and arise from fertilized eggs.

Diagnosis of most intestinal roundworms is usually by examination of a small bit of feces under the microscope and finding characteristic shelled embryos or juveniles. However, pinworm eggs are not often found in the feces because the female deposits them on the skin around the anus. The "Scotch tape method" is more effective. The sticky side of cellulose tape is applied around the anus to collect the shelled embryos, then the tape is placed on a glass slide and examined under the microscope. Several drugs are effective against this parasite, but all members of a family should be treated at the same time, since the worm easily spreads through a household.

Filarial Worms

At least eight species of filarial nematodes infect humans, and some of these are major causes of diseases. Some 250 million people in tropical countries are infected with *Wuchereria bancrofti* (named for Otto Wucherer) or *Brugia malayi* (named for S. L. Brug), which places these species among the scourges of humanity. The worms live in the lymphatic system, and the females are as long as 100 mm. The disease symptoms are associated with inflammation and obstruction of the lymphatic system. The females release live young, the tiny microfilariae, into the blood and lymph. The microfilariae are picked up by mosquitos as the insects feed, and they develop in the mosquitos to the infective stage. They escape from the mosquito when it is feeding again on a human and penetrate the wound made by the mosquito bite.

The dramatic manifestations of elephantiasis are occasionally produced after long and repeated exposure to the worms. The condition is marked by an excessive growth of connective tissue and enormous swelling of affected parts, such as the scrotum, legs, arms, and more rarely, the vulva and breasts (Figure 8-13).

Another filarial worm causes river blindness (onchocerciasis) and is carried by blackflies. It infects more than 30 million people in parts of Africa, Arabia, Central America, and South America.

The most common filarial worm in the United States is probably the dog heartworm, *Dirofilaria immitis* (Figure 8-14). Carried by mosquitos, it also can infect other canids, cats, ferrets, sea lions, and occasionally humans. Along the Atlantic and Gulf Coast states and northward along the Mississippi River throughout the midwestern states, prevalance in dogs is up to 45%. It occurs in other states at a lower prevalence. This worm causes a very serious disease among dogs, and no responsible owner should fail to provide "heartworm pills" for a dog during mosquito season.

Phylum Nematomorpha

The popular name for the Nematomorpha (nem′a-to-mor′fa) (Gr. *nema, nematos,* thread, + *morphē,* form) is "horsehair worms," based on an old superstition that the worms arise from horsehairs that happen to fall into the water; and indeed they resemble hairs from a horse's tail. They were long included with the nematodes, with which they share the structure of the cuticle, presence of epidermal cords, longitudinal muscles only, and pattern of nervous system. However, since the early larval form of some species has a striking resemblance to the Priapulida, it is impossible to say to what group the nematomorphs are most closely related.

About 250 species of horsehair worms have been named. Worldwide in distribution, horsehair worms are free-living as adults and parasitic in arthropods as juveniles. Adults have a vestigial digestive tract and do not feed, but they can live almost anywhere in wet or moist surroundings if oxygen is adequate. Juveniles do not emerge from the arthropod host unless there is water nearby, and adults are often seen wriggling slowly about in ponds or streams. Juveniles of freshwater forms use various terrestrial insects as hosts, while marine forms use certain crabs.

Phylum Acanthocephala

The members of the phylum Acanthocephala (a-kan′tho-sef′a-la) (Gr. *akantha,* spine or thorn, + *kephalē,* head) are commonly known as "spiny-headed worms." The phylum derives

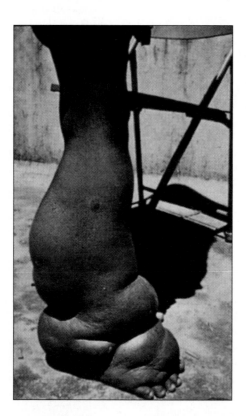

FIGURE 8-13

Elephantiasis of leg caused by adult filarial worms of *Wuchereria bancrofti,* which live in lymph passages and block the flow of lymph. Tiny juveniles, called microfilariae, are picked up in a blood meal of a mosquito, where they develop to the infective stage and are transmitted to a new host.

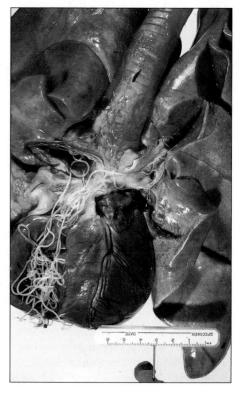

FIGURE 8-14

Dirofilaria immitis in a dog's heart. This nematode is a major menace to the health of dogs in North America. The adults live in the heart, and the juveniles circulate in the blood where they are picked up and transmitted by mosquitos.

its name from one of its most distinctive features, a cylindrical invaginable **proboscis** (Figure 8-15) bearing rows of recurved spines, by which the worms attach themselves to the intestine of their hosts. All acanthocephalans are endoparasitic, living as adults in the intestines of vertebrates.

Over 500 species are known, most of which parasitize fishes, birds, and mammals, and the phylum is worldwide in distribution. However, no species is normally a parasite of humans, although rarely humans are infected with species that usually occur in other hosts.

Various species range in size from less than 2 mm to over 1 m in length, with the females of a species usually being larger than the males. In life, the body is usually bilaterally flattened, with numerous transverse wrinkles. The worms are usually cream color but may be yellowish or brown from absorption of pigments from the intestinal contents.

The body wall is syncytial, and its surface is punctuated by minute crypts 4 to 6 μm deep, which greatly increase the surface area of the tegument. About 80% of the thickness of the tegument is the radial fiber zone, which contains a **lacunar system** of ramifying fluid-filled canals (Figure 8-15). The function of the lacunar system is unclear, but it may serve in distribution of nutrients to the peculiar, tubelike muscles in the body wall of these organisms.

Excretion is across the body wall in most species. In one family there is a pair of **protonephridia** with flame cells that unite to form a common tube that opens into the sperm duct or uterus.

Because acanthocephalans have no digestive tract, they must absorb all nutrients through their tegument. They can absorb various molecules by specific membrane transport mechanisms, and their tegument can carry out pinocytosis.

Sexes are separate. The male has a protrusible penis, and at copulation the sperm travel up the genital duct and escape into the pseudocoel of the female. The zygotes develop into shelled acanthor larvae. The shelled larvae escape from the vertebrate host in the feces, and, if eaten by a suitable arthropod, they hatch and work their way into the hemocoel where they grow to juvenile acanthocephalans. Either development ceases until the arthropod is eaten by a suitable host, or they may pass through several transport hosts in which they encyst until eaten.

Phylum Entoprocta

Entoprocta (en'to-prok'ta) (Gr. *entos,* within, + *proktos,* anus) is a small phylum of less than 100 species of tiny, sessile animals that, superficially, look much like hydroid cnidarians, but their tentacles are ciliated and tend to roll inward (Figure 8-16). Most entoprocts are microscopic, and none is more than 5 mm long. They are all stalked and sessile forms; some are colonial, and some are solitary. All are ciliary feeders.

With the exception of one genus all entoprocts are marine forms that have a wide distribution from the polar regions to the tropics. Most marine species are restricted to coastal and brackish waters and often grow on shells and algae. Some are commensals on marine annelid worms. Freshwater entoprocts (Figure 8-16) occur on the underside of rocks in running water.

In December, 1995, P. Funch and R. M. Kristensen reported that they had found some very strange little creatures clinging to the mouthparts of the Norway lobster (Nephrops norvegicus), *so strange that they did not fit into any known phylum (Nature* **378:** *711–714). Funch and Kristensen concluded that the organisms, only 0.35 mm long, represented a new phylum, which was named* **Cycliophora.** *The name refers to a crown of compound cilia, reminiscent of rotifers, with which the organisms feed. They were described as "acoelomate," although whether they might have a pseudocoel is unclear, and they do have a cuticle. Their life cycle seems bizarre. The sessile feeding stages on the lobster's mouthparts undergo internal budding to produce motile stages: (1) larvae containing new feeding stages; (2) dwarf males, which become attached to feeding stages that contain developing females; and (3) females, which also attach to the lobster's mouthparts, then produce dispersive larvae and degenerate.*

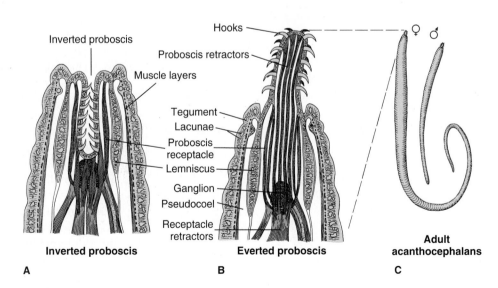

Inverted proboscis

Hooks

Proboscis retractors

Muscle layers

Tegument
Lacunae
Proboscis receptacle
Lemniscus
Ganglion
Pseudocoel
Receptacle retractors

Inverted proboscis
A

Everted proboscis
B

♀ ♂

Adult acanthocephalans
C

FIGURE 8-15

Structure of a spiny-headed worm (phylum Acanthocephala). **A** and **B,** Eversible spiny proboscis by which the parasite attaches to the intestine of the host, often doing great damage. Because they lack a digestive tract, food is absorbed through the tegument. **C,** Male is typically smaller than female.

Whether the proposed new phylum will withstand the scrutiny of further research is unknown, and its possible relationships to other phyla are quite unclear. Funch and Kristensen think that the organisms are protostomes and see affinities with the Entoprocta and Ectoprocta. Little short of astonishing, however, is their abundance on the mouthparts of a host as well known as the Norway lobster. How could biologists have failed to notice them before? At a time when habitat destruction drives many species to extinction every year, we wonder if there are phyla suffering the same fate. S. Conway Morris pondered the possibility of further undiscovered phyla (Nature 378:661–662), suggesting you may need a couple of zoology textbooks and a decent microscope when you next dine at your favorite seafood restaurant:"Who knows what might be found lurking under the lettuce?"

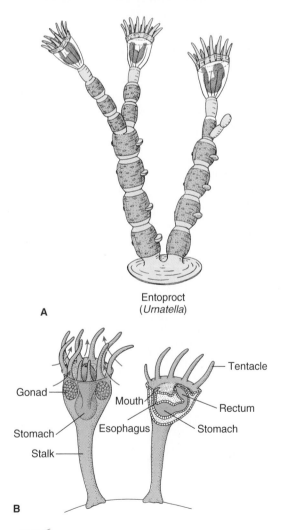

A

B

Entoproct
(*Urnatella*)

Tentacle

Gonad

Mouth

Rectum

Stomach

Esophagus

Stomach

Stalk

FIGURE 8-16

A, *Urnatella,* a freshwater entoproct, forms small colonies of two or three stalks from a basal plate. **B,** *Loxosomella,* a solitary entoproct. Both solitary and colonial entoprocts can reproduce asexually by budding, as well as sexually.

The body of entoprocts is cup shaped, bears a crown, or circle, of ciliated tentacles, and attaches to a substratum by a stalk in solitary species. In colonial species, the stalks of the individuals connect to the stolon of the colony. The tentacles and stalk are continuations of the body wall.

The gut is ∪-shaped and ciliated, and both the mouth and anus open within the circle of tentacles. They capture food particles in the current created by the tentacular cilia and then pass the particles along the tentacles to the mouth. Entoprocts have a pair of protonephridia but no circulatory or respiratory organs.

Some species are monoecious, some are dioecious, and some seem to be protandrous hermaphrodites; that is, the gonad at first produces sperm and then eggs. Cleavage is modified spiral and mosaic, and a trochophore-like larva is produced, similar to that of some molluscs and annelids (p. 159).

Phylogeny and Adaptive Radiation

Phylogeny

Hyman[2] grouped the Rotifera, Gastrotricha, Kinorhyncha, Priapulida, Nematoda, and Nematomorpha into a single phylum (Aschelminthes). All of these phyla share a certain combination of characteristics, and Hyman contended that the evidences of relationships were so concrete and specific that they could not be disregarded. Nevertheless, most authors now consider that differences between the groups are sufficient to merit phylum status for each, although some accept the concept of the Aschelminthes as a superphylum. The phyla may well have been derived originally from the protostome line via an acoelomate common ancestor resembling certain flatworms in many features.

Loriciferans bear some similarity to kinorhynchs, larval Priapulida, larval nematomorphs, rotifers, and tardigrades (p. 241). Although the loriciferans are poorly known, cladistic analysis suggests that they form a sister group to the kinorhynchs and that these two phyla together are a sister group of the priapulids (Figure 8-17).

Acanthocephalans are highly specialized parasites with a unique structure and have doubtless been so for millions of years. Any ancestral species or related group that would shed a clue to the phyletic relationships of the Acanthocephala probably went extinct long ago. Like the cestodes, acanthocephalans have no digestive tract and must absorb all nutrients across the tegument, but the tegument of the two groups is quite different in structure.

The entoprocts were once included with the phylum Ectoprocta in a phylum called Bryozoa, but the ectoprocts are true coelomate animals, and many zoologists prefer to place them in a separate group. Some biologists still refer to ecto-

[2]Hyman, L. H. 1951. The invertebrates, vol. III. Acanthocephala, Aschelminthes, and Entoprocta, the pseudocoelomate Bilateria. New York, McGraw-Hill Book Company.

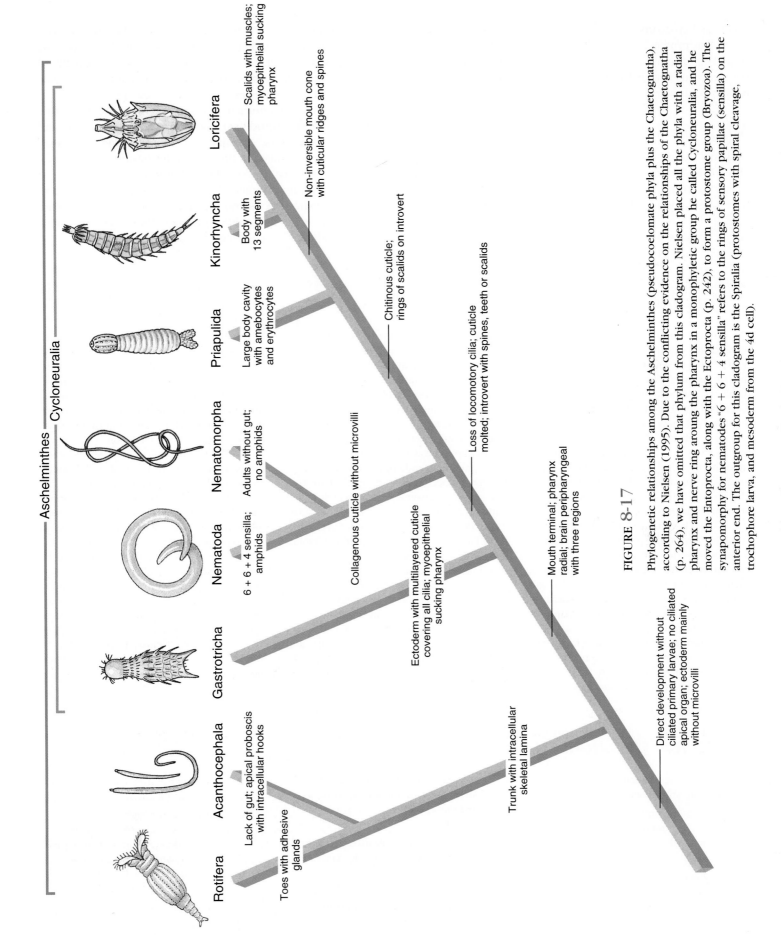

Rotifera — Toes with adhesive glands

Acanthocephala — Lack of gut; apical proboscis with intracellular hooks

Gastrotricha

Nematoda — 6 + 6 + 4 sensilla; amphids

Nematomorpha — Adults without gut; no amphids

Priapulida — Large body cavity with amebocytes and erythrocytes

Kinorhyncha — Body with 13 segments

Loricifera — Scalids with muscles; myoepithelial sucking pharynx

Aschelminthes

Cycloneuralia

Trunk with intracellular skeletal lamina

Direct development without ciliated primary larvae; no ciliated apical organ; ectoderm mainly without microvilli

Ectoderm with multilayered cuticle covering all cilia; myoepithelial sucking pharynx

Collagenous cuticle without microvilli

Mouth terminal; pharynx radial; brain peripharyngeal with three regions

Loss of locomotory cilia; cuticle molted; introvert with spines, teeth or scalids

Chitinous cuticle; rings of scalids on introvert

Non-inversible mouth cone with cuticular ridges and spines

FIGURE 8-17

Phylogenetic relationships among the Aschelminthes (pseudocoelomate phyla plus the Chaetognatha), according to Nielsen (1995). Due to the conflicting evidence on the relationships of the Chaetognatha (p. 264), we have omitted that phylum from this cladogram. Nielsen placed all the phyla with a radial pharynx and nerve ring around the pharynx in a monophyletic group he called Cycloneuralia, and he moved the Entoprocta, along with the Ectoprocta (p. 242), to form a protostome group (Bryozoa). The synapomorphy for nematodes "6 + 6 + 4 sensilla" refers to the rings of sensory papillae (sensilla) on the anterior end. The outgroup for this cladogram is the Spiralia (protostomes with spiral cleavage, trochophore larva, and mesoderm from the 4d cell).

151

procts as bryozoans. The Entoprocta may be distantly related to the Ectoprocta, but there is little evidence of close relationship. The entoprocts may have arisen as an early offshoot of the same line that led to the ectoprocts. These relationships remain controversial.

Adaptive Radiation

Certainly the most impressive adaptive radiation in this group of phyla is shown by the nematodes. They are by far the most numerous in terms of both individuals and species, and they have been able to adapt to almost every habitat available to animal life. Their basic pseudocoelomate body plan, with the cuticle, hydrostatic skeleton, and longitudinal muscles, has proved generalized and plastic enough to adapt to an enormous variety of physical conditions. Free-living lines gave rise to parasitic forms on at least several occasions, and virtually all potential hosts have been exploited. All types of life cycle occur: from the simple and direct to the complex, with intermediate hosts; from normal dioecious reproduction to parthenogenesis, hermaphroditism, and alternation of free-living and parasitic generations. A major factor contributing to the evolutionary opportunism of the nematodes has been their extraordinary capacity to survive suboptimal conditions, for example, the developmental arrests in many free-living and animal parasitic species and the ability to undergo cryptobiosis (survival in harsh conditions by assuming a very low metabolic rate) in many free-living and plant parasitic species.

Summary

The phyla covered in this chapter possess a body cavity called a pseudocoel, which is derived from the embryonic blastocoel, rather than a secondary cavity in the mesoderm (coelom). Several of the groups exhibit eutely, a constant number of cells or nuclei in adult individuals of a given species.

The phylum Rotifera is composed of small, mostly freshwater organisms with a ciliated corona, which creates currents of water to draw planktonic food toward the mouth. The mouth opens into a muscular pharynx, or mastax, that is equipped with jaws.

The Gastrotricha, Kinorhyncha, and Loricifera are small phyla of tiny, aquatic pseudocoelomates. Gastrotrichs move by cilia or adhesive glands, and kinorhynchs anchor and then pull themselves by the spines on their head. Loriciferans can withdraw their bodies into the lorica. Priapulids are marine burrowing forms.

By far the largest and most important of this group of phyla are the nematodes, of which there may be as many as 500,000 species in the world. They are more or less cylindrical, tapering at the ends, and covered with a tough, secreted cuticle. The body-wall muscles are longitudinal only, and to function well in locomotion, such an arrangement must enclose a volume of fluid in the pseudocoel at high hydrostatic pressure. This fact of nematode life has a profound effect on most of their other physiological functions, for example, ingestion of food, egestion of feces, excretion, and copulation. Most nematodes are dioecious, and there are four juvenile stages, each separated by a molt of the cuticle. Almost all invertebrate and vertebrate animals and many plants have nematode parasites, and many other nematodes are free-living in soil and aquatic habitats. Some parasitic nematodes have part of their life cycle free-living, some undergo a tissue migration in their host, and some have an intermediate host in their life cycle.

The Nematomorpha or horsehair worms are related to the nematodes and have parasitic juvenile stages in arthropods, followed by a free-living, aquatic, nonfeeding adult stage.

Acanthocephalans are all parasitic in the intestines of vertebrates as adults, and their juvenile stages develop in arthropods. They have an anterior, invaginable proboscis armed with spines, which they embed in the intestinal wall of their host. They do not have a digestive tract and so must absorb all nutrients across their tegument.

The Entoprocta are small, sessile, aquatic animals with a crown of ciliated tentacles encircling both the mouth and anus.

The Rotifera, Gastrotricha, Kinorhyncha, Priapulida, Nematoda, and Nematomorpha have been included by some workers in one phylum, but most biologists believe that the groups are not sufficiently related to be encompassed by a single phylum. It is possible that they are derived from a common ancestor in the protostome line. Phylogenetic relationships of the Acanthocephala and Entoprocta are even more obscure. Of all these phyla, the Nematoda have achieved enormous evolutionary success and undergone great adaptive radiation.

Review Questions

1. Explain the difference between a true coelom and a pseudocoel.

2. What is the normal size of a rotifer; where is it found; and what are its major body features?

3. Explain the difference between mictic and amictic eggs of rotifers, and tell the adaptive value of each.

4. What is eutely?

5. What are the approximate lengths of loriciferans, priapulids, gastrotrichs, and kinorhynchs? Where are they found?

6. A skeleton is a supportive structure. Explain how a hydrostatic skeleton supports an animal.

7. What feature of body-wall muscles in nematodes requires a high hydrostatic pressure in the pseudocoelomic fluid for efficient function?

8. Explain how the high pseudocoelomic pressure affects feeding and defecation in nematodes. How could ameboid sperm be an adaptation to the high hydrostatic pressure in the pseudocoel?

9. Explain the interaction of the cuticle, body-wall muscles, and pseudocoelomic fluid in the locomotion of nematodes.

10. Outline the life cycle of each of the following: *Ascaris lumbricoides*, hookworm, *Enterobius vermicularis*, *Trichinella spiralis*, *Wuchereria bancrofti*.

11. Where in the human body are the adults of each species in question 10 found?

12. Where are juveniles and adults of nematomorphs found?

13. The evolutionary ancestry of acanthocephalans is particularly obscure. Describe some characters of

acanthocephalans that support that statement.

14. How do acanthocephalans get food?

15. What characteristics distinguish entoprocts among the pseudocoelomates?

Selected References

See also general references on page 395.

Bird, A. F., and J. Bird. 1991. The structure of nematodes, ed. 2. New York, Academic Press. *The most authoritative reference available on nematode morphology. Highly recommended.*

Bundy, D. A. P., and E. S. Cooper. 1989. *Trichuris* and trichuriasis in humans. In J. R. Baker and R. Muller (eds.), Advances in parasitology, vol. 28. London, Academic Press. *A recent estimate of people in the world infected with whipworm* (Trichuris) *is 687 million. This number is exceeded among nematodes only by hookworms (932 million) and* Ascaris *(1.26 billion).*

Despommier, D. D. 1990. *Trichinella spiralis:* the worm that would be virus. Parasitol. Today **6**:193–196. *The first-stage juveniles of* Trichinella *are among the largest of all intracellular parasites.*

Duke, B. O. L. 1990. Onchocerciasis (river blindness)—can it be eradicated? Parasitol. Today **6**:82–84. *Despite the introduction of a very effective drug, the author predicts that this parasite will not be eradicated in the foreseeable future.*

Ogilvie, B. M., M. E. Selkirk, and R. M. Maizels. 1990. The molecular revolution and nematode parasitology: yesterday, today, and

tomorrow. J. Parasitol. **76**:607–618. *Modern molecular biology has wrought enormous changes in investigations on nematodes.*

Poinar, G. O., Jr. 1983. The natural history of nematodes. Englewood Cliffs, New Jersey, Prentice-Hall, Inc. *Contains a great deal of information about these fascinating creatures, including free-living and plant and animal parasites.*

Roberts, L. 1990. The worm project. Science **248**:1310–1313. *A free-living nematode,* Caenorhabditis elegans, *has been of great value in studies of development and genetics.*

Links to the Internet

Visit this textbook's web site at http://www.mhhe.com/zoology to find live Internet links for each of the references listed below.

1. Introduction to the Aschelminth Phyla. University of California at Berkeley Museum of Paleontology. This site contains many SEM images of microscopic nematodes, as well as links to other aschelminth phyla.

Horsehair Worms
2. Horsehair worms in Illinois. General information on nematomorphs, supported by the Illinois Natural Survey, with many links to economic entomology, aquatic ecology, wildlife ecology, and biodiversity.

Rotifers
3. Rotifer Biology Project. The Department of Biological Sciences at the University of Texas at El Paso supports this site, which covers rotifer systematics, ecology, and links to other sites.

Nematodes
4. Society of Nematologists. This page has much information and links to other sites containing information on nematodes.
5. *Caenorhabditis elegans* WWW server. This site contains information on this nematode, links, meetings, the genome project, and a literature search.
6. *Caenorhabditis Genetics.* An Introduction to *C. elegans.* A very thorough introduction to the biology of this significant nematode.
7. Department of Nematology. This page, supported by the University of Nebraska Nematology Department, contains much information on plant and parasitic nematodes. This home page has many links, news, images, research on nematodes, as well as a link to a course in nematology.
8. Phylum Nematoda. University of Michigan site covering members of

phylum Nematoda. Links to the various taxa in this group, including many individual species.

9. *Dracunculiasis* (Guinea Worm Disease). This CDC site is linked to sites with more information on the guinea worm, as well as MMWR reports on the subject.
10. *Dracunculiasis* (Guinea Worm Disease). A CDC fact sheet on dracunculiasis.
11. Hookworm. This CDC site is linked to information on hookworms, and contains references on hookworms.
12. Ascariasis (Intestinal Roundworms). This CDC site is linked to information on human intestinal roundworms
13. Pinworms. This CDC site has information on a very common nematode infection: pinworms.
14. Trichinosis. This CDC site has information on trichinosis (caused by the trichina worm).

Molluscs

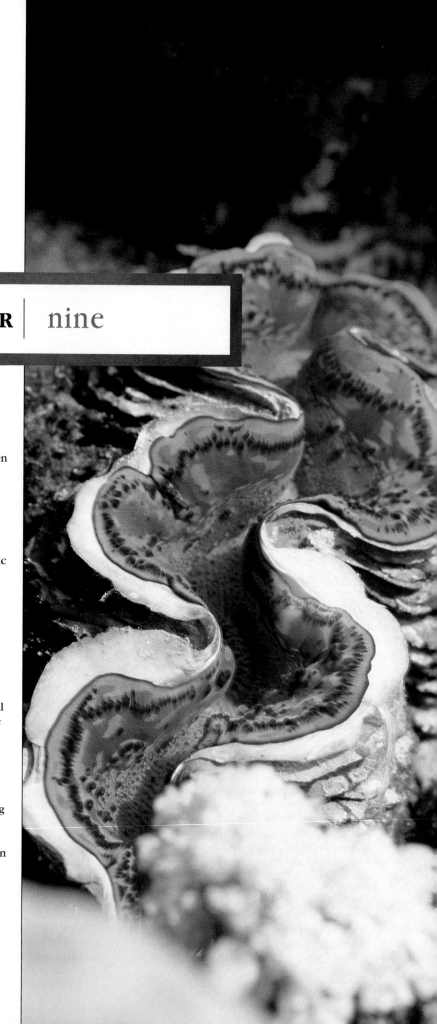

CHAPTER | nine

A Significant Space

L ong ago in the Precambrian era, the most complex animals
populating the seas were acoelomate. They must have been
inefficient burrowers, and they were unable to exploit the rich
subsurface ooze. Any that developed fluid-filled spaces within
the body would have had a substantial advantage because these
spaces could serve has a hydrostatic skeleton and improve
burrowing efficiency.

The simplest, and probably the first, mode of achieving a
fluid-filled space within the body was retention of the embryonic
blastocoel, as in the pseudocoelomates. This evolutionary
solution was not ideal because the organs lay loose in the body
cavity. Improved efficiency of the pseudocoel as a hydrostatic
skeleton depended on increasingly high hydrostatic pressure, a
condition that severely limited the potential for adaptive
radiation.

Some descendants of the Precambrian acoelomate
organisms evolved a more elegant arrangement: a fluid-filled
space within the mesoderm, the **coelom.** This space was lined
with mesoderm, and the organs were suspended by mesodermal
membranes, the **mesenteries.** Not only could the coelom serve
as an efficient hydrostatic skeleton, with circular and
longitudinal body-wall muscles acting as antagonists, but a more
stable arrangement of organs resulted in less crowding. The
mesenteries provided an ideal location for networks of blood
vessels, and the alimentary canal could become more muscular,
more highly specialized, and more diversified without interfering
with other organs.

Development of the coelom was a major step in the
evolution of larger and more complex forms. All major groups in
the chapters to follow are coelomates.

Next to Arthropoda the phylum Mollusca (mol-lus'ka) (L. *mollusca,* soft) has the most named species in the animal kingdom—probably about 50,000 living species, not to mention some 35,000 fossil species discovered to date. The name Mollusca indicates one of their distinctive characteristics, a soft body.

This very diverse group includes organisms as different as chitons, snails, clams, and octopuses (Figure 9-1). The group ranges from fairly simple organisms to some of the most complex of invertebrates, and in size from almost microscopic to the giant squid *Architeuthis harveyi* (Gr. *archi,* primitive, + *teuthis,* squid). The body of this huge species may grow up to 18 m long if its tentacles are extended. It may weigh up to 454 kg (1000 pounds). The shells of some of the giant clams *Tridacna gigas* (Gr. *tridaknos,* eaten at three bites) (see Figure 9-21), which inhabit the Indo-Pacific coral reefs, reach 1.5 m in length and weigh over 225 kg. These are extremes, however, since probably 80% of all molluscs are less than 5 cm in maximum shell size.

The enormous variety, great beauty, and availability of the shells of molluscs have made shell collecting a popular pastime. However, many amateur shell collectors, although able to name hundreds of the shells that grace our beaches, know very little about the living animals that created those shells and once lived in them. The largest classes of molluscs are the Gastropoda (snails and their relatives), Bivalvia (clams, oysters, and others), Polyplacophora (chitons), and Cephalopoda (squids, octopuses, nautiluses). The Monoplacophora, Scaphopoda (tusk shells), Caudofoveata, and Solenogastres are much smaller classes.

Ecological Relationships

Molluscs are found in a great range of habitats, from the tropics to polar seas, at altitudes exceeding 7000 m, in ponds, lakes, and streams, on mudflats, in pounding surf, and in open ocean from the surface to the abyssal depths. Most of them live in the sea, and they represent a variety of life-styles, including bottom feeders, burrowers, borers, and pelagic forms. The phylum includes some of the most sluggish and some of the swiftest and most active of the invertebrates. It includes herbivorous grazers, predaceous carnivores, and ciliary filter feeders.

According to the fossil evidence, the molluscs originated in the sea, and most of them have remained there. Much of their evolution occurred along shores, where food was abundant and habitats were varied. Only the bivalves and gastropods moved on to brackish and freshwater habitats. As filter feeders, the bivalves were unable to leave aquatic surroundings; however, the snails (gastropods) actually invaded the land and may have been the first animals to do so. Terrestrial snails are limited in range by their need for humidity, shelter, and the presence of calcium in the soil.

Economic Importance

A group as large as the molluscs would naturally affect humans in some way. A wide variety of molluscs are used as food. Pearls, both natural and cultured, are produced in the shells of clams and oysters, most of them in a marine oyster, found around eastern Asia (Figure 9-4B).

Position in Animal Kingdom

1. The molluscs are one of the major groups of true **coelomate** animals.
2. They belong to the **protostome** branch, or schizocoelous coelomates, and have spiral cleavage and mosaic development.
3. Many molluscs have a **trochophore larva** similar to the trochophore larva of marine annelids and other marine protostomes. Developmental evidence thus indicates that molluscs and annelids share a common ancestor.
4. Because molluscs are not metameric, they must have diverged from their common ancestor with the annelids before the advent of metamerism.
5. All the **organ systems** found in more derived invertebrates are present and well developed.

Biological Contributions

1. In molluscs gaseous exchange occurs not only through the body surface as in phyla discussed previously, but also in specialized **respiratory organs** in the form of **gills** or a **lung.**
2. Most classes have an **open circulatory system** with pumping **heart,** vessels, and blood sinuses. In most cephalopods the circulatory system is closed.
3. The efficiency of the respiratory and circulatory systems in the cephalopods has made greater body size possible. Invertebrates reach their largest size in some of the cephalopods.
4. They have a fleshy **mantle** that in most cases secretes a shell and is variously modified for a number of functions. Other features unique to the phylum are the **radula** and the muscular **foot.**
5. The highly developed direct **eye** of cephalopods is similar to the indirect eye of the vertebrates but arises as a skin derivative in contrast to the brain eye of vertebrates.

Characteristics of Phylum Mollusca

1. Body bilaterally symmetrical (bilateral asymmetry in some); unsegmented; usually with definite head

2. Ventral body wall specialized as a **muscular foot,** variously modified but used chiefly for locomotion

3. Dorsal body wall forms the **mantle,** which encloses the **mantle cavity,** is modified into **gills** or a **lung,** and secretes the **shell** (shell absent in some)

4. Surface epithelium usually ciliated and bearing mucous glands and sensory nerve endings

5. Coelom mainly limited to area around heart

6. Complex digestive system; rasping organ (**radula**) usually present (Figure 9-3); anus usually emptying into mantle cavity

7. **Open circulatory system** (mostly closed in cephalopods) of heart (usually three chambered, two in most gastropods), blood vessels, and sinuses; respiratory pigments in blood

8. Gaseous exchange by gills, lung, mantle, or body surface

9. Usually one or two kidneys (**metanephridia**) opening into the pericardial cavity and usually emptying into the mantle cavity

10. Nervous system of paired cerebral, pleural, pedal, and visceral ganglia, with nerve cords and subepidermal plexus; ganglia centralized in nerve ring in polyplacophorans, gastropods, and cephalopods

11. Sensory organs of touch, smell, taste, equilibrium, and vision (in some); eyes highly developed in cephalopods

A

B

C

D

E

FIGURE 9-1

Molluscs: a diversity of life forms. The basic body plan of this ancient group has become variously adapted for different habitats. **A,** A chiton *(Tonicella lineata),* class Polyplacophora. **B,** A marine snail *(Calliostoma annulata),* class Gastropoda. **C,** A nudibranch *(Chromodoris kuniei),* class Gastropoda. **D,** Pacific giant clam *(Panope abrupta),* with siphons to the left, class Bivalvia. **E,** An octopus *(Octopus briareus),* class Cephalopoda, forages at night on a Caribbean coral reef.

Some molluscs are destructive. The burrowing ship-worms (Figure 9-24), a kind of clam, do great damage to wooden ships and wharves. To prevent the ravages of ship-worms, wharves must be either creosoted or built of concrete. Snails and slugs often damage garden and other vegetation. In addition, many snails serve as intermediate hosts for serious parasites. A certain boring snail, the oyster drill, is second only to the sea star in destroying oysters.

Pearl production is the by-product of a protective device used by the animal when a foreign object, such as a grain of sand or a parasite, becomes lodged between the shell and mantle. The mantle secretes many layers of nacre around the irritating object (Figure 9-4). Pearls are cultured by inserting small spheres, usually made from pieces of the shells of freshwater clams, in the mantle of a certain species of oyster and by maintaining the oysters in enclosures. The oyster deposits its own nacre around the "seed" in a much shorter time than would be required to form a pearl normally.

Form and Function

Body Plan

Reduced to its simplest dimensions, the mollusc body may be said to consist of a **head-foot** portion and a **visceral mass** portion (Figure 9-2). The head-foot is the more active area, containing the feeding, cephalic sensory, and locomotor organs. It depends primarily on muscular action for its function. The visceral mass is the portion containing digestive, circulatory, respiratory, and reproductive organs, and it depends primarily on ciliary tracts for its functioning. Two folds of skin, outgrowths of the dorsal body wall, make up a protective **mantle,** which encloses a space between the mantle and body wall called the **mantle cavity.** The mantle cavity houses the **gills** or lung, and in some molluscs the mantle secretes a protective **shell** over the visceral mass and head-foot. Modifications of the structures that make up the head-foot and the visceral mass produce the great profusion of different patterns making up this major group of animals.

Head-Foot

Most molluscs have a well-developed head, which bears the mouth and some specialized sensory organs. Photosensory receptors range from fairly simple to the complex eyes of the cephalopods. Tentacles are often present. Within the mouth is a structure unique to molluscs, the radula, and usually posterior to the mouth is the chief locomotor organ, or foot.

Radula The radula is a rasping, protrusible, tonguelike organ found in all molluscs except the bivalves and some nudibranchs and their relatives. It is a ribbonlike membrane on which are mounted rows of tiny teeth that point backward (Figure 9-3). Complex muscles move the radula and its supporting cartilages (**odontophore**) in and out while the membrane is partly rotated over the tips of the cartilages. There may be a few or as many as 250,000 teeth, which, when protruded, can scrape, pierce, tear, or cut particles of food material, and the radula may serve as a rasping file for carrying the particles in a continuous stream toward the digestive tract.

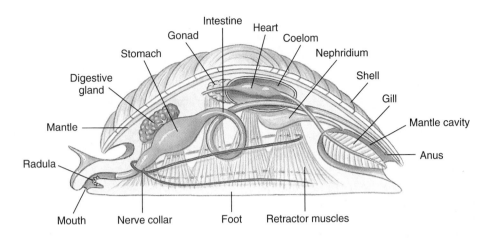

FIGURE 9-2

Generalized mollusc. Although this construct is often presented as a "hypothetical ancestral mollusc" (HAM), most experts now agree that it never actually existed. It is an abstraction used to facilitate description of the general body plan of molluscs.

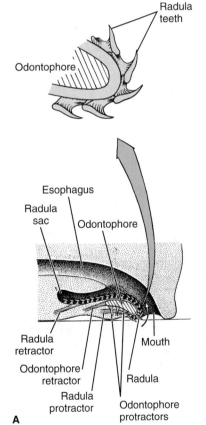

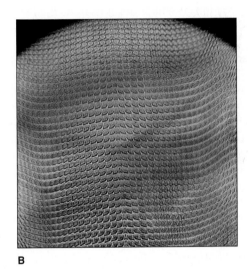

FIGURE 9-3

A, Diagrammatic longitudinal section of gastropod head showing the radula and radula sac. The radula moves back and forth over the odontophore cartilage. As the animal grazes, the mouth opens, the odontophore is thrust forward, the radula gives a strong scrape backward bringing food into the pharynx, and the mouth closes. The sequence is repeated rhythmically. As the radula ribbon wears out anteriorly, it is continually replaced posteriorly. **B,** Radula of a snail prepared for microscopic examination.

Foot The molluscan foot may be variously adapted for locomotion, for attachment to a substratum, or for a combination of functions. It is usually a ventral, solelike structure in which waves of muscular contraction effect a creeping locomotion. However, there are many modifications, such as the attachment disc of the limpets, the laterally compressed "hatchet foot" of the bivalves, or the siphon for jet propulsion in the squids and octopuses. Secreted mucus is often used as an aid to adhesion or as a slime track by small molluscs that glide on cilia.

Visceral Mass

Mantle and Mantle Cavity   The mantle is a sheath of skin extending from the visceral hump that hangs down on each side of the body, protecting the soft parts and creating between itself and the visceral mass the space called the mantle cavity. The surface of the mantle secretes the shell externally or internally.

The mantle cavity plays an enormous role in the life of the mollusc. It usually houses the respiratory organs (gills or lung), which develop from the mantle, and the mantle's own exposed surface serves also for gaseous exchange. The products from the digestive, excretory, and reproductive systems empty into the mantle cavity. In aquatic molluscs a continuous current of water, kept moving by surface cilia or by muscular pumping, brings in oxygen, and in some forms, food; flushes out wastes; and carries reproductive products out to the environment. In aquatic forms the mantle usually has sensory receptors for sampling the environmental water. In cephalopods (squids and octopuses) the muscular mantle and its cavity create the jet propulsion used in locomotion.

Shell   The shell of the mollusc, when present, is secreted by the mantle and is lined by it. Typically there are three layers (Figure 9-4). The **periostracum** is the outer horny layer, composed of an organic substance called conchiolin, which is a resistant protein. It helps protect the underlying calcareous layers from erosion by boring organisms. It is secreted by a fold of the mantle edge, and growth occurs only at the margin of the shell. On the older parts of the shell the periostracum often becomes worn away. The middle **prismatic layer** is composed of densely packed prisms of calcium carbonate laid down in a protein matrix. It is secreted by the glandular margin of the mantle, and increase in shell size occurs at the shell margin as the animal grows. The inner **nacreous layer** of the shell is composed of calcium carbonate sheets laid down over a thin protein matrix. This layer is secreted continuously by the mantle surface, so that it becomes thicker during the life of the animal.

Freshwater molluscs usually have a thick periostracum that gives some protection against the acids produced in the water by the decay of leaf litter. In some marine molluscs the periostracum is thick, but in some it is relatively thin or

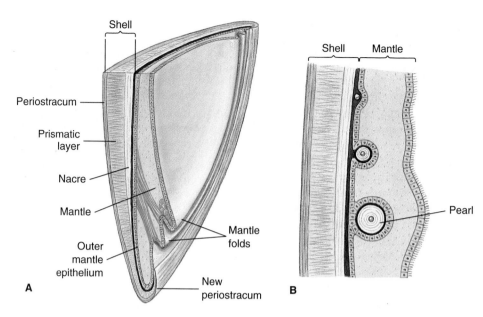

FIGURE 9-4

A, Diagrammatic vertical section of shell and mantle of a bivalve. The outer mantle epithelium secretes the shell; the inner epithelium is usually ciliated. **B,** Formation of pearl between mantle and shell as a parasite or bit of sand under the mantle becomes covered with nacre.

absent. There is a great range in variation in shell structure. Calcium for the shell comes from the environmental water or soil or from food. In most molluscs the first shell appears during the larval period and grows continuously throughout life.

Internal Structure and Function

In the molluscs, oxygen-carbon dioxide exchange occurs not only through the body surface, particularly that of the mantle, but in specialized respiratory organs such as gills or lungs, which are derivatives of the mantle. There is an **open circulatory system** with a pumping **heart,** blood vessels, and blood sinuses (rather than capillaries), which permeate the organs. Most cephalopods have a closed blood system with heart, vessels, and capillaries. The digestive tract is complex and highly specialized according to the feeding habits of the various molluscs. Most molluscs have a pair of **kidneys (metanephridia),** which connect with the coelom; the ducts of the kidneys in many forms serve also for the discharge of eggs and sperm. The **nervous system** consists of several pairs of ganglia with connecting nerve cords. There are various types of highly specialized sense organs.

Most molluscs are dioecious, although some of the gastropods are hermaphroditic. Many aquatic molluscs pass through free-swimming **trochophore** (Figure 9-5) and **veliger** (Figure 9-6) larval stages. The veliger is the free-swimming larva of most marine snails, tusk shells, and bivalves. It develops from the trochophore and has the beginning of a foot, shell, and mantle.

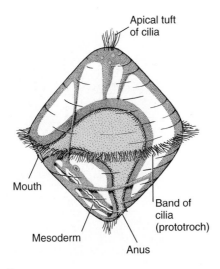

FIGURE 9-5

A generalized trochophore larva. Molluscs and annelids with primitive embryonic development have trochophore larvae, as do several other phyla.

The trochophore larva (Figure 9-5) is minute, translucent, more or less top shaped, and has a prominent circlet of cilia (prototroch) and sometimes one or two accessory circlets. It is found in molluscs and annelids with primitive embryonic development and is considered one of the evidences for common phylogenetic origin of the two phyla. Some form of trochophore-like larva also occurs in marine turbellarians, nemertines, brachiopods, phoronids, sipunculids, and echiurids, and it probably reflects some phylogenetic relationship among all these phyla.

Classes Caudofoveata and Solenogastres

The caudofoveates and the solenogasters (see Figure 9-35, p. 176) are often united in the class Aplacophora, and they are both wormlike and shell-less, with calcareous scales or spicules in their integument, with reduced head, and without nephridia. In contrast to the caudofoveates, the solenogasters usually have no true gills, and they are hermaphroditic. The caudofoveates are burrowing marine animals, feeding on microorganisms and detritus, whereas the solenogasters live freely on the bottom and often feed on cnidarians. The caudofoveates may have more features closer to those of the ancestral mollusc than do any other living groups.

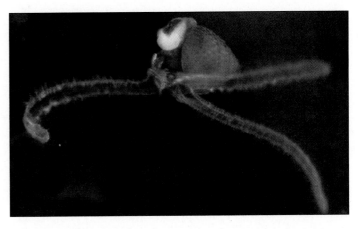

FIGURE 9-6

Veliger of a snail, *Pedicularia*, swimming. The adults are parasitic on corals. The ciliated process (velum) develops from the prototroch of the trochophore (Figure 9-5).

Class Monoplacophora

Until 1952 the Monoplacophora (mon-o-pla-kof′o-ra) were known only from Paleozoic shells. However, in that year living specimens of *Neopilina* (Gr. *neo*, new, + *pilos*, felt cap) were dredged up from the ocean bottom near the west coast of Costa Rica. These molluscs are small and have a low, rounded shell and a creeping foot (Figure 9-7). They have a superficial resemblance to the limpets, but unlike most other molluscs, a number of organs are serially repeated. Serial repetition occurs to a more limited extent in the chitons. Some authors have considered the monoplacophorans truly metameric (p. 50), indicating that molluscs descended from a metameric, annelid-like ancestor and that metamerism was lost secondarily in other molluscs. Others believe that *Neopilina* shows only pseudometamerism and that molluscs did not have a metameric ancestor. However, the phylogenetic affinity of annelids is strongly supported by embryological evidence.

Class Polyplacophora: Chitons

The chitons are somewhat flattened and have a convex dorsal surface that bears eight articulating limy **plates**, or **valves**, which give them their name (Figures 9-8 and 9-9). The term Polyplacophora means "bearing many plates," in contrast to the Monoplacophora, which bear one shell (*mono*, single). The plates overlap posteriorly and are usually dull colored to match the rocks to which the chitons cling.

Most chitons are small (2 to 5 cm); the largest rarely exceeds 30 cm. They commonly occur on rocky surfaces in intertidal regions, although some live at great depths. Chitons are stay-at-home organisms, straying only very short distances for feeding. In feeding, a sensory subradular organ protrudes from the mouth to explore for algae or colonial organisms. When

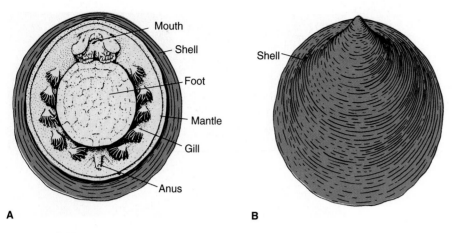

A   **B**

FIGURE 9-7

Neopilina, class Monoplacophora. Living specimens range from 3 mm to about 3 cm in length. **A,** Ventral view. **B,** Dorsal view.

some are found, the radula projects to scrape algae off the rocks. A chiton clings tenaciously to its rock with the broad flat foot. If detached, it can roll up like an armadillo for protection.

The mantle forms a **girdle** around the margin of the plates, and in some species mantle folds cover part or all of the plates. On each side of the broad ventral foot and lying between the foot and the mantle is a row of gills suspended from the roof of the mantle cavity. With the foot and the mantle margin adhering tightly to the substrate, these grooves become closed chambers, open only at the ends. Water enters the grooves anteriorly, flows across the gills, and leaves posteriorly, thus bringing a continuous supply of oxygen to the gills.

Blood pumped by the three-chambered heart reaches the gills by way of an aorta and sinuses. Two kidneys carry waste from the pericardial cavity to the exterior. Two pairs of longitudinal nerve cords are connected in the buccal region. Sense organs include shell eyes on the surface of the shell (in some) and a pair of **osphradia** (chemosensory organs for sampling water).

Sexes are separate in chitons. Sperm shed by males in the excurrent water enter the gill grooves of the females by incurrent openings. Eggs are shed into the sea singly or in strings or masses of jelly. The trochophore larva metamorphoses directly into a juvenile, without an intervening veliger stage.

Class Scaphopoda

The Scaphopoda (ska-fop′o-da), commonly called the tusk shells or tooth shells, are sedentary marine molluscs that have a slender body covered with a mantle and a tubular shell open at both ends. Here the molluscan body plan has taken a new direction, with the mantle wrapped around the viscera and fused to form a tube. Most scaphopods are 2.5 to 5 cm long, although they range from 4 mm to 25 cm long.

FIGURE 9-8

Mossy chiton, *Mopalia muscosa.* The upper surface of the mantle, or "girdle," is covered with hairs and bristles, an adaptation for defense.

The foot, which protrudes through the larger end of the shell, functions in burrowing into mud or sand, always leaving the small end of the shell exposed to the water above (Figure 9-10). Respiratory water circulates through the mantle cavity both by movements of the foot and by ciliary action. Gaseous exchange occurs in the mantle. Most of the food is detritus and protozoa from the substrate. Cilia of the foot or on the mucus-covered, ciliated knobs of long tentacles catch the food.

Class Gastropoda

Among the molluscs the class Gastropoda (gas-trop′o-da) (Gr. *gastēr,* stomach, + *pous, podos,* foot) is by far the largest and most diverse, containing about 40,000 living and 15,000 fossil species. Its members differ so widely that there is no single general term in our language that can apply to them as a group. They include snails, limpets, slugs, whelks, conchs, periwinkles, sea slugs, sea hares, sea butterflies, and others. They range from some marine molluscs with many primitive characters to highly evolved, air-breathing snails and slugs.

Gastropods are often sluggish, sedentary animals because most of them have heavy shells and slow locomotor organs. When present, the shell is almost always of one piece (univalve) and may be coiled or uncoiled. Some snails have an

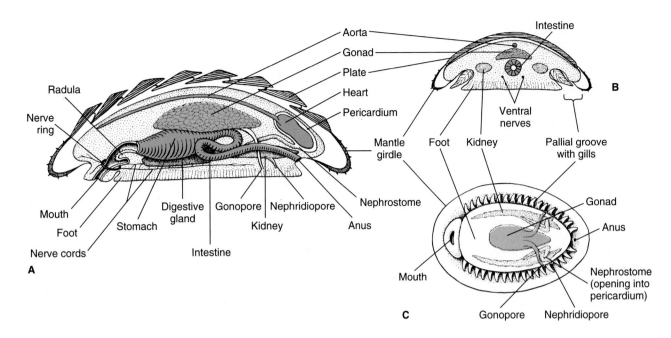

FIGURE 9-9

Anatomy of a chiton (class Polyplacophora). **A,** Longitudinal section. **B,** Transverse section. **C,** External ventral view.

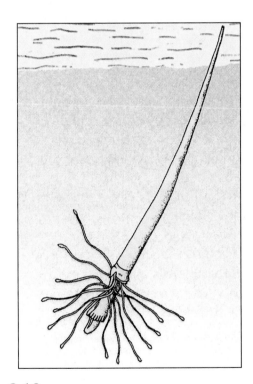

FIGURE 9-10

The tusk shell, *Dentalium,* a scaphopod. It burrows into soft mud or sand and feeds by means of its prehensile tentacles. Respiratory currents of water are drawn in by ciliary action through the small open end of the shell, then expelled through the same opening by muscular action.

operculum, a horny plate that covers the shell aperture when the body withdraws into the shell. It protects the body and prevents water loss. These animals are basically bilaterally symmetrical, but because of **torsion,** a twisting process that occurs in the veliger stage, the visceral mass has become asymmetrical.

Form and Function

Torsion

Of all the molluscs, only gastropods undergo torsion. Torsion is a peculiar phenomenon that moves the mantle cavity to the front of the body, thus twisting the visceral organs as well through a 90- to 180-degree rotation. Torsion occurs during the veliger stage, and in some species the first part may take only a few seconds. The second 90 degrees typically takes a longer period. Before torsion occurs, the embryo is bilaterally symmetrical with an anterior mouth and a posterior anus and mantle cavity (Figure 9-11). The change comes about by an uneven growth of the right and left muscles that attach the shell to the head-foot.

After torsion, the anus and mantle cavity become anterior and open above the mouth and head. The left gill, kidney, and heart auricle are now on the right side, whereas the original right gill, kidney, and heart auricle (lost in most modern gastropods) are now on the left, and the nerve cords have been twisted into a figure eight. Because of the space available in the mantle cavity, the animal's sensitive head end can now be withdrawn into the protection of the shell, with the tougher foot forming a barrier to the outside.

The curious arrangement that results from torsion poses a serious sanitation problem by creating the possibility of wastes being washed back over the gills (**fouling**) and causes us to wonder what evolutionary factors favored such a strange realignment of the body. Several explanations have been proposed, none entirely satisfying. For example, sense organs of the mantle cavity (osphradia) would better sample water when turned in the direction of travel, and as mentioned already, the head could be withdrawn into the shell. Certainly the consequences of torsion and the resulting need to avoid fouling have been very important in the subsequent evolution of gastropods. We cannot explore these consequences, however, until we describe another unusual feature of gastropods—coiling.

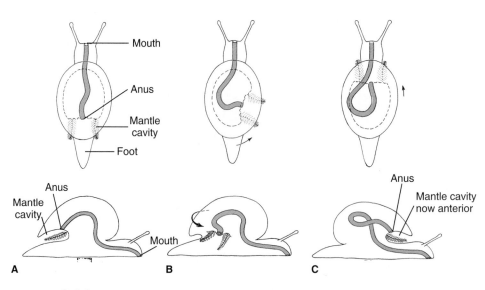

FIGURE 9-11

Torsion in gastropods. **A,** Ancestral condition before torsion. **B,** Intermediate condition. **C,** Early gastropod, torsion complete; direction of crawling now tends to carry waste products back into the mantle cavity, resulting in fouling.

Coiling

The coiling, or spiral winding, of the shell and visceral hump is not the same as torsion. Coiling may occur in the larval stage at the same time as torsion, but the fossil record shows that coiling was a separate evolutionary event and originated in gastropods earlier than torsion did. Nevertheless, all living gastropods have descended from coiled, torted ancestors, whether or not they now show these characteristics.

Early gastropods had a bilaterally symmetrical shell with all the whorls lying in a single plane (Figure 9-12A). Such a shell was not very compact, since each whorl had to lie completely outside the preceding one. Curiously, a few modern species have secondarily returned to that form. The compactness problem of the planospiral shell was solved by a shape in which each succeeding whorl was at the side of the preceding one (Figure 9-12B). However, this shape clearly was unbalanced, hanging as it did with much weight over to one side. They achieved better weight distribution by shifting the shell upward and posteriorly, with the shell axis oblique to the longitudinal axis of the foot (Figure 9-12C and D). The weight and bulk of the main body whorl, the largest whorl of the shell, pressed on the right side of the mantle cavity, however, and apparently interfered with the organs on that side. Accordingly, the gill, auricle, and kidney of the right side have been lost in all except a few living gastropods, leading to a condition of **bilateral asymmetry.**

Adaptations to Avoid Fouling

Although the loss of the right gill was probably an adaptation to the mechanics of carrying the coiled shell, that condition made possible a way to avoid fouling, which is displayed in most modern gastropods. Water is brought into the left side of

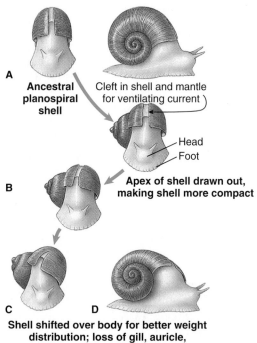

FIGURE 9-12

Evolution of shell in gastropods. **A,** Earliest coiled shells were planospiral, each whorl lying completely outside the preceding whorl. Interestingly, the shell has become planospiral secondarily in some living forms. **B,** Better compactness was achieved by snails in which each whorl lay partially to the side of the preceding whorl. **C** and **D,** Better weight distribution resulted when shell was moved upward and posteriorly.

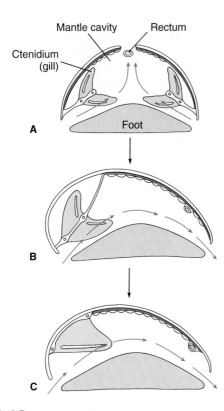

Ctenidium (gill)

Mantle cavity   Rectum

A Foot

B

C

FIGURE 9-13

Evolution of the gills in gastropods. **A,** Primitive condition in prosobranchs with two gills and excurrent water leaving the mantle cavity by a dorsal slit or hole. **B,** Condition after one gill had been lost. **C,** Advanced condition in prosobranchs, in which filaments on one side of remaining gill are lost, and axis is attached to mantle wall.

the mantle cavity and out the right side, carrying with it the wastes from the anus and nephridiopore, which lie near the right side (Figure 9-13). Some gastropods with primitive characteristics (those with two gills, such as abalone) (Figure 9-14A) avoid fouling by venting the excurrent water through a dorsal slit or hole in the shell above the anus (Figure 9-12). The opisthobranchs (nudibranchs and others) have evolved an even more curious "twist"; after undergoing torsion as larvae, they develop various degrees of *detorsion* as adults. The pulmonates (most freshwater and terrestrial snails) have lost the gill altogether, and the vascularized mantle wall has become a lung. The anus and nephridiopore open near the opening of the lung to the outside (pneumostome), and waste is expelled forcibly with air or water from the lung.

Feeding Habits

Feeding habits of gastropods are as varied as their shapes and habitats, but all include the use of some adaptation of the radula. Many gastropods are herbivorous, rasping off particles of algae. Some herbivores are grazers, some are browsers, some are planktonic feeders. The abalone (Figure 9-14) holds seaweed with the foot and breaks off pieces with

the radula. Some snails are scavengers, living on dead and decayed flesh; others are carnivorous, tearing their prey apart with their radular teeth. Some, such as the oyster borer and the moon snail (Figure 9-14B), have an extensible proboscis for drilling holes in the shells of the bivalves whose soft parts they find delectable. Some even have a spine for opening the shells. Most of the pulmonates (air-breathing snails) (Figure 9-20, p. 167) are herbivorous, but some live on earthworms and other snails.

> *Among the most interesting predators are the poisonous cone shells (Figure 9-15), which feed on vertebrates or other invertebrates, depending on the species. When Conus senses the presence of its prey, a single radular tooth slides into position at the tip of the proboscis. When the proboscis strikes the prey, it expels the tooth like a harpoon, and the poison tranquilizes or kills the prey at once. Some species can deliver very painful stings, and the sting of several species is lethal to humans. The venom consists of a series of toxic peptides, and each Conus species carries peptides (**conotoxins**) that are specific for the neuroreceptors of its preferred prey.*

Some of the sessile gastropods, such as the slipper shells, are ciliary feeders that use the gill cilia to draw in particulate matter, which they roll into a mucous ball and carry to their mouth. Some sea butterflies secrete a mucous net to catch small planktonic forms and then draw the web into the mouth.

After maceration by the radula or by some grinding device, such as the so-called gizzard in sea hares (Figure 9-16) and in others, digestion is usually extracellular in the lumen of the stomach or digestive glands. In ciliary feeders the stomachs are sorting regions and most of the digestion is intracellular in the digestive gland.

Internal Form and Function

Respiration in most gastropods is carried out by a gill (two gills in a few), although some aquatic forms lack gills and depend on the skin. The pulmonates have a lung. Freshwater pulmonates must surface to expel a bubble of gas from the lung and curl the edge of the mantle around the **pneumostome** (pulmonary opening in the mantle cavity) (Figure 9-20B) to form a siphon for taking in air.

Most gastropods have a single nephridium (kidney). The circulatory and nervous systems are well developed (Figure 9-17). The nervous system includes three pairs of ganglia connected by nerves. Sense organs include eyes, statocysts, tactile organs, and chemoreceptors.

There are both dioecious and hermaphroditic gastropods. During copulation in hermaphroditic species there is sometimes an exchange of **spermatophores** (bundles of

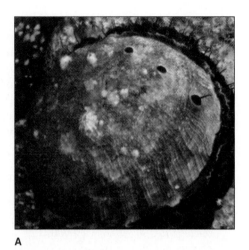

A

B

FIGURE 9-14

A, Red abalone, *Haliotus rufescens.* This huge, limpetlike snail is prized as food and extensively marketed. Abalones are strict vegetarians, feeding especially on sea lettuce and kelp. **B,** Moon snail, *Polinices lewisii.* A common inhabitant of West Coast sand flats, the moon snail is a predator of clams and mussels. It uses its radula to drill neat holes through its victim's shell, through which the proboscis is then extended to eat the bivalve's fleshy body.

FIGURE 9-15

Conus extends its long, wormlike proboscis. When the fish attempts to consume this tasty morsel, the *Conus* stings it in the mouth and kills it. The snail engulfs the fish with its distensible stomach, then regurgitates the scales and bones some hours later.

sperm), so that self-fertilization is avoided. Many forms perform courtship ceremonies. Most land snails lay their eggs in holes in the ground or under logs. Some aquatic gastropods lay their eggs in gelatinous masses; others enclose them in gelatinous capsules or in parchment egg cases. Most marine gastropods go through a free-swimming veliger larval stage during which torsion and coiling occur. Others develop directly into a juvenile within the egg capsule.

Major Groups of Gastropods

Traditional classification of the class Gastropoda recognized three subclasses: Prosobranchia, much the largest subclass, almost all of which are marine; Opisthobranchia, an assemblage including sea slugs, sea hares, nudibranchs, and canoe shells—all marine; and Pulmonata, containing most freshwater and terrestrial species. Currently, gastropod taxonomy is in a state of flux, and some workers regard any attempt to present a classfication of the class as premature.[1] Nevertheless, present evidence suggests that the Prosobranchia is paraphyletic. The Opisthobranchia may or may not be paraphyletic, but the Opisthobranchia and Pulmonata together apparently form a monophyletic grouping.

---

[1]Bieler, R. 1992. Ann. Rev. Ecol. Syst. **23**:311–338.

Rhinophore   Oral tentacle

A

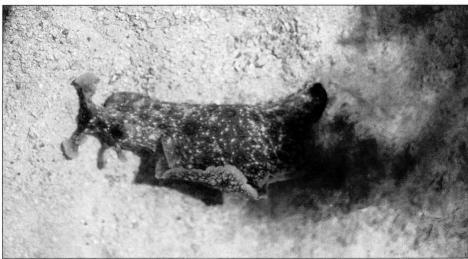

B

FIGURE 9-16

A, The sea hare, *Aplysia dactylomela,* crawls and swims across a coral reef, assisted by large, winglike parapodia, here curled above the body. **B,** When attacked, the sea hare squirts a copious protective secretion from its "purple gland" in the mantle cavity.

Familiar examples of marine gastropods are periwinkles, limpets (Figure 9-18A), whelks, conchs, abalones (Figure 9-14A), slipper shells, oyster borers, rock shells, and cowries.

At present 8 to 12 groups of opisthobranchs are recognized. Some have a gill and a shell, although the latter may be vestigial, and some have no shell or true gill. The large sea hare *Aplysia* (Figure 9-16) has large earlike anterior tentacles and a vestigial shell. Nudibranchs have no shell as adults and rank among the most beautiful and colorful of molluscs (Figure 9-19). Having lost the gill, the body surface of some nudibranchs is often increased for gaseous exchange by small projections **(cerata),** or a ruffling of the mantle edge.

The third major group (Pulmonata) contains most land and freshwater snails and slugs. Usually lacking gills, their mantle cavity has become a lung, which fills with air by contraction of the mantle floor. Aquatic and a few terrestrial species have one pair of nonretractile tentacles, at the base of which are the eyes; land forms usually have two pairs of tentacles, with the posterior pair bearing the eyes (Figures 9-17 and 9-20). The few nonpulmonate species of gastropods that live in fresh water usually can be distinguished from pulmonates because they have an operculum, which is lacking in pulmonates.

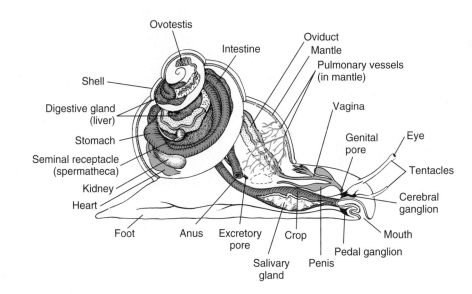

FIGURE 9-17

Anatomy of a pulmonate snail.

Class Bivalvia (Pelecypoda)

The Bivalvia (bi-val've-a) are also known as Pelecypoda (pel-e-sip'o-da) (Gr. *pelekus,* hatchet, + *pous, podus,* foot). They are bivalved (two-shelled) molluscs that include the mussels, clams, scallops, oysters, and shipworms and range in size from tiny seed shells 1 to 2 mm in length to the giant, South Pacific clams *Tridacna,* mentioned previously (Figure 9-21). Most bivalves are sedentary **suspension feeders** that depend on ciliary currents produced by the gills to bring in food materials. Unlike the gastropods, they have no head, no radula, and very little cephalization (Figure 9-22).

Most bivalves are marine, but many live in brackish water and in streams, ponds, and lakes.

Form and Function

Shell

Bivalves are laterally compressed, and their two shells **(valves)** are held together dorsally by a **hinge ligament** that causes the valves to gape ventrally. **Adductor muscles** work in opposition to the hinge ligament and draw the valves together (Figure 9-23C and D). Projecting above the hinge ligament on each valve is the **umbo,** which is the oldest part of the shell. The valves function largely for protection, but those of shipworms (Figure 9-24) have microscopic teeth for rasping wood, and rock borers use spiny valves for boring into rock. A few bivalves such as scallops (Figure 9-25) use their shells for locomotion by clapping the valves together so that they move in spurts.

Body and Mantle

The **visceral mass** is suspended from the dorsal midline, and the muscular foot is attached to the visceral mass anteroventrally. The gills hang down on each side, each covered by a fold of the mantle. The posterior edges of the mantle folds form dorsal excurrent and ventral incurrent openings (Figures 9-23A, and 9-26). In some marine bivalves part of the mantle is drawn out into long muscular siphons to allow the clam to burrow into the mud or sand and extend the siphons to the water above. Cilia on the gills and inner surface of the mantle direct the flow of water over the gills.

Locomotion

Most bivalves move by extending the slender muscular foot between the valves (Figure 9-23A). They pump blood into the foot, causing it to swell and to act as an anchor in the mud or sand, then longitudinal muscles contract to shorten the foot and pull the animal forward. In most bivalves the foot is used

A   B

FIGURE 9-18

A, Keyhole limpet, *Diodora aspera,* a prosobranch gastropod with a hole in the apex through which the water leaves the shell. **B,** The flamingo tongue, *Cyphoma gibbosum,* is a showy inhabitant of Caribbean coral reefs, where it is associated with gorgonians. This snail has a smooth creamy orange to pink shell that is normally covered by the brightly marked mantle.

FIGURE 9-19

An aeolid nudibranch, *Flabellina iodinea.* Its long, dorsal cerata contain nematocysts, which the animal obtains from its cnidarian diet.

for burrowing, but a few creep. Some bivalves are sessile: oysters attach their shells to a surface by secreting cement, and mussels (Figure 9-27) attach themselves by secreting a number of slender **byssal threads.**

Feeding and Digestion

Most bivalves are suspension feeders. The respiratory currents bring both oxygen and organic materials to the gills where ciliary tracts direct them to the tiny pores of the gills.

Pneumostome

FIGURE 9-20

A, Pulmonate land snail. Note two pairs of tentacles; the second larger pair bears the eyes. **B,** Banana slug, *Ariolimax columbianus.*

FIGURE 9-21

Giant clam, *Tridacna gigas,* lies buried in coral rock with only the richly colored fluted siphonal edge visible. These bivalves bear enormous numbers of symbiotic single-cell algae (zooxanthellae) that provide much of the clam's nutrition.

Gland cells on the gills and labial palps secrete copious amounts of mucus, which entangles particles suspended in the water going through gill pores. Ciliary tracts move the particle-laden mucus to the mouth (Figure 9-22).

In the stomach the mucus and food particles are kept whirling by a rotating gelatinous rod, called a **crystalline style.** Solution of layers of the rotating style frees digestive enzymes for extracellular digestion. Ciliated ridges of the stomach sort food particles and direct suitable particles to the **digestive gland** for intracellular digestion.

Shipworms (Figure 9-24) feed on the particles they excavate as they burrow in wood. Symbiotic bacteria live in a special organ in these bivalves and produce cellulase to digest the wood. Other bivalves such as giant clams gain much of their nutrition from the photosynthetic products of symbiotic algae living in their mantle tissue (Figure 9-21).

Internal Features and Reproduction

Bivalves have a three-chambered heart that pumps blood through the gills and mantle for oxygenation and to the kidneys for waste elimination (Figure 9-28). They have three pairs of widely separated ganglia and poorly developed sense organs. A few bivalves have ocelli. The steely blue eyes of some scallops (Figure 9-25), located around the mantle edge, are remarkably complex, equipped with cornea, lens, and retina.

Freshwater clams were once abundant and diverse in streams throughout the eastern United States, but they are now easily the most jeopardized group of animals in the country. Of the more than 300 species once present, 12 are extinct, 42 are listed as threatened or endangered, and as many as 88 more may be listed soon. A combination of causes is responsible, of which a decline in water quality is among the most important. Pollution and sedimentation from mining, industry, and agriculture are among the culprits. Habitat destruction by altering natural water courses and damming is an important factor. Poaching to supply the Japanese cultured pearl industry is partially to blame (see note on p. 156). And in addition to everything else, the prolific zebra mussels (see next note) attach in great numbers to native clams, exhausting food supplies (phytoplankton) in the surrounding water.

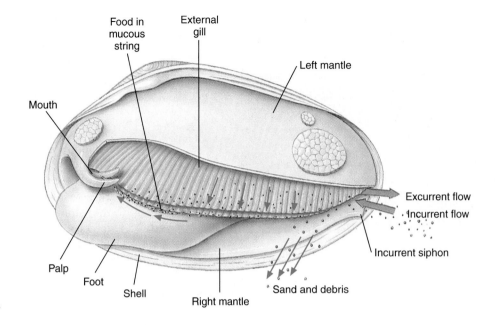

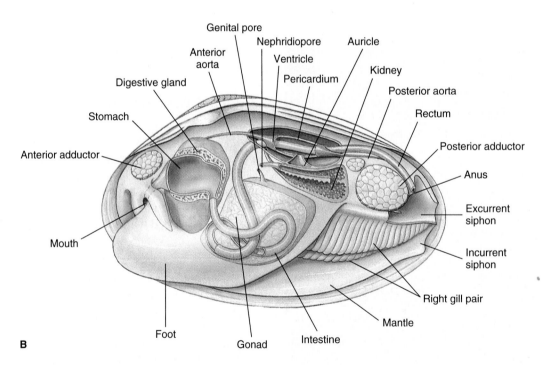

FIGURE 9-22

A, Feeding mechanism of freshwater clam. Left valve and mantle are removed. Water enters the mantle cavity posteriorly and is drawn forward by ciliary action to the gills and palps. As water enters the tiny openings of the gills, food particles are sieved out and caught up in strings of mucus that are carried by cilia to the palps and directed to the mouth. Sand and debris drop into the mantle cavity and are removed by cilia. **B,** Clam anatomy.

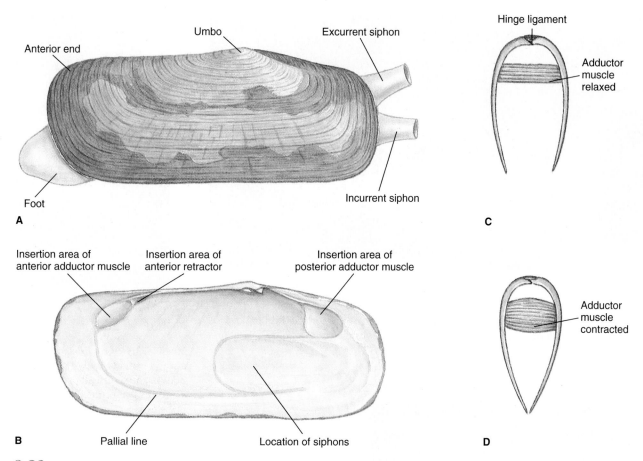

FIGURE 9-23

Tagelus plebius, the stubby razor clam (class Bivalvia). **A,** External view of left valve. **B,** Inside of right shell showing scars where muscles were attached. The mantle was attached to the pallial line. **C** and **D,** Sections showing function of adductor muscles and hinge ligament. In **C** the adductor muscle is relaxed, allowing the hinge ligament to pull the valves apart. In **D** the adductor muscle is contracted, pulling the valves together.

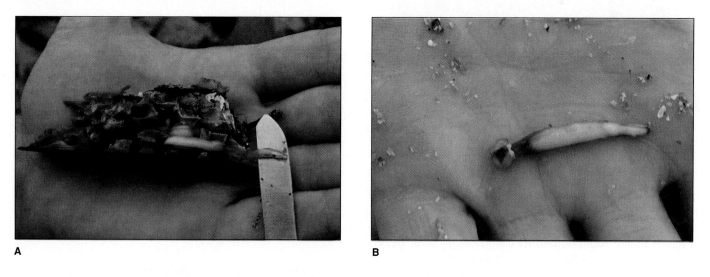

FIGURE 9-24

A, Shipworms are bivalves that burrow in wood, causing great damage to unprotected wooden hulls and piers. **B,** The two small, anterior valves, seen at left, are used as rasping organs to extend the burrow.

FIGURE 9-25

Representing a group that has evolved from burrowing ancestors, the surface-dwelling bay scallop *Argopecten irradians* has developed sensory tentacles and a series of blue eyes along its mantle edges.

FIGURE 9-27

Mussels, *Mytilus edulis,* occur in northern oceans around the world; they form dense beds in the intertidal zone. A host of marine creatures live protected beneath attached mussels.

FIGURE 9-26

In the northwest ugly clam, *Entodesma navicula,* the incurrent and excurrent siphons are clearly visible.

Zebra mussels, Dreissena polymorpha, *are a recent and potentiallly disastrous biological introduction into North America. They were apparently picked up as veligers with ballast water by one or more ships in freshwater ports in northern Europe and then expelled between Lake Huron and Lake Erie in 1986. This 4 cm bivalve spread throughout the Great Lakes by 1990, and by 1994 it was as far south on the Mississippi as New Orleans, as far north as Duluth, Minnesota, and as far east as the Hudson River in New York. It attaches to any firm surface and filter feeds on phytoplankton. Large numbers build up rapidly. They foul water intake pipes of municipal and industrial plants, impede intake of water for municipal supplies, and have far-reaching effects on the ecosystem (see preceding note). Zebra mussels may cost $5 billion or more to control by the end of the century.*

Sexes are separate, and fertilization is usually external. Marine embryos usually go through three free-swimming larval stages—**trochophore, veliger larva,** and young **spat**—before reaching adulthood. In freshwater clams fertilization is internal, and some of the gill tubes become temporary brood chambers. There the zygotes develop into specialized veligers called **glochidia,** which are discharged with the excurrent flow (Figure 9-29). If the larvae come in contact with a passing fish, they hitchhike a ride as parasites in the fish's gills for the next 20 to 70 days before sinking to the bottom to become sedentary adults.

Class Cephalopoda

The Cephalopoda (sef-a-lop′o-da) are the most complex of the molluscs—in fact, in some respects they are the most complex of all invertebrates. They include the squids, octopuses, nautiluses, and cuttlefishes. All are marine, and all are active predators.

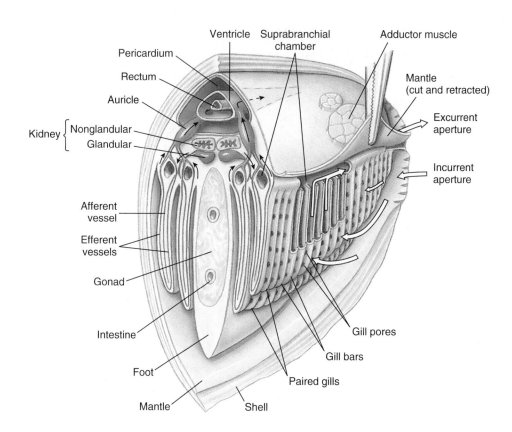

Ventricle Suprabranchial
chamber

Pericardium

Rectum

Auricle

Kidney { Nonglandular
Glandular

Afferent
vessel

Efferent
vessels

Gonad

Intestine

Foot

Mantle

Shell

Paired gills

Gill bars

Gill pores

Adductor muscle

Mantle
(cut and retracted)

Excurrent
aperture

Incurrent
aperture

FIGURE 9-28

Section through heart region of a freshwater clam to show relation of circulatory and respiratory systems. Respiratory water currents: water is drawn in by cilia, enters gill pores, and then passes up water tubes to suprabranchial chambers and out excurrent aperture. Blood in gills exchanges carbon dioxide and oxygen. Blood circulation: ventricle pumps blood forward to sinuses of foot and viscera, and posteriorly to mantle sinuses. Blood returns from mantle to auricles; it returns from viscera to the kidney, and then goes to the gills, and finally to the auricles.

Cephalopods (Gr. *kephalē,* head, + *pous, podos,* foot) have a modified foot that is concentrated in the head region. It takes the form of a funnel for expelling water from the mantle cavity, and the anterior margin is drawn out into a circle or crown of arms or tentacles.

Cephalopods range in size from 2 to 3 cm up to the giant squid, *Architeuthis,* which is the largest invertebrate known. The squid *Loligo* (L., cuttlefish) is about 30 cm long (Figure 9-30 A).

The enormous giant squid, Architeuthis, *is very poorly known because no one has ever been able to study a living specimen. The anatomy has been studied from stranded animals, from those captured in the nets of fishermen, and from specimens found in the stomach of sperm whales. The mantle length is 5 to 6 m, and the head is up to one meter long. They have the largest eyes in the animal kingdom: up to 25 cm (10 inches) in diameter. They apparently eat fish and other squids, and they are an important food item for sperm whales. They are thought to live on or near the bottom at a depth of 1000 m, but some have been seen swimming at the surface.*

Fossil records of cephalopods go back to Cambrian times. The earliest shells were straight cones; others were curved or coiled, culminating in the coiled shell similar to that of the modern *Nautilus* spp. (Gr. *nautilos,* sailor)—the only remain-

ing members of the once flourishing nautiloids (Figure 9-31). Cephalopods without shells or with internal shells (such as octopuses and squids) probably evolved from a straight-shelled ancestor. Ammonoids were widely prevalent in the Mesozoic era but became extinct by the end of the Cretaceous period. They had chambered shells analogous to nautiloids, but the septa were more complex. The reasons for their extinction remain a mystery. Present evidence suggests that they were gone before the asteroid bombardment at the end of the Cretaceous period (p. 31), and some nautiloids, which some ammonoids closely resembled, survive to the present.

Form and Function

Shell

Although early nautiloid shells were heavy, they were made buoyant by a series of **gas chambers,** as is that of the *Nautilus* (Figure 9-31B), enabling the animal to swim while carrying its shell. The shell of *Nautilus,* although coiled, is quite different from that of a gastropod. Transverse septa divide the shell into internal chambers (Figure 9-31B). The living animal inhabits only the largest, last chamber. As it grows it moves forward, secreting behind it a new septum. A cord of living tissue (the **siphuncle**), which extends from the visceral mass, extends through the septa and connects the chambers. Cuttlefishes also have a small coiled or curved shell, but it is entirely enclosed by the mantle. In squids most of the shell has disappeared, leaving only a thin, horny strip called a **pen,** which the mantle encloses. In *Octopus* (Gr. *oktos,* eight, + *pous, podos,* foot) the shell is very reduced.

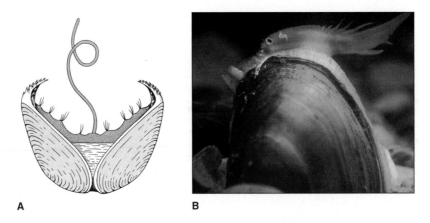

FIGURE 9-29

A, Glochidium, or larval form, for some freshwater clams. When the larva is released from brood pouch of mother, it may become attached to a fish's gill by clamping its valves closed. It remains as a parasite on the fish for several weeks. Its size is approximately 0.3 mm. **B,** Some clams have adaptations that help their glochidia find a host. The mantle edge of this female pocketbook mussel (*Lampsilis ovata*) mimics a small minnow, complete with eye. When a smallmouth bass comes to dine, it gets doused with glochidia.

After Nautilus *secretes a new septum, the new chamber is filled with fluid similar in ionic composition to that of the* Nautilus' *blood (and of seawater). Fluid removal involves the active secretion of ions into tiny intercellular spaces in the siphuncular epithelium, so that a very high local osmotic pressure is produced, and the water is drawn out of the chamber by osmosis. The gas in the chamber is only the respiratory gas from the siphuncle tissue that diffuses into the chamber as the fluid is removed. Thus the gas pressure in the chamber is 1 atmosphere or less because it is in equilibrium with the gases dissolved in the seawater surrounding the* Nautilus, *which are in turn in equilibrium with air at the surface of the sea, despite the fact that the* Nautilus *may be swimming at 400 m beneath the surface. That the shell can withstand implosion by the surrounding 41 atmospheres (about 600 pounds per square inch), and that the siphuncle can remove water against this pressure are marvelous feats of natural engineering!*

Locomotion

Most cephalopods swim by forcefully expelling water from the mantle cavity through a ventral **funnel**—a sort of jet propulsion method. The funnel is mobile and can be pointed forward or backward to control direction; the force of water expulsion determines speed.

Squids and cuttlefishes are excellent swimmers. The squid body is streamlined and built for speed (Figure 9-30).

Cuttlefishes swim more slowly (Figure 9-32). Both squids and cuttlefishes have lateral fins that can serve as stabilizers, but they are held close to the body for rapid swimming. The gas-filled chambers of *Nautilus* keep the shell upright. Although not as fast as squids, they move surprisingly well.

Octopus has a rather globular body and no fins (Figure 9-1E). Octopuses can swim backwards by spurting jets of water from their funnel, but they are better adapted to crawling about over the rocks and coral, using the suction discs on their arms to pull or to anchor themselves. Some deep-water octopods have fins and the arms webbed like an umbrella; they swim in a sort of medusa fashion.

External features

During embryonic development of cephalopods, the head and foot become indistinguishable. The ring around the mouth, which bears the arms and tentacles, is apparently derived from the anterior margin of the head.

In *Nautilus* the head with its 60 to 90 or more tentacles can be extruded from the opening of the body compartment of the shell (Figure 9-31). Its tentacles have no suckers but adhere to prey by secretions. The tentacles search for, sense, and grasp food. Beneath the head is the funnel. The shell shelters the mantle, mantle cavity, and visceral mass. Two pairs of gills are located in the mantle cavity.

Cephalopods other than nautiloids have only one pair of gills. Octopods have 8 arms with suckers; squids and cuttlefishes (decapods) have 10 arms: 8 arms with suckers and a pair of long retractile tentacles. The thick mantle covering the trunk fits loosely at the neck region allowing intake of water into the mantle cavity. When the mantle edges contract closely about the neck, water is expelled through the funnel. The

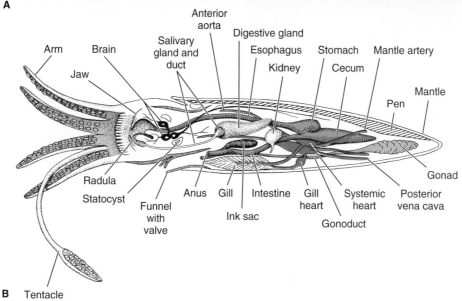

FIGURE 9-30

A, Squid *Sepioteuthis lessoniana.* **B,** Lateral view of squid anatomy, with the left half of the mantle removed.

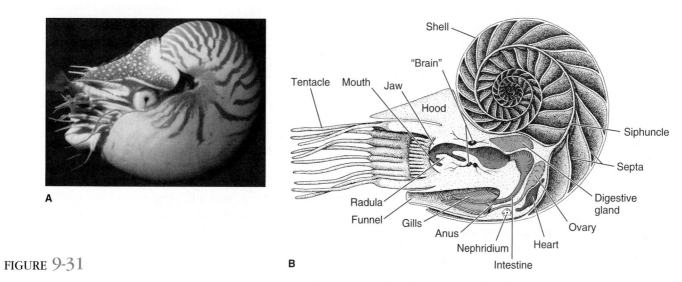

FIGURE 9-31

Nautilus, a cephalopod. **A,** Live *Nautilus,* feeding on a fish. **B,** Longitudinal section, showing gas-filled chambers of shell, and diagram of body structure.

FIGURE 9-32

A cuttlefish, *Sepia latimanus.* Cuttlefishes have an internal shell familiar to keepers of caged birds as "cuttlebone."

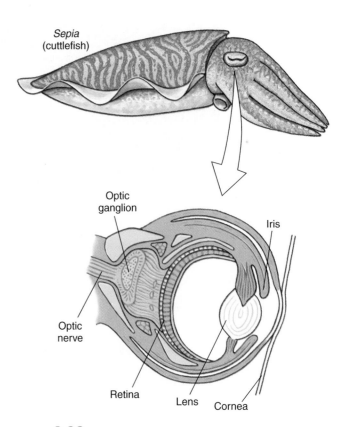

FIGURE 9-33

Eye of a cuttlefish (*Sepia*). The structure of cephalopod eyes shows a high degree of convergent evolution with the eyes of vertebrates.

water current thus created provides oxygenation for the gills in the mantle cavity, jet power for locomotion, and a means of carrying wastes and sexual products away from the body.

Color Changes

There are special pigment cells called **chromatophores** in the skin of most cephalopods, which by expanding and contracting produce color changes. They are controlled by the nervous system and perhaps by hormones. Some color changes are protective to agree with background hues; most are behavioral and are associated with alarm or with courtship. Many deep-sea squids are also bioluminescent.

Ink Production

Most cephalopods other than nautiloids have an ink sac that empties into the rectum. The sac contains an ink gland that secretes a dark fluid containing the pigment melanin. When the animal is alarmed, it releases a cloud of ink through the anus to form a "smokescreen" to confuse the enemy.

Feeding and Nutrition

Cephalopods are predaceous, feeding chiefly on small fishes, molluscs, crustaceans, and worms. Their arms, which are used in food capture and handling, have a complex musculature and are capable of delicately controlled movements. They are highly mobile and swiftly seize prey and bring it to the mouth. Strong, beaklike **jaws** grasp the prey, and the **radula** tears off pieces of flesh (Figure 9-30B). Octopods and cuttlefishes have salivary glands that secrete a poison for immobilizing prey. Digestion is extracellular and occurs in the stomach and cecum.

Internal Features and Reproduction

The active habits of cephalopods are reflected in their internal anatomy, particularly their respiratory, circulatory, and nervous systems. They have the most complex brain among the invertebrates (Figure 9-30B). Except for *Nautilus,* which has relatively simple eyes, cephalopods have elaborate eyes with cornea, lens, chambers, and retina (Figure 9-33)—similar to the camera-type eye of vertebrates.

Ciliary propulsion would not circulate enough water over the gills for an active animal, and cephalopods ventilate their gills by muscular action of the mantle wall. They have a closed circulatory system with a network of vessels, and blood flows through the gills via capillaries. The plan of circulation of the ancestral mollusc places the entire systemic circulation before the blood reaches the gills, which means that the blood pressure at the gills is too low for rapid gaseous exchange. This functional problem has been solved by the evolutionary development of **accessory** or **branchial hearts** (Figure 9-30B).

Sexes are separate in cephalopods. In the male seminal vesicle the spermatozoa are encased in spermatophores and stored in a sac that opens into the mantle cavity. During copulation one arm of the adult male plucks a spermatophore from

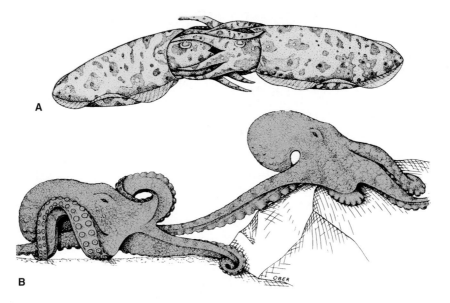

FIGURE 9-34

Copulation in cephalopods. **A,** Mating cuttlefishes. **B,** Male octopus uses modified arm to deposit spermatophores in female mantle cavity to fertilize her eggs. Octopuses often tend their eggs during development.

his own mantle cavity and inserts it into the mantle cavity of the female near the oviduct opening (Figure 9-34). Before copulation males often undergo color displays, apparently directed against rival males and for courtship of females. Eggs are fertilized as they leave the oviduct and are usually attached to stones or other objects to develop. Some octopods tend their eggs.

Phylogeny and Adaptive Radiation

On the basis of such shared features as spiral cleavage, mesoderm from the 4d blastomere, and trochophore larva, most zoologists have accepted the Mollusca as protostomes, allied with the annelid-arthropod line. Opinions differ, however, as to whether the molluscs were derived from a flatworm ancestor independent of the annelids, from the annelid-arthropod line after the advent of the coelom, or from a protoannelid after metamerism arose. This last hypothesis is strengthened if *Neopilina* (class Monoplacophora) is metameric, as some scientists have contended. This question is yet to be resolved. Many zoologists suggest that the replication of body parts found in the monoplacophorans is pseudometamerism. In this case molluscs would have branched off from the annelid-arthropod line after the coelom arose but before the advent of metamerism.

The ancestral mollusc was a more or less wormlike, dorsoventrally flattened organism with a ventral gliding surface and a dorsal mantle with a chitinous cuticle and calcare-

ous scales (Figure 9-35). It had a posterior mantle cavity with two gills, a radula, a ladderlike nervous system, and an open circulatory system with a heart. Among living molluscs the primitive condition is most nearly approached by the caudofoveates, although the foot is reduced to an oral shield in this class. The solenogasters have lost the gills, and the foot is represented by the ventral groove. Both these classes probably branched off from a common ancestor before the development of a solid shell and a ventral muscularized foot. The polyplacophorans probably also branched off early from the main lines of molluscan evolution. Some workers believe that the shells of polyplacophorans are not homologous to the shells of other molluscs because they differ structurally and developmentally. Thus the Polyplacophora and the remaining classes are sister groups (Figure 9-36).

Present evidence suggests that the Gastropoda is a monophyletic group and that the Gastropoda and Cephalopoda form a sister group to the Monoplacophora (see Figure 9-36). Both the gastropods and cephalopods have a greatly expanded visceral mass. Torsion brought the mantle cavity toward the head in gastropods, but in cephalopods the mantle cavity was extended ventrally. The evolution of a chambered shell in cephalopods was a very important contribution to their freedom from the substratum and their ability to swim. The development of their respiratory, circulatory, and nervous systems is correlated with their predatory and swimming habits.

Scaphopods and bivalves have an expanded mantle cavity that essentially envelops the body. Adaptations for burrowing characterize this clade: the spatulate foot and reduction of the head and sense organs.

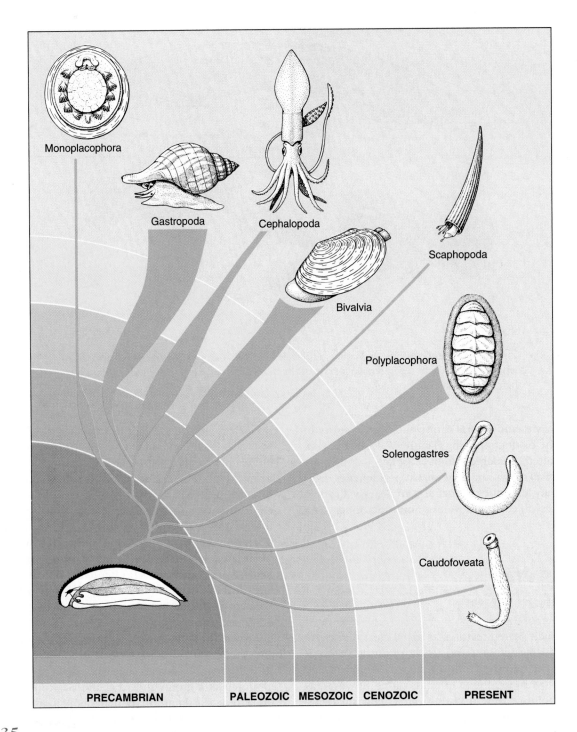

FIGURE 9-35

The classes of Mollusca, showing their derivations and relative abundance.

Most of the diversity among the molluscs is related to their adaptation to different habitats and modes of life and to a wide variety of feeding methods, ranging from sedentary filter feeding to active predation. There are many adaptations for food gathering within the phylum and an enormous variety in radular structure and function, particularly among the gastropods.

The versatile glandular mantle has probably shown more plastic adaptative capacity than any other molluscan structure. Besides secreting the shell and forming the mantle cavity, it is variously modified into gills, lungs, siphons, and apertures, and it sometimes functions in locomotion, in the feeding processes, or in a sensory capacity. The shell, too, has undergone a variety of evolutionary adaptations.

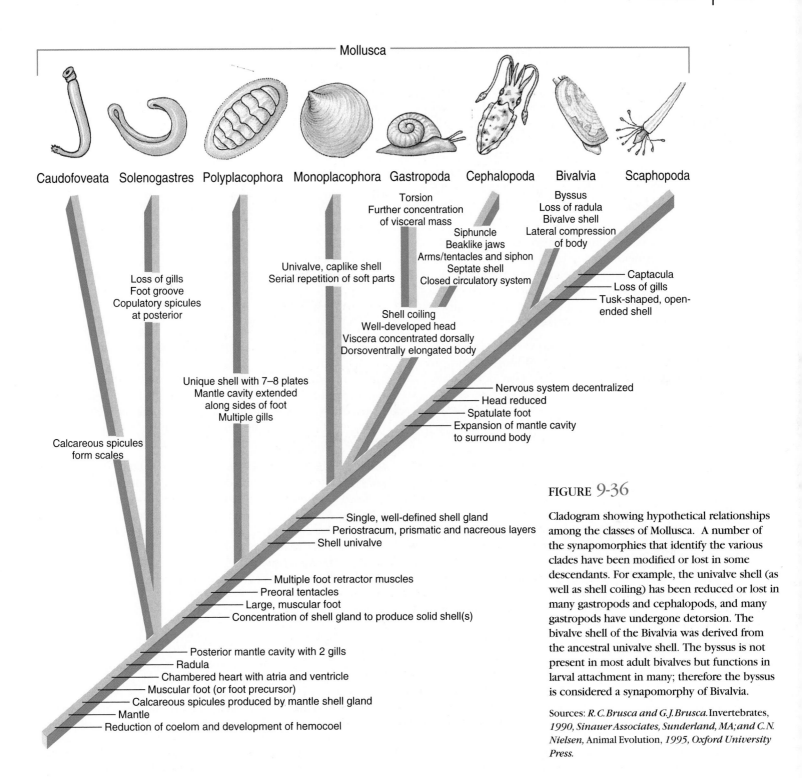

FIGURE 9-36

Cladogram showing hypothetical relationships among the classes of Mollusca. A number of the synapomorphies that identify the various clades have been modified or lost in some descendants. For example, the univalve shell (as well as shell coiling) has been reduced or lost in many gastropods and cephalopods, and many gastropods have undergone detorsion. The bivalve shell of the Bivalvia was derived from the ancestral univalve shell. The byssus is not present in most adult bivalves but functions in larval attachment in many; therefore the byssus is considered a synapomorphy of Bivalvia.

Sources: *R. C. Brusca and G. J. Brusca.* Invertebrates, *1990, Sinauer Associates, Sunderland, MA; and C. N. Nielsen,* Animal Evolution, *1995, Oxford University Press.*

Summary

The Mollusca is one of the largest and most diverse of all phyla, its members ranging in size from very small organisms to the largest of invertebrates. Their basic body divisions are the head-foot and the visceral mass, usually covered by a shell. The majority are marine, but some are fresh water, and a few are terrestrial. They occupy a variety of niches; a number are economically important, and a few are medically important as hosts of parasites.

The molluscs are coelomate (have a coelom), although their coelom is limited to the area around the heart. The evolutionary development of a coelom was important because it enabled better organization of visceral organs and, in many of the animals that have it, an efficient hydrostatic skeleton. The mantle and mantle cavity are important charac-

Classification of Phylum Mollusca

Class Caudofoveata (kaw′do-fo-ve-at′a) (L. *cauda*, tail, + *fovea*, small pit): **caudofoveates.** Wormlike; shell, head, and excretory organs absent; radula usually present; mantle with chitinous cuticle and calcareous scales; oral pedal shield near anterior mouth; mantle cavity at posterior end with pair of gills; sexes separate; often united with solenogasters in class Aplacophora. Examples: *Chaetoderma, Limifossor.*

Class Solenogastres (so-len′o-gas′trez) (Gr. *solēn*, pipe, + *gaster*, stomach): **solenogasters.** Wormlike; shell, head, and excretory organs absent; radula usually present; mantle usually covered with scales or spicules; mantle cavity posterior, without true gills, but sometimes with secondary respiratory structures; foot represented by long, narrow, ventral pedal groove; hermaphroditic. Example: *Neomenia.*

Class Monoplacophora (mon′o-pla-kof′o-ra) (Gr. *monos*, one, + *plax*, plate, + *phora*, bearing): **monoplacophorans.** Body bilaterally symmetrical with a broad flat foot; a single limpetlike shell; mantle cavity with five or six pairs of gills; large coelomic cavities; radula present; six pairs of nephridia, two of which are gonoducts; separate sexes. Example: *Neopilina* (Figure 9-7).

Class Polyplacophora (pol′y-pla-kof′o-ra) (Gr. *polys*, many, several, + *plax*, plate, + *phora*, bearing): **chitons.** Elongated, dorsoventrally flattened body with reduced head; bilaterally symmetrical; radula present; shell of eight dorsal plates; foot broad and flat; gills multiple, along sides of body between foot and mantle edge; sexes usually separate, with a trochophore but no veliger larva. Examples: *Mopalia* (Figure 9-8), *Chaetopleura.*

Class Scaphopoda (ska-fop′o-da) (Gr. *skaphē*, trough, boat, + *pous, podos*, foot): **tusk shells.** Body enclosed in a one-piece tubular shell open at both ends; conical foot; mouth with radula and tentacles; head absent; mantle for respiration; sexes separate; trochophore larva. Example: *Dentalium* (Figure 9-10).

Class Gastropoda (gas-trop′o-da) (Gr. *gaster*, belly, + *pous, podos*, foot): **snails and relatives.** Body asymmetrical; usually in a coiled shell (shell uncoiled or absent in some); head well developed, with radula; foot large and flat; dioecious or monoecious, some with trochophore, typically with veliger, some without larva. Examples: *Busycon, Polinices* (Figure 9-14B), *Physa, Helix, Aplysia* (Figure 9-16).

Class Bivalvia (bi-val′ve-a) (L. *bi*, two, + *valva*, folding door, valve) **(Pelecypoda): bivalves.** Body enclosed in a two-lobed mantle; shell of two lateral valves of variable size and form, with dorsal hinge; head greatly reduced but mouth with labial palps; no radula; no cephalic eyes; gills platelike; foot usually wedge shaped; sexes usually separate, typically with trochophore and veliger larvae. Examples: *Mytilus* (Figure 9-27), *Venus, Bankia* (Figure 9-24).

Class Cephalopoda (sef′a-lop′o-da) (Gr. *kephalē*, head, + *pous, podos*, foot): **squids and octopuses.** Shell often reduced or absent; head well developed with eyes and a radula; head with arms or tentacles; foot modified into a funnel; nervous system of well-developed ganglia, centralized to form a brain; sexes separate, with direct development. Examples: *Loligo* (Figure 9-30), *Octopus* (Figure 9-1E), *Sepia* (Figure 9-32), *Nautilus* (Figure 9-31).

---

teristics of molluscs. The mantle secretes the shell and overlies a part of the visceral mass to form a cavity housing the gills. The mantle cavity has been modified into a lung in some molluscs. The foot is usually a ventral, solelike, locomotory organ, but it may be variously modified, as in the cephalopods, where it has become a funnel and arms. Most molluscs except bivalves have a radula, which is a protrusible, tonguelike organ with teeth used in feeding. Except in the cephalopods, which have a closed circulatory system, the circulatory system of molluscs is open, with a heart and blood sinuses. Molluscs usually have a pair of nephridia connecting with the coelom and a complex nervous system with a variety of sense organs. The primitive larva of molluscs is the trochophore, and most marine molluscs have a more advanced larva, the veliger.

The classes Caudofoveata and Solenogastres are small groups of wormlike molluscs with no shell. The Scaphopoda is a slightly larger class with a tubular shell, open at both ends, and the mantle wrapped around the body.

The class Monoplacophora is a tiny, univalve marine group showing pseudometamerism or vestiges of true metamerism. The Polyplacophora are more common and diverse marine organisms with shells in the form of a series of eight plates. They are rather sedentary animals with a row of gills along each side of their foot.

The Gastropoda is the most successful and largest class of molluscs. Its interesting evolutionary history includes torsion, or the twisting of the posterior end to the anterior, so that the anus and head are at the same end, and coiling, an elongation and spiraling of the visceral mass. Torsion led to the survival problem of fouling, which is the release of excreta over the head and in front of the gills, and this problem was solved in various ways among different gastropods. Among the solutions to fouling were bringing water into one side of the mantle cavity and out the other (many gastropods), some degree of detorsion (opisthobranchs and pulmonates), and conversion of the mantle cavity into a lung (pulmonates).

Members of the class Bivalvia are marine and fresh water, and their shell is divided into two valves joined by a dorsal ligament and held together by an adductor muscle. Most of them are suspension feeders, drawing water through their gills by ciliary action.

Members of the class Cephalopoda are the most complex molluscs; they are all predators and many can swim rapidly. Their tentacles capture prey by adhesive secretions or by suckers. They swim by forcefully expelling water from their mantle cavity through a funnel.

There is strong embryological evidence that the molluscs are related to the annelids, although the molluscs are not metameric. The enormous diversity of molluscs can be derived from a hypothetical ancestor that showed the basic molluscan body plan.

Review Questions

1. How does the coelom develop embryologically? Why was the evolutionary development of the coelom important?

2. Members of the phylum Mollusca are extremely diverse, yet the phylum clearly constitutes a monophyletic clade. What evidence can you cite in support of this statement?

3. How are molluscs important to humans?

4. Distinguish among the following classes of molluscs: Polyplacophora, Gastropoda, Bivalvia, Cephalopoda.

5. Define the following: radula, odontophore, periostracum, prismatic layer, nacreous layer, trochophore, veliger, glochidium, osphradium.

6. Briefly describe the habitat and habits of a typical chiton.

7. Define the following with respect to gastropods: operculum, torsion, fouling, bilateral asymmetry.

8. Torsion in gastropods created a selective disadvantage: fouling. Suggest one or more potential selective advantages that could have offset the disadvantage. What are ways that gastropods have evolved to avoid fouling?

9. Distinguish among opisthobranchs and pulmonates.

10. Briefly describe how a typical bivalve feeds and how it burrows.

11. What is the function of the siphuncle of cephalopods?

12. Describe how cephalopods swim and how they eat.

13. Cephalopods are actively swimming predators, but they evolved from a slow-moving, probably grazing ancestor. Describe evolutionary modifications of the ancestral plan that make the cephalopod life-style possible.

14. To what other major invertebrate groups are molluscs related, and what is the nature of the evidence for the relationship?

15. Briefly describe the characteristics of the hypothetical ancestral mollusc, and tell how each class of molluscs differs from the primitive condition with respect to each of the following: shell, radula, foot, mantle cavity and gills, circulatory system, head.

Selected References

See also general references on page 395.

Barinaga, M. 1990. Science digests the secrets of voracious killer snails. Science **249**:250-251. *Describes research on the toxins produced by cone snails.*

Kuznik, F. 1993. America's aching mussels. National Wildlife (Oct./Nov):34-39. *Details the miserable status of freshwater clams (or mussels) in the United States.*

Morris, P. A. (W. J. Clench [editor]). 1973. A field guide to shells of the Atlantic and Gulf Coasts and the West Indies, ed. 3. Boston, Houghton Mifflin Company. *An excellent revision of a popular handbook.*

Moynihan, M. 1985. Communication and noncommunication by cephalopods.

Bloomington, Indiana University Press. *Readable summarization of our understanding of communication in this remarkable group of molluscs.*

Roper, C. R. E., and K. J. Boss. 1982. The giant squid. Sci. Am. **246**:96-105 (April). *Many mysteries remain about the deep-sea squid, Architeuthis, because it has never been studied alive. It can reach a weight of 1000 pounds and a length of 18 m, and its eyes are as large as automobile headlights.*

Ross, J. 1994. An aquatic invader is running amok in U.S. waterways. Smithsonian **24**(11):40-50 (Feb.). *A small bivalve, the zebra mussel, apparently introduced into the Great Lakes with ballast water from ships, is clogging up intake pipes and municipal water supplies. It will take billions of dollars to control.*

Ward, P., L. Greenwald, and O. E. Greenwald, 1980. The buoyancy of the chambered nautilus. Sci Am. **243**:190-203 (Oct.). *Reviews discoveries on how the nautilus removes the water from a chamber after screening a new septum.*

Yonge, C. M. 1975. Giant clams. Sci. Am. **232**:96-105 (Apr.). *Details the fascinating morphological adaptations of this bivalve for life with its mutualistic zooxanthellae.*

Links to the Internet

Visit this textbook's web site at http://www.mhhe.com/zoology to find live Internet links for each of the references listed below.

1. Biosis/Mollusca. This site provides information on the molluscan classes, including systematics, phylogeny, and bibliographies.

2. Phylum Mollusca. This site provides information on the molluscan classes.

3. Mollusca. Arizona's Tree of Life Web Page. Pictures, characteristics, phylogenetic relationships, references on molluscs. Pictures and references, and links to polyplacophorans, gastropods, bivalves, and cephalopods. Most sites have reference lists.

Gastropods and Bivalves

4. Molluscs. Keys to Marine Invertebrates of the Woods Hole Region. Descriptive information, definition of terminology, and keys to the bivalves and gastropods of the Woods Hole Region.

5. Aplysia Hometank. Information on the use of the shell-less gastropod, *Aplysia,* in neuroscience research.

Cephalopods

6. Cephalopod Page. Information and images of cephalopods.

Segmented Worms: The Annelids

CHAPTER ten

Dividing Up the Body

Although a spacious, fluid-filled coelom provided an efficient hydrostatic skeleton for burrowing, precise control of body movements was not possible in the earliest coelomates. The force of muscle contraction in one area was carried throughout the body by the fluid in the undivided coelom. This defect was remedied when a series of partitions (septa) evolved in the ancestral annelids. When the septa divided the coelom into a series of compartments, components of most other body systems, such as circulatory, nervous, and excretory, were repeated in each segment. This body plan is known as *metamerism.*

The evolutionary advent of metamerism was highly significant because it made possible development of much greater complexity in structure and function. Metamerism not only increased the efficiency of burrowing, it made possible independent and separate movements by the separate segments. The need for fine control of movements led, in turn, to the evolution of a more sophisticated nervous system. Moreover, repetition of body parts gave the organisms a built-in redundancy, as in some human-made systems. This redundancy provided a safety factor: if one segment should fail, the others could still function. Thus an injury to one part would not necessarily be fatal.

The evolutionary potential of the metameric body plan is amply demonstrated by the large and diverse phylum Arthropoda. Metamerism also arose independently in the deuterostome line, which includes the numerous and adaptively diverse vertebrates.

Annelida (an-nel'i-da) (L. *annellus,* little ring, + *-ida,* suffix) consists of the segmented worms. It is a large phylum, numbering approximately 15,000 species, the most familiar of which are earthworms and freshwater worms (class Oligochaeta) and leeches (class Hirudinea). However, approximately two-thirds of the phylum is composed of marine worms (class Polychaeta), which are less familiar to most people. Among the latter are many curious members; some are strange, even grotesque, whereas others are graceful and beautiful. They include clam worms, plumed worms, parchment worms, scaleworms, lugworms, and many others. The annelids are true coelomates and belong to the protostome branch, with spiral and mosaic cleavage. They are a highly developed group in which the nervous system is more centralized and the circulatory system more complex than those of the phyla we have studied so far.

Annelids are sometimes called "bristle worms" because, with the exception of leeches, most annelids bear tiny chitinous bristles called **setae** (L. *seta,* hair or bristle). Short needlelike setae help anchor the somites during locomotion to prevent backward slipping; long, hairlike setae aid aquatic forms in swimming. Because many annelids are either burrowers or live in secreted tubes, the stiff setae also aid in preventing the worm from being pulled out or washed out of its home. Robins know from experience how effective the earthworms' setae are.

Ecological Relationships

Annelids are worldwide in distribution, occurring in the sea, fresh water, and terrestrial soil. Some marine annelids live quietly in tubes or burrow into bottom mud or sand. Some of these feed on organic matter in the mud through which they burrow; others feed on suspended particles with elaborate ciliary or mucous devices for trapping food. Many are predators, either pelagic or hiding in crevices of coral or rock except when hunting. Freshwater annelids burrow in mud or sand, live among vegetation, or swim about freely. The most familiar annelids are the terrestrial earthworms, which move about through the soil. Some leeches are bloodsuckers, and others are carnivores; most of them live in fresh water.

Economic Importance

Much of the economic importance of annelids is indirect, deriving from their ecological roles. Many are members of grazing food chains or detritus food chains, serving as prey for other organisms of more direct interest to humans, such as fishes. Consequently, a lively market in some polychaetes and oligochaetes as fish bait thrives. The burrows of earthworms increase drainage and aeration of soils, and the migrations of worms help mix the soil and distribute organic matter to deeper layers (p. 186). Some marine annelids that burrow serve an analogous role in the sea; the lugworm (*Arenicola,* Figure 10-10) is sometimes called the "earthworm of the sea."

Position in Animal Kingdom

1. Annelids belong to the **protostome** branch of the animal kingdom and have **spiral cleavage** and **mosaic development,** characters shared with and indicating relationship to the molluscs and primitive arthropods.
2. Annelids, together with the molluscs and arthropods, form a sister group with the flatworms.
3. Annelids as a group show a metamerism with comparatively few differences between the different somites.
4. Characters shared with arthropods include an outer secreted cuticle and a similar nervous system.

Biological Contributions

1. The introduction of **metamerism** by the group represents the greatest innovation in this phylum. An homologous but more highly specialized metamerism is seen in the arthropods.
2. A true coelomic cavity reaches a high stage of development in this group.
3. Specialization of the head region into differentiated organs, such as the tentacles, palps, and eyespots of the polychaetes, is carried further in some annelids than in other invertebrates so far considered.
4. There are modifications of the **nervous system,** with cerebral ganglia (brain), two closely fused ventral nerve cords with giant fibers running the length of the body, and various ganglia with their lateral branches.
5. The circulatory system is much more complex than any we have so far considered. It is a closed system with muscular blood vessels and aortic arches ("hearts") for propelling the blood.
6. The appearance of the fleshy **parapodia,** with their respiratory and locomotor functions, introduces a suggestion of the paired appendages and specialized gills found in the more highly organized arthropods.
7. The well-developed **nephridia** in most of the somites have reached a differentiation that involves removal of waste from the blood as well as from the coelom.
8. Annelids are the most highly organized animals capable of complete regeneration. However, this ability varies greatly within the group.

Characteristics of Phylum Annelida

1. Body **metamerically segmented;** symmetry bilateral
2. Body wall with outer circular and inner longitudinal muscle layers; outer transparent moist cuticle secreted by epithelium
3. **Chitinous setae,** often present on fleshy appendanges called **parapodia;** setae absent in leeches
4. **Coelom (schizocoel) well developed** and divided by septa, except in leeches; coelomic fluid supplies turgidity and functions as hydrostatic skeleton
5. Blood system closed and **segmentally arranged;** respiratory pigments (hemoglobin, hemerythrin, or chlorocruorin) often present; amebocytes in blood plasma
6. Digestive system complete and not metamerically arranged
7. Respiratory gas exchange through skin, gills, or parapodia
8. Excretory system typically a pair of nephridia for each metamere
9. Nervous system with a double ventral nerve cord and a pair of ganglia with lateral nerves in each metamere; brain, a pair of dorsal cerebral ganglia with connectives to cord
10. Sensory system of tactile organs, taste buds, statocysts (in some), photoreceptor cells, and eyes with lenses (in some)
11. Hermaphroditic or separate sexes; larvae, if present, are trochophore type; asexual reproduction by budding in some; spiral and mosaic cleavage; mesoderm from 4d blastomere (see Figure 3-11)

New medical uses for leeches (p. 191) have led to a revival of the market in blood-sucking leeches and establishment of "leech farms" where these organisms are raised in captivity.

Body Plan

The annelid body typically has a head or prostomium, a segmented body, and a terminal portion bearing the anus. New segments form just in front of the terminal portion; thus, the oldest segments are at the anterior end and the youngest segments are at the posterior end. Neither the prostomium nor the terminal portion is considered a metamere.

The body wall is made up of strong circular and longitudinal muscles adapted for swimming, crawling, and burrowing and is covered with epidermis and a thin, outer layer of nonchitinous cuticle (Figure 10-1).

In most annelids the coelom develops embryonically as a split in the mesoderm on each side of the gut (**schizocoel**), forming a pair of coelomic compartments in each segment. Each compartment is surrounded by **peritoneum** (a layer of mesodermal epithelium), which lines the body wall, forms dorsal and ventral **mesenteries** (double-membrane partitions that support the gut), and covers all the organs (Figure 10-1). Where peritonea of adjacent segments meet, **septa** are formed. These are perforated by the gut and longitudinal blood vessels. Not only is the coelom metamerically arranged, but practically every body system is affected in some way by this segmental arrangement.

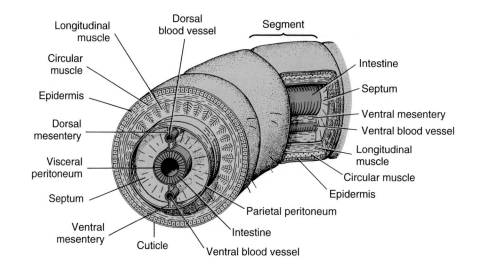

FIGURE 10-1

Annelid body plan.

Except in the leeches, the coelom is filled with fluid and serves as a **hydrostatic skeleton.** Because the volume of the fluid is essentially constant, contraction of the longitudinal body wall muscles causes the body to shorten and become larger in diameter, whereas contraction of the circular muscles causes it to lengthen and become thinner. Separation of the hydrostatic skeleton into a metameric series of coelomic cavities by septa increases its efficiency greatly because the force of local muscle contraction is not transferred throughout the length of the worm. Widening and elongation can occur in restricted areas. Crawling motions are effected by alternating waves of contraction by longitudinal and circular muscles passing down the body (**peristaltic contraction).**

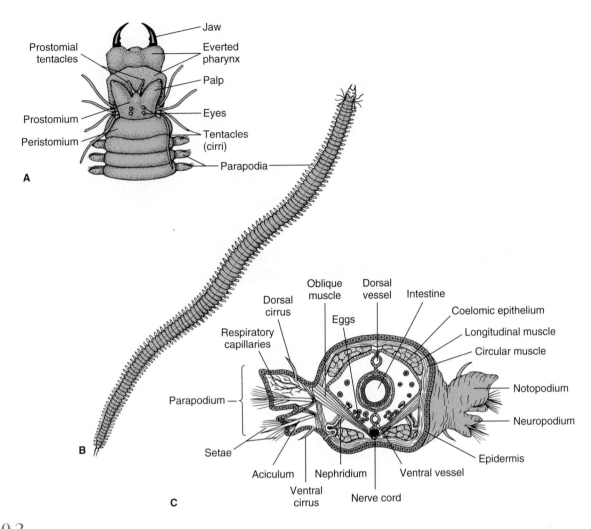

FIGURE 10-2

Nereis virens, a polychaete. **A,** Anterior end, with pharynx everted. **B,** External structure. **C,** Generalized transverse section through region of the intestine.

Segments in which longitudinal muscles contract, widen and anchor themselves against burrow walls or other substrata while other segments, in which circular muscles contract, elongate and stretch forward. Forces powerful enough for burrowing as well as locomotion can thus be generated. Swimming forms use undulatory rather than peristaltic movements in locomotion.

Class Polychaeta

The polychaetes (Gr. *polys,* many, + *chaitē,* long hair) are the largest class of annelids, with more than 10,000 described species, most of them marine. Although the majority of them are from 5 to 10 cm long, some are less than a millimeter, and others may be as long as 3 m. Some are brightly colored in reds and greens; others are dull or iridescent. Some are picturesque, such as the "featherduster" worms (Figure 10-3).

Polychaetes live under rocks, in coral crevices, or in abandoned shells, or they burrow into mud or sand; some build their own tubes on submerged objects or in bottom material; some adopt the tubes or homes of other animals; some are pelagic, making up a part of the planktonic population. They are extremely abundant in some areas; for example, a square meter of mudflat may contain thousands of polychaetes. They play a significant part in marine food chains, since they are eaten by fish, crustaceans, hydroids, and many others.

Polychaetes differ from other annelids in having a well-differentiated head with specialized sense organs; paired, paddlelike appendages (**parapodia**) on most segments; and no clitellum (Figure 10-2). As their name implies, they have many setae, usually arranged in bundles on the parapodia. They show a pronounced differentiation of some body somites and a specialization of sensory organs practically unknown among clitellates (oligochaetes and leeches).

A polychaete typically has a head, or **prostomium,** which may or may not be retractile and which often bears

A

B

FIGURE 10-3

Tube-dwelling sedentary polychaetes. **A,** The Christmas-tree worm, *Spirobranchus giganteus,* has a double crown of radioles and lives in a calcareous tube. **B,** Sabellid polychaetes, *Bispira brunnea,* live in leathery tubes.

eyes, antennae, and sensory palps (Figure 10-2). The first segment **(peristomium)** surrounds the mouth and may bear setae, palps, or, in predatory forms, chitinous jaws. Ciliary feeders may bear a tentacular crown that may be opened like a fan or withdrawn into the tube (Figure 10-3).

The trunk is segmented and most segments bear parapodia, which may have lobes, cirri, setae, and other parts on them (Figure 10-2C). Parapodia are composed of two main parts—a dorsal **notopodium** and ventral **neuropodium**—either of which may be prominent or reduced in a given species. The parapodia are used in crawling, swimming, or anchoring in tubes. They usually serve as the chief respiratory organs, although some polychaetes also may have gills. *Amphitrite,* for example, has three pairs of branched gills and long extensible tentacles (Figure 10-4). *Arenicola,* the lugworm (Figure 10-10), which burrows through the sand leaving characteristic castings at the entrance to its burrow, has paired gills on certain somites.

Sense organs are more highly developed in polychaetes than in oligochaetes and include eyes and statocysts. Eyes, when present, may range from simple eyespots to well-developed organs. Usually the eyes are retinal cups, with rod-like photoreceptor cells lining the cup wall and directed toward the lumen of the cup.

In contrast to clitellates, polychaetes have no permanent sex organs, possess no permanent ducts for their sex cells, and usually have separate sexes. Gonads appear as temporary swellings of the peritoneum and shed their gametes into the coelom. Gametes are carried outside through gonoducts, through nephridia, or by rupture of the body wall. Fertilization is external, and development is indirect with a trochophore larva.

Some polychaetes are free-moving pelagic forms, some are active burrowers and crawlers, and some are sedentary, living in tubes or burrows that they rarely (or never) leave. An example of an active, predatory worm is *Nereis* (Greek mythology, a Nereid, or daughter of Nereus, ancient sea god), the clam worm (Figure 10-2). *Nereis* has a muscular, eversible pharynx equipped with jaws that can be thrust out with surprising speed and dexterity for capturing prey. Scale worms (Figure 10-5) often live as commensals with other invertebrates, and fireworms (Figure 10-6) feed on gorgonians and stony corals.

Most sedentary tube and burrow dwellers are particle feeders, using ciliary or mucoid methods of obtaining food. The principal food source is plankton and detritus. Some, like *Amphitrite* (Greek mythology, sea goddess) (Figure 10-4), with head peeping out of the mud, send out long extensible tentacles over the surface. Cilia and mucus on the tentacles entrap particles found on the sea bottom and move them toward the mouth (deposit feeding).

The fanworms, or "featherduster" worms, are beautiful tubeworms, fascinating to watch as they emerge from their secreted tubes and unfurl their lovely tentacular crowns to feed. A slight disturbance, sometimes even a passing shadow, causes them to duck quickly into the safety of the homes they have built. Food attracted to the feathery arms, or **radioles,** by ciliary action is trapped in mucus and is carried down ciliated food grooves to the mouth (Figure 10-7). Particles too large for the food grooves are carried along the margins and dropped off. Further sorting may occur near the mouth where only the small particles of food enter the mouth, and sand grains are stored in a sac to be used later in enlarging the tube.

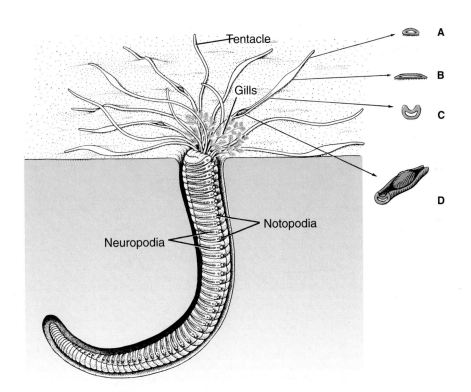

FIGURE 10-4

Amphitrite, which builds its tubes in mud or sand, extends long grooved tentacles out over the mud to pick up bits of organic matter. The smallest particles are moved along food grooves by cilia and larger particles by peristaltic movement. Its plumelike gills are blood red. **A,** Section through exploratory end of a tentacle. **B,** Section through a tentacle in an area adhering to substratum. **C,** Section showing ciliary groove. **D,** Particle being carried toward mouth.

Some polychaetes live most of the year as sexually unripe animals called atokes, *but during the breeding season a portion of the body develops into a sexually ripe worm called an* epitoke, *which is swollen with gametes (Figure 10-8). One example is the palolo worm, which lives in burrows among the coral reefs of the South Seas. During the reproductive cycle, the posterior somites become swollen with gametes. During the swarming period, which occurs at the beginning of the last quarter of the October-November moon, these epitokes break off and swim to the surface. Just before sunrise, the sea is literally covered with them, and at sunrise they burst, freeing the eggs and sperm for fertilization. The anterior portions of the worms regenerate new posterior sections. A related form swarms in the Atlantic in the third quarter of the June-July moon. Swarming is of great adaptive value because the synchronous maturation of all the epitokes ensures the maximum number of fertilized eggs. However, it is very hazardous; many types of predators have a feast. In the meantime, the atoke remains safe in its burrow to produce another epitoke at the next cycle!*

FIGURE 10-5

The scale worm *Hesperonoe adventor* normally lives as a commensal in the tubes of *Urechis* (phylum Echiura, p. 238).

FIGURE 10-6

The fireworm *Hermodice carunculata* feeds on gorgonians and stony corals. Its setae are like tiny glass fibers and serve to ward off predators.

Some worms, such as *Chaetopterus* (Gr. *chaitē,* hair or mane, + *pteron,* wing), secrete mucous filters through which they pump water to collect edible particles (Figure 10-9). The lugworm *Arenicola* (L. *arena,* sand, + *colere,* to inhabit) lives in an L-shaped burrow in which, by peristaltic movements, it causes water to flow. Food particles are filtered out by the sand at the front of its burrow, and it ingests the food-laden sand (Figure 10-10).

Tube dwellers secrete many types of tubes. Some are parchmentlike; some are firm, calcareous tubes attached to rocks or other surfaces (Figure 10-3A); and some are simply grains of sand or bits of shell or seaweed cemented together with mucous secretions. Many burrowers in sand and mud flats simply line their burrows with mucus (Figure 10-10).

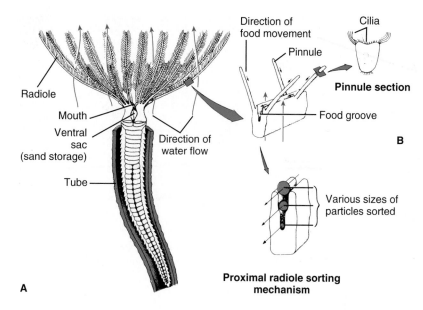

FIGURE 10-7

Sabella, a polychaete ciliary feeder. **A,** Anterior view of the crown. Cilia direct small food particles along grooved radioles to the mouth and discard larger particles. Sand grains are directed to storage sacs and later used in tube building. **B,** Distal portion of radiole showing ciliary tracts of pinnules and food grooves.

FIGURE 10-8

Eunice viridis, the Samoan palolo worm. The posterior somites make up the epitokal region, consisting of segments packed with gametes. Each segment has one eyespot on the ventral side. Once a year the worms swarm, and the epitokes detach, rise to the surface, and discharge their ripe gametes, leaving the water milky. By the next breeding season, the epitokes are regenerated.

Class Oligochaeta

The more than 3000 species of oligochaetes (Gr. *oligos,* few, + *chaitē,* long hair) are found in a great variety of sizes and habitats. They include the familiar earthworms and many species that live in fresh water. Most are terrestrial or freshwater forms, but some are parasitic, and a few live in marine or brackish water.

Oligochaetes, with few exceptions, bear setae, which may be long or short, straight or curved, blunt or needlelike, or arranged singly or in bundles. Whatever the type, they are less numerous in oligochaetes than in polychaetes. Aquatic forms usually have longer setae than do earthworms.

Earthworms

The most familiar of oligochaetes are earthworms ("night crawlers"), which burrow in moist, rich soil, emerging at night to feed on surface detritus and vegetation and to breed. In damp, rainy weather they stay near the surface, often with mouth or anus protruding from the burrow. In very dry weather they may burrow several feet underground, coil up in a slime chamber, and become dormant. *Lumbricus terrestris* (L. *lumbricum,* earthworm), the form commonly studied in school laboratories, is approximately 12 to 30 cm long (Figure 10-11). Giant tropical earthworms may have from 150 to 250 or more segments and may grow to as much as 3 m in length. They usually live in branched and interconnected tunnels.

Aristotle called earthworms the "intestines of the soil." Some 22 centuries later Charles Darwin published his observations in his classic The Formation of Vegetable Mould Through the Action of Worms. *He showed how worms enrich the soil by bringing subsoil to the surface and mixing it with the topsoil. An earthworm can ingest its own weight in soil every 24 hours, and Darwin estimated that from 10 to 18 tons of dry earth per acre pass through their intestines annually, thus bringing up potassium and phosphorus from the subsoil and also adding to the soil nitrogenous products from their own metabolism. They expose the mold to the air and sift it into small particles. They also drag leaves, twigs, and organic substances into their burrows, closer to the roots of plants. Their activities are important in aerating the soil. Darwin's views were at odds with his contemporaries, who thought earthworms were harmful to plants. But recent research has amply confirmed Darwin's findings, and earthworm management is now practiced in many countries.*

FIGURE 10-9

Chaetopterus, a sedentary polychaete, lives in a U-shaped tube in the sea bottom. It pumps water through the parchment-like tube (of which one-half has been cut away here) with its three pistonlike fans. The fans beat 60 times per minute to keep water currents moving. The winglike notopodia of the twelfth segment continuously secrete a mucous net that strains out food particles. As the net fills with food, the food cup rolls it into a ball, and when the ball is large enough (about 3 mm), the food cup bends forward and deposits the ball in a ciliated groove to be carried to the mouth and swallowed.

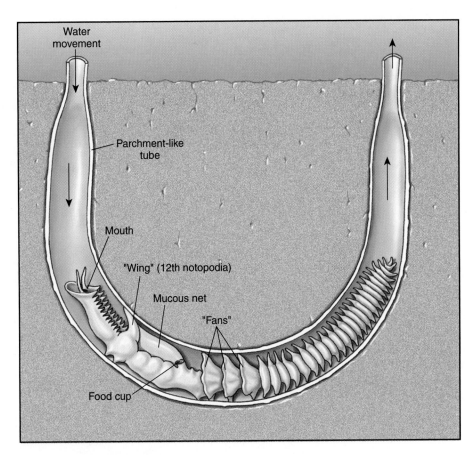

Form and Function

In earthworms a fleshy prostomium overhangs the mouth at the anterior end, and the anus is on the terminal end (Figure 10-11B). In most earthworms, each segment bears four pairs of chitinous setae (Figure 10-11C), although in some oligochaetes each segment may have up to 100 or more setae. Each seta is a bristlelike rod set in a sac within the body wall and moved by tiny muscles (Figure 10-12). The setae project through small pores in the cuticle to the outside. In locomotion and burrowing, setae anchor parts of the body to prevent slipping. Earthworms move by peristaltic movement. Contraction of circular muscles in the anterior end lengthens the body, thus pushing the anterior end forward where it is anchored by setae; contractions of longitudinal muscles then shorten the body, pulling the posterior end forward. As these waves of contraction pass along the entire body, it gradually is moved forward.

As in other annelids, the digestive tract is unsegmented and extends the length of the worm. The food of the earthworm is mainly decayed organic matter and bits of vegetation drawn in by the muscular **pharynx** (Figure 10-11A). **Chloragogen tissue** is found in the typhlosole and around the intestine. Chloragogen cells synthesize glycogen and fat and can break free to distribute these nutrients through the coelom. Chloragogen cells also serve an excretory function.

Annelids have a double transport system—the coelomic fluid and circulatory system. Food, wastes, and respir-

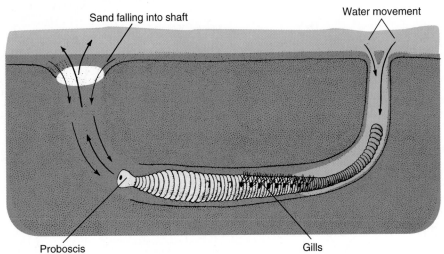

FIGURE 10-10

Arenicola, the lugworm, lives in an L-shaped burrow in intertidal mudflats. It burrows by successive eversions and retractions of its proboscis. By peristaltic movements it keeps water coming in the open end of the tube and filtering through the sand at the head end. The worm then ingests the food-laden sand.

atory gases are carried by both in varying degrees. The blood is in a closed system of blood vessels, including capillary systems in the tissues. There are five main longitudinal trunks, of which the **dorsal blood vessel** is the main pumping organ (Figure 10-11A and C). The blood contains colorless ameboid cells and a dissolved respiratory pigment, **hemoglobin.** The

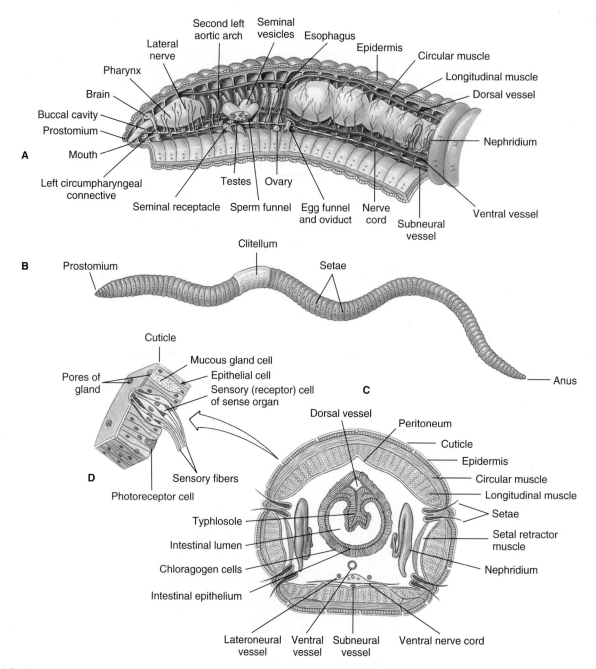

FIGURE 10-11

Earthworm anatomy. **A,** Internal structure of anterior portion of worm. **B,** External features, lateral view. **C,** Generalized transverse section through region posterior to clitellum. **D,** Portion of epidermis showing sensory, glandular, and epithelial cells.

blood of other annelids may have respiratory pigments other than hemoglobin.

The organs of excretion are the **nephridia,** a pair of which is found in each somite except the first three and the last one. Each one occupies parts of two successive somites (Figure 10-13). A ciliated funnel, known as the **nephrostome,** lies just anterior to an intersegmental septum and leads by a small ciliated tubule through the septum into the somite behind, where it connects with the main part of the nephridium. This part of the nephridium is made up of sev-

eral loops of increasing size, which finally terminate in a bladderlike structure leading to an aperture, the **nephridiopore,** which opens to the outside near the ventral row of setae. Cilia draw wastes from the coelom into the nephrostome and tubule. In the tubule, water and salts are resorbed, forming a dilute urine that discharges to the outside through the nephridiopore.

The nervous system in oligochaetes is typical of all annelids. There is a pair of **cerebral ganglia** (brain) above the pharynx joined to the ventral nerve cord by a pair of connec-

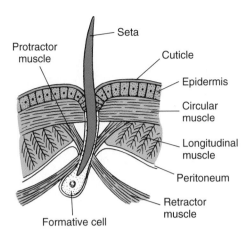

FIGURE 10-12

Seta with its muscle attachments showing relation to adjacent structures. Setae lost by wear and tear are replaced by new ones, which develop from formative cells.

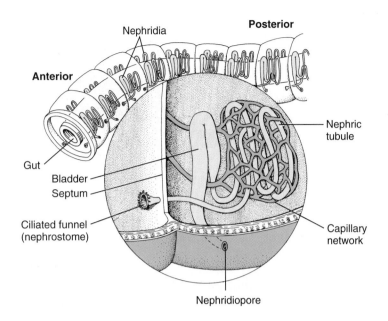

FIGURE 10-13

Nephridium of earthworm. Wastes are drawn into the ciliated nephrostome in one segment, then passed through the loops of the nephridium, and are expelled through the nephridiopore of the next segment.

tives around the pharynx (Figure 10-11 A). The double nerve cord has a pair of ganglia in each somite, giving off segmental nerves containing both sensory and motor fibers. For rapid escape movements most annelids are provided with one or more very large axons commonly called **giant axons,** or giant fibers, in the ventral nerve cord. The speed of conduction in these giant nerve fibers is much greater than that in small axons.

> *In the dorsal median giant fiber of* Lumbricus, *which is 90 to 160 μm in diameter, the speed of conduction has been estimated at 20 to 45 m/second, several times faster than in ordinary neurons of this species. This is also much faster than in polychaete giant fibers, probably because in the earthworms the giant fibers are enclosed in myelinated sheaths. The speed of conduction may be altered by changes in temperature.*

FIGURE 10-14

Two earthworms in copulation. Their anterior ends point in opposite directions as their ventral surfaces are held together by mucous bands secreted by the clitella. Mutual insemination occurs during copulation. After separation each worm secretes a cocoon to receive its eggs and sperm.

Earthworms are hermaphroditic and exchange sperm during copulation, which usually occurs at night. When mating, the worms extend their anterior ends from their burrows and bring their ventral surfaces together (Figure 10-14). They are held together by mucus secreted by the **clitellum** and by special ventral setae, which penetrate each other's bodies in the regions of contact. Sperm are discharged and travel to the seminal receptacles of the other worm in its seminal grooves. After copulation each worm secretes around its clitellum, first a mucous tube and then a tough, chitinlike band that forms a **cocoon** (Figure 10-15). As the cocoon passes forward, eggs from the oviducts, albumin

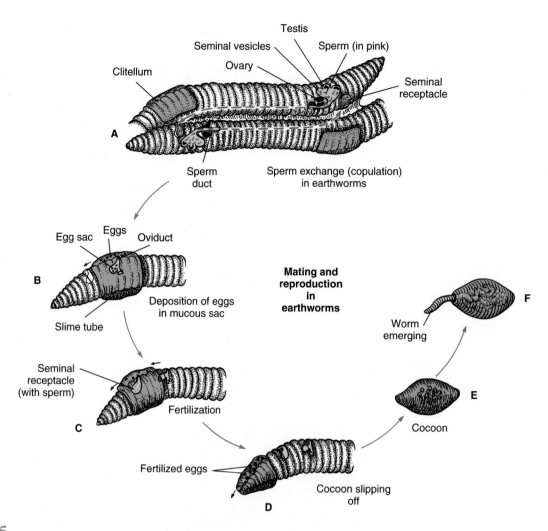

FIGURE **10-15**

Earthworm copulation and formation of egg cocoons. **A,** Mutual insemination occurs during copulation; sperm from the genital pore (somite 15) pass along seminal grooves to seminal receptacles (somites 9 and 10) of each mate. **B** and **C,** After worms separate, a slime tube formed over the clitellum passes forward to receive eggs from oviducts and sperm from seminal receptacles. **D,** As the cocoon slips off over the anterior end, its ends close and seal. **E,** The cocoon is deposited near the burrow entrance. **F,** Young worms emerge in two to three weeks.

from the skin glands, and sperm from the mate (stored in the seminal receptacles) are poured into it. Fertilization of the eggs now takes place within the cocoon. When the cocoon leaves the worm, its ends close, producing a lemon-shaped body. Embryonic development occurs within the cocoon, and the form that hatches from the egg is a young worm similar to the adult. It does not develop a clitellum until it is sexually mature.

Freshwater Oligochaetes

Freshwater oligochaetes usually are smaller and have more conspicuous setae than do the earthworms. They are more mobile than earthworms and tend to have better developed sense organs. They are generally benthic forms that creep about on the bottom or burrow into the soft mud. Aquatic oligochaetes provide an important food source for fishes. A few are ectoparasitic.

Some aquatic forms have **gills.** In some the gills are long, slender projections from the body surface. Others have ciliated posterior gills (Figure 10-16D), which they extend from their tubes and use to keep the water moving. Most forms respire through the skin as do the earthworms.

The chief foods are algae and detritus, which worms may pick up by extending a mucus-coated pharynx. Burrowers swallow mud and digest the organic material. Some, such as *Aeolosoma,* are ciliary feeders that use currents produced by cilia at the anterior end of the body to sweep food particles into the mouth (Figure 10-16B).

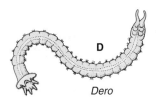
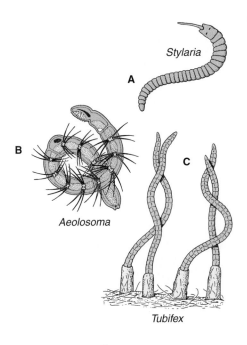

FIGURE 10-16

Some freshwater oligochaetes. **A,** *Stylaria* has the prostomium drawn out into a long snout. **B,** *Aeolosoma* uses cilia around the mouth to sweep in food particles, and it buds off new individuals asexually. **C,** *Tubifex* lives head down in long tubes. **D,** *Dero* has ciliated anal gills.

FIGURE 10-17

The world's largest leech, *Haementeria ghilianii,* on the arm of Dr. Roy K. Sawyer, who found it in French Guiana, South America.

Class Hirudinea

Leeches, numbering over 500 species, are found predominantly in freshwater habitats, but a few are marine, and some have even adapted to terrestrial life in moist, warm areas. Most leeches are between 2 and 6 cm in length, but some are smaller, and some reach 20 cm or more (Figure 10-17). They are found in a variety of patterns and colors—black, brown, red, or olive green. They are usually flattened dorsoventrally (Figure 10-18).

Form and Function

Leeches have a fixed number of segments, usually 34, and typically have both an anterior and a posterior sucker. They have no parapodia and, except in one genus (*Acanthobdella*), they have no setae. Another primitive character of *Acanthobdella* is its five anterior coelomic compartments separated by septa; septa have disappeared in all other leeches. The coelom has become filled with connective tissue and muscle, substantially reducing its effectiveness as a hydrostatic skeleton.

Many leeches live as carnivores on small invertebrates; some are temporary parasites, sucking blood from vertebrates; and some are permanent parasites, never leaving their host. Most leeches have a muscular, protrusible proboscis or a muscular pharynx with three jaws armed with teeth. They feed on the body juices of their prey, penetrating its surface with their proboscis or jaws and sucking the fluids with their powerful, muscular pharynx. Blood-sucking leeches (Figure 10-19) secrete an anticoagulant in their saliva. Predatory leeches feed frequently, but those that feed on blood of vertebrates consume large meals (up to several times their body weight) and digest the food slowly. The slow digestion of their meals results from the lack of secretion of amylases, lipase, or endopeptidases by their gut. In fact, they apparently depend mostly on bacteria in their gut for digestion of the blood meal.

For centuries the "medicinal leech" (Hirudo medicinalis) *was used for blood letting because of the mistaken idea that bodily disorders and fevers were caused by an excess of blood. The use of leeches has an advantage over the lancing of veins; the operation is painless because of anesthetic components in the leech's saliva. A 10 to 12 cm long leech can extend to a much greater length when distended with blood, and the amount of blood it can suck is considerable. Leech collecting and leech culture in ponds were practiced in Europe on a commercial scale during the nineteenth century. Wordsworth's poem "The Leech-Gatherer" was based on this use of the leech.*

Leeches are once again being used medicinally. When fingers or toes are severed, microsurgeons often can reconnect arteries but not the more delicate veins. Leeches are used to relieve congestion until the veins can grow back into the healing digit.

Leeches are hermaphroditic but practice cross-fertilization during copulation. Sperm are transferred by a penis or by hypodermic impregnation. Leeches have a clitellum, but it is evident only during the breeding season. After copulation, the clitellum secretes a cocoon that receives the eggs and sperm. Cocoons are buried in bottom mud, attached to submerged objects or, in terrestrial species, placed in damp soil. Development is similar to that of oligochaetes.

> *Leeches are highly sensitive to stimuli associated with the presence of a prey or host. They are attracted by and will attempt to attach to an object smeared with appropriate host substances, such as fish scales, oil secretions, or sweat. Those that feed on the blood of mammals are attracted by warmth, and the terrestrial haemadipsid leeches of the tropics will converge on a person standing in one place.*

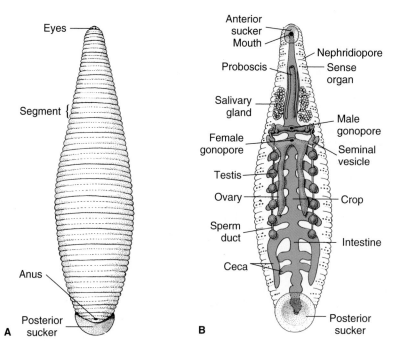

FIGURE 10-18

Structure of a leech, *Placobdella*. **A,** External appearance, dorsal view. **B,** Internal structure, ventral view.

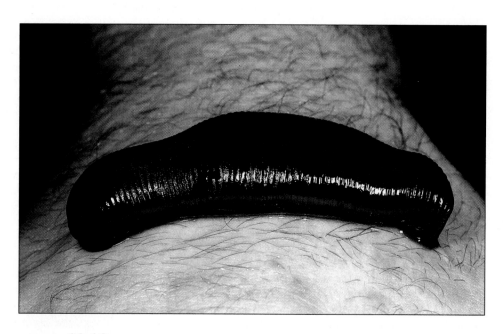

FIGURE 10-19

Hirudo medicinalis feeding on blood from a human arm.

Phylogeny and Adaptive Radiation

Phylogeny

There are so many similarities in the early development of the molluscs, annelids, and primitive arthropods that there seems little doubt about their close relationship. These three phyla apparently form a sister group to the flatworms. Many marine annelids and molluscs have an early embryogenesis typical of protostomes, shared with some marine flatworms, suggesting it is an ancestral trait. Annelids share with the arthropods an outer secreted cuticle and a similar nervous system, and there is a similarity between the lateral appendages (parapodia) of many marine annelids and the appendages of certain arthropods with other primitive characters. The most important resemblance, however, probably lies in the segmented plan of the annelid and the arthropod body structure.

What can we infer about the common ancestor of the annelids? This subject has been a long and continuing debate. Most hypotheses of annelid origin have assumed that metamerism arose in connection with the development of lateral appendages (parapodia) resembling those of the polychaetes. However, the oligochaete body is adapted to vagrant burrowing in the substratum with a peristaltic movement that is highly benefited by a metameric coelom. On the other hand, polychaetes

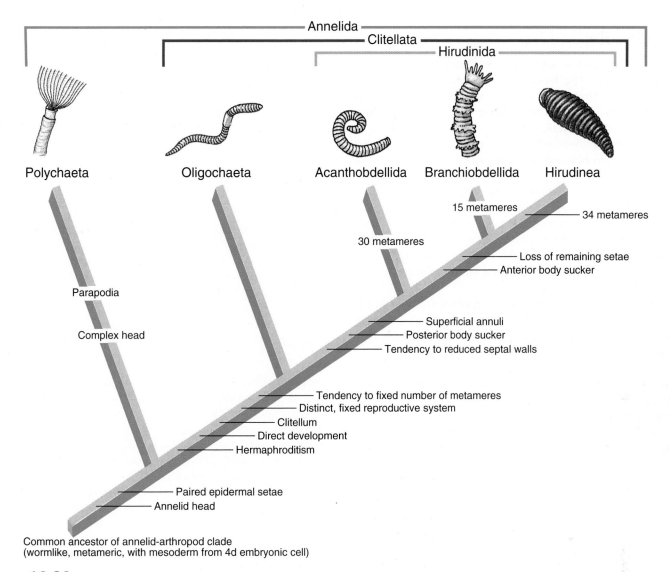

Annelida
Clitellata
Hirudinida

Polychaeta   Oligochaeta   Acanthobdellida   Branchiobdellida   Hirudinea

15 metameres — 34 metameres

30 metameres

Loss of remaining setae
Anterior body sucker

Parapodia

Complex head

Superficial annuli
Posterior body sucker
Tendency to reduced septal walls

Tendency to fixed number of metameres
Distinct, fixed reproductive system
Clitellum
Direct development
Hermaphroditism

Paired epidermal setae
Annelid head

Common ancestor of annelid-arthropod clade
(wormlike, metameric, with mesoderm from 4d embryonic cell)

FIGURE 10-20

Cladogram of the annelids, showing the appearance of shared derived characters that specify the five monophyletic groups. The Acanthobdellida and the Branchiobdellida are two small groups discussed briefly in the note on this page. Brusca and Brusca place both groups, together with the Hirudinea ("true" leeches), within a single taxon, the Hirudinida. This clade has several synapomorphies: tendency toward reduction of septal walls, the appearance of a posterior sucker, and the subdivision of body segments by superficial annuli. Note also that, according to this scheme, the Oligochaeta have no defining synapomorphies; that is, they are defined solely by primitive characters that they retain, and thus might be paraphyletic.

(Source: R. C. Brusca and G. J. Brusca, Invertebrates, *1990, Sinauer Associates, Sunderland, MA.)*

with well-developed parapodia are generally adapted to swimming and crawling in a medium too fluid for effective peristaltic locomotion. Although the parapodia do not prevent such locomotion, they do little to further it, and it seems likely that they evolved as an adaptation for swimming. Although the polychaetes are more primitive in some characters, such as in their reproductive system, some authorities have argued that the ancestral annelids were more similar to the oligochaetes in overall body plan and that those of the polychaetes and leeches are more evolutionarily derived. The leeches are closely related to the oligochaetes but have diverged from them in connection with a swimming existence and the abandonment of a burrowing mode of life. This relationship is indicated by the cladogram in Figure 10-20.

The Branchiobdellida, a group of small annelids that are parasitic or commensal on crayfish and show similarities to both oligochaetes and leeches, are here placed with the oligochaetes, but they are considered a separate class by some authorities. They have 14 or 15 segments and bear a head sucker.

One genus of leech, Acanthobdella, *has some characteristics of leeches and some of oligochaetes; it is sometimes separated from the other leeches into a special class, Acanthobdellida, that characteristically has 27 somites, setae on the first five segments, and no anterior sucker.*

Adaptive Radiation

Annelids are an ancient group that has undergone extensive adaptive radiation. The basic body structure, particularly of the polychaetes, lends itself to great modification. As marine worms, polychaetes have a wide range of habitats in an environment that is not physically or physiologically demanding. In contrast, the environment of earthworms imposes strict physical and physiological demands, which has limited their radiation.

A basic adaptive feature in the evolution of annelids is their septal arrangement, resulting in fluid-filled coelomic compartments. Fluid pressure in these compartments is used as a hydrostatic skeleton in precise movements such as burrowing and swimming. Powerful circular and longitudinal muscles have been adapted for flexing, shortening, and lengthening the body.

There is wide variation in feeding adaptations, from the sucking pharynx of the oligochaetes and the chitinous jaws of carnivorous polychaetes to the specialized tentacles and radioles of particle feeders.

In polychaetes the parapodia have been adapted in many ways and for a variety of functions, chiefly locomotion and respiration.

In leeches many adaptations, such as suckers, cutting jaws, pumping pharynx, distensible gut, and the secretion of anticoagulants, are related to their predatory and blood-sucking habits.

Classification of Phylum Annelida

Class Polychaeta (pol′e-ke′ta) (Gr. *polys,* many, +*chaitē,* long hair). Mostly marine; head distinct and bearing eyes and tentacles; most segments with parapodia (lateral appendages) bearing tufts of many setae; clitellum absent; sexes usually separate; gonads transitory; asexual budding in some; trochophore larva usually; mostly marine. Examples: *Nereis* (Figure 10-2), *Aphrodita, Glycera, Arenicola* (Figure 10-10), *Chaetopterus* (Figure 10-9), *Amphitrite* (Figure 10-4).

Class Oligochaeta (ol′i-go-ke′ta) (Gr. *oligos,* few, +*chaitē,* long hair). Body with conspicuous segmentation; number of segments variable; setae few per metamere; no parapodia; head absent; coelom spacious and usually divided by intersegmental septa; hermaphroditic; development direct, no larva; chiefly terrestrial and freshwater. Examples: *Lumbricus* (Figure 10-11), *Stylaria* (Figure 10-16A), *Aeolosoma* (Figure 10-16B), *Tubifex* (Figure 10-16C).

Class Hirudinea (hir′u-din′e-a) (L. *hirudo,* leech, + *-ea,* characterized by): **leeches.** Body with fixed number of segments (usually 34) with many annuli; body usually with anterior and posterior suckers; clitellum present; no parapodia; setae absent (except *Acanthobdella*); coelom closely packed with connective tissue and muscle; development direct; hermaphroditic; terrestrial, freshwater, and marine. Examples: *Hirudo, Placobdella* (Figure 10-18), *Macrobdella*.

Summary

The phylum Annelida is a large, cosmopolitan group containing the marine polychaetes, the earthworms and freshwater oligochaetes, and the leeches. Certainly the most important structural advancement of this group is metamerism, the division of the body into a series of similar segments, each of which contains a repeated arrangement of many organs and systems. The coelom is also highly developed in the annelids and has, together with the septal arrangement of fluid-filled compartments and a well-developed body-wall musculature, provided the annelids with an effective hydrostatic skeleton for precise burrowing and swimming. The primitive metamerism of the annelids lays the groundwork for the much more specialized metamerism of the arthropods to be considered in the next chapter.

Polychaetes are the largest class of annelids and are mostly marine. They have on each somite many setae, which are borne on paired parapodia. Parapodia show a wide variety of adaptations among polychaetes, including specialization for swimming, respiration, crawling, maintaining position in a burrow, pumping water through a burrow, and as accessory feeding organs. Some polychaetes are mostly predaceous and have an eversible pharynx with jaws. Other polychaetes rarely leave the burrows or tubes in which they live. Several styles of deposit and suspension feeding are shown among the members of this group. Polychaetes are dioecious with a primitive reproductive system and no clitellum. They practice external fertilization, and their larva is a trochophore.

The class Oligochaeta contains the earthworms and many freshwater forms; they have a small number of setae per segment (compared to the Polychaeta) and no parapodia. The circulatory system is closed, and the dorsal blood vessel is the main pumping organ. There is a pair of nephridia in most somites. Earthworms contain the typical annelid nervous system: dorsal cerebral ganglia connected to a double, ventral nerve cord with segmental ganglia, running the length of the worm. Oligochaetes are hermaphroditic and practice cross-fertilization. The clitellum plays an important role in reproduction, including secretion of mucus to surround the worms during copulation and secretion of a cocoon to receive the eggs and sperm and in which embryonic development occurs. A small, juvenile worm hatches from the cocoon.

The leeches (class Hirudinea) are mostly fresh water, although a few are marine and a few are terrestrial. They feed mostly on fluids; many are predators, some are temporary parasites, and a few are permanent parasites. The hermaphroditic leeches reproduce in a fashion similar to oligochaetes, with cross-fertilization and cocoon formation by the clitellum.

Embryological evidence supports a phylogenetic relationship of the annelids with the molluscs and arthropods.

Review Questions

1. Name the major characteristics that distinguish the phylum Annelida.

2. Distinguish among the classes of the phylum Annelida.

3. Describe the annelid body plan, including the body wall, segments, coelom and its compartments, and coelomic lining.

4. Explain how the hydrostatic skeleton of the annelids helps them to burrow. How is the efficiency for burrowing increased by metamerism?

5. Describe at least three ways that various polychaetes obtain food.

6. Define each of the following: prostomium, peristomium, radioles, parapodium, neuropodium, notopodium.

7. Explain the function of each of the following in earthworms: pharynx, calciferous glands, crop, gizzard, typhlosole, chloragogen tissue.

8. Describe the main features of each of the following in earthworms: circulatory system, nervous system, excretory system.

9. Describe the function of the clitellum and the cocoon.

10. How are freshwater oligochaetes generally different from earthworms?

11. Describe the ways in which leeches obtain food.

12. What are the main differences in reproduction and development among the three classes of annelids?

13. What was the evolutionary significance of metamerism and the coelom to its earliest possessors?

14. What are the phylogenetic relationships between the molluscs, annelids, and arthropods? What is the evidence for these relationships? What do you think is the most important synapomorphy shared by the annelids and arthropods and not the molluscs?

Selected References

See also general references on page 395.

Clark, R. B. 1964. Dynamics in metazoan evolution. The origin of the coelom and segments. Oxford, Clarendon Press. *An important treatise giving the author's hypotheses on the subject.*

Dales, R. P. 1967. Annelids, ed. 2. London, The Hutchinson Publishing Group, Ltd. *A concise account of the annelids.*

Kingman, J., and P. Kingman. 1993. The dance of the luminescent threadworms. Underwater Naturalist **22**(2):36. *Describes the spectacular swarming of the epitokes of Odontosyllis enopla off Belize during July and August, the third and fourth nights after a full moon.*

Lent, C. M., and M. H. Dickinson. 1988. The neurobiology of feeding in leeches.

Sci. Am. **258**:98-103 (June). *Feeding behavior in leeches is controlled by a single neurotransmitter (serotonin).*

Nicholls, J. C., and D. Van Essen. 1974. The nervous system of the leech. Sci. Am. **230**:38-48 (Jan.). *Because the leech has large nerve cells and only a few neurons perform a given function, its nervous system is particularly appropriate for experimental studies.*

Links to the Internet

Visit this textbook's web site at http://www.mhhe.com/zoology to find live Internet links for each of the references listed below.

1. Introduction to the Annelida. University of California at Berkeley Museum of Paleontology. This site provides an introduction to the annelids. It contains information on annelids in the fossil record and annelid life histories, systematics, and morphology, as well as many links.

2. Phylum Annelida. University of Michigan site describing annelids. Links to the three classes of annelids.

3. BIOSIS Annelida. A great deal of general information on annelids and specific information on the annelid classes can be found here.

4. Annelida. Arizona's Tree of Life Web Page. Pictures, characteristics, phylogenetic relationships, references on annelids.

Oligochaetes

5. FAQs about Earthworms. This site contains photographs of earthworms and cocoons, answers to frequently asked questions about earthworms, and a description of earthworm sampling techniques.

Polychaetes

6. Chaetozone. This site provides a polychaete newsletter and polychaete research news.

7. Introduction to the Polychaeta. University of California at Berkeley Museum of Paleontology. Links to the fossil record, life history and ecology, systematics, and more on morphology. Each site typically includes at least one picture of a member of the taxon.

Hirudineans

8. Leeches. This site provides interesting information on the uses of leeches in anticoagulant research. In addition, it answers many frequently asked questions about leeches.

9. Leeches. A diverse site focusing on leeches that includes the history of leeching, many pictures of leeches, and current research on this group.

Arthropods

CHAPTER eleven

A Winning Combination

T unis, Algeria—Treating it as an invading army, Tunisia, Algeria, and Morocco have mobilized to fight the most serious infestation of locusts in over 30 years. Billions of the insects have already caused extensive damage to crops and are threatening to inflict great harm to the delicate economies of North Africa.

Source: *New York Times,* April 20, 1988

The staggering losses occasionally inflicted by the billions of locusts in Africa serve as only one reminder of our ceaseless struggle with the dominant group of animals on earth today: the insects in the phylum Arthropoda. With nearly 1 million species recorded, and probably as many yet remaining to be classified, arthropods far outnumber all the other species of animals in the world combined. Numbers of individuals are equally enormous. Some scientists have estimated that there are 200 million insects for every single human alive today! Insects have an unmatched ability to adapt to all terrestrial environments and to virtually all climates, and crustaceans have radiated likewise in aquatic environments. Having originally evolved as land animals, insects developed wings and invaded the air 150 million years before flying reptiles, birds, or mammals. Many insects have exploited freshwater habitats, where they are now widely prevalent. In the sea the numbers of insects are more limited, but there are vast numbers of crustaceans in marine habitats.

How can we account for the enormous success of these creatures? Arthropods have a combination of valuable structural and physiological adaptations, including a versatile exoskeleton, metamerism, an efficient respiratory system, and highly developed sensory organs. In addition, many have a waterproofed cuticle and have extraordinary abilities to survive adverse environmental conditions. We will describe these adaptations and others in this chapter.

The phylum Arthropoda (arthrop'o-da) (Gr. *arthron*, joint, + *pous, podos*, foot) embraces the largest assemblage of living animals on earth. It includes the spiders, scorpions, ticks, mites, crustaceans, millipedes, centipedes, insects, and some smaller groups. In addition there is a rich fossil record extending back to the mid-Cambrian period (Figure 11-1).

Arthropods are eucoelomate protostomes with well-developed organ systems, and their cuticular exoskeleton containing chitin is a prominent characteristic. Like the annelids, they are conspicuously segmented; their primitive body pattern is a linear series of similar somites, each with a pair of jointed appendages. However, unlike the annelids, the arthropods have embellished the segmentation theme: variation occurs in the pattern of somites and appendages in the phylum. Often the somites are combined or fused into functional groups, called **tagmata,** for specialized purposes. Appendages, too, are frequently differentiated and specialized for walking, swimming, flying, or eating.

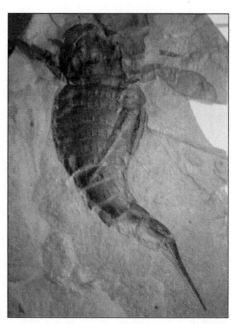

FIGURE 11-1

Fossils of early arthropods. **A,** Trilobite fossils, dorsal view. These animals were abundant in the mid-Cambrian period. **B,** Eurypterid fossil; eurypterids flourished in Europe and North America from Ordovician to Permian periods.

Few arthropods exceed 60 cm in length, and most are far below this size. The largest is a Japanese crab (*Macrocheira kaempferi*), which has approximately a 3.7 m span; the smallest is a parasitic mite, which is less than 0.1 mm long.

Arthropods are usually active, energetic animals. However we judge them, whether by their great diversity or their wide ecological distribution or their vast numbers of species, the answer is the same: they are the most abundant and diverse of all animals.

Although arthropods compete with us for food supplies and spread serious diseases, they are essential in pollination of many food plants, and they also serve as food, yield drugs and dyes, and create such products as silk, honey, and beeswax.

Ecological Relationships

Arthropods are found in all types of environment from low ocean depths to very high altitudes and from the tropics far into both north and south polar regions. Some species are adapted for life in the air; others for life on land or in fresh, brackish, and marine waters; others live in or on the bodies of plants and other animals. Some live in places where no other animal could survive.

Although all types—carnivorous, omnivorous, and herbivorous—occur in this vast group, the majority are herbivorous. Most aquatic arthropods depend on algae for their nourishment, and the majority of land forms live chiefly on

plants. There are many parasites. In diversity of ecological distribution the arthropods have no rivals.

Why Have Arthropods Achieved Such Great Diversity and Abundance?

Arthropods have achieved a great diversity, number of species, wide distribution, variety of habitats and feeding habits, and power of adaptation to changing conditions. These are some of the structural and physiological patterns that have been helpful to them:

1. **A versatile exoskeleton.** The arthropods possess an exoskeleton that is highly protective without sacrificing mobility. The skeleton is the **cuticle,** an outer covering secreted by the underlying epidermis.

The cuticle consists of an inner and thicker **procuticle** and an outer, relatively thin **epicuticle.** The procuticle is divided into the **exocuticle,** which is secreted before a molt, and **endocuticle,** which is secreted after molting. Both layers of the procuticle contain **chitin** bound with protein. Chitin is a tough, resistant, nitrogenous polysaccharide that is insoluble in water, alkalis, and weak acids. Thus the procuticle not only is flexible and lightweight but also affords protection, particularly against dehydration. In most crustaceans, the procuticle in some areas is also impregnated with **calcium salts,** which reduce its flexibility. In the hard shells of lobsters and crabs, for instance, this calcification is extreme. The outer

Characteristics of Phylum Arthropoda

1. Bilateral symmetry; metameric body, **tagmata** of head and trunk; head, thorax, and abdomen; or cephalothorax and abdomen

2. **Appendages jointed;** primitively, one pair to each somite (metamere), but number often reduced; appendages often modified for specialized functions

3. **Exoskeleton of cuticle** containing protein, lipid, chitin, and often calcium carbonate secreted by underlying epidermis and shed (molted) at intervals

4. Muscular system complex, with exoskeleton for attachment; striated muscles for rapid action; smooth muscles for visceral organs; **no cilia**

5. Coelom reduced; most of body cavity consisting of **hemocoel** (sinuses, or spaces, in the tissues) filled with blood

6. Complete digestive system; mouthparts modified from appendages and adapted for different methods of feeding

7. **Circulatory system open,** with dorsal contractile heart, arteries, and hemocoel

8. Respiration by body surface, gills, tracheae (air tubes), or book lungs

9. Paired excretory glands called coxal, antennal, or maxillary glands present in some, homologous to metameric nephridial system of annelids; some with other excretory organs, called Malpighian tubules

10. Nervous system of annelid plan, with dorsal brain connected by a ring around the gullet to a double nerve chain of ventral ganglia; fusion of ganglia in some species; well-developed sensory organs

11. Sexes usually separate, with paired reproductive organs and ducts; usually internal fertilization; oviparous or ovoviviparous; often with metamorphosis; parthenogenesis in a few forms; **growth with ecdysis**

epicuticle is composed of protein and often lipids. The protein is stabilized and hardened by a chemical process called tanning, adding further protection. Both the procuticle and the epicuticle are laminated, that is, composed of several layers each.

The cuticle may be soft and permeable or may form a veritable coat of armor. Between body segments and between the segments of appendages it is thin and flexible, permitting free movement of the joints. In crustaceans and insects the cuticle forms ingrowths for muscle attachment. It may also line the foregut and hindgut, line and support the trachea, and be adapted for a variety of purposes.

The nonexpansible cuticular exoskeleton does, however, impose important conditions on growth. To grow, an arthropod must shed its outer covering at intervals and grow a larger one—a process called **ecdysis, or molting.** Arthropods molt from four to seven times before reaching adulthood, and some continue to molt after that. Much of an arthropod's physiology centers on molting, particularly in young animals—preparation, molting itself, and then all the processes that must be completed in the postmolt period.

An exoskeleton is also relatively heavy and becomes proportionately heavier with increasing size. Weight of the exoskeleton tends to limit the ultimate body size.

2. **Segmentation and appendages for more efficient locomotion.** Typically each somite has a pair of jointed appendages, but this arrangement is often modified, with both segments and appendages specialized for adaptive functions. The limb segments are essentially hollow levers that are moved by muscles, most of which are striated for rapid action. The jointed appendages are equipped with sensory hairs and are variously modified for sensory functions, food handling, and swift and efficient walking or swimming.

3. **Air piped directly to cells.** Most land arthropods have a highly efficient tracheal system of air tubes, which delivers oxygen directly to tissues and cells and makes a high metabolic rate possible. Aquatic arthropods breathe mainly by some form of gill.

4. **Highly developed sensory organs.** Sensory organs are found in great variety, from the compound (mosaic) eye to the senses of touch, smell, hearing, balancing, and chemical reception. Arthropods are keenly alert to what goes on in their environment.

5. **Complex behavior patterns.** Arthropods exceed most other invertebrates in complexity and organization of their activities. Innate (unlearned) behavior unquestionably controls much of what they do, but learning also plays an important part in the lives of many arthropods.

6. **Reduced competition through metamorphosis.** Many arthropods pass through metamorphic changes, including a larval form quite different from the adult in structure. The larval form is often adapted for eating a different kind of food from that of the adult and occupies a different space, resulting in less competition within a species.

Subphylum Trilobita

The trilobites (Figure 11-1A) probably had their beginnings a million or more years before the Cambrian period in which they flourished. They have been extinct some 200 million years, but were abundant during the Cambrian and Ordovician periods. Their name refers to the trilobed shape of the body, caused by a pair of longitudinal grooves. They were bottom dwellers, probably scavengers. Most of them could roll up like pill bugs.

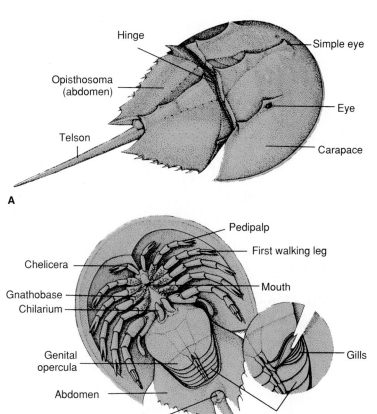

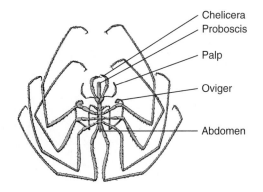

FIGURE 11-3

Pycnogonid, *Nymphon* sp. In this genus all the anterior appendages (chelicerae, palps, and ovigers) are present in both sexes, although ovigers are often not present in females of other genera.

had some resemblances to the marine horseshoe crabs (Figure 11-2) and to the scorpions, their terrestrial counterparts.

Subclass Xiphosurida: Horseshoe Crabs

The xiphosurids are an ancient marine group that dates from the Cambrian period. There are only three genera (five species) living today. *Limulus* (L. *limus*, sidelong, askew) (Figure 11-2), which lives in shallow water along the North American Atlantic coast, goes back practically unchanged to the Triassic period. Horseshoe crabs have an unsegmented, horseshoe-shaped **carapace** (hard dorsal shield) and a broad abdomen, which has a long spinelike **telson,** or tailpiece. On some of the abdominal appendages **book gills** (flat leaflike gills) are exposed. Horseshoe crabs can swim awkwardly by means of the abdominal plates and can walk on their walking legs. They feed at night on worms and small molluscs and are harmless to humans.

Class Pycnogonida: Sea Spiders

Pycnogonids are curious little marine animals that are much more common than most of us realize. They stalk about on their four pairs of long, thin walking legs, sucking juices from hydroids and soft-bodied animals with their large suctorial proboscis (Figure 11-3). They often have a pair of ovigerous legs **(ovigers)** with which males carry the egg masses. Their odd appearance is enhanced by the much reduced abdomen attached to the elongated cephalothorax. Most are only a few millimeters long, although some are much larger. They are common in all oceans.

Class Arachnida

Arachnids (Gr. *arachnē,* spider) are a numerous and diverse group, with over 50,000 species described so far.

FIGURE 11-2

A, Dorsal view of horseshoe crab *Limulus* (class Merostomata). They grow to 0.5 m in length. **B,** Ventral view.

Subphylum Chelicerata

Chelicerate arthropods are a very ancient group that includes the eurypterids (extinct), horseshoe crabs, spiders, ticks and mites, scorpions, sea spiders, and others. They are characterized by having six pairs of appendages that include a pair of **chelicerae,** a pair of **pedipalps,** and **four pairs of walking legs** (a pair of chelicerae and five pairs of walking legs in horseshoe crabs). They have **no mandibles** and **no antennae.** Most chelicerates suck liquid food from their prey.

Class Merostomata

Subclass Eurypterida

The eurypterids, or giant water scorpions (Figure 11-1B), lived 200 to 500 million years ago and some were perhaps the largest of all arthropods, reaching a length of 3 m. They

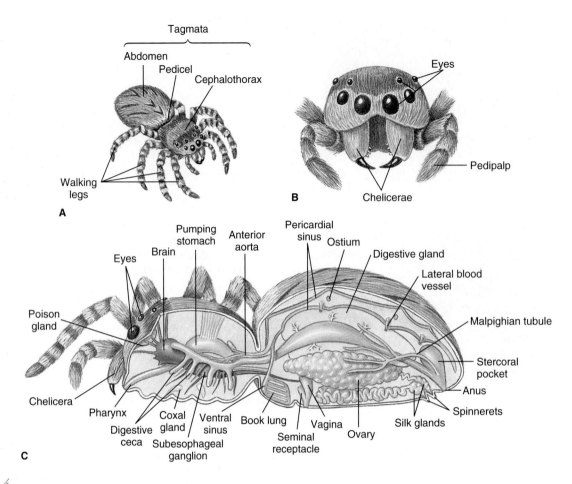

FIGURE 11-4

A, External anatomy of a jumping spider. **B,** Anterior view of head. **C,** Internal anatomy of a spider.

They include the spiders, scorpions, pseudoscorpions, whip scorpions, ticks, mites, harvestmen (daddy longlegs), and others. The arachnid tagmata are a cephalothorax and an abdomen.

Order Araneae: Spiders

The spiders are a large group of 35,000 recognized species, distributed all over the world. The cephalothorax and abdomen show no external segmentation, and the tagmata are joined by a narrow, waistlike **pedicel** (Figure 11-4).

All spiders are predaceous and feed largely on insects (Figure 11-5). Their chelicerae function as fangs and bear ducts from their poison glands, with which they effectively dispatch their prey. Some spiders chase their prey, others ambush them, and many trap them in a net of silk. After the spider seizes its prey with its chelicerae and injects venom, it liquefies the tissues with a digestive fluid and sucks up the resulting broth into the stomach. Spiders with teeth at the bases of the chelicerae crush or chew up the prey, aiding digestion by enzymes from the mouth. Many spiders provision their young with previously captured prey.

Spiders breathe by means of **book lungs** or **tracheae** or both. Book lungs, which are unique in spiders, consist of many parallel air pockets extending into a blood-filled chamber (Figure 11-4). Air enters the chamber by a slit in the body wall. The tracheae are a system of air tubes that carry air directly to the tissues from openings called **spiracles.** The tracheae are similar to those in insects (p. 219), but are much less extensive.

Spiders and insects have a unique excretory system of **Malpighian tubules** (Figure 11-4), which work in conjunction with specialized rectal glands. Potassium and other solutes and waste materials are secreted into the tubules, which drain the fluid, or "urine," into the intestine. The rectal glands reabsorb most of the potassium and water, leaving behind such wastes as uric acid. By this cycling of water and potassium, species living in dry environments conserve body fluids, producing a nearly dry mixture of urine and feces. Many spiders also have **coxal glands,** which are modified nephridia that open at the coxa, or base, of the first and third walking legs

Spiders usually have eight **simple eyes,** each provided with a lens, optic rods, and a retina (Figure 11-4B). Chiefly they perceive moving objects, but some, such as those of the hunting and jumping spiders, may form images. Because vision is usually poor, a spider's awareness of its environment depends especially on its hairlike **sensory setae.** Every seta on its surface is useful in communicating some information about the surroundings, air currents, or changing tensions in the spider's web. By sensing the vibrations of its web, the spider can judge the size and activity of its entangled prey or can receive the message tapped out on a silk thread by a prospective mate.

Web-Spinning Habits The ability to spin silk is an important factor in the lives of spiders, as it is in some other arachnids. Two or three pairs of spinnerets containing hundreds of microscopic tubes connect to special abdominal **silk glands** (Figure 11-4C). A protein secretion emitted as a liquid hardens on contact with air to form the silk thread. Spiders' silk threads are stronger than steel threads of the same diameter and are said to be second in tensional strength only to fused quartz fibers. The threads will stretch one-fifth of their length before breaking.

The spider web used for trapping insects is familiar to most people. The webs of some species consist merely of a few strands of silk radiating out from a spider's burrow or place of retreat. Other species spin beautiful, geometric orb webs. However, spiders use silk threads for many purposes besides web making. They use silk threads to line their nests; form sperm webs or egg sacs; build draglines; make bridge lines, warning threads, molting threads, attachment discs, or nursery webs; or to wrap up prey securely (Figure 11-6). Not all spiders spin webs for traps. Some, such as the wolf spiders, jumping spiders (Figure 11-5B), and fisher spiders (Figure 11-7), simply chase and catch their prey.

Reproduction Before mating, the male spins a small web, deposits a drop of sperm on it, and then picks the sperm up and stores it in special cavities of his pedipalps. When he mates, he inserts the pedipalps into the female genital opening to store the sperm in his mate's seminal receptacles. A courtship ritual usually precedes mating. The female lays her fertilized eggs in a silken cocoon, which she may carry about or may attach to a web or plant. A cocoon may contain hundreds of eggs, which hatch in approximately two weeks. The young usually remain in the egg sac for a few weeks and molt once before leaving it. Several molts occur before adulthood.

A **B**

FIGURE 11-5

A, A camouflaged crab spider, *Misumenoides* sp., awaits its insect prey. Its coloration matches the petals among which it lies, thus deceiving insects that visit the flowers in search of pollen or nectar. **B,** A jumping spider, *Eris aurantius.* This species has excellent vision and stalks an insect until it is close enough to leap with unerring precision, fixing its chelicerae into its prey.

FIGURE 11-6

A grasshopper, snared and helpless in the web of a golden garden spider *(Argiope aurantia),* is wrapped in silk while still alive. If the spider is not hungry, the prize is saved for a later meal.

Are Spiders Really Dangerous? It is truly amazing that such small and helpless creatures as spiders have generated so much unreasoned fear in the human heart. Spiders are timid creatures, which, rather than being dangerous enemies to humans, are actually allies in our continuing conflict with insects. The venom produced to kill prey is usually harmless to humans. Even the most poisonous spiders bite only when threatened or when defending their eggs or young. American

FIGURE 11-7

A fisher spider, *Dolomedes triton,* feeds on a minnow. This handsome spider feeds mostly on aquatic and terrestrial insects but occasionally captures small fishes and tadpoles. It pulls its paralyzed victim from the water, pumps in digestive enzymes, then sucks out the predigested contents.

tarantulas (Figure 11-8), despite their fearsome appearance, are *not* dangerous. They rarely bite, and their bite is not considered serious.

Two genera in the United States can give severe or even fatal bites: *Latrodectus* (L. *latro,* robber, + *dektes,* biter), and *Loxosceles* (Gr. *loxos,* crooked, + *skelos,* leg). The most important species are *Latrodectus mactans,* the **black widow,** and *Loxosceles reclusa,* the **brown recluse.** The black widow is moderate to small in size and shiny black, with a bright orange or red "hourglass" on the underside of the abdomen (Figure 11-9A). The venom is neurotoxic; that is, it acts on the nervous system. About four or five out of each 1000 bites reported were fatal.

The brown recluse, which is smaller than the black widow, is brown, and bears a violin-shaped dorsal stripe on its cephalothorax (Figure 11-9B). Its venom is hemolytic rather than neurotoxic, destroying tissues and skin surrounding the bite. Its bite can be mild to serious and occasionally fatal.

Some spiders in other parts of the world are dangerous, for example, the funnel-web spider *Atrax robustus* in Australia. Most dangerous of all are certain ctenid spiders in South America, for example, *Phoneutria fera.* In contrast to most spiders, these are quite aggressive.

Order Scorpionida: Scorpions

Although scorpions are more common in tropical and subtropical regions, some occur in temperate zones. Scorpions are generally secretive, hiding in burrows or under objects by day and feeding at night. They feed largely on insects and

FIGURE 11-8

A tarantula, *Rhechostica hentzi.*

spiders, which they seize with clawlike pedipalps and tear up with jawlike chelicerae.

The scorpion's body consists of a rather short cephalothorax, which bears the appendages and from one to six pairs of eyes and a clearly segmented abdomen. The abdomen is divided into a broader **preabdomen** and tail-like **postabdomen,** which ends in a stinging apparatus used to inject venom (Figure 11-10A). The venom of most species is not harmful to humans, although that of certain species of *Androctonus* in Africa and *Centruroides* in Mexico, Arizona, and New Mexico can be fatal unless antivenom is available.

Scorpions bear living young, which their mother carries on her back until after the first molt.

Order Opiliones: Harvestmen

Harvestmen, often known as "daddy longlegs," are common in the United States and other parts of the world (Figure 11-10B). These curious creatures are easily distinguished from spiders by the fact that their abdomen and cephalothorax are broadly joined, without constriction of the pedicel, and by the presence of external segmentation of the abdomen. They have four pairs of long, spindly legs, and without apparent ill effect, they can cast off one or more legs if they are grasped by a predator (or human hand). The ends of their chelicerae are pincerlike, and they feed much more as scavengers than do most arachnids.

Order Acari: Ticks and Mites

Acarines differ from all other arachnids in having their cephalothorax and abdomen completely fused, with no sign of external division or segmentation (Figure 11-11). Their mouthparts are carried on a little anterior projection, the

capitulum. They are found almost everywhere—in both fresh and salt water, on vegetation, on the ground, and parasitic on vertebrates and invertebrates. Over 25,000 species have been described, many of which are of importance to humans, but this is probably only a fraction of the species that exist.

Many species of mites are entirely free living. *Dermatophagoides farinae* (Gr. *dermatos,* skin, + *phago,* to eat, + *eidos,* likeness of form) (Figure 11-12) and related species are denizens of house dust all over the world, sometimes causing allergies and dermatoses. Some mites are marine, but most aquatic species are found in fresh water. They have long, hairlike setae on their legs for swimming, and their larvae may be parasitic on aquatic invertebrates. Such abundant organisms must be important ecologically, but many acarines have more direct effects on our food supply and health. The spider mites (family Tetranychidae) are serious agricultural pests on fruit trees, cotton, clover, and many other plants. Larvae of the genus *Trombicula* are called chiggers or redbugs. They feed on the dermal tissues of terrestrial vertebrates, including humans, and cause an irritating dermatitis; some species of chiggers transmit a disease called Asiatic scrub typhus. The hair-follicle mite, *Demodex* (Figure 11-13), is apparently nonpathogenic in humans; it infects most of us although we are unaware of it. Other species of *Demodex* and other genera of mites cause mange in domestic animals.

FIGURE 11-9

A, A black widow spider, *Latrodectus mactans,* suspended on her web. Note the orange "hourglass" on the ventral side of her abdomen. **B,** The brown recluse spider, *Loxosceles reclusa,* is a small venomous spider. Note the small violin-shaped marking on its cephalothorax. The venom is hemolytic and dangerous.

FIGURE 11-10

A, Scorpion (order Scorpionida) with young, which stay with the mother until the first molt. **B,** A harvestman (order Opiliones). Harvestmen run rapidly on their stiltlike legs. They are especially noticeable during the harvesting season, hence the common name.

The inflamed welt and intense itching that follows a chigger bite is not the result of the chigger burrowing into the skin, as is popularly believed. Rather the chigger bites through the skin with its chelicerae and injects a salivary secretion containing powerful enzymes that liquefy skin cells. Human skin responds defensively by forming a hardened tube that the larva uses as a sort of drinking straw and through which it gorges itself with host cells and fluid. Scratching usually removes the chigger but leaves the tube, which is a source of irritation for several days.

FIGURE 11-11

A wood tick, *Dermacentor variabilis* (order Acarina).

Ticks are usually larger than mites. They pierce the skin of vertebrates and suck blood until enormously distended;

FIGURE 11-12

Scanning electron microgaph of house dust mite, *Dermatophagoides farinae.*

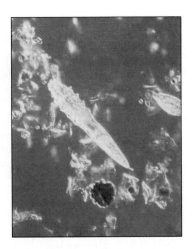

FIGURE 11-13

Demodex follicuorum, the human follicle mite. This tiny mite (100 to 400 μm) lives in follicles, particularly around the nose and eyes. Its prevalence ranges from about 20% in persons 20 years of age or younger to nearly 100% in the aged.

then they drop off and digest their meal. After molting, they are ready for another meal. In addition to disease conditions that they themselves cause, ticks are among the world's premier disease vectors, ranking second only to mosquitos. They carry a greater variety of infectious agents than any other arthropods; such agents include protozoan, rickettsial, viral, bacterial, and fungal organisms. Species of *Ixodes* carry the most common arthropod-borne infection in the United States, Lyme disease (see note below). Species of *Dermacentor* (Figure 11-11) and other ticks transmit Rocky Mountain spotted fever, a poorly named disease because most cases occur in the eastern United States. *Dermacentor* also transmits tularemia and the agents of several other diseases. Texas cattle fever, also called red-water fever, is caused by a protozoan parasite transmitted by the cattle tick *Boophilus annulatus.* Many more examples could be cited.

In the 1970s people in the town of Lyme, Connecticut, experienced an epidemic of arthritis. Subsequently known as Lyme disease, it is caused by a bacterium and carried by ticks of the genus Ixodes. *Now thousands of cases are reported each year in Europe and North America, and other cases have been reported from Japan, Australia, and South Africa. Many people bitten by infected ticks recover spontaneously or do not suffer any ill effects. Others, if not treated at an early stage, develop a chronic, disabling disease. Lyme disease is now the leading arthropod-borne disease in the United States.*

Subphylum Crustacea

Crustaceans traditionally have been included as a class in the subphylum Mandibulata, along with insects and myriapods. Members of all of these groups have, at least, a pair of antennae, mandibles, and maxillae on the head. Whether the Mandibulata constitutes a monophyletic grouping has been debated, and we discuss this question further on page 232.

The 30,000 or more species of Crustacea (L. *crusta,* shell) include lobsters, crayfishes, shrimp, crabs, water fleas, copepods, and barnacles. It is the only arthropod class that is primarily aquatic; they are mainly marine, but many freshwater and a few terrestrial species are known. The majority are free living, but many are sessile, commensal, or parasitic. Crustaceans are often very important components of aquatic ecosystems, and several have considerable economic importance.

Crustaceans are the only arthropods with **two pairs of antennae** (Figure 11-14). In addition to the antennae and **mandibles,** they have **two pairs of maxillae** on the head, followed by a pair of appendages on each body segment (although appendages on some somites are absent in some groups). All appendages, except perhaps the first antennae (antennules), are primitively **biramous** (two main branches), and at least some of the appendages of all present-day adults show that condition. Organs specialized for respiration, if present, are in the form of gills. Crustaceans lack Malpighian tubules.

Crustaceans primitively have 60 segments or more, but most tend to have between 16 and 20 somites and increased tagmatization. The major tagmata are the **head, thorax,** and **abdomen,** but these are not homologous throughout the subphylum (or even within some classes) because of varying

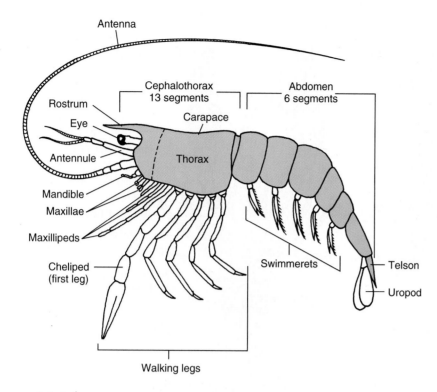

FIGURE 11-14

Archetypical plan of the Malacostraca. Note that the maxillae and maxillipeds have been separated diagrammatically to illustrate general plan. Typically in the living animal only the third maxilliped is visible externally. In the order Decapoda the carapace covers the cephalothorax, as shown here.

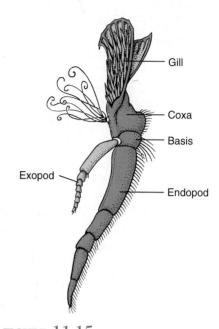

FIGURE 11-15

Parts of a biramous crustacean appendage (third maxilliped of a crayfish).

degrees of fusion of somites, for example, as in the cephalothorax.

In many crustaceans, the dorsal cuticle of the head extends posteriorly and around the sides of the animal to cover or fuse with some or all of the thoracic and abdominal somites. This covering is the **carapace.** In some groups the carapace forms clamshell-like valves that cover most or all of the body. In the decapods (including lobsters, shrimp, crabs, and others) the carapace covers the entire cephalothorax but not the abdomen.

Form and Function

Appendages

Some of the modifications of crustacean appendages may be illustrated by those of crayfishes and lobsters (class Malacostraca, order Decapoda, p. 212). The **swimmerets,** or abdominal appendages, retain the primitive biramous condition. Such an appendage consists of inner and outer branches, called the **endopod** and **exopod,** which are attached to one or more basal segments collectively called the **protopod** (Figure 11-15).

There are many modifications of this plan. In the primitive character state for crustaceans, all of the trunk appendages are rather similar in structure and adapted for swimming. The evolu-

tionary trend, shown in the crayfishes, has been toward reduction in number of appendages and toward a variety of modifications that fit them for many functions. Some are foliaceous (flat and leaflike), as are the maxillae; some are biramous, as are the swimmerets, maxillipeds, uropods, and antennae; some have lost one branch and are **uniramous,** as are the walking legs.

> *The terminology applied by various workers to crustacean appendages has not been blessed with uniformity. At least two systems are in wide use. Alternative terms to those we use, for example, are protopodite, exopodite, endopodite, basipodite, coxopodite, and epipodite. The first and second pairs of antennae may be called antennules and antennae, and the first and second maxillae are often called maxillules and maxillae. A rose by any other name....*

In crayfishes we find the first three pairs of thoracic appendages, called **maxillipeds,** serving along with the two pairs of maxillae as food handlers; the other five pairs of appendages are lengthened and strengthened for walking and defense (Figure 11-16). The first pair of walking legs, called

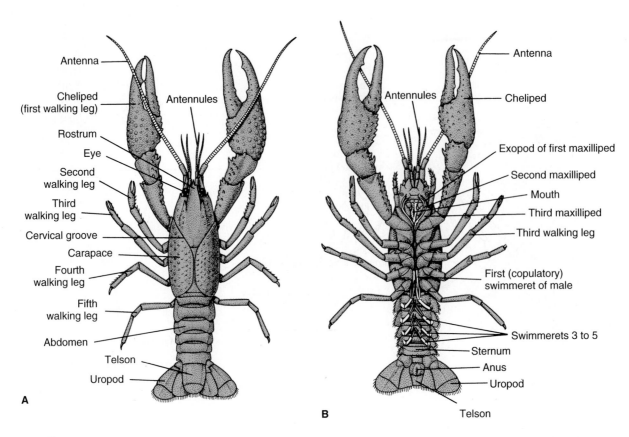

FIGURE 11-16

External structure of crayfishes. **A,** Dorsal view. **B,** Ventral view.

chelipeds, are enlarged with a strong claw, or chela, for defense. The abdominal swimmerets serve not only for locomotion, but in the male the first pair is modified for copulation, and in the female they all serve as a nursery for attached eggs and young. The last pair of appendages, the **uropods,** are wide and serve as paddles for swift backward movements, and, with the telson, they form a protective device for eggs or young on the swimmerets.

Ecdysis

The problem of growth despite a restrictive exoskeleton is solved in crustaceans, as in other arthropods, by ecdysis (Gr. *ekdysis,* to strip off), the periodic shedding of the old cuticle and the formation of a larger new one. Molting occurs most frequently during larval stages and less often as the animal reaches adulthood. Although the actual shedding of the cuticle is periodic, the molting process and the preparations for it, involving the storage of reserves and changes in the integument, are a continuous process going on during most of the animal's life.

During each **premolt** period the old cuticle becomes thinner as inorganic salts are withdrawn from it and stored in the tissues. Other reserves, both organic and inorganic, also accumulate and are stored. The underlying epidermis begins

to grow by cell division; it secretes first a new inner layer of epicuticle and then enzymes that digest away the inner layers of old endocuticle (Figure 11-17). Gradually a new cuticle forms inside the degenerating old one. Finally the actual ecdysis occurs as the old cuticle ruptures, usually along the mid-dorsal line, and the animal backs out (Figure 11-18). By taking in air or water the animal swells to stretch the new larger cuticle to its full size. During the **postmolt** period the cuticle thickens, the outer layer hardens by tanning, and the inner layer is strengthened as salvaged inorganic salts and other constituents are redeposited. Usually the animal is very secretive during the postmolt period when its defenseless condition makes it particularly vulnerable to predation.

That ecdysis is under hormonal control has been demonstrated in both crustaceans and insects, but the process is often initiated by a stimulus perceived by the central nervous system. The action of the stimulus in decapods is to decrease the production of a **molt-inhibiting hormone** from neurosecretory cells in the **X-organ** of the eyestalk. The sinus gland, also in the eyestalk, releases the hormone. When the level of molt-inhibiting hormone drops, the **Y-organs** near the mandibles produce **molting hormone.** This hormone initiates the processes leading to premolt. The Y-organs are homologous to the prothoracic glands of insects, which produce ecdysone.

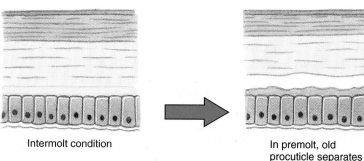

Intermolt condition

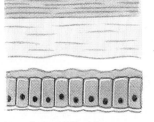

In premolt, old
procuticle separates
from epidermis, which
secretes new epicuticle

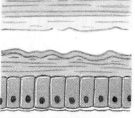

As new exocuticle is
secreted, molting fluid
dissolves old endocuticle,
and solution products
are reabsorbed

> *Neurosecretory cells are modified nerve cells that secrete hormones.*
> *They are widespread in invertebrates and also occur in vertebrates. Cells*
> *in the vertebrate hypothalamus and in the posterior pituitary are good*
> *examples.*

Other Endocrine Functions

Body color of crustaceans is largely a result of pigments in special branched cells **(chromatophores)** in the epidermis. The chromatophores change color by concentrating the pigment granules in the center of the cells, which causes a lightening effect, or by dispersing pigment throughout the cells, which causes darkening. Neurosecretory cells in the eyestalk control pigment behavior. Neurosecretory hormones also control pigment in the eyes for light and dark adaptation, and other neurosecretory hormones control the rate and amplitude of the heartbeat.

Androgenic glands, which are not neurosecretory, occur in male malacostracans, and their secretion stimulates expression of male sexual characteristics. If androgenic glands are artificially implanted in a female, her ovaries transform to testes and begin to produce sperm, and her appendages begin to take on male characteristics at the next molt.

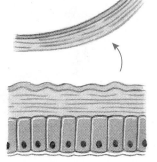

At ecdysis, the
old epicuticle and
exocuticle are discarded

Feeding Habits

Feeding habits and adaptations for feeding vary greatly among crustaceans. Many forms can shift from one type of feeding to another depending on environment and food availability, but fundamentally the same set of mouthparts is used by all. The mandibles and maxillae are involved in the actual ingestion; maxillipeds hold and crush food. In predators the walking legs, particularly the chelipeds, serve in food capture.

Many crustaceans, both large and small, are predatory, and some have interesting adaptations for killing their prey. One shrimplike form, *Lygiosquilla,* has on one of its walking legs a specialized digit that can be drawn into a groove and released suddenly to pierce a passing prey. The pistol shrimp *Alpheus* has one enormously enlarged chela that can be cocked like the hammer of a gun and snapped with a force that stuns its prey.

The food of crustaceans ranges from plankton, detritus, and bacteria, used by **suspension feeders,** to larvae, worms, crustaceans, snails, and fishes, used by **predators,** and dead animal and plant matter, used by **scavengers.** Suspension feeders, such as fairy shrimps, water fleas, and barnacles, use their legs, which bear a thick fringe of setae, to create water currents that sweep food particles through the setae. The mud shrimp *Upogebia* uses long setae on its first two pairs of thoracic

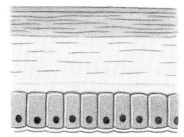

In postmolt, new
cuticle is stretched and
unfolded, and endocuticle
is secreted

FIGURE 11-17

Cuticle secretion and reabsorption in ecdysis.

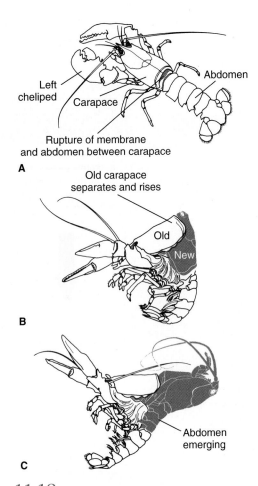

Left
cheliped
Carapace
Abdomen
Rupture of membrane
and abdomen between carapace

A

Old carapace
separates and rises

Old
New

B

Abdomen
emerging

C

FIGURE 11-18

Molting sequence in the northern lobster, *Homarus americanus.*
A, Membrane between the carapace and abdomen ruptures, and the
carapace begins a slow elevation. This step may take up to two
hours. **B** and **C,** Head, thorax, and finally abdomen are withdrawn.
This process usually takes no more than 15 minutes. Immediately
after ecdysis, the chelipeds are desiccated and the body is very soft.
The lobster now begins rapid absorption of water so that within 12
hours the body increases about 20% in length and 50% in weight.
Water will be replaced by living tissue in succeeding weeks.

appendages to strain food material from water circulated
through its burrow by movements of its swimmerets.

Crayfishes have a two-part stomach. The first contains a
gastric mill in which food, already torn up by the mandibles,
can be ground up further by three calcareous teeth into particles fine enough to pass through a filter of setae in the second
part of the stomach; the food particles then pass into the intestine for chemical digestion.

Respiration, Excretion, and Circulation

The **gills** of crustaceans vary in shape—treelike, leaflike, or
filamentous—all provided with blood vessels or sinuses.
They usually are attached to appendages and kept ventilated by
the movement of appendages in the water. The overlapping

carapace usually protects the **branchial chambers.** Some
smaller crustaceans breathe through the general body surface.

Excretory and **osmoregulatory** organs in crustaceans
are paired glands located in the head, with excretory pores
opening at the base of either the antennae or the maxillae,
thus **antennal glands** or **maxillary glands,** respectively
(Figure 11-19). The antennal glands of decapods are also called
green glands. They resemble the coxal glands of the chelicerates. The waste product is mostly ammonia with some
urea and uric acid. Some wastes diffuse through the gills as
well as through the excretory glands.

Circulation, as in other arthropods, is an **open system**
consisting of a heart, either compact or tubular, and arteries,
which transport blood to different areas of the hemocoel.
Some smaller crustaceans lack a heart. An open circulatory
system depends less on heartbeats for circulation because the
movement of organs and limbs circulates the blood more
effectively in open sinuses than in capillaries. The blood may
contain as respiratory pigments either hemocyanin or hemoglobin (hemocyanin in decapods), and it has the property of
clotting to prevent loss of blood in minor injuries.

Nervous and Sensory Systems

A cerebral ganglion above the esophagus sends nerves to the
anterior sense organs and connects to a subesophageal ganglion by a pair of connectives around the esophagus. A double
ventral nerve cord has a ganglion in each segment that sends
nerves to the viscera, appendages, and muscles (Figure 11-19).
Giant fiber systems are common among the crustaceans.

Sensory organs are well developed. There are two types of
eyes—the median, or nauplius, eye and compound eyes. The
median eye consists usually of a group of three pigment cups
containing retinal cells; the eye may or may not have a lens.
Median eyes are found in nauplius larvae and in some adult
forms, and they may be the only adult eye, as in copepods.

Most crustaceans have **compound eyes** similar to
insect eyes. In crabs and crayfishes they also are on the ends of
movable eyestalks (Figure 11-19). Compound eyes are precise
instruments, different from vertebrate eyes, yet especially
adept at detecting motion; they are able to analyze polarized
light. The convex corneal surface gives a wide visual field, particularly in the stalked eyes where the surface may cover an
arc of 200 degrees or more.

Compound eyes are composed of many tapering units
called **ommatidia** set close together (Figure 11-20). The facets,
or corneal surfaces, of the ommatidia give the surface of the
eye the appearance of a fine mosaic. Most crustacean eyes are
adapted either to bright or to dim light, depending on their
diurnal or nocturnal habits, but some are able, by means of
screening pigments, to adapt, to some extent at least, to both
bright and dim light. The number of ommatidia varies from a
dozen or two in some small crustaceans to 15,000 or more in a
large lobster. Some insects have approximately 30,000.

Other sensory organs include statocysts, tactile setae on
the cuticle of most of the body, and chemosensitive setae,
especially on the antennae, antennules, and mouthparts.

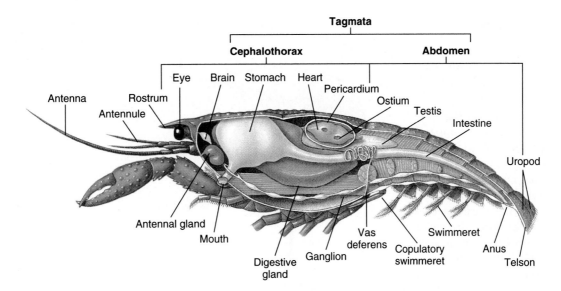

FIGURE 11-19

Internal structure of a male crayfish.

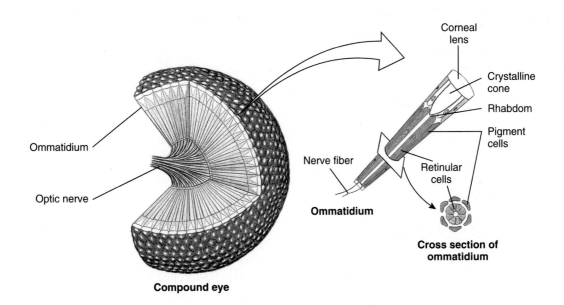

FIGURE 11-20

Compound eye of an insect. A single ommatidium is shown enlarged to the right.

Reproduction and Life Cycles

Most crustaceans have separate sexes, and numerous specializations for copulation occur among the different groups. Barnacles are monoecious but generally practice cross-fertilization. In some ostracods males are scarce, and reproduction is usually parthenogenetic. Most crustaceans brood their eggs in some manner—branchiopods and barnacles have special brood chambers, copepods have egg sacs attached to the sides of the abdomen (see Figure 11-24), and malacostracans usually carry eggs and young attached to their appendages.

The organism that hatches from an egg of a crayfish is a tiny juvenile with the same form as the adult and a complete set of appendages and somites. However, most crustaceans produce larvae that must go through a series of changes, either gradual or abrupt over the course of the series of molts, to assume the adult form (metamorphosis). The primitive larva of the crustaceans is the **nauplius** (Figure 11-21). It has an unsegmented body, a frontal eye, and three pairs of appendages, representing the two pairs of antennae and the mandibles. The form of the developmental stages and postlarvae of different groups of Crustacea are varied and have special names.

Class Branchiopoda

Members of the class Branchiopoda (bran′kee-op′o-da) (Gr. *branchia*, gills, + *pous, podos,* foot) have several primitive characteristics. Four orders are recognized: **Anostraca** (fairy shrimp and brine shrimp), which lack a carapace; **Notostraca** (tadpole shrimp such as *Triops*), whose carapace forms a large dorsal shield covering most of the trunk somites; **Conchostraca** (clam shrimp such as *Lynceus*), whose carapace is bivalved and usually encloses the entire body; and **Cladocera** (water fleas such as *Daphnia,* Figure 11-22), with a carapace typically covering the entire body but not the head. Branchiopods have reduced first antennae and second maxillae. Their legs are flattened and leaflike **(phyllopodia)** and are the chief respiratory organs (hence, the name branchiopods). The legs also are used in suspension feeding in most branchiopods, and in groups other than the cladocerans, they are used for locomotion as well. The most important and diverse order is the Cladocera, which often forms a large segment of the freshwater zooplankton.

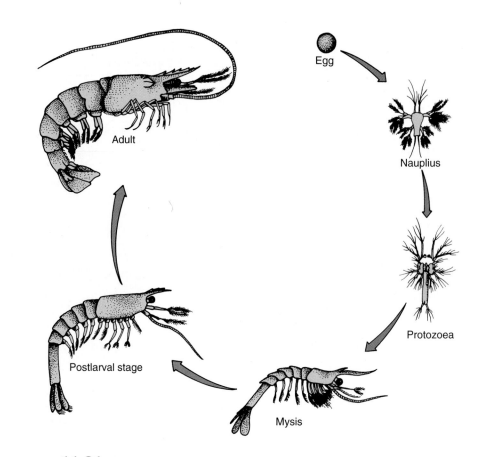

FIGURE 11-21

Life cycle of the Gulf shrimp *Penaeus.* Penaeids spawn at depths of 40 to 90 m. The young larval forms make up part of the plankton fauna and work their way inshore to water of lower salinity to develop as juveniles. Older shrimp return to deeper water offshore.

Class Maxillopoda

The class Maxillopoda includes a number of crustacean groups traditionally considered classes themselves. Specialists have recognized evidence that these groups descended from a common ancestor and thus form a clade within the Crustacea. They basically have five cephalic, six thoracic, and usually four abdominal somites plus a telson, but reductions are common. No typical appendages occur on the abdomen. The eye of the nauplius (when present) has a unique structure and is referred to as a **maxillopodan eye.**

Members of the subclass **Ostracoda** (os-trak′o-da) (Gr. *ostrakodes,* testaceous, that is, having a shell) are, like conchostracans, enclosed in a bivalved carapace and resemble tiny clams, 0.25 to 8 mm long (Figure 11-23). Ostracods show considerable fusion of trunk somites, and numbers of thoracic appendages are reduced to two or none.

The subclass **Copepoda** (ko-pep′o-da) (Gr. *kōpē,* oar, + *pous, podos,* foot) is an important group of Crustacea, second only to Malacostraca in number of species. Copepods are small (usually a few millimeters or less in length), rather elongate, tapering toward the posterior end, lacking a carapace, and retaining the simple, median, nauplius eye in the adult (Figure 11-24). They have four pairs of rather flattened, biramous, thoracic

swimming appendages, and a fifth, reduced pair. The abdomen bears no legs. Many symbiotic as well as free-living species are known. Many of the parasites are highly modified, and the adults may be so highly modified (and may depart so far from the description just given) that they can hardly be recognized as arthropods. Ecologically, free-living copepods are of extreme importance, often dominating the primary consumer level (herbivore) in aquatic communities.

The subclass **Branchiura** (bran-kee-u′ra) (Gr. *branchia,* gills, + *ura,* tail) is a small group of primarily fish parasites, which, despite its name, has no gills (Figure 11-25). Members of this group are usually between 5 and 10 mm long and may be found on marine or freshwater fish. They typically have a broad, shieldlike carapace, compound eyes, four biramous thoracic appendages for swimming, and a short, unsegmented abdomen. The second maxillae have become modified as suction cups.

The subclass **Cirripedia** (sir-i-ped′i-a) (L. *cirrus,* curl of hair, + *pes, pedis,* foot) includes barnacles, which are usually enclosed in a shell of calcareous plates, as well as three smaller orders of burrowing or parasitic forms. Barnacles are sessile as adults and may be attached to the substrate by a stalk (gooseneck barnacles) (Figure 11-26B) or directly (acorn barnacles)

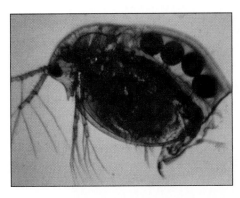

FIGURE **11-22**

A water flea, *Daphnia* (order Cladocera), photographed with polarized light. These microscopic forms occur in great numbers in northern lakes and are an important component of the food chain leading to fishes.

FIGURE **11-23**

An ostracod (subclass Ostracoda, class Maxillopoda).

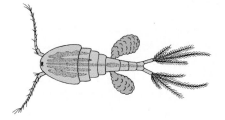

FIGURE **11-24**

A copepod with attached ovisacs (subclass Copepoda, class Maxillopoda).

FIGURE **11-25**

Fish louse (subclass Branchiura, class Maxillopoda).

A   **B**

FIGURE **11-26**

A, Acorn barnacles, *Semibalanus cariosus* (subclass Cirripedia) are found on rocks along the Pacific Coast of North America. **B,** Common gooseneck barnacles, *Lepas anatifera.* Note the feeding legs, or cirri, on *Lepas.* Barnacles attach themselves to a variety of firm substrates, including rocks, pilings, and boat bottoms.

(Figure 11-26A). Typically, the carapace (mantle) surrounds the body and secretes a shell of calcareous plates. The head is reduced, the abdomen absent, and the thoracic legs are long, many-jointed cirri with hairlike setae. The cirri are extended through an opening between the calcareous plates to filter from the water the small particles on which the animal feeds (Figure 11-26B).

> *Barnacles frequently foul ship bottoms by settling and growing there. So great may be their number that the speed of the ship may be reduced 30% to 40%, necessitating expensive drydocking of the ship to clean them off.*

Class Malacostraca

The class Malacostraca (mal′a-kos′tra-ka) (Gr. *malakos,* soft, + *ostrakon,* shell) is the largest class of Crustacea and shows great diversity. We will mention only 4 of its 12 to 13 orders. The trunk of malacostracans usually has eight thoracic and six abdominal somites, each with a pair of appendages. There are many marine and freshwater species.

The **Isopoda** (i-sop′o-da) (Gr. *isos,* equal, + *pous, podos,* foot) are commonly dorsoventrally flattened, lack a carapace, and have sessile compound eyes. Their abdominal appendages bear gills. Common land forms are sow bugs or pill bugs (*Porcellio* and *Armadillidium,* Figure 11-27A), which live under stones and in damp places. *Asellus* is common in fresh water, and *Ligia* is abundant on sea beaches and rocky shores. Some isopods are parasites of other crustaceans or of fish (Figure 11-28).

The **Amphipoda** (am-fip′o-da) (Gr. *amphis,* on both sides, + *pous, podos,* foot) resemble isopods in that the members have no carapace and have sessile compound eyes. However, they are usually compressed laterally, and their gills are in the thoracic position, as in other malacostracans. There are many marine amphipods (Figure 11-29), such as the beach flea, *Orchestia,* and numerous freshwater species.

A

FIGURE 11-27

A, Four pill bugs, *Armadillidium vulgare* (order Isopoda), common terrestrial forms. **B,** Freshwater sow bug, *Caecidotea* sp., an aquatic isopod.

B

FIGURE 11-28

An isopod parasite (*Anilocra* sp.) on a coney (*Cephalopholis fulvus*) inhabiting a Caribbean coral reef.

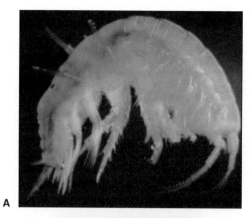

A

B

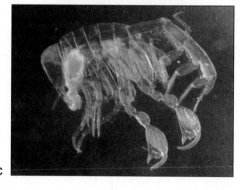

C

FIGURE 11-29

Marine amphipods. **A,** Free-swimming amphipod, *Anisogammarus* sp. **B,** Skeleton shrimp, *Caprella* sp., shown on a bryozoan colony, resemble praying mantids. **C,** *Phronima,* a marine pelagic amphipod, takes over the tunic of a salp (subphylum Urochordata, Chapter 14). Swimming by means of its abdominal swimmerets, which protude from the opening of the barrel-shaped tunic, the amphipod maneuvers to catch its prey.

The **Euphausiacea** (yu-faws′i-a′se-a) (Gr. *eu,* well, + *phausi,* shining bright, + *acea,* L. suffix, pertaining to) is a group of only about 90 species, but they are important as the oceanic plankton known as "krill." They are about 3 to 6 cm long (Figure 11-30) and commonly occur in great oceanic swarms, where they are eaten by baleen whales and many fishes.

The **Decapoda** (de-cap′o-da) (Gr. *deka,* ten, + *pous, podos,* foot) have five pairs of walking legs of which the first is often modified to form pincers **(chelae)** (Figures 11-14 and 11-16). These are the lobsters, crayfishes (see Figure 11-14), shrimps (Figure 11-21), and crabs, the largest of the crustaceans (Figure 11-31). True crabs differ from the others in having a broader carapace and a much reduced abdomen

(Figure 11-31A and C). Familiar examples are the fiddler crabs *Uca,* which burrow in sand just below the high-tide level (Figure 11-31C), decorator crabs, which cover their carapaces with sponges and sea anemones for camouflage, and spider crabs, such as *Libinia.* Hermit crabs (Figure 11-31B) have become adapted to live in snail shells; their abdomens, which lack a hard exoskeleton, are protected by the snail shell.

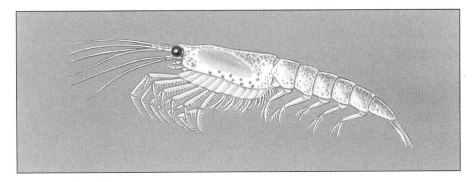

FIGURE 11-30

Meganyctiphanes, order Euphausiacea, the northern krill.

A B C

D E

FIGURE 11-31

Decapod crustaceans. **A,** The bright orange tropical rock crab, *Grapsus grapsus,* is a conspicuous exception to the rule that most crabs bear cryptic coloration. **B,** The hermit crab, *Elassochirus gilli,* which has a soft abdominal exoskeleton, lives in a snail shell that it carries about and into which it can withdraw for protection. **C,** The male fiddler crab, *Uca* sp., uses its enlarged cheliped to wave territorial displays and in threat and combat. **D,** The red night shrimp, *Rhynchocinetes rigens,* prowls caves and overhangs of coral reefs, but only at night. **E,** Spiny lobster *Panulirus argus.*

Subphylum Uniramia

Appendages of members of the Uniramia (yu′ni-ra′me-a) (L. *unus,* one, + *ramus,* a branch) are unbranched, as the name implies. The subphylum includes the insects and the myriapods. The term **myriapod** (Gr. *myrias,* a myriad, + *podos,* foot) refers to several classes that have evolved a pattern of two tagmata—head and trunk—with paired appendages on most or all trunk somites. Myriapods include the Chilopoda (centipedes), Diplopoda (millipedes), Pauropoda (pauropods), and Symphyla (symphylans).

Insects have evolved a pattern of three tagmata—head, thorax, and abdomen—with appendages on the head and thorax but greatly reduced or absent from the abdomen. The common ancestor of insects probably resembled myriapods in general body form.

The head of myriapods and insects resembles the crustacean head but has only **one pair of antennae,** instead of two. It also has **mandibles** and two pairs of **maxillae** (one pair of maxillae in millipedes). The legs are all **uniramous.**

Respiratory exchange is by body surface and tracheal systems, although juveniles, if aquatic, may have gills.

Class Chilopoda: Centipedes

Centipedes are active predators with a preference for moist places such as under logs or stones, where they feed on earthworms, insects, etc. Their bodies are somewhat flattened dorsoventrally, and they may contain from a few to 177 somites (Figure 11-32). Each somite, except the one behind the head and the last two, bears one pair of appendages. Those of the first body segment are modified to form poison claws, which they use to kill their prey. Most species are harmless to humans.

The head bears a pair of eyes, each consisting of a group of ocelli (simple eyes). Respiration is by tracheal tubes with a pair of spiracles in each somite. Sexes are separate, and all species are oviparous. The young are similar to the adult. The common house centipede *Scutigera,* with 15 pairs of legs, and *Scolopendra* (Figure 11-32), with 21 pairs of legs, are familiar genera.

Class Diplopoda: Millipedes

Diplopods, or "double-footed" arthropods, are commonly called millipedes, which literally means "thousand feet" (Figure 11-33). Although they do not have a thousand legs, they do have a great many. Their cylindrical bodies are made up of 25 to 100 segments. The four thoracic segments bear only one pair of

FIGURE 11-32

A, A centipede, *Scolopendra* (class Chilopoda) from the Amazon Basin, Peru. Most segments have one pair of appendages each. First segment bears a pair of poison claws, which in some species can inflict serious wounds. Centipedes are carnivorous. **B,** Head of centipede.

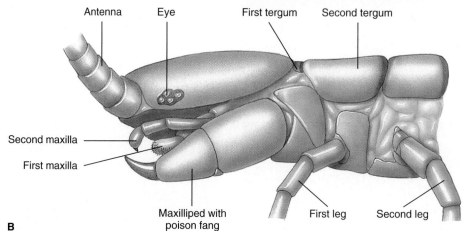

legs each, but the abdominal segments each have two pairs, a condition that may have evolved from fusion of somites. Two pairs of spiracles occur on each abdominal somite, each opening into an air chamber that gives off tracheal tubes.

Millipedes are less active than centipedes and are generally herbivorous, living on decayed plant and animal matter and sometimes living plants. They prefer dark moist places under stones and logs. The female lays her eggs in a nest and guards them carefully. The larval forms have only one pair of legs to a somite.

Class Insecta: Insects

Insects are the most numerous and diverse of all the groups of arthropods (Figure 11-34). There are more species of insects than species in all the other classes of animals combined. The number of insect species named has been estimated at close to 1 million, with thousands, perhaps millions, of other species yet to be discovered and classified.

It is difficult to appreciate fully the significance of this extensive group and its role in the biological pattern of animal life. The study of insects **(entomology)** occupies the time and resources of thousands of skilled men and women all over the world. The struggle between humans and their insect competitors seems to be an endless one, yet paradoxically, insects are so interwoven into the economy of nature in so many roles that we would have a difficult time without them.

Insects differ from other arthropods in having **three pairs of legs** and usually **two pairs of wings** on the thoracic

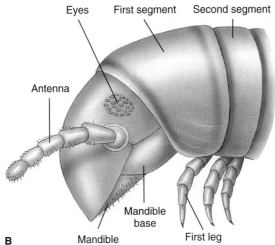

A

B

FIGURE 11-33

A, A tropical millipede with warning coloration. Note the typical doubling of appendages on most segments, hence diplosegments. **B,** Head of millipede.

region of the body (Figure 11-35), although some have one pair of wings or none. In size insects range from less than 1 mm to 20 cm in length, the majority being less than 2.5 cm long.

Distribution and Adaptability

Insects have spread into practically all habitats that can support life, but only a relatively few are marine. They are common in brackish water, in salt marshes, and on sandy beaches. They are abundant in fresh water, soils, forests, and plants, and they are found even in deserts and wastelands, on mountaintops, and as parasites in and on the bodies of plants and animals, including other insects.

Their wide distribution is made possible by their powers of flight and their highly adaptable nature. In many cases they can easily surmount barriers that are impassable to many other animals. Their small size and well-protected eggs allow them to be carried great distances by wind, water, and other animals.

The amazing adaptability of insects is evidenced by their wide distribution and enormous diversity of species. Such diversity enables this vigorous group to take advantage of all available resources of food and shelter.

Much of the success of insects is due to the adaptive qualities of the cuticular exoskeleton, as is the case in other arthropods. However, the great exploitation of terrestrial environments by insects has been made possible by the array of adaptations they possess to withstand its rigors. For example, their epicuticle has a waxy and a varnish layer and they can close their spiracles; both characteristics minimize evaporative water loss. They extract the utmost in fluid from food and fecal material, and many can retain the water produced in oxidative metabolism. Many can enter a resting stage (diapause) and lie dormant during inhospitable conditions.

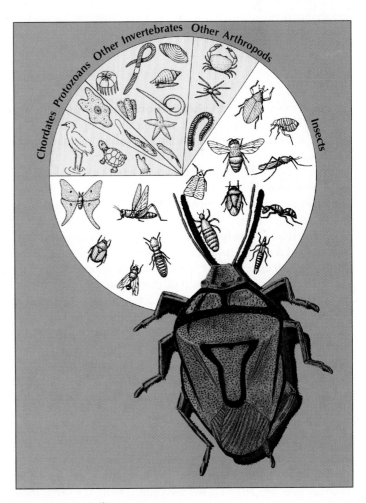

FIGURE 11-34

Pie diagram indicating relative numbers of species of insects to the rest of the animal kingdom and animal-like protistans.

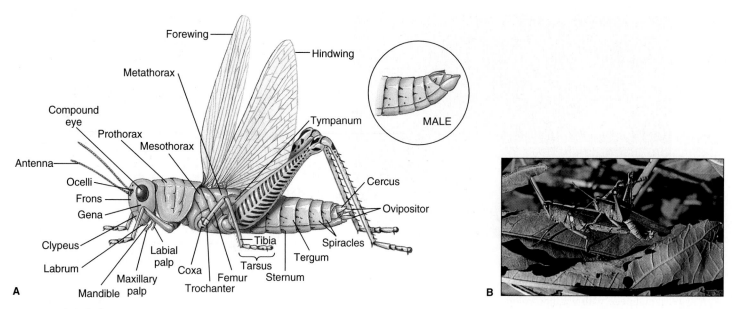

FIGURE 11-35

A, External features of a female grasshopper. The terminal segment of a male with external genitalia is shown in inset. **B,** A pair of grasshoppers, *Schistocerca obscura* (order Orthoptera), copulating. The African desert locust mentioned in the chapter prologue (p. 196) is *Schistocerca gregaria.*

External Features

The insect tagmata are the **head, thorax,** and **abdomen.** The cuticle of each body segment is typically composed of four plates **(sclerites),** a dorsal notum **(tergum),** a ventral **sternum,** and a pair of lateral **pleura.** The pleura of abdominal segments are membranous rather than sclerotized.

The head usually bears a pair of relatively large compound eyes, a pair of antennae, and usually three ocelli. Mouthparts typically consist of a **labrum,** a pair each of **mandibles** and **maxillae,** a **labium,** and a tonguelike **hypopharynx.** The type of mouthparts an insect possesses determines how it feeds. We will discuss some of these modifications later.

The thorax is composed of three somites: **prothorax, mesothorax,** and **metathorax,** each bearing a pair of legs (Figure 11-35). In most insects the mesothorax and metathorax each bear a pair of wings. Wings consist of a double membrane that contains veins of thicker cuticle, which serve to strengthen the wing. Although these veins vary in their patterns among the different species, they are constant within a species and serve as one means of classification and identification.

Legs of insects are often modified for special purposes. Terrestrial forms have walking legs with terminal pads and claws as in beetles. These pads may be sticky for walking upside down, as in house flies. The hindlegs of grasshoppers and crickets are adapted for jumping (Figure 11-35B). Mole crickets have the first pair of legs modified for burrowing in the ground. Water bugs and many beetles have paddle-shaped appendages for swimming. For grasping its prey, the forelegs of the praying mantis are long and strong (Figure 11-36).

Wings and the Flight Mechanism

Insects share the power of flight with birds and flying mammals. However, their wings have evolved in a different manner from that of the limb buds of birds and mammals and are not homologous with them. Insect wings are formed by outgrowth from the body wall of the mesothoracic and metathoracic segments and are composed of cuticle.

Most insects have two pairs of wings, but the Diptera (true flies) have only one pair (Figure 11-37), the hindwings being represented by a pair of small **halteres** (balancers) that vibrate and are responsible for equilibrium during flight. Males in the order Strepsiptera have only the hind pair of wings and an anterior pair of halteres. The males of scale insects also have one pair of wings but no halteres. Some insects are wingless. Ants and termites, for example, have wings only on males, and on females during certain periods; workers are always wingless. Lice and fleas are always wingless.

Wings may be thin and membranous, as in flies and many others (Figure 11-37); thick and horny, as in the forewings of beetles (see Figure 11-48); parchmentlike, as in the forewings of grasshoppers; covered with fine scales, as in butterflies and moths; or with hairs, as in caddis flies.

Wing movements are controlled by a complex of muscles in the thorax. **Direct flight muscles** are attached to a part of the wing itself. **Indirect flight muscles** are not attached to the wing and cause wing movement by altering the shape of the thorax. The wing is hinged at the thoracic tergum and also slightly laterally on a pleural process, which acts as a fulcrum (Figure 11-38). In all insects, the upstroke of the wing is effected by contracting indirect muscles that

FIGURE 11-37

Horse fly *Tabanus* sp. (order Diptera). Adult female tabanids feed on the blood of birds and mammals, and their bite is painful. They can be a significant pest of livestock.

FIGURE 11-36

A, Praying mantis (order Orthoptera), feeding on an insect. **B,** Praying mantis laying eggs.

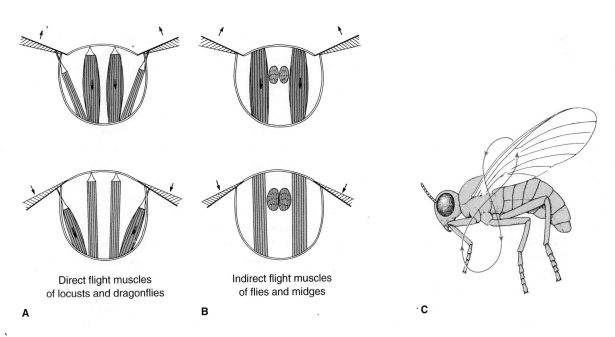

Direct flight muscles of locusts and dragonflies

A

Indirect flight muscles of flies and midges

B

C

FIGURE 11-38

A, Flight muscles of insects such as cockroaches, in which upstroke is by indirect muscles and downstroke is by direct muscles. **B,** In insects such as flies and bees, both upstroke and downstroke are by indirect muscles. **C,** The figure-eight path followed by the wing of a flying insect during the upstroke and downstroke.

pull the tergum down toward the sternum (Figure 11-38A). Dragonflies and cockroaches accomplish the downstroke by contracting direct muscles attached to the wings lateral to the pleural fulcrum. In Hymenoptera and Diptera all flight muscles are indirect. The downstroke occurs when the sternotergal muscles relax and longitudinal muscles of the thorax arch the tergum (Figure 11-38B), pulling the tergal articulations upward relative to the pleura. The downstroke in beetles and grasshoppers involves both direct and indirect muscles.

Contraction of flight muscles has two basic types of neural control: **synchronous** and **asynchronous.** Larger insects such as dragonflies and butterflies have synchronous muscles, in which a single volley of nerve impulses stimulates a muscle contraction and thus one wing stroke. Asynchronous muscles are found in the more specialized insects. Their mechanism of action is complex and depends on the storage of potential energy in resilient parts of the thoracic cuticle. As one set of muscles contracts (moving the wing in one direction), they stretch the antagonistic set of muscles, causing

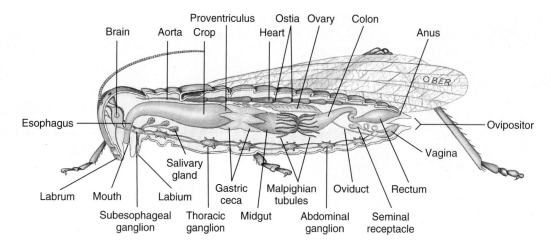

FIGURE 11-39

Internal structure of female grasshopper.

them to contract (and move the wing in the other direction). Because the muscle contractions are not phase-related to nervous stimulation, only occasional nerve impulses are necessary to keep the muscles responsive to alternating stretch activation. Thus extremely rapid wing beats are possible. For example, butterflies (with synchronous muscles) may beat as few as four times per second. Insects with asynchronous muscles, such as flies and bees, may vibrate at 100 beats per second or more. The fruit fly *Drosophila* (Gr. *drosos,* dew, + *philos,* loving) can fly at 300 beats per second, and midges have been clocked at more than 1000 beats per second!

Obviously flying entails more than a simple flapping of wings; a forward thrust is necessary. As the indirect flight muscles alternate rhythmically to raise and lower the wings, the direct flight muscles alter the angle of the wings so that they act as lifting airfoils during both the upstroke and the downstroke, twisting the leading edge of the wings downward during the downstroke and upward during the upstroke. This modulation produces a figure-eight movement (Figure 11-38C) that aids in spilling air from the trailing edges of the wings. The quality of the forward thrust depends, of course, on several factors, such as variations in wing venation, how much the wings are tilted, and how they are feathered.

Flight speeds vary. The fastest flyers usually have narrow, fast-moving wings with a strong tilt and a strong figure-eight component. Sphinx moths and horse flies are said to achieve approximately 48 km (30 miles) per hour and dragonflies approximately 40 km (25 miles) per hour. Some insects are capable of long continuous flights. The migrating monarch butterfly *Danaus plexippus* (Gr. after Danaus, mythical king of Arabia) (see Figure 11-42) travels south for hundreds of miles in the fall, flying at a speed of approximately 10 km (6 miles) per hour.

Internal Form and Function

Nutrition The digestive system (Figure 11-39) consists of a **foregut** (mouth with salivary glands, **esophagus, crop** for storage, and **gizzard** for grinding), **midgut** (stomach and gastric ceca), and **hindgut** (intestine, rectum, and anus). The foregut and hindgut are lined with cuticle, so absorption of food is confined largely to the midgut, although some absorption may take place in all sections. The majority of insects feed on plant juices and plant tissues. Such a food habit is called **phytophagous.** Some insects feed on specific plants; others, such as grasshoppers, can eat almost any plant. The caterpillars of many moths and butterflies eat the foliage of only certain plants. Certain species of ants and termites cultivate fungus gardens as a source of food.

Many beetles and the larvae of many insects live on dead animals **(saprophagous).** A number of insects are **predaceous,** catching and eating other insects as well as other types of animals.

Many insects, adults as well as larvae, are **parasitic.** Adult fleas, for instance, live on the blood of birds and mammals, all stages of biting lice and sucking lice are parasites of birds and mammals, and the larvae of many varieties of wasps live on spiders and caterpillars. In turn, many insects are parasitized by other insects. Some of the latter are beneficial to humans by controlling the numbers of injurious insects. When parasitic insects are themselves parasitized by other insects, the condition is known as **hyperparasitism.**

The feeding habits of insects are determined to some extent by their mouthparts, which are highly specialized for each type of feeding.

Biting and **chewing mouthparts,** such as those of the grasshopper and many herbivorous insects, are adapted for seizing and crushing food (Figure 11-40A). The mandibles of chewing insects are strong, toothed plates whose edges can bite or tear while the maxillae hold the food and pass it toward the mouth. Enzymes secreted by the salivary glands add chemical action to the chewing process.

Sucking mouthparts are greatly varied. House flies and fruit flies have no mandibles; the labium is modified into two soft lobes containing many small tubules that sponge up liquids with a capillary action much as the holes of a commercial sponge do (Figure 11-40D). Horse flies, however, are fitted not only to sponge up surface liquids but to bite into the skin with slender, tapering mandibles and then sponge up blood. Mosquitos combine **piercing** by means of needlelike stylets and sucking through a food channel (Figure 11-40B). In honeybees the labium forms a flexible and contractile "tongue" covered with many hairs. When the bee plunges its proboscis into nectar, the tip of the tongue bends upward and moves back and forth rapidly. Liquid enters the tube by capillarity and is drawn up the tube continuously by a pumping pharynx. In butterflies and moths mandibles are usually absent, and the maxillae are modified into a long sucking proboscis (Figure 11-40C) for drawing nectar from flowers. At rest, the proboscis is coiled up into a flat spiral. In feeding it extends, and pharyngeal muscles pump up the fluid.

Gas Exchange Terrestrial animals require efficient respiratory systems that permit rapid oxygen–carbon dioxide exchange but at the same time restrict water loss. In insects this is the function of the **tracheal system,** an extensive network of thin-walled tubes that branch into every part of the body (Figure 11-41). The tracheal trunks open to the outside by paired **spiracles,** usually two on the thorax and seven or eight on the abdomen. A spiracle may be merely a hole in the integument, as in primatively wingless insects, but it is usually provided with a valve or other closing mechanism that cuts down water loss. The evolution of such a device must have been very important in enabling insects to move into drier habitats. The **tracheae** are composed of a single layer of cells and are lined with cuticle that is shed, along with the outer cuticle, during the molt. They are supported by spiral thickenings

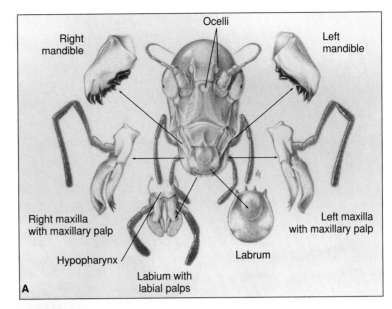

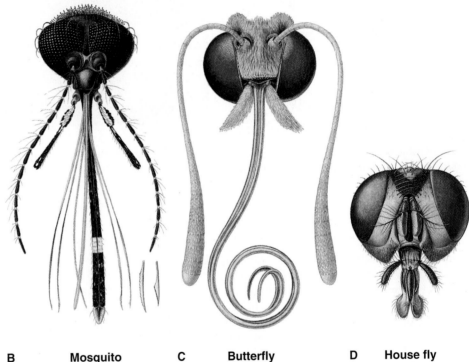

B Mosquito   **C Butterfly**   **D House fly**

FIGURE 11-40

Four types of insect mouthparts.

of the cuticle, the **taenidia,** that prevent their collapse. The tracheae branch out into smaller tubes, ending in very fine, fluid-filled tubules called **tracheoles** (not lined with cuticle), which branch into a fine network over the cells. Scarcely any living cell is located more than a few micrometers away from a tracheole. In fact, the ends of some tracheoles actually indent the membranes of the cells they supply, so that they terminate close to the mitochondria. The tracheal system affords an efficient system of transport without the use of oxygen-carrying pigments in the blood.

In some very small insects gas transport occurs entirely by diffusion along a concentration gradient. As oxygen is used, a partial vacuum develops in the tracheae, and air is sucked in through the spiracles. Larger or more active insects employ some ventilation device for moving air in and out of the tubes. Usually muscular movements in the abdomen perform the pumping action that draws air in or expels it.

The tracheal system is primarily adapted for breathing air, but many insects (nymphs, larvae, and adults) live in water. In small, soft-bodied aquatic nymphs, gaseous exchange may

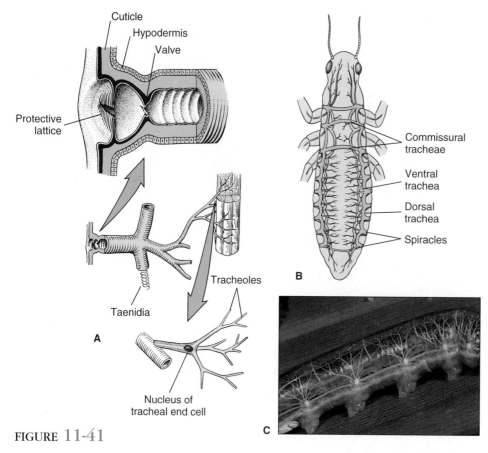

FIGURE **11-41**

A, Relationship of spiracle, tracheae, taenidia (chitinous bands that strengthen the tracheae), and tracheoles (diagrammatic). **B,** Generalized arrangement of insect tracheal system (diagrammatic). Air sacs and tracheoles not shown. **C,** Spiracles and tracheae of a caterpillar visible through its transparent cuticle.

Although the diving beetle Dytiscus (Gr. dytikos, able to swim) can fly, it spends most of its life in the water as an excellent swimmer. It uses an "artificial gill" in the form of a bubble of air held under its wing covers. The bubble is kept stable by a layer of hairs on top of the abdomen and is in contact with the spiracles on the abdomen. Oxygen from the bubble diffuses into the tracheae and is replaced by diffusion of oxygen from the water. Thus the bubble can last for several hours before the beetle must surface to replace it. Mosquito larvae are not good swimmers but live just below the surface, putting out short breathing tubes like snorkels to the surface for air. Spreading oil on the water, a favorite method of mosquito control, clogs the tracheae with oil and so suffocates the larvae. "Rattailed maggots" of the syrphid flies have an extensible tail that can stretch as much as 15 cm to the water surface.

occur by diffusion through the body wall, usually into and out of a tracheal network just under the integument. The aquatic nymphs of stoneflies and mayflies are equipped with **tracheal gills,** which are thin extensions of the body wall containing a rich tracheal supply. The gills of dragonfly nymphs are ridges in the rectum (rectal gills) where gas exchange occurs as water moves in and out of the rectum.

Excretion and Water Balance Malpighian tubules (Figure 11-39) are typical of most insects. As in the spiders (p. 200), Malpighian tubules are very efficient, both as excretory organs and as a means of conserving body fluids—an important factor in the success of terrestrial animals.

Since water requirements vary among different types of insects, this ability to cycle water and salts is very important. Insects living in dry environments may resorb nearly all water from the rectum, producing a nearly dry mixture of urine and feces. Leaf-feeding insects take in and excrete quantities of fluid. Freshwater larvae need to excrete water and conserve salts. Insects that feed on dry grains need to conserve water and excrete salt.

Nervous System The nervous system in general resembles that of the larger crustaceans, with a similar tendency toward fusion of ganglia (Figure 11-39). Some insects have a giant fiber system. There is also a visceral nervous system that corresponds in function with the autonomic nervous system of vertebrates. Neurosecretory cells located in various parts of the brain have an endocrine function, but, except for their role in molting and metamorphosis, little is known of their activity.

Sense Organs The sensory perceptions of insects are usually keen. Organs receptive to mechanical, auditory, chemical, visual, and other stimuli are well developed. They are scattered over the body but are especially numerous on the appendages.

Photoreceptors include both ocelli and compound eyes. The compound eyes are large and constructed of ommatidia, like those of crustaceans (p. 208). Apparently, visual acuity in insect eyes is much lower than that of human eyes, but most flying insects rate much higher than humans in flicker-fusion tests. Flickers of light become fused in the human eye at a frequency of 45 to 55 per second, but in bees and blow flies they do not fuse until 200 to 300 per second. This would be an advantage in analyzing a fast-changing landscape.

Most insects have three ocelli on the head, and they also have dermal light receptors on the body surface, but not much is known about them.

Sounds may be detected by sensitive hairlike **sensilla** or by tympanic organs sensitive to sonic or ultrasonic sound. Sensilla are modifications in the cuticular surface for reception of sensory stimuli other than light and are supplied with one or more neurons. Tympanic organs, found in grasshoppers (Figure 11-35A), crickets, cicadas, butterflies, and moths, involve a number of sensory cells extending to a thin tympanic membrane that encloses an air space in which vibrations can be detected.

Chemoreceptive sensilla, which are peglike or setae, are especially abundant on the antennae, mouthparts, or legs. Mechanical stimuli, such as contact pressure, vibrations, and tension changes in the cuticle, are picked up by sensilla or by sensory cells in the epidermis. Insects also sense temperature, humidity, body position (proprioception), gravity, etc.

Reproduction   Sexes are separate in insects, and fertilization is usually internal. Insects have various means of attracting mates. The female moth gives off a chemical (pheromone) that can be detected for a great distance by the male (males may be able to detect the pheromone several miles from the female). Fireflies use flashes of light; some insects find each other by means of sounds or color signals and by various kinds of courtship behavior.

Sperm are usually deposited in the vagina of the female at the time of copulation (Figure 11-35B). In some orders the sperm are encased in spermatophores that may be transferred at copulation or deposited on the substratum to be picked up by the female. A male silverfish deposits a spermatophore on the ground, then spins signal threads to guide the female to it. During the evolutionary transition of mandibulates from aquatic to terrestrial life, spermatophores were widely used, with copulation evolving much later.

Usually the sperm are stored in the seminal receptacle of the female in numbers sufficient to fertilize more than one batch of eggs. Many insects mate only once during their lifetime, and none mates more than a few times.

Insects usually lay a great many eggs. A queen honeybee, for example, may lay more than 1 million eggs during her lifetime. On the other hand, some flies are ovoviviparous and bring forth only a single offspring at a time. Forms that make no provision for the care of the young may lay many more eggs than those that provide for the young or those that have a very short life cycle.

Most species lay their eggs in a particular type of place to which they are guided by visual, chemical, or other clues. Butterflies and moths lay their eggs on the specific kind of plant on which their caterpillars must feed. A tiger moth may look for a pigweed, a sphinx moth for a tomato or tobacco plant, and a monarch butterfly for a milkweed plant (Figure 11-42). Insects whose immature stages are aquatic lay their eggs in water. A tiny braconid wasp lays her eggs on a caterpillar of the sphinx moth where they will feed and pupate in tiny white cocoons (Figure 11-43). An ichneumon wasp, with unerring accuracy, seeks out a certain kind of larva in which her young will live as internal parasites. Her long ovipositors may have to penetrate 1 to 2 cm of wood to find and deposit her eggs in the larva of a wood wasp or a wood-boring beetle (Figure 11-44).

Metamorphosis and Growth

Although many animals undergo a metamorphosis, insects illustrate it more dramatically than any other group. The transformation of a caterpillar into a beautiful moth or butterfly is indeed an astonishing change.

Early development occurs within the eggshell, and the hatching young escape from the capsule in various ways. During postembryonic development most insects change in form; that is, they undergo **metamorphosis.** A number of molts are necessary during the growth period, and each stage of the insect between molts is called an **instar.**

Approximately 88% of insects go through **holometabolous (complete) metamorphosis** (Gr. *holo,* complete, + *metabolē,* change) (Figure 11-42), which separates the physiological processes of growth (**larva**) from those of differentiation (**pupa**) and reproduction (**adult**). Each stage functions efficiently without competition with the other stages, because the larvae often live in entirely different surroundings and eat different foods from the adults. The wormlike larvae, which usually have chewing mouthparts, are known by various common names, such as caterpillars, maggots, bagworms, fuzzy worms, and grubs. After a series of instars during which the wings are developing internally, the larva forms a case or cocoon about itself and becomes a pupa, or chrysalis, a nonfeeding stage in which many insects pass the winter. When the final molt occurs a full-grown adult emerges (Figure 11-42), pale and with wings wrinkled. In a short time the wings expand and harden, and the insect is on its way. The stages, then, are **egg, larva** (several instars), **pupa,** and **adult.** The adult undergoes no further molting.

Some insects undergo **hemimetabolous (gradual,** or **incomplete) metamorphosis** (Gr. *hemi,* half, + *metabolē,* change) (Figure 11-45). These include insects such as bugs, scale insects, lice, and grasshoppers, which have terrestrial young, and mayflies, stoneflies (Figure 11-46A), and dragonflies (Figure 11-46B), which lay their eggs in water. The young are called **nymphs** (Figure 11-47), and their wings develop externally as budlike outgrowths in the early instars and increase in size as the animal grows by successive molts and becomes a winged adult. Aquatic nymphs have tracheal gills or other modifications for aquatic life. The stages are **egg, nymph** (several instars), and **adult.**

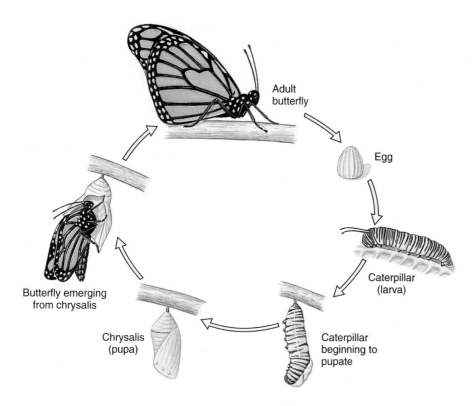

FIGURE 11-42

Holometabolous (complete) metamorphosis in a butterfly, *Danaus plexippus* (order Lepidoptera). Eggs hatch to produce first of several larval instars. The last larval instar molts to become a pupa. The adult emerges at the pupal molt.

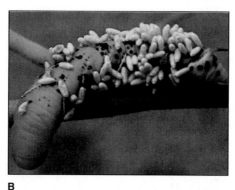

A

B

FIGURE 11-43

A, Hornworm, larval stage of a sphinx moth (order Lepidoptera). The more than 100 species of North American sphinx moths are strong fliers and mostly nocturnal feeders. Their larvae, called hornworms because of the large fleshy posterior spine, are often pests of tomatoes, tobacco, and other plants. **B,** Hornworm parasitized by a tiny wasp, *Apanteles*, which laid its eggs inside the caterpillar. The wasp larvae have emerged, and their pupae are on the caterpillar's skin. Young wasps emerge in 5 to 10 days, and the caterpillar usually dies.

FIGURE 11-44

An ichneumon wasp with the end of the abdomen raised to thrust her long ovipositor into wood to find a tunnel made by the larva of a wood wasp or wood-boring beetle. She can bore 13 mm or more into the wood to lay her eggs in the larva of the wood-boring beetle, which will become host for the ichneumon larvae. Other ichneumon species attack spiders, moths, flies, crickets, caterpillars, and other insects.

The biological meaning of the word "bug" is a great deal more restrictive than in common English usage. People often refer to all insects as "bugs," even extending the word to include such nonanimals as bacteria, viruses, and glitches in computer programs. Strictly speaking, however, a bug is a member of the order Hemiptera and nothing else.

A few insects, such as silverfish (see Figure 11-57) and springtails, undergo direct development. The young, or juveniles, are similar to the adults except in size and sexual maturation. The stages are egg, juvenile, and adult. Such insects include the primitively wingless insects.

Hormones control and regulate metamorphosis in insects. Three major endocrine organs are involved in development through the juvenile instars and eventually the emergence of the adult.

FIGURE 11-45

Life history of a hemimetabolous insect.

FIGURE 11-46

A, A stonefly, *Perla* sp. (order Plecoptera). **B,** A ten-spot dragonfly, *Libellula pulchella* (order Odonata). **C,** Nymph (larva) of a dragonfly. Both stoneflies and dragonflies have aquatic larvae that undergo gradual metamorphosis.

These organs and the hormones they produce are the **brain (ecdysiotropin), ecdysial (prothoracic) glands (ecdysone),** and **corpora allata (juvenile hormone).** Hormonal control of molting and metamorphosis is the same in holometabolous and hemimetabolous insects.

Diapause Many animals can enter a state of dormancy during adverse conditions, and there are periods in the life cycle of many insects when a particular stage can remain dormant for a long time because external climatic conditions are too harsh for normal activity. Most insects enter such a stage facultatively when some factor of the environment,

FIGURE 11-47

A, Ecdysis in a cicada, *Tibicen davisi* (order Homoptera). The old cuticle splits along a dorsal midline as a result of increased blood pressure and of air forced into the thorax by muscle contraction. The emerging insect is pale, and its new cuticle is soft. The wings will be expanded by blood pumped into veins, and the insect enlarges by taking in air. **B,** An adult *Tibicen davisi*.

such as temperature, becomes unfavorable, and the state continues until conditions again become favorable.

However, some species have a prolonged arrest of growth that is internally programmed and is usually seasonal. This type of dormancy is called **diapause** (di′a-poz) (Gr. *dia*, through, dividing into two parts, + *pausis*, a stopping), and it is an important adaptation to survive adverse environmental conditions. Diapause usually is triggered by some external signal, such as shortening day length. Diapause always occurs at the end of an active growth stage of the molting cycle so that, when the diapause period is over, the insect is ready for another molt.

Behavior and Communication

The keen sensory perceptions of insects make them extremely responsive to many stimuli. The stimuli may be internal (physiological) or external (environmental), and the responses are governed by both the physiological state of the animal and the pattern of nerve pathways involved. Many of the responses are simple, such as orientation toward or away from the stimulus, for example, attraction of a moth to light, avoidance of light by a cockroach, or attraction of carrion flies to the odor of dead flesh.

Much of the behavior of insects, however, is not a simple matter of orientation but involves a complex series of responses. A pair of tumble bugs, or dung beetles, chew off a bit of dung, roll it into a ball, and roll the ball laboriously to where they intend to bury it, after laying their eggs in it (Figure 11-48). A female cicada slits the bark of a twig and then lays an egg in each of the slits. A female potter wasp *Eumenes* scoops up clay into pellets, carries them one by one to her building site, and fashions them into dainty little narrow-necked clay pots, into each of which she lays an egg. Then she hunts and paralyzes a number of caterpillars, pokes them into the opening of a pot, and closes up the opening with clay. Each egg, in its own protective pot, hatches to find a well-stocked larder of food awaiting it.

Some insects can memorize and perform in sequence tasks involving multiple signals in various sensory areas. Worker honeybees have been trained to walk through mazes that involved five turns in sequence, using such clues as the color of a marker, the distance between two spots, or the angle of a turn. The same is true of ants. Workers of one species of Formica *learned a six-point maze at a rate only two or three times slower than that of laboratory rats. The foraging trips of ants and bees often wind and loop about in a circuitous route, but once the forager has found food, the return trip is relatively direct. One investigator suggested that the continuous series of calculations necessary to figure the angles, directions, distance, and speed of the trip and to convert it into a direct return could involve a stopwatch, a compass, and integral vector calculus. How the insect does it is unknown.*

FIGURE 11-48

Tumble bugs, or dung beetles, *Canthon pilularis* (order Coleoptera), chew off a bit of dung, roll it into a ball, and then roll it to where they will bury it in soil. One beetle pushes while the other pulls. Eggs are laid in the ball, and the larvae feed on the dung. Tumble bugs are black, an inch or less in length, and common in pastures.

Much of such behavior is "innate," that is, entire sequences of actions apparently have been genetically programmed. However, a great deal more learning is involved than we once believed. A potter wasp, for example, must learn where she has left her pots if she is to return to fill them with caterpillars one at a time. Social insects, which have been studied extensively, are capable of most of the basic forms of learning used by mammals. The exception is insight learning. Apparently insects, when faced with a new problem, cannot reorganize their memories to construct a new response.

Insects communicate with other members of their species by means of chemical, visual, auditory, and tactile signals. **Chemical signals** take the form of **pheromones,** which are substances secreted by one individual that affect the behavior or physiological processes of another individual. Examples of pheromones include sex attractants, releasers of certain behavior patterns, trail markers, alarm signals, and territorial markers. Like hormones, pheromones are effective in minute quantities. Social insects, such as bees, ants, wasps, and termites, can recognize a nestmate—or an alien in the nest—by means of identification pheromones. Pheromones determine caste in termites, and to some extent in ants and bees. In fact, pheromones are probably a primary integrating force in populations of social insects. Many insect pheromones have been extracted and chemically identified.

Sound production and **reception** (phonoproduction and phonoreception) in insects have been studied extensively, and although a sense of hearing is not present in all insects, this means of communication is meaningful to insects that use it. Sounds serve as warning devices, advertisement of territorial claims, or courtship songs. The sounds of crickets and grasshoppers seem to be concerned with courtship and aggression.

Male crickets scrape the rough edges of the forewings together to produce their characteristic chirping. Male cicadas produce the long, drawn-out sound of their recruitment call by vibrating membranes in a pair of organs located on the ventral side of the basal abdominal segment.

There are many forms of **tactile communication,** such as tapping, stroking, grasping, and antennae touching, which evoke responses varying from recognition to recruitment and alarm. Certain kinds of flies, springtails, and beetles manufacture their own **visual signals** in the form of **bioluminescence.** The best known of the luminescent beetles are fireflies, or lightning bugs (which are neither flies nor bugs, but beetles), in which the flash of light is a means of locating a prospective mate. Each species has its own characteristic flashing rhythm produced on the ventral side of the last abdominal segments. Females flash an answer to the species-specific pattern to attract the males. This interesting "love call" has been adopted by species of *Photuris,* which prey on the male fireflies of other species they attract (Figure 11-49).

FIGURE 11-49

Firefly femme fatale, *Photuris versicolor,* eating a male *Photinus tanytoxus,* which she has attracted with false mating signals.

FIGURE 11-50

An ant (order Hymenoptera) tending a group of aphids (order Homoptera). The aphids feed copiously on plant juices and excrete the excess as a clear liquid rich in carbohydrates ("honey-dew"), which is cherished as a food by ants.

Social Behavior Insects rank very high in the animal kingdom in their organization of social groups, and cooperation within the more complex groups depends heavily on chemical and tactile communication. Social communities are not all complex, however. Some community groups are temporary and uncoordinated, as are the hibernating associations of carpenter bees or the feeding gatherings of aphids (Figure 11-50). Some are coordinated for only brief periods, such as the tent caterpillars *Malacosoma,* that join in building a home web and a feeding net. However, all these are still open communities with social behavior.

In the true societies of some orders, such as Hymenoptera (honeybees and ants) and Isoptera (termites), a complex social life is necessary for the perpetuation of the species. Such societies are closed. In them all stages of the life cycle are involved, the communities are usually permanent, all activities are collective, and there is reciprocal communication. There is a high degree of efficiency in the division of labor. Such a society is essentially a family group in which the mother or perhaps both parents remain with the young, sharing the duties of the group in a cooperative manner. The society usually demonstrates polymorphism, or **caste** differentiation.

Honeybees have one of the most complex social organizations in the insect world. Instead of lasting one season, their organization continues for a more or less indefinite period. As many as 60,000 to 70,000 honeybees may live in a single hive. Of these, there are three castes—a single sexually mature female, or **queen,** a few hundred **drones,** which are sexually mature males, and thousands of **workers,** which are sexually inactive genetic females (Figure 11-51).

Workers take care of the young, secrete wax with which they build the six-sided cells of the honeycomb, gather the nectar from flowers, manufacture honey, collect pollen, and ventilate and guard the hive. One drone, sometimes more, fertilizes the queen during the mating flight, at which time enough sperm are stored in her seminal receptacle to last her lifetime.

Castes are determined partly by fertilization and partly by what is fed to the larvae. Drones develop from unfertilized eggs (and consequently are haploid); queens and workers develop from fertilized eggs (and thus are diploid; see haplodiploidy, p. 147). Female larvae that will become queens are fed **royal jelly,** a secretion from the salivary glands of the nurse workers. Royal jelly differs from the "worker jelly" fed to ordinary larvae, but the components in it that are essential for queen determination have not yet been identified. Honey and pollen are added to the worker diet about the third day of larval life. Pheromones in the "queen substance," which is produced by the queen's mandibular glands, prevent the female workers from maturing sexually. Workers produce royal jelly only when the level of "queen substance" pheromone in the colony drops. This drop occurs when the queen becomes too old, dies, or is removed. Then the workers start enlarging a larval cell and feeding the larva the royal jelly that produces a new queen.

Honeybees have evolved an efficient system of communication by which, through certain body movements, their scouts inform the workers of the location and quantity of food sources.

Termite colonies contain several castes, consisting of fertile individuals, both males and females, and sterile individuals (Figure 11-52). Some of the fertile individuals may have wings and may leave the colony, mate, lose their wings, and as **king** and **queen** start a new colony. Wingless fertile individuals may under certain conditions substitute for the king or queen. Sterile members are wingless and become **workers** and **soldiers.** Soldiers have large heads and mandibles and serve for the

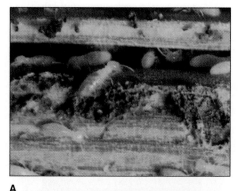

A

B

FIGURE 11-51

Queen bee surrounded by her court. The queen is the only egg layer in the colony. The attendants, attracted by her pheromones, constantly lick her body. As food is transferred from these bees to others, the queen's presence is communicated throughout the colony.

FIGURE 11-52

A, Termite workers, *Reticulitermes flavipes* (order Isoptera), eating yellow pine. Workers are wingless sterile adults that tend the nest and care for the young. **B,** The termite queen becomes a distended egg-laying machine. The queen and several workers and soldiers are shown here.

FIGURE 11-53

A weaver ant nest in Australia.

defense of the colony. As in bees and ants, extrinsic factors cause caste differentiation. Reproductive individuals and soldiers secrete inhibiting pheromones that pass throughout the colony to the nymphs through a mutual feeding process, called **trophallaxis,** so that they become sterile workers. Workers also produce pheromones, and if the level of "worker substance" or "soldier substance" falls, as might happen after an attack by marauding predators, for example, the next generation produces compensating proportions of the appropriate caste.

Ants also have highly organized societies. Superficially, they resemble termites, but they are quite different (belong to a different order) and can be distinguished easily. In contrast to termites, ants are usually dark in color, are hard bodied, and have a constriction between the thorax and abdomen.

In ant colonies males die soon after mating and the queen either starts her own new colony or joins some established colony and does the egg laying. The sterile females are wingless workers and soldiers that do the work of the colony—gather food, care for the young, and protect the colony. In many larger colonies there may be two or three types of individuals within each caste.

Ants have evolved some striking patterns of "economic" behavior, such as making slaves, farming fungi, herding "ant cows" (aphids or other homopterans, Figure 11-50), sewing their nests together with silk (Figure 11-53), and using tools.

Insects and Human Welfare

Beneficial Insects Although most of us think of insects primarily as pests, humanity would have great difficulty in surviving if all insects were suddenly to disappear. Insects are necessary for the cross-fertilization of many crops. Bees pollinate almost $10 billion worth of food crops per year in the United States alone, and this value does not include pollination of forage crops for livestock or pollination by other insects. In addition, some insects produce useful materials: honey and beeswax from bees, silk from silkworms, and shellac from a wax secreted by the lac insects.

Very early in their evolution insects and flowering plants formed a relationship of mutual adaptations that have been to each other's advantage. Insects exploit flowers for food, and flowers exploit insects for pollination. Each floral development of petal and sepal arrangement is correlated with the sensory adjustment of certain pollinating insects. Among these mutual adaptations are amazing devices of allurements, traps, specialized structure, and precise timing.

Many predaceous insects, such as tiger beetles, aphid lions, ant lions, praying mantids, and ladybird beetles, destroy harmful insects (Figure 11-54A, B). Some insects control harmful ones by parasitizing them or by laying their eggs where their young, when hatched, may devour the host (Figure 11-54C). Dead animals are quickly consumed by maggots hatched from eggs laid on carcasses.

Insects and their larvae serve as an important source of food for many birds, fish, and other animals.

Harmful Insects Harmful insects include those which eat and destroy plants and fruits, such as grasshoppers, chinch bugs, corn borers, boll weevils, grain weevils, San Jose scale, and scores of others (Figure 11-55). Practically every cultivated crop has some insect pest. Lice, bloodsucking flies, warble flies, bot flies, and many others attack humans or domestic animals or both. Malaria, carried by the *Anopheles* mosquito (Figure 11-56), is still one of the world's worst killers; mosquitos also transmit yellow fever and filariasis. Fleas carry plague, which at many times in history has almost wiped out whole human populations. House flies are the vector for typhoid and lice for typhus fever; tsetse flies carry African sleeping sickness; and certain bloodsucking bugs are carriers of Chagas' disease. In addition there is a tremendous destruction of food, clothing, and property by weevils, cockroaches, ants, clothes moths, termites, and carpet beetles. Not the least of the insect pests are bedbugs, *Cimex*, bloodsucking hemipterous insects that humans contracted, probably early in their evolution, from bats that shared their caves.

Control of Insects Because all insects are an integral part of the ecological communities to which they belong, their total destruction would probably do more harm than good. Food chains would be disturbed, some of our most loved birds would disappear, the biological cycles by which dead animal and plant matter disintegrates and returns to enrich the soil would be seriously impeded, and we would lose many flowering plants, including many food-crop plants. We often have overlooked the beneficial role of insects in our environment, and in our zeal to control the pests we have indiscriminately sprayed the landscape

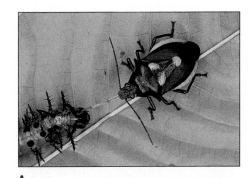

A

B

C

FIGURE 11-54

Some beneficial insects. **A,** A predaceous stink bug (order Hemiptera) feeds on a caterpillar. Note the sucking proboscis of the bug. **B,** A ladybird beetle ("ladybug," order Coleoptera). Adults (and larvae of most species) feed voraciously on plant pests such as mites, aphids, scale insects, and thrips. **C,** A parasitic wasp (*Larra bicolor*) attacking a mole cricket. The wasp drives the cricket from its burrow, then stings and paralyzes it. After the wasp deposits her eggs, the mole cricket recovers and resumes an active life—until it is killed by the developing wasp larvae.

A

B

C

FIGURE 11-55

Insect pests. **A,** Japanese beetles, *Popillia japonica* (order Coleoptera) are serious pests of fruit trees and ornamental shrubs. They were introduced into the United States from Japan in 1917. **B,** Longtailed mealybug, *Pseudococcus longispinus* (order Homoptera). Many mealybugs are pests of commercially valuable plants. **C,** Corn ear worms, *Heliothis zea* (order Lepidoptera). An even more serious pest of corn is the infamous corn borer, an import from Europe in 1908 or 1909.

with extremely effective "broad-spectrum" insecticides that eradicate beneficial, as well as harmful, insects. We have also found, to our chagrin, that many of the chemicals we have used persist in the environment and accumulate as residues in the bodies of animals higher up in food chains. Also, many

Classification of Class Insecta

Insects are divided into orders chiefly on the basis of morphology and developmental features. Entomologists do not all agree on the names of the orders or on the limits of each order. Some tend to combine and others to divide the groups. However, the following synopsis of the major orders is one that is rather widely accepted.

Order Protura (pro-tu′ra) (Gr. *protos,* first, + *oura,* tail). Minute (1 to 1.5 mm); no eyes or antennae; appendages on abdomen as well as thorax; live in soil and dark, humid places; slight, gradual metamorphosis.

Order Diplura (dip-lu′ra) (Gr. *diploos,* double, + *oura,* tail): **japygids.** Usually less than 10 mm; pale, eyeless; a pair of long terminal filaments or pair of caudal forceps; live in damp humus or rotting logs; development direct.

Order Collembola (col-lem′bo-la) (Gr. *kolla,* glue, + *embolon,* peg, wedge): **springtails** and **snow fleas.** Small (5 mm or less); no eyes; respiration by trachea or body surface; a springing organ folded under the abdomen for leaping; abundant in soil; sometimes swarm on pond surface film or on snowbanks in spring; development direct.

Order Thysanura (thy-sa-nu′ra) (Gr. *thysanos,* tassel, + *oura,* tail): **silverfish** (Figure 11-57) and **bristletails.** Small to medium size; large eyes; long antennae; three long terminal cerci; live under stones and leaves and around human habitations; development direct.

Order Ephemeroptera (e-fem-er-op′ter-a) (Gr. *ephēmeros,* lasting but a day, + *pteron,* wing): **mayflies** (Figure 11-58). Wings membranous; forewings larger than hindwings; adult mouthparts vestigial; nymphs aquatic, with lateral tracheal gills, hemimetabolous development.

Order Odonata (o-do-na′ta) (Gr. *odontos,* tooth, + *ata,* characterized by): **dragonflies** (Figure 11-46B), **damselflies.** Large; membranous wings are long, narrow, net veined, and similar in size; long and slender body; aquatic nymphs with aquatic gills and prehensile labium for capture of prey; hemimetabolous development.

Order Orthoptera (or-thop′ter-a) (Gr. *orthos,* straight, + *pteron,* wing): **grasshoppers, locusts, crickets, cockroaches, walkingsticks, praying mantids** (Figures 11-35 and 11-36). Wings when present, with forewings thickened and hindwings folded like a fan under forewings; chewing mouthparts; hemimetabolous development.

Order Isoptera (i-sop′ter-a) (Gr. *isos,* equal, + *pteron,* wing): **termites** (Figure 11-52). Small; membranous, narrow wings similar in size with few veins; wings shed at maturity; erroneously called "white ants"; distinguishable from true ants by broad union of thorax and abdomen; complex social organization; hemimetabolous development.

Order Mallophaga (mal-lof′a-ga) (Gr. *mallos,* wool, + *phagein,* to eat): **biting lice** (Figure 11-59). As large as 6 mm; wingless; chewing mouthparts; legs adapted for clinging to host; live on birds and mammals; hemimetabolous development.

Order Anoplura (an-o-plu′ra) (Gr. *anoplos,* unarmed, + *oura,* tail): **sucking lice** (Figure 11-60). Depressed body; as large as 6 mm; wingless; mouthparts for piercing and sucking; adapted for clinging to warm-blooded host; includes the head louse, body louse, crab louse, others; hemimetabolous development.

Order Hemiptera (he-mip′ter-a) (Gr. *hemi,* half + *pteron,* wing) **(Heteroptera): true bugs** (Figure 11-54A). Size 2 to 100 mm; wings present or absent; forewings with basal portion leathery, apical portion membranous; hindwings membranous; at rest, wings held flat over abdomen; piercing-sucking mouthparts; many with odorous scent

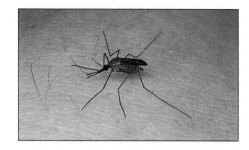

FIGURE **11-56**

A mosquito, *Anopheles quadrimaculatus* (order Diptera). *Anopheles* spp. are vectors of malaria.

FIGURE **11-57**

Silverfish *Lepisma* (order Thysanura) is often found in homes.

glands; include water scorpions, water striders, bedbugs, squash bugs, assassin bugs, chinch bugs, stinkbugs, plant bugs, lace bugs, others; hemimetabolous.

Order Homoptera (ho-mop′ter-a) (Gr. *homos,* same, + *pteron,* wing): **cicadas** (Figure 11-47), **aphids** (Figure 11-50), **scale insects, mealybugs** (Figure 11-55B), **leafhoppers, treehoppers** (Figure 11-61). (Often included as suborder under Hemiptera.) If winged, either membranous or thickened forewings and membranous hindwings; wings held rooflike over body; piercing-sucking mouthparts; all plant eaters; some destructive; a few serving as source of shellac, dyes, etc.; some with complex life histories; hemimetabolous.

Order Neuroptera (neu-rop′ter-a) (Gr. *neuron,* nerve, + *pteron,* wing): **dobsonflies, ant lions** (Figure 11-62), **lacewings.** Medium to large size; similar, membranous wings with many cross veins; chewing mouthparts; dobsonflies with greatly enlarged mandibles in males, and with aquatic larvae; ant lion larvae (doodlebugs) make craters in sand to trap ants; holometabolous development.

Order Coleoptera (ko-le-op′ter-a) (Gr. *koleos,* sheath, + *pteron,* wing): **beetles** (Figures 11-54B; 11-55A), **fireflies** (Figure 11-49), **weevils.** The largest order of animals in the world; forewings (elytra) thick, hard, opaque; membranous hindwings folded under forewings at rest; mouthparts for biting and chewing; includes ground beetles, carrion beetles, whirligig beetles, darkling beetles, stag beetles, dung beetles (Figure 11-48), diving beetles, boll weevils, fireflies, ladybird beetles (ladybugs), others; holometabolous.

Order Lepidoptera (lep-i-dop′ter-a) (Gr. *lepidos,* scale, + *pteron,* wing): **butterflies** and **moths** (Figures 11-42 and 11-55C). Membranous wings covered with overlapping scales, wings coupled at base; mouthparts a sucking tube, coiled when not in use; larvae (caterpillars) with chewing mandibles for plant eating, stubby prolegs on the abdomen, and silk glands for spinning cocoons; antennae knobbed in butterflies and usually plumed in moths; holometabolous.

Order Diptera (dip′ter-a) (Gr. *dis,* two, + *pteron,* wing): **true flies.** Single pair of wings, membranous and narrow; hindwings reduced to inconspicuous balancers (halteres); sucking mouthparts or adapted for sponging or lapping or piercing; legless larvae called maggots or, when aquatic, wigglers; include crane flies, mosquitos (Figure 11-56), moth flies, midges, fruit flies, flesh flies, house flies, horseflies (Figure 11-37), bot flies, blow flies, gnats, and many others; holometabolous.

Order Trichoptera (tri-kop′ter-a) (Gr. *trichos,* hair, + *pteron,* wing): **caddisflies.** Small, soft bodied; wings well-veined and hairy, folded rooflike over hairy body; chewing mouthparts; aquatic larvae construct cases of leaves, sand, gravel, bits of shell, or plant matter, bound together with secreted silk or cement; some make silk feeding nets attached to rocks in streams; holometabolous.

Order Siphonaptera (si-fon-ap′ter-a) (Gr. *siphon,* a siphon, + *apteros,* wingless): **fleas** (Figure 11-63). Small; wingless; bodies laterally compressed; legs adapted for leaping; no eyes; ectoparasitic on birds and mammals; larvae legless and scavengers; holometabolous.

Order Hymenoptera (hi-men-op′ter-a) (Gr. *hymen,* membrane, + *pteron,* wing): **ants, bees** (Figure 11-51), **wasps.** Very small to large; membranous, narrow wings coupled distally; subordinate hindwings; mouthparts for biting and lapping up liquids; ovipositor sometimes modified into stinger, piercer, or saw; both social and solitary species; most larvae legless, blind, and maggotlike; holometabolous.

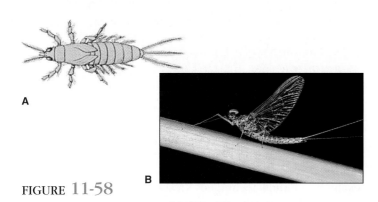

A

B

FIGURE 11-58

Mayfly (order Ephemeroptera). **A,** Nymph. **B,** Adult.

strains of insects have developed resistance to insecticides in common use.

In recent years, methods of control other than chemical insecticides have been under intense investigation and experimentation. Economics, concern for the environment, and consumer demand are causing thousands of farmers across the United States to use alternatives to strict dependence on chemicals.

Several types of biological controls have been developed and are under investigation. All of these areas present problems but also show great possibilities. One is the use of bacterial, fungal, and viral pathogens. A bacterium, *Bacillus thuringiensis,* is quite effective in control of lepidopteran pests (cabbage looper, imported cabbage worm, tomato

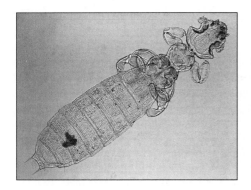

FIGURE 11-59

Gliricola porcelli (order Mallophaga), a chewing louse of guinea pigs. Antennae are normally held in the deep grooves on the sides of the head.

FIGURE 11-60

The head and body louse of humans *Pediculus humanus* (order Anoplura) feeding.

FIGURE 11-61

Oak treehoppers *Platycotis vittata* (order Homoptera).

worm, gypsy moth). Other strains of *B. thuringiensis* attack insects in other orders, and the species range of target insects is being widened by techniques of genetic engineering. Genes coding for the toxin produced by *B. thuringiensis* are also being introduced into other bacteria and even into the plants themselves, which makes the plants resistant to insect attack.

A number of viruses and fungi that have potential as insecticides have been isolated. Difficulties and expense in rearing these agents are being overcome in certain cases, and some are nearing commercial production.

Introduction of natural predators or parasites of the insect pests has met with some success. In the United States the vedalia beetle from Australia helps control the cottony-cushion scale on citrus plants, and numerous instances of control by use of insect parasites have been recorded.

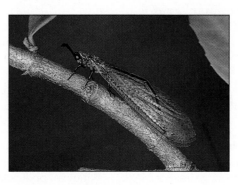

FIGURE 11-62

Adult ant lion (order Neuroptera).

FIGURE 11-63

Female human flea, *Pulex irritans* (order Siphonaptera).

complex compounds that are produced in such minute amounts. Nevertheless, pheromones are likely to play an important role in biological pest control in the future.

> The gypsy moth, introduced into the United States in 1858 in an ill-advised attempt to breed a better silkworm, has spread throughout the northeast as far south as southern Virginia. It defoliates oak forests in years when there are outbreaks. In 1981, it defoliated 13 million acres in 17 northeastern states.

Another approach to biological control is to interfere with the reproduction or behavior of insect pests with sterile males or with naturally occurring organic compounds that act as hormones or pheromones. Such research, although very promising, is slow because of our limited understanding of insect behavior and the problems of isolating and identifying

> The sterile male approach has been used effectively in eradicating screwworm flies, a livestock pest. Large numbers of male insects, sterilized by irradiation, are introduced into the natural population; females that mate with the sterile flies lay infertile eggs.

A systems approach referred to as **integrated pest management** is being increasingly practiced. This approach involves integrated utilization of all possible, practical techniques to contain pest infestations at a tolerable level, for example, cultural techniques (resistant plant varieties, crop rotation, tillage techniques, timing of sowing, planting, or harvesting), use of biological controls, and sparing use of insecticides.

Classification of Phylum Arthropoda

Subphylum Trilobita (tri′lo-bi′ta) (Gr. *tri-*, three, + *lobos*, lobe): **trilobites.** All extinct forms; Cambrian to Carboniferous; body divided by two longitudinal furrows into three lobes; distinct head, thorax, and abdomen; biramous appendages.

Subphylum Chelicerata (ke-lis′e-ra′ta) (Gr. *chēle*, claw, + *keratos*, a horn): **eurypterids, horseshoe crabs, spiders, ticks.** First pair of appendages modified to form chelicerae; pair of pedipalps and four pairs of legs; no antennae, no mandibles; cephalothorax and abdomen often with segments fused.

Class Merostomata (mer′o-sto′ma-ta) (Gr. *meros*, thigh, + *stomatos*, mouth): **aquatic chelicerates.** Cephalothorax and abdomen; compound lateral eyes; appendages with gills; sharp telson; **subclasses Eurypterida** (all extinct) and **Xiphosurida,** the horseshoe crabs.

Class Pycnogonida (pik′no-gon′i-da) (Gr. *pyknos*, compact, + *gonia*, knee, angle): **sea spiders.** Small (3 to 4 mm), but some reach 500 mm; body chiefly cephalothorax; tiny abdomen; usually four pairs of long walking legs (some with five or six pairs); one pair of subsidiary legs (ovigers) for egg bearing; mouth on long proboscis; four simple eyes; no respiratory or excretory system. Example: *Pycnogonum.*

Class Arachnida (ar-ack′ni-da) (Gr. *arachnē*, spider): **scorpions, spiders, mites, ticks, harvestmen.** Four pairs of legs; segmented or unsegmented abdomen with or without appendages and generally distinct from cephalothorax; respiration by gills, tracheae, or book lungs; excretion by Malpighian tubules or coxal glands; dorsal bilobed brain connected to ventral ganglionic mass with nerves; simple eyes; sexes separate; chiefly oviparous; no true metamorphosis. Examples: *Argiope, Centruroides.*

Subphylum Crustacea (crus-ta′she-a) (L. *crusta*, shell, + *acea*, group suffix): **crustaceans.** Mostly aquatic, with gills; cephalothorax usually with dorsal carapace; biramous appendages, modified for various functions; head appendages consisting of two pairs of antennae, one pair of mandibles, and two pairs of maxillae; sexes usually separate; development primitively with nauplius stage.

Class Branchiopoda (bran′kee-op′o-da) (Gr. *branchia*, gills, + *pous, podos*, foot): **branchiopods.** Flattened, leaflike swimming appendages (phyllopodia) with respiratory function. Examples: *Triops, Lynceus, Daphnia.*

Class Maxillopoda (maks′i-lop′o-da) (L. *maxilla*, jawbone, + Gr. *pous, podos*, foot): **ostracods, copepods, branchiurans, barnacles.** Five cephalic, six thoracic, and usually four abdominal somites; no typical appendages on abdomen; unique maxillopodan eye. Examples: *Cypris, Cyclops, Ergasilus, Argulus, Balanus.*

Class Malacostraca (mal′a-kos′tra-ka) (Gr. *malakos*, soft, + *ostrakon*, shell): **shrimps, crayfishes, lobsters, crabs.** Usually with eight thoracic and six abdominal somites, each with a pair of appendages. Examples: *Armadillidium, Gammarus, Megacytiphanes, Grapsus, Homarus, Panulirus.*

Subphylum Uniramia (yu-ni-ra′me-a) (L. *unus*, one, + *ramus*, a branch): **insects** and **myriapods.** All appendages uniramous; head appendages consisting of one pair of antennae, one pair of mandibles, and one or two pairs of maxillae.

Class Diplopoda (di-plop′o-da) (Gr. *diploos*, double, + *pous, podos*, foot): **millipedes.** Subcylindrical body; head with short antennae and simple eyes; body with variable number of somites; short legs, usually two pairs of legs to a somite; separate sexes. Examples: *Julus, Spirobolus.*

Class Chilopoda (ki-lop′o-da) (Gr. *cheilos*, lip, + *pous, podos*, foot): **centipedes.** Dorsoventrally flattened body; variable number of somites, each with one pair of legs; one pair of long antennae; separate sexes. Examples: *Cermatia, Lithobius, Geophilus.*

Class Pauropoda (pau-rop′o-da) (Gr. *pauros*, small, + *pous, podos*, foot): **pauropods.** Minute (1 to 1.5 mm), cylindrical body consisting of double segments and bearing nine or ten pairs of legs; no eyes. Example: *Pauropus.*

Class Symphyla (sim′fi-la) (Gr. *syn*, together, + *phylon*, tribe): **garden centipedes.** Slender (1 to 8 mm) with long, threadlike antennae; body consisting of 15 to 23 segments with 10 to 12 pairs of legs; no eyes. Example: *Scutigerella.*

Class Insecta (in-sek′ta) (L. *insectus*, cut into): **insects.** Body with distinct head, thorax, and abdomen; pair of antennae; mouthparts modified for different food habits; head of six fused somites; thorax of three somites; abdomen with variable number, usually 11 somites; thorax with two pairs of wings (sometimes one pair or none) and three pairs of jointed legs; separate sexes; usually oviparous; gradual or abrupt metamorphosis. (Insect orders on pp. 228–229).

Phylogeny and Adaptive Radiation

Phylogeny

The similarities between annelids and arthropods give strong support to the hypothesis that both phyla originated from a line of coelomate segmented protostomes, these in time diverged to form a protoannelid line with laterally located parapodia and one or more protoarthropod lines with more ventrally located parapodia. Some authors have contended that the Arthropoda is polyphyletic and that some or all of the present subphyla are derived from different annelidlike ancestors that have undergone "arthropodization." The crucial development is the hardening of the cuticle to form an arthropod exoskeleton, and most of the features that distinguish arthropods from annelids result from the stiffened exoskeleton. For example, the vital role in annelids of the coelomic compartments as a hydrostatic skeleton was gone; therefore intersegmental septa were unnecessary, as was a closed circulatory system. Jointed appendages, of course, are required if the external surface is hard, and the body wall muscles of the annelid type could be converted and inserted on the considerable inner surfaces of the cuticle for efficient movement of body parts. Compared with annelids, there was a great restriction in permeable surfaces for respiration and excretion, and such adaptations as tracheal systems and Malpighian tubules were necessary. Thus arthropods *could* have evolved more than once. However, other zoologists argue strongly that the similarities of the arthropods outweigh their differences, and the phylum is monophyletic. The phyla Onycophora and Tardigrada (Chapter 12) may be sister taxa to the arthropods. A cladogram depicting possible relationships is in Chapter 12 (p. 246).

The crustaceans traditionally have been allied with the insects and myriapods in a group known as Mandibulata because they all have mandibles, as contrasted with chelicerae. Critics of this traditional grouping have argued that the mandibles in each group were so different that they could not have been inherited from a common ancestor. However, advocates of the "mandibulate hypothesis" maintain, on the basis of present evidence, that the crustacean mandible and the insect/myriapod mandible are indeed homologous. They also point out numerous other similarities between the groups, such as the basic structure of the ommatidia, the tripartite brain, and the head primitively of five somites, each with a pair of appendages. Figure 11-64 is a cladogram depicting this mandibulate hypothesis.

Most zoologists agree that the insects and myriapods share a number of important characteristics and that they probably evolved from a common ancestor (Figure 11-64). The ancestor probably had a head and trunk of many similar somites, a primitive character retained by the myriapods.

Evolution of the insects involved specialization of the first three postcephalic somites to become the locomotor segments (thorax) and a loss or reduction of appendages on the rest of the body (abdomen). The wingless orders traditionally have been regarded as having the most primitive characteristics, but placing them in a single subclass (Apterygota) creates a paraphyletic taxon. Three of the apterygote orders (Diplura, Collembola, Protura) have their mandibles and first maxillae located deeply in pouches in the head, a condition known as **endognathy.** They share other primitive and derived characters, and there are many similarities between the endognathous insects and the myriapods. All other insects are **ectognathous,** including the wingless order Thysanura. Ectognathous insects do not have their mandibles and maxillae in pouches, and they share other synapomorphies. Endognathous and ectognathous insects form sister groups, and the Thysanura diverged from the common ancestor of ectognathous insects before the advent of flight, which unites the remaining ectognathous orders.

The ancestral winged insect gave rise to three lines, which differed in their ability to flex their wings. Two of these (Odonata and Ephemeroptera) have outspread wings. The other line has wings that can fold back over the abdomen. It branched into three groups: one with hemimetabolous metamorphosis and chewing mouthparts (Orthoptera, Dermaptera, Isoptera, Embioptera); one group with hemimetabolous metamorphosis and a tendency to have sucking mouthparts (Thysanoptera, Hemiptera, Homoptera, Mallophaga, Anoplura); and a group with holometabolous metamorphosis. Insects in the last group have the most specialized life history and apparently form a clade.

Adaptive Radiation

Annelids show little specialization or fusion of somites and relatively little differentiation of appendages. However, in arthropods the adaptive trend has been toward tagmatization of the body by differentiation or fusion of somites, giving rise in more derived groups to such tagmata as head and trunk; head, thorax, and abdomen; or cephalothorax (fused head and thorax) and abdomen. A series of similar appendages, one pair on each trunk somite, is the primitive character state, still retained by some crustaceans and the myriapods. More derived forms have appendages specialized for specific functions, and some appendages are lost entirely.

Much of the amazing diversity in arthropods seems to have developed because of modification and specialization of their cuticular exoskeleton and their jointed appendages, thus resulting in a wide variety of locomotor and feeding adaptations. Whether it be in the area of habitat, feeding adaptations, means of locomotion, reproduction, or general mode of living, the adaptive achievements of the arthropods are truly remarkable.

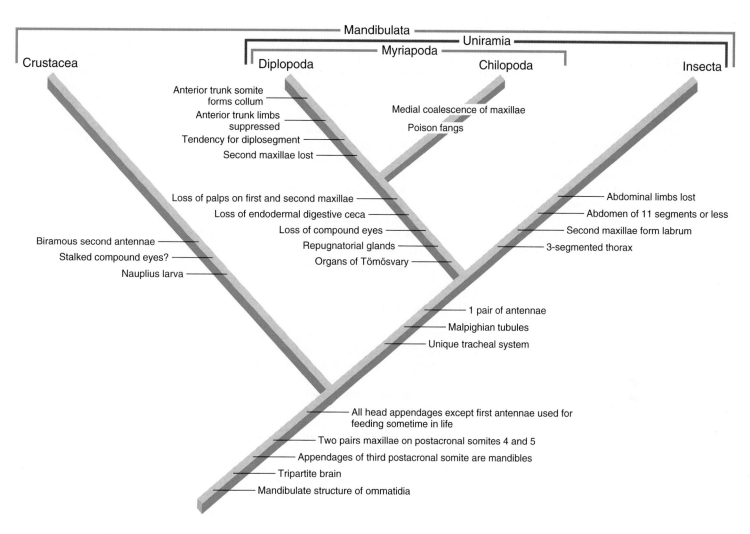

FIGURE 11-64

Cladogram showing "mandibulate hypothesis," the uniramians and crustaceans being sister groups evolving from a common ancestor defined by numerous synapomorphies. The character followed by a question mark may be a primitive rather than derived feature. The **acron** corresponds to the prostomium of annelids and is the anterior closure, not counted as a somite. The myriapods and insects also would be sister groups. Organs of Tömösvary are unique sensory organs opening at the bases of the antennae, and repugnatorial glands, located on certain somites or legs, secrete an obnoxious substance for defense. The gnathochilarium is formed in diplopods by fusion of the first maxillae, and the collum is the collarlike tergite of the first trunk segment. Outgroup for this cladogram would be the chelicerates.

Source: R. C. Brusca and G. J. Brusca, Invertebrates, *1990. Sinauer Associates, Sunderland, MA.*

Summary

The Arthropoda is the largest, most abundant and diverse phylum in the world. Arthropods are metameric, coelomate protostomes with well-developed organ systems. Most show marked tagmatization. They are extremely diverse and occur in all habitats capable of supporting life. Perhaps more than any other single factor, the success of the arthropods is explained by the adaptations made possible by their cuticular exoskeleton. Other important elements in their success are jointed appendages, tracheal respiration, efficient sensory organs, complex behavior, metamorphosis, and the ability to fly.

Members of the subphylum Chelicerata have no antennae, and their main feeding appendages are chelicerae. In addition, they have a pair of pedipalps (which may be similar to walking legs) and four pairs of walking legs. The great majority of living chelicerates are in the class Arachnida: spiders (order Araneae), scorpions (order Scorpionida), harvestmen (order Opiliones), and ticks and mites (order Acarina).

The tagmata of spiders (cephalothorax and abdomen) show no external segmentation and join by a waistlike pedicel. Chelicerae of spiders have poison glands for paralyzing or killing their prey. Spiders can spin silk, which they use for a variety of purposes.

The cephalothorax and abdomen of ticks and mites are completely fused, and the anterior capitulum bears the mouthparts. Ticks and mites are the most numerous of any arachnids; some are important disease carriers, and others are serious plant pests.

The Crustacea is a large, primarily aquatic subphylum of arthropods. Crustaceans bear two pairs of antennae, mandibles, and two pairs of maxillae on the head. Their appendages are primitively biramous, and the major tagmata are the head, thorax, and abdomen. Many have a carapace and respire by means of gills.

All arthropods must periodically cast off their cuticle (ecdysis) and grow larger before the newly secreted cuticle hardens. Premolt and postmolt periods are hormonally controlled, as are several other structures and functions.

There are many predators, scavengers, filter feeders, and parasites among the Crustacea. Respiration is through the body surface or by gills, and excretory organs take the form of maxillary or antennal glands. Circulation, as in other arthropods, is through an open system of sinuses (hemocoel), and a dorsal, tubular heart is the chief pumping organ. Most crustaceans have compound eyes composed of units called ommatidia.

Members of the class Maxillopoda, subclass Copepoda, lack a carapace and abdominal appendages. They are abundant and are among the most important of the primary consumers in many freshwater and marine ecosystems. The Malacostraca are the largest crustacean class, and the most important orders are the Isopoda, Amphipoda, Euphausiacea, and Decapoda. All have both abdominal and thoracic appendages. Decapods include crabs, shrimp, lobster, crayfish, and others; they have five pairs of walking legs (including the chelipeds) on their thorax.

Members of the subphylum Uniramia have uniramous appendages and bear one pair of antennae, a pair of mandibles, and two pairs of maxillae (one pair of maxillae in millipedes) on the head. The tagmata are the head and trunk in the myriapods and head, thorax, and abdomen in the insects.

The Insecta is the largest class of the world's largest phylum. Insects are easily recognized by the combination of their tagmata and the possession of three pairs of thoracic legs.

The radiation and abundance of insects are largely explained by several features allowing them to exploit terrestrial habitats, such as waterproof cuticle and other mechanisms to minimize water loss and the ability to become dormant during adverse conditions.

Feeding habits vary greatly among insects, and there is an enormous variety of specialization of mouthparts reflecting the particular feeding habits of a given insect. Insects breathe by means of a tracheal system, which is a system of tubes that opens by spiracles on the thorax and abdomen. Excretory organs are Malpighian tubules.

Sexes are separate in insects, and fertilization is usually internal. Almost all insects undergo metamorphosis during development. In hemimetabolous (gradual) metamorphosis, the larval instars are called nymphs, and the adult emerges at the last nymphal molt. In holometabolous (complete) metamorphosis, the last larval molt gives rise to a nonfeeding stage (pupa). A winged adult emerges at the final, pupal, molt. Both types of metamorphosis are hormonally controlled.

Insects are important to human welfare, particularly because they pollinate food crop plants, control populations of other, harmful insects by predation and parasitism, and serve as food for other animals. Many insects are harmful to human interests because they feed on crop plants, and many are carriers of important diseases affecting humans and domestic animals.

Arthropods share a common ancestor with annelids, and many zoologists believe that the Arthropoda is monophyletic. Present evidence suggests that the crustaceans and uniramians form a mandibulate clade. The endognathous insects have a number of primitive characters and show similarities with the myriapods. Wings, hemimetabolous metamorphosis, and holometabolous metamorphosis evolved among the ectognathous insects.

Adaptive radiation of the arthropods has been enormous, and they are extremely abundant.

Review Questions

1. Give the characteristics of arthropods that most clearly distinguish them from the Annelida.

2. Name the subphyla of arthropods, and give a few examples of each.

3. Much of the success of the arthropods has been attributed to their cuticle. Why do you think this is so? Describe some other factors that probably contributed to their success.

4. What is a trilobite?

5. What appendages are characteristic of chelicerates?

6. Briefly describe the appearance of each of the following: eurypterids, horseshoe crabs, pycnogonids.

7. Tell the mechanism of each of the following with respect to spiders: feeding, excretion, sensory reception, webspinning, reproduction.

8. Distinguish each of the following orders from each other: Araneae, Scorpionida, Opiliones, Acarina.

9. People fear spiders and scorpions, but ticks and mites are far more important medically and economically. Why? Give examples.

10. What are the tagmata and the appendages on the head of crustaceans? What are some other important characteristics of Crustacea?

11. Of the classes of Crustacea, the Branchiopoda, Maxillopoda, and Malacostraca are the most important. Distinguish them from each other.

12. Distinguish among the subclasses Ostracoda, Copepoda, Branchiura, and Cirripedia of the crustacean class Maxillopoda.

13. Copepods sometimes have been called "insects of the sea" because marine planktonic copepods probably are the most abundant animals in the world. What is their ecological importance?

14. Define each of the following: swimmeret, endopod, exopod, maxilliped, cheliped, uropod, nauplius.

15. Describe the molting process in Crustacea, including the action of the hormones.

16. Explain the mechanism of each of the following with respect to crustaceans: feeding, respiration, excretion, circulation, sensory reception, reproduction.

17. Distinguish the following from each other: Diplopoda, Chilopoda, Insecta.

18. Define each of the following with respect to insects: sclerite, notum, tergum, sternum, pleura, labrum, labium, hypopharynx, haltere, instar, diapause.

19. Explain why wings powered by indirect flight muscles can beat much

more rapidly than those powered by direct flight muscles.

20. What different modes of feeding are found in insects, and how are these reflected in their mouthparts?

21. Describe each of the following with respect to insects: respiration, excretion and water balance, sensory reception, reproduction.

22. Explain the difference between holometabolous and hemimetabolous metamorphosis in insects, including the stages in each.

23. Describe and give an example of each of the four ways insects communicate with each other.

24. What are the castes found in honeybees and in termites, and what is the function of each? What is trophallaxis?

25. Name several ways in which insects are beneficial to humans and several ways they are detrimental.

26. For the past 50 or more years, people have relied on toxic insecticides for control of harmful insects. What problems have arisen resulting from

such reliance on insecticides? What are the alternatives? What is integrated pest management?

27. Some biologists suggest that the Arthropoda is polyphyletic. Explain why this could be so despite the characteristics shared by all arthropods.

28. We believe that the earliest insects were wingless, that is, the lack of wings is the primitive condition, and this was observed in the traditional subclass Apterygota. We now consider the Apterygota paraphyletic. Why?

Selected References

See also general references on page 395.

Berenbaum, M. R. 1995. Bugs in the system. Reading, Massachusetts, Addison-Wesley Publishing Company. *How insects impact human affairs. Well written for a wide audience, highly recommended.*

Blum, M. S., ed. 1985. Fundamentals of insect physiology. New York, John Wiley & Sons. *Good, multiauthored text on insect physiology. Recommended.*

Borror, D. J., D. M. Delong, and C. A. Triplehorn. 1989. An introduction to the study of insects, ed. 6. Philadelphia, Saunders College Publishing. *A good entomology text.*

Cronin, T. W., N. J. Marshall, and M. F. Land. 1994. The unique visual system of the mantis shrimp. Am. Sci. **82:**356-365. *The ancestors of mantis shrimps diverged from other crustaceans about 400 million years ago. Accuracy in the raptorial strike of these aggressive predators requires a highly refined visual system.*

Foelix, R. F. 1982. Biology of spiders. Cambridge, Massachusetts, Harvard University Press. *Attractive, comprehensive book with extensive references; of interest to both amateurs and professionals.*

Hadley, N. F. 1986. The arthropod cuticle. Sci. Am. **234:**100-107 (March). *Modern studies on the chemistry and structure of arthropod cuticle help to explain its remarkable properties.*

Heinrich, B., and H. Esch. 1994. Thermoregulation in bees. Am. Sci. **82:**164-170. *Fascinating behavioral and physiological adaptations for increasing and decreasing body temperature allow bees to function in a surprisingly wide range of environmental temperatures.*

Hölldobler, B. K., and E. O. Wilson. 1990. The ants. Cambridge, Massachusetts, Harvard University Press. *The fascinating story of social organization in ants.*

Kaston, B. J. 1978. How to know the spiders, ed. 3. Dubuque, Iowa, William C. Brown Company, Publishers. *Spiral-bound identification manual.*

Lane, R. P., and R. W. Crosskey, eds. 1993. Medical insects and arachnids. London, Chapman and Hall. *This is the best book currently available on medical entomology.*

McDaniel, B. 1979. How to know the ticks and mites. Dubuque, Iowa, William C. Brown Company, Publishers. *Useful, well-illustrated keys to genera and higher categories of ticks and mites in the United States.*

McMasters, J. H. 1989. The flight of the bumblebee and related myths of entomological engineering. Am. Sci. **77:**164-169. *There is a popular myth about an aerodynamicist who "proved" that a bumblebee cannot fly—but his assumptions were wildly wrong.*

Moffat, A. S. 1991. Research on biological pest control moves ahead. Science **252:**211-212. *A report on the current status of biological pest control, including the contributions of genetic engineering.*

Polis, G. A., ed. 1990. The biology of scorpions. Stanford, California, Stanford University Press. *The editor brings together a readable summary of what is known about scorpions.*

Schram, F. R. 1986. Crustacea. New York, Oxford University Press. *The most recent comprehensive account.*

Shear, W. A. 1994. Untangling the evolution of the web. Am. Sci. **82:**256-266. *Fossil spider webs are nonexistent. Evolution of the web must be studied by comparing modern spider webs to each other and correlating studies of spider anatomy.*

Topoff, H. 1990. Slave-making ants. Am. Sci. **78:**520-528. *An amazing type of social parasitism in which certain species of ants raid the colonies of related species, abduct their pupae, then exploit them to do all the work in the host colony.*

Vollrath, F. 1992. Spider webs and silks. Sci. Am. **266:**70-76 (Mar.). *Spider web design and silk must obey the same constraints as materials used in human structural engineering; we can learn useful lessons from spiders.*

Wooley, T. A. 1988. Acarology. Mites and human welfare. New York, John Wiley.

Wootton, R. J. 1990. The mechanical design of insect wings. Sci. Am. **263:**114-120 (Nov.). *The ingenious architecture of insect wings and how they are adapted to flight.*

Links to the Internet

Visit this textbook's web site at http://www.mhhe.com/zoology to find live Internet links for each of the references listed below.

1. Arthropoda. Arizona's Tree of Life Web Page. Pictures, characteristics, phylogenetic relationships, references on arthropods. Links to specific arthropod groups.

2. Phylum Arthropoda. University of Michigan site describing arthropods. Links to chelicerates (Merostomata, Pycnogonida, Arachnida), crustaceans (remipedians, cephalocarids, branchiopods, maxillopodans, malacostracans), and uniramids (chilopodans, diplopodans, and insects). Each web link has pictures, descriptive material, links, references, and some have other web resources as well.

Chelicerates

3. Introduction to the Cheliceramorpha. University of California at Berkeley Museum of Paleontology. Links to the fossil records, life history and ecology, systematics, and more on morphology. Each site typically includes at least one picture of a member of the taxon.

4. Trilobite Link Page. Information on trilobites, classification, and many links.

5. Arachnology. Many links to information on arachnids are found here. Learn about arachnid societies, scientists, taxonomy, paleontology, and myths. This site also provides links to photographs of arachnids and even a test of your level of arachnophobia for fun.

6. BIOSIS Arachnida. This site provides many links to information on the paleontology, taxonomy, life history, and ecology of members of most arachnid groups.

7. Arachnida. Arizona's Tree of Life Web Page. Pictures, characteristics, phylogenetic relationships, references on arachnids. Links to Araneae, Acari, Scorpionida, and further links from these sites.

8. Spotlight on Lyme Disease. A CDC site with links to fact sheets, definitions, distribution with the United States, public information, plus pictures of the ticks that transmit Lyme disease.

Crustaceans

9. BIOSIS Crustacea. This site provides links to information on the paleontology, taxonomy, life history, and ecology of members of most crustacean groups.

10. Crustacea. Arizona's Tree of Life Web Page. Pictures, characteristics, phylogenetic relationships, references on crustaceans. Links to remipedians, branchipodans, maxillopodans, and malacostracans.

11. Freshwater Crayfish Home Page. This site includes links to many other sites on crayfish, threatened species, diseases, and recipes.

Myriapods

12. Introduction to the Myriapoda. University of California at Berkeley Museum of Paleontology. Links to the fossil record, life history and ecology, systematics, and more on morphology of the Chilopoda, Diplopoda, Symphyla, and Pauropoda. Each site typically includes at least one picture of a member of the taxon.

Insects

13. Introduction to the Uniramia. University of California at Berkeley Museum of Paleontology. Links to the fossil record, life history and ecology, systematics, and more on morphology. Each site typically includes at least one picture of a member of the taxon.

14. Systematics of the Uniramia 2. University of California at Berkeley Museum of Paleontology. Links to many individual orders of insects, such as siphonapterans, coleopterans, lepidopterans, and dipterans. Links to the fossil record, life history and ecology, systematics, and more on morphology. Each site typically has at least one picture of a member of the taxon.

15. Introduction to the Parainsecta. Springtails and the proturans. University of California at Berkeley Museum of Paleontology. Links to the fossil record, life history and ecology, systematics, and more on morphology. Each site typically includes at lease one picture of a member of the taxon.

16. The Virtual Insectary. This site has information on insect feeding habits

and habitats, as well as many images of insects.

17. Insecta. Arizona's Tree of Life Web Page. Pictures, characteristics, phylogenetic relationships, references on insects. Links to paleopterygous and neopterygous insects.

18. Out in the Web. Links for more entomological information.

19. University of Florida Book of Insect Records. Thirty-eight separate chapters on fascinating insects ranging from the longest migration, the longest copulation, the smallest eggs, to the smallest adult.

20. Hexapoda. Arizona's Tree of Life Web Page. Pictures, characteristics, phylogenetic relationships, references on hexapods, with links to springtails, proturans, diplurans, and insects.

21. UD General Key Insects. This interactive key from the University of Delaware combines a dichotomous key with beautiful photographs and photomicrographs to allow keying of insects to the ordinal level. When the insect is identified, information on the order is given, complete with further links for more information, as well as gorgeous pictures of the insect.

22. Neoptera. Arizona's Tree of Life Web Page. Pictures, characteristics, phylogenetic relationships, references on neoptera. Links to other insect orders.

23. Beetles. This is a site devoted entirely to beetles. In addition to seeing photographs, you learn how beetles fly and navigate.

24. Coleopterists Society. Links to information about the Society, web sites, beetles conservation, and more.

25. Smithsonian Institution's Colepteran Web Site. Information on systematics, organizations, jobs, etc..

26. Monarch Watch. Information on the monarch, sightings of monarchs on their migration, and links to other sites.

27. Smithsonian Institution's Dipteran Web Site. Information on systematics, organizations, jobs, etc.

28. Iowa State University's Tasty Insect Recipes. Learn to cook banana worm bread, bug blox, and chocolate chirpie chip cookies. Insects can be nutritious and delicious!

Lesser Protostomes and Lophophorates

Some Evolutionary Experiments

The early Cambrian period, about 570 million years ago, was the most fertile time in evolutionary history. For 3 billion years before this period, evolution had forged little more than bacteria and blue-green algae. Then, within the space of a few million years, all of the major phyla, and probably all of the smaller phyla, became established. This was the Cambrian explosion, the greatest evolutionary "bang" the world has known. In fact, the fossil record suggests that more phyla existed in the Paleozoic era than exist now, but some disappeared during major extinction events that punctuated the evolution of life on earth. The greatest of these disruptions was the Permian extinction about 230 million years ago. Thus evolution has led to many "experimental models." Some of these models failed because they were unable to survive in changing conditions. Others gave rise to abundant and dominant species and individuals that inhabit the world today. Still others radiated but little, with small numbers of species persisting, whereas others were formerly more abundant but are now in decline.

The great evolutionary flow that began with the appearance of the coelom and led to the three huge phyla of molluscs, annelids, and arthropods produced other lines as well. Those that have survived are small and lack great economic and ecological importance; they are usually grouped together as "lesser protostomes." They probably diverged at different times from different ancestors, but in all likelihood each shares a common ancestor with annelids or arthropods or both.

Three phyla—Phoronida, Ectoprocta, and Brachiopoda— are included in this chapter mainly for convenience. They are apparently related to each other because they all possess a crown of cilated tentacles, called a lophophore, used in food capture and respiration. The brachiopods were abundant in the Paleozoic but began to decline thereafter. The exception to the common theme of this chapter is the phylum Ectoprocta. It arose in the Cambrian, became widespread in the Paleozoic, and remains a prevalent group today.

This chapter includes brief discussions of nine coelomate phyla whose position in the phylogenetic lines of the animal kingdom is somewhat problematic. Other than having derived from the main protostome line, we can say little of the relationships of six of the phyla. The Sipuncula, Echiura, and Pogonophora have some annelid-like characters. The Pentastomida, Onychophora, and Tardigrada show some arthropod-like characteristics; they often have been grouped together and called Pararthropoda because they have unjointed limbs with claws (at some stage) and a cuticle that undergoes molting. The relationships of the lophophorate phyla are also obscure, but cladistic analysis places them with the deuterostomes.

The Lesser Protostomes

Three of the lesser protostome phyla are benthic (bottom-dwelling) marine worms. The Pentastomida is an entirely parasitic group, and the onychophorans are terrestrial (but limited to damp areas). Tardigrades are found in marine, freshwater, and terrestrial habitats.

Phylum Sipuncula

The phylum Sipuncula (sy-pun′kyu-la) (L. *sipunculus,* little siphon) consists of about 330 species of benthic marine worms, predominantly littoral or sublittoral. Sometimes called "peanut worms," they live sedentary lives in burrows in mud or sand (Figure 12-1), occupy borrowed snail shells, or live in coral crevices or among vegetation. Some species construct their own rock burrows by chemical and perhaps mechanical means. More than half the species are restricted to tropical zones. Some are tiny, slender worms, but the majority range from 15 to 30 cm in length.

Sipunculans are not metameric, nor do they possess setae. Their head is in the form of an **introvert,** which is crowned by ciliated tentacles surrounding the mouth (Figure 12-1). They are largely deposit feeders, extending the introvert and tentacles from their burrow to explore and feed. They have a cerebral ganglion, nerve cord, and pair of nephridia; the coelomic fluid contains red blood cells bearing a respiratory pigment known as hemerythrin.

Sipunculan larvae are trochophores (Figure 9-5, p. 159), and their early embryological development indicates affinities to the Annelida and Echiura. They appear to have diverged from a common ancestor of the three phyla before metamerism evolved.

Phylum Echiura

The phylum Echiura (ek-ee-yur′a) (Gr. *echis,* viper, + *oura,* tail) consists of marine worms that burrow into mud or sand

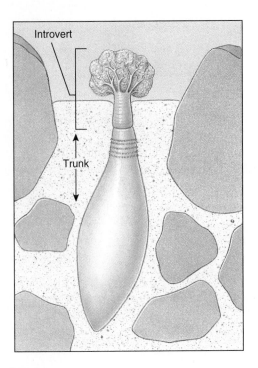

FIGURE 12-1

Themiste, a sipunculan.

or live in empty snail shells, sand dollar tests, or rocky crevices. They are found in all oceans, most commonly in littoral zones of warm waters. They vary in length from a few millimeters to 40 to 50 cm.

Although there are only about 140 species, echiurans are more diverse than sipunculans. Their bodies are cylindrical. Anterior to the mouth is a flattened, extensible proboscis, which, unlike the introvert of sipunculans, cannot be retracted into the trunk. Echiurans are often called "spoonworms" because of the shape of the contracted proboscis in some of them. The proboscis has a ciliated groove leading to the mouth. While the animal lies buried, the proboscis can extend out over the mud for exploration and deposit feeding (Figure 12-2). *Urechis* (Gr. *oura,* tail, + *echis,* viper), however, secretes a mucous net in a U-shaped burrow through which it pumps water and strains out food particles. *Urechis* is sometimes called the "fat innkeeper" because it has characteristic species of commensals living with it in its burrow, including a crab, fish, clam, and polychaete annelid.

Echiurans, with the exception of *Urechis,* have a **closed circulatory system** with a contractile vessel; most have one to three pairs of nephridia (some have many pairs), and all have a nerve ring and ventral nerve cord. A pair of anal sacs arises from the rectum and opens into the coelom; they are probably respiratory in function and possibly accessory nephridial organs.

Early cleavage and trochophore stages are very similar to those of annelids and sipunculans.

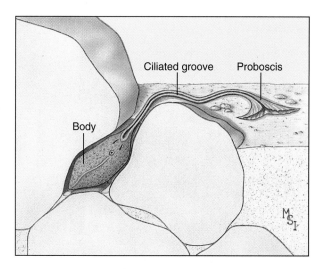

FIGURE 12-2

Bonellia (phylum Echiura) is a detritus feeder. Lying in its burrow, it explores the surface with its long proboscis, which picks up organic particles and carries them along a ciliated groove to the mouth.

In some species of echiurans sexual dimorphism is pronounced, with the female being much the larger of the two. Bonellia has an extreme sexual dimorphism, and sex is determined in a very interesting way. At first the free-swimming larvae are sexually undifferentiated. Those that come into contact with the proboscis of a female become tiny males (1 to 3 mm long) that migrate to the female uterus. About 20 males are usually found in a single female. Larvae that do not contact a female proboscis metamorphose into females. The stimulus for development into males is apparently a pheromone produced by the female proboscis.

Phylum Pogonophora

The phylum Pogonophora (po-go-nof′e-ra) (Gr. *pōgōn*, beard, + *pherō*, to bear), or beardworms, was entirely unknown before the twentieth century. The first specimens to be described were collected from deep-sea dredgings in 1900 off the coast of Indonesia. They have since been discovered in several seas, including the western Atlantic off the eastern coast of the United States. Some 80 species have been described so far.

Most pogonophores live in the bottom ooze on the ocean floor, usually at depths of more than 200 m. Their usual length varies from 5 to 85 cm, with a diameter usually of less than a millimeter. They are sessile and secrete very long chitinous tubes in which they live, probably extending the anterior end only for feeding.

The body has a short forepart, a long, very slender trunk, and a small, metameric opisthosoma (Figure 12-3). They are covered with a cuticle similar in structure to that of annelids and sipunculans, and they bear setae on the trunk and opisthosoma similar to those of annelids. A series of coelomic compartments divides the body. The forepart bears from one to many tentacles.

Pogonophores are remarkable in having no mouth or digestive tract, making their mode of nutrition a puzzling matter. They absorb some nutrients dissolved in seawater, such as glucose, amino acids, and fatty acids, through the pinnules and microvilli of their tentacles. They apparently derive most of their energy, however, from a mutualistic association with chemoautotrophic bacteria. These bacteria oxidize hydrogen sulfide to provide the energy to produce organic compounds from carbon dioxide. An organ called the **trophosome** bears the bacteria. The trophosome develops embryonically from the midgut (all traces of the foregut and hindgut are absent in the adult).

There is a well-developed, closed, blood vascular system. Sexes are separate.

Pogonophores have photoreceptor cells very similar to those of annelids (oligochaetes and leeches), and the structure of the cuticle, the makeup of the setae, and the segmentation of the opisthosoma all point strongly toward a common ancestor with the annelids.

Among the most amazing animals found in the deep-water, Pacific rift communities are the giant pogonophorans, Riftia pachyptila. Much larger than any pogonophores reported before, they measure up to 1.5 m in length and 4 cm in diameter. Some authors consider them a separate phylum, the **Vestimentifera.** *The trophosome of other pogonophores is confined to the posterior part of the trunk, which is buried in sulfide-rich sediments, but the trophosome of Riftia occupies most of its large trunk. It has a much larger supply of hydrogen sulfide, enough to nourish its large body, in the effluent of the hydrothermal vents.*

Phylum Pentastomida

The wormlike Pentastomida (pen-ta-stom′i-da) (Gr. *pente*, five, + *stoma*, mouth) are parasites, 2.5 to 12 cm long, that are found in the lungs and nasal passages of carnivorous vertebrates—most commonly in reptiles (Figure 12-4). Some human infections have been found in Africa and Europe. The intermediate host is usually a vertebrate that is eaten by the final host.

They have four clawlike appendages at their anterior end, and their body is covered by a chitinous cuticle, which they periodically molt during juvenile stages. Pentastomids

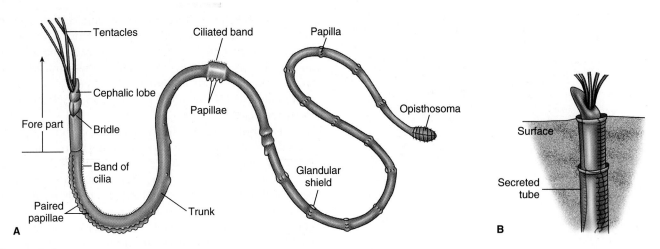

FIGURE 12-3

Diagram of a typical pogonophoran. **A,** External features. The body, in life, is much more elongated than shown in this diagram. **B,** Position in its tube.

show arthropod affinities, but there is little agreement as to where they fit in that phylum. On the basis of the structure of their spermatozoa, pentastomids seem to be closest to the crustacean subclass Branchiura. Some authorities consider Pentastomida a subphylum of Arthropoda. Their extensive modifications for parasitic life make their ancestry difficult to determine.

Phylum Onychophora

Members of the phylum Onychophora (on-i-kof'o-ra) (Gr. *onyx,* claw, + *pherō,* to bear) are called "velvet worms" or "walking worms." They are about 70 species of caterpillar-like animals, 1.4 to 15 cm long, that live in rain forests and other tropical and semitropical leafy habitats.

The fossil record of the onychophorans shows that they have changed little in their 500-million-year history. They were originally marine animals and were probably far more common than they are now. Today they are all terrestrial and are extremely retiring, coming out only at night or when the air is nearly saturated with moisture.

Onychophorans are covered by a soft cuticle, which contains chitin and protein. Their wormlike bodies are carried on 14 to 43 pairs of stumpy, unjointed legs, each ending with a flexible pad and two claws (Figure 12-5). The head bears a pair of flexible antennae with annelid-like eyes at the base.

They are air breathers, using a **tracheal system** that connects with pores scattered over the body. The tracheal system, although similar to that of arthropods, probably evolved independently. Other arthropod-like characteristics include an open circulatory system with a tubular heart, a hemocoel for a body cavity, and a large brain. Annelid-like characteristics include segmentally arranged nephridia, a muscular body wall, and pigment-cup ocelli.

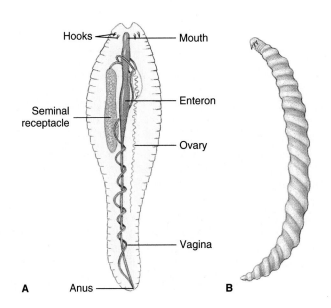

FIGURE 12-4

Two pentastomids. **A,** *Linguatula,* found in the nasal passages of carnivorous mammals. The female is shown with some internal structures. **B,** Female *Armillifer,* a pentastomid with pronounced body rings. In parts of Africa and Asia, humans are parasitized by immature stages; adults (10 cm long or more) live in the lungs of snakes. Human infection may occur from eating snakes or from contaminated food and water.

Onychophorans are dioecious. In some species there is a placental attachment between mother and young, and the young are born as juveniles (viviparous); in others the young are also not encased in a shell when released from the mother, but they develop in the uterus without attachment (ovoviviparous). Two Australian genera are oviparous and lay shell-covered eggs in moist places.

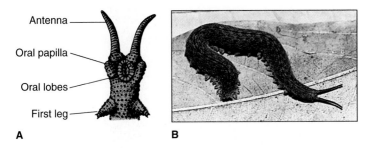

Antenna

Oral papilla

Oral lobes

First leg

A

B

FIGURE 12-5

Peripatus, a caterpillar-like onychophoran that has characteristics in common with both annelids and arthropods. **A,** Ventral view of head. **B,** In natural habitat.

Phylum Tardigrada

Tardigrada (tar-di-gray′da) (L. *tardus,* slow, + *gradus,* step), or "water bears," are minute forms usually less than a millimeter in length. Most of the 300 to 400 species are terrestrial forms that live in the water film that surrounds mosses and lichens. Some live in fresh water, and a few are marine.

The body bears eight short, **unjointed legs,** each with claws (Figure 12-6). Unable to swim, they creep about awkwardly, clinging to the substrate with their claws. A pair of sharp stylets and a sucking pharynx adapt them for piercing and sucking plant cells or small prey such as nematodes and rotifers.

There is a body covering of nonchitinous **cuticle** that is molted several times during the life cycle. As in arthropods, muscle fibers are attached to the cuticular exoskeleton, and the body cavity is a hemocoel.

The annelid-type nervous system is surprisingly complex, and in some species there is a pair of eyespots. Circulatory and respiratory organs are lacking.

Females may deposit their eggs in the old cuticle as they molt or attach them to a substrate. Embryonic formation of the coelom is enterocoelous, a deuterostome characteristic. Nevertheless, their numerous arthropod-like characteristics strongly suggest a common ancestry with the Arthropoda (see Figure 12-13).

One of the most intriguing features of terrestrial tardigrades is their capacity to enter a state of suspended animation, called **cryptobiosis,** during which metabolism is virtually imperceptible. Under gradual drying conditions they reduce the water content of their body from 85% to only 3%, movement ceases, and the body becomes barrel shaped. In a cryptobiotic state tardigrades can withstand harsh environmental conditions: temperature extremes, ionizing radiation, oxygen deficiency, etc., and many survive for years. Activity resumes when moisture is again available.

The Lophophorates

The three lophophorate phyla may appear to have little in common. The **phoronids** (phylum Phoronida) are wormlike

FIGURE 12-6

Scanning electron micrograph of *Echiniscus maucci,* phylum Tardigrada. This species is 300 to 500 µm long. Unable to swim, it clings to moss or water plants with its claws, and if the environment dries up, it goes into a state of suspended animation and "sleeps away" the drought.

marine forms that live in secreted tubes in sand or mud or attached to rocks or shells. The **ectoprocts** (phylum Ectoprocta) are minute forms, mostly colonial, whose protective cases often form encrusting masses on rocks, shells, or plants. The **brachiopods** (phylum Brachiopoda) are bottom-dwelling marine forms that superficially resemble molluscs because of their bivalved shells. All have a free-swimming larval stage but are sessile as adults.

One may wonder why these three apparently widely different types of animals are considered together. They are all coelomate; all have some deuterostome and some protostome characteristics; all are sessile; and none has a distinct head. But these characteristics are also shared by other phyla. What really sets them apart from other phyla is the common possession of a ciliary feeding device called a **lophophore** (Gr. *lophas,* crest or tuft, + *phorein,* to bear).

A lophophore is a unique arrangement of ciliated tentacles borne on a ridge (a fold of the body wall), which surrounds the mouth but not the anus. The lophophore with its crown of tentacles contains within it an extension of the coelom, and the thin, ciliated walls of the tentacles not only constitute an efficient feeding device but also serve as a respiratory surface for exchange of gases between environmental water and coelomic fluid. The lophophore can usually be extended for feeding or withdrawn for protection.

In addition, all three phyla have a U-shaped alimentary canal, with the anus placed near the mouth but outside the lophophore. The coelom is divided primitively into three compartments, the **protocoel,** the **mesocoel** and the **metacoel,** and the mesocoel extends into the hollow tentacles of the lophophore. The protocoel, where present, forms a cavity in a flap over the mouth, the **epistome.** The portion of the body that contains the mesocoel is known as the **mesosome,** and that containing the metacoel is the **metasome.**

Phylum Phoronida

The phylum Phoronida (fo-ron′i-da) (Gr. *phoros,* bearing, + L. *nidus,* nest) comprises approximately 10 species of small wormlike animals that live on the bottom of shallow coastal waters, especially in temperate seas. The phylum name refers to the tentacled lophophore. Phoronids range from a few millimeters to 30 cm in length. Each worm secretes a leathery or chitinous tube in which it lies free, but which it never leaves (Figure 12-7). The tubes may be anchored singly or in a tangled mass on rocks, shells, or pilings or buried in the sand. The tentacles on the lophophore are thrust out for feeding, but if the animal is disturbed it can withdraw completely into its tube.

The lophophore has two parallel ridges curved in a horseshoe shape, the bend located ventrally and the mouth lying between the two ridges. The cilia on the tentacles direct a water current toward a groove between the two ridges, which leads toward the mouth. Plankton and detritus caught in this current become entangled in mucus and are carried by the cilia to the mouth.

Mesenteric partitions divide the coelomic cavity into proto-, meso-, and metacoel, similar to the compartments of deuterostomes. Phoronids have a closed system of contractile blood vessels but no heart; the red blood contains hemoglobin. There is a pair of metanephridia. A nerve ring sends nerves to the tentacles and body wall.

There are both monoecious (the majority) and dioecious species of Phoronida, and at least one species reproduces asexually. Cleavage seems to be related to both spiral and radial types.

Phylum Ectoprocta

The Ectoprocta (ek′to-prok′ta) (Gr. *ektos,* outside, + *proktos,* anus) have long been called bryozoans (Gr. *bryon,* moss, + *zoōn,* animal), or moss animals, a term that originally included the Entoprocta also.

Of the 4000 or so species of ectoprocts, few are more than 0.5 mm long; all are aquatic, both freshwater and marine, but they largely occur in shallow waters; and most, with very few exceptions, are colony builders. Ectoprocts, unlike the other phyla considered in this chapter, were abundant and widespread in the past and remain so today. They left a rich fossil record since the Ordovician era. Modern marine forms exploit all kinds of firm surfaces, such as shells, rock, large brown algae, mangrove roots, and ship bottoms. They are one of the most important groups of fouling organisms on boat hulls; they decrease efficiency of the hull passing through the water and make periodic scraping of the hull necessary.

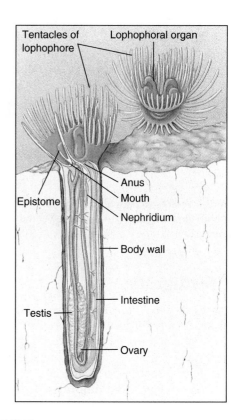

FIGURE 12-7

Internal structure of *Phoronis* (phylum Phoronida), in diagrammatic vertical section.

Each member of a colony lives in a tiny chamber, called a **zoecium,** which is secreted by its epidermis (Figure 12-8A). Each individual, or **zooid,** consists of a feeding **polypide** and a case-forming **cystid.** The polypide includes the lophophore, digestive tract, muscles, and nerve centers. The cystid is the body wall of the animal, together with its secreted exoskeleton. The exoskeleton, or zoecium, may, according to the species, be gelatinous, chitinous, or stiffened with calcium and possibly also impregnated with sand. The shape may be boxlike, vaselike, oval, or tubular.

Some colonies form limy encrustations on seaweed, shells, and rocks (Figure 12-9); others form fuzzy or shrubby growths or erect, branching colonies that look like seaweed. Some ectoprocts might easily be mistaken for hydroids but can be distinguished under a microscope by the presence of ciliated tentacles (Figure 12-10). In some freshwater forms individuals are borne on finely branching stolons that form delicate tracings on the underside of rocks or plants. Other freshwater ectoprocts are embedded in large masses of gelatinous material. Although the zooids are minute, the colonies may be several centimeters in diameter; some encrusting colonies may be a meter or more in width, and erect forms may reach 30 cm or more in height.

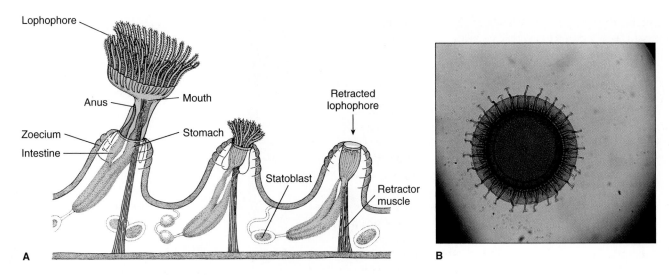

FIGURE 12-8

A, Small portion of freshwater colony of *Plumatella* (phylum Ectoprocta), which grows on the underside of rocks. These tiny individuals disappear into their chitinous zoecia when disturbed. **B,** Statoblast of a freshwater ectoproct, *Cristatella.* Statoblasts are a kind of bud that survives over winter when the colony dies in the autumn. This one is about 1 mm in diameter and bears hooked spines.

The polypide lives in a type of jack-in-the-box existence, popping up to feed and then quickly withdrawing into its little chamber, which often has a tiny trapdoor (operculum) that shuts to conceal its inhabitant. To extend the tentacular crown, certain muscles contract, which increases the hydrostatic pressure within the body cavity and pushes out the lophophore. Other muscles can contract to withdraw the crown to safety with great speed.

The lophophore ridge tends to be circular in marine ectoprocts and U-shaped in freshwater species. When feeding, the animal extends the lophophore, and the tentacles spread out to a funnel shape (Figure 12-10). Cilia on the tentacles draw water into the funnel and out between the tentacles. Food particles trapped in mucus in the funnel are drawn into the mouth, both by the pumping action of the muscular pharynx and by the action of cilia in the pharynx.

Respiratory, vascular, and excretory organs are absent. Gaseous exchange occurs through the body surface, and, since the ectoprocts are small, the coelomic fluid is adequate for internal transport. Coelomocytes engulf and store waste materials. There is a ganglionic mass and a nerve ring around the pharynx, but no sense organs are present. A septum divides the coelom into an anterior mesocoel in the lophophore and a larger posterior metacoel. The protocoel and epistome are present only in freshwater ectoprocts. Pores in the walls between adjoining zooids permit exchange of materials by way of the coelomic fluid.

Feeding individuals make up most colonies, but polymorphism also occurs. One type of modified zooid resembles a bird beak that snaps at small invading organisms that

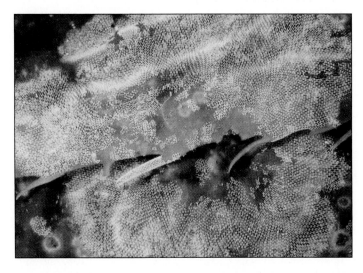

FIGURE 12-9

Skeletal remains of a colony of *Membranipora*, a marine encrusting form of Ectoprocta. Each little oblong zoecium is the calcareous former home of a tiny ectoproct.

might foul a colony. Another type has a long bristle that sweeps away foreign particles.

Most ectoprocts are hermaphroditic. Some species shed eggs into the seawater, but most brood their eggs, some within the coelom and some externally in a special zoecium in which the embryo develops. Cleavage is radial but apparently determinate.

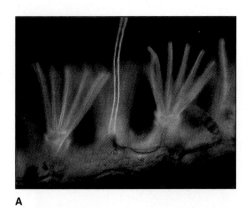

Lingula
(inarticulate)
A

Terebratella
(articulate)
B

FIGURE 12-11

Brachipods. **A,** *Lingula,* an inarticulate brachiopod that normally occupies a burrow. The contractile pedicel can withdraw the body into the burrow. **B,** An articulate brachiopod, *Terebratella.* The valves have a tooth-and-socket articulation and a short pedicel projects through one valve to attach to the substratum (pedicel shown in Figure 12-12).

FIGURE 12-10

A, Ciliated lophophore of *Electra pilosa,* a marine ectoproct. **B,** *Plumatella repens,* a freshwater ectoproct. It grows on the underside of rocks and vegetation in lakes, ponds, and streams.

Brooding is often accompanied by degeneration of the lophophore and gut of the adults, the remains of which contract into minute dark balls, or **brown bodies.** Later, new internal organs regenerate in the old chambers. The brown bodies may remain passive or may be taken up and eliminated by the new digestive tract—an unusual kind of storage excretion.

Freshwater species reproduce both sexually and asexually. Asexual reproduction is by budding or by means of **stato-blasts,** which are hard, resistant capsules containing a mass of germinative cells that form during the summer and fall (Figure 12-8B). When the colony dies in late autumn, the statoblasts are released, and in spring they can give rise to new polypides and eventually to new colonies.

Phylum Brachiopoda

The Brachiopoda (brak'i-op'o-da) (Gr. *brachiōn,* arm, + *pous, podos,* foot), or lamp shells, is an ancient group. Compared with the fewer than 300 species now living, some 30,000 fossil species, which flourished in the Paleozoic and Mesozoic seas, have been described. The brachiopods were once very abundant, but they are now apparently in decline. Some modern forms have changed little from the early ones. The genus *Lingula* (L., little tongue) (Figure 12-11A) has existed virtually unchanged for over 400 million years. Most modern brachiopod shells range between 5 and 80 mm, but some fossil forms reached 30 cm in length.

Brachiopods are all attached, bottom-dwelling, marine forms that mostly prefer shallow water. Externally bra-

chiopods resemble the bivalved molluscs in having two calcareous shell valves secreted by the mantle. They were, in fact, classed with the molluscs until the middle of the nineteenth century, and their name refers to the arms of the lophophore, which were thought homologous to the mollusc foot. Brachiopods, however, have **dorsal** and **ventral valves** instead of right and left lateral valves as do the bivalve molluscs and, unlike the bivalves, most of them are attached to a substrate either directly or by means of a fleshy stalk called a **pedicel** (or pedicle).

In most brachiopods the ventral (pedicel) valve is slightly larger than the dorsal (brachial) valve, and one end projects in the form of a short pointed beak that is perforated where the fleshy stalk passes through (Figure 12-11B). In many the shape of the pedicel valve is that of the classical oil lamp of Greek and Roman times, so that the brachiopods came to be known as the "lamp shells."

There are two classes of brachiopods based on shell structure. The shell valves of Articulata are connected by a hinge with an interlocking tooth-and-socket arrangement (articular process); those of Inarticulata lack the hinge and are held together by muscles only.

The body occupies only the posterior part of the space between the valves (Figure 12-12A), and extensions of the body wall form mantle lobes that line and secrete the shell. The large horseshoe-shaped lophophore in the anterior mantle cavity bears long ciliated tentacles used in respiration and feeding. Ciliary water currents carry food particles between

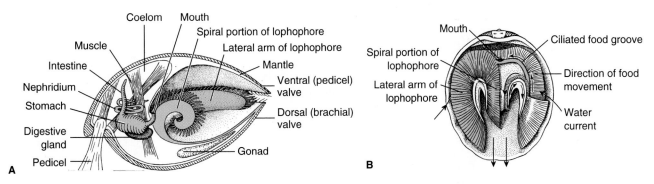

FIGURE 12-12

Brachiopod anatomy. **A,** An articulate brachiopod (longitudinal section). **B,** Feeding and respiratory currents. The large arrows show water flow over the lophophore; the small arrows indicate food movement toward the mouth in the ciliated food groove.

the gaping valves and over the lophophore. Food is caught in mucus on the tentacles and carried in a ciliated food groove along the arm of the lophophore to the mouth (Figure 12-12).

There is no cavity in the epistome of articulates, but in inarticulates there is a protocoel in the epistome that opens into the mesocoel. As in the other lophophorates, the posterior metacoel bears the viscera. One or two pairs of nephridia open into the coelom and empty into the mantle cavity. There is an open circulatory system with a contractile heart. There is a nerve ring with a small dorsal and a larger ventral ganglion.

Sexes are separate and paired gonads discharge gametes through the nephridia. The development of brachiopods is similar in some ways to that of the deuterostomes, with radial, mostly equal, holoblastic cleavage and the coelom forming enterocoelically in the articulates. The free-swimming larva of the articulates resembles a trochophore.

Phylogeny

The early embryological development of sipunculans, echiurans, and annelids is almost identical, showing a very close relationship among the three phyla. Sipunculans and echiurans are not metameric and thus are more primitive in that characteristic than annelids. They probably represent collateral evolutionary lines that branched from protoannelid stock before the origin of metamerism.

Several characters suggest relationship of the Pogonophora to the Annelida, as we noted previously.

The phylogenetic affinities of the Pentastomida are uncertain. Most modern taxonomists align them with the arthropods, however, and evidence is accumulating that they are most closely related to the crustacean subclass Branchiura (p. 210). This evidence includes similarities in morphology of their sperm and in base sequences of ribosomal RNA. If the pentastomids really are close to the branchi-

urans, then their status as a phylum should be revoked, and they should be classified as crustacean arthropods.

Onychophorans share a number of characteristics with the annelids: metamerically arranged nephridia, muscular body wall, pigment cup ocelli, and ciliated reproductive ducts. Characteristics shared with arthropods include the cuticle, tubular heart and hemocoel with open circulatory system, presence of tracheae (probably not homologous), and large size of the brain. Unique characteristics include oral papillae, slime glands, body tubercles, and suppression of external segmentation. Some authors believe the onychophorans should be included with the arthropods, but that would require redefining the phylum Arthropoda. Most authors believe that the differences seem to warrant keeping them in a separate phylum (Figure 12-13).

The affinities of tardigrades are among the most puzzling of all animal groups. They have some similarities to rotifers, particularly in their reproduction and their cryptobiotic tendencies, and some authors have called them pseudocoelomates. Their embryogenesis, however, would seem to put them among the coelomates. The enterocoelic origin of the mesoderm is a deuterostome characteristic. Other authors identify several important synapomorphies that suggest an ancestor in common with the arthropods (Figure 12-13).

The phylogenetic position of the lophophorates has been the subject of much controversy and debate. Sometimes they have been considered protostomes with some deuterostome characters, and at other times deuterostomes with some protostome characters. Brusca and Brusca[1] contended that there is overwhelming evidence that they are a monophyletic clade and are deuterostomes. Their common possession of a lophophore is a unique synapomorphy. Other features, such as the U-shaped digestive tract, metanephridia (except in ectoprocts), and tendency to secrete outer casings may be homologous within the clade, but they are convergent with many other taxa.

[1]Brusca, R. C., and G. J. Brusca. 1990. *Invertebrates.* Sunderland, Massachusetts, Sinauer Associates.

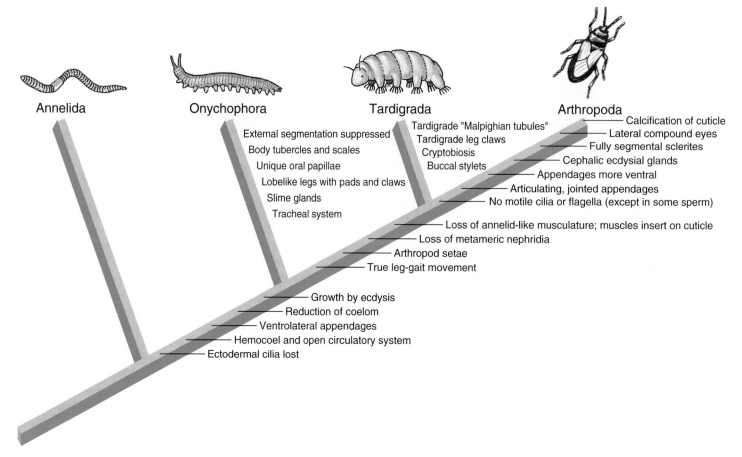

FIGURE 12-13

Cladogram depicting hypothetical relationships of Onychophora and Tardigrada to annelids and arthropods. Onychophorans diverged from the arthropod line after the development of such synapomorphies as hemocoel and growth by ecdysis, but they share several primitive characters with the annelids, such as the metameric arrangement of the nephridia. Note that the tracheal system of onychophorans is not homologous to that of arthropods but represents a convergence. The phylogenetic relationships of other phyla covered in this chapter are too difficult to evaluate to permit construction of a cladogram that includes them.

Summary

The nine small, coelomate phyla considered in this chapter are grouped together here for convenience. The Sipuncula, Echiura, and Pogonophora probably share an ancestor with annelids; the Pentastomida, Onychophora, and Tardigrada apparently have a common ancestor with the arthropods. The lophophorates (Phoronida, Ectoprocta, and Brachiopoda) have some characteristics of both protostomes and deuterostomes.

Sipunculans and echiurans are burrowing marine worms; neither shows metamerism. Pogonophorans live in tubes on the deep ocean floor and have no digestive tract. They are metameric.

Although the Pentastomida have certain arthropod-like characteristics, their specialized modifications complicate determination of their phylogenetic relationships. They may be related to the crustacean subclass Branchiura. Onychophora are caterpillar-like animals that are metameric and show some annelid and arthropod characteristics. Tardigrades are minute, mostly terrestrial animals that have a hemocoel, as do arthropods.

The Phoronida, Ectoprocta, and Brachiopoda all bear a lophophore, which is a crown of ciliated tentacles surrounding the mouth but not the anus and containing an extension of the mesocoel. They are also sessile as adults, have a U-shaped digestive tract, and have free-swimming larvae. The lophophore functions both as a respiratory and a feeding structure.

Phoronida are the least abundant of the lophophorates, living in tubes mostly in shallow coastal waters.

Ectoprocts are abundant in marine habitats, living on a variety of submerged substrata, and a number of species are common in fresh water. Ectoprocts are colonial, and although each individual is quite small, the colonies are commonly several centimeters or more in width or height.

Brachiopods were a widely prevalent phylum in the Paleozoic era but have been declining since the early Mesozoic era. They have a dorsal and a ventral shell and usually attach to the substrate directly or by means of a pedicel.

Because of their possession of a unique synapomorphy, the lophophore, the lophophorates appear to form a clade.

Review Questions

1. Give the main distinguishing characteristics of the following, and tell where each lives: Sipuncula, Echiura, Pogonophora, Pentastomida, Onychophora, Tardigrada.

2. What do the members of each of the aforementioned groups eat?

3. What is the evidence that the Sipuncula and Echiura share an ancestor with the annelids?

4. What is the largest pogonophoran known? Where is it found, and how is it nourished?

5. Some investigators regard the Onychophora as a "missing link" between the Annelida and the Arthopoda. Give evidence for and against this hypothesis.

6. What is the survival value of cryptobiosis in tardigrades?

7. Some investigators consider the lophophorates a monophyletic clade within the deuterostomes. What are some synapomorphies that distinguish this clade, and what are some synapomorphies that diagnose each phylum of lophophorates?

8. Define each of the following: lophophore, zoecium, zooid, polypide, cystid, brown bodies, statoblasts.

9. What are the coelomic compartments found in the lophophorates?

10. Brachiopods superficially resemble bivalve molluscs. How would you explain the difference to a layperson?

11. The lophophorates are sometimes placed in a phylogenetic position between the protostomes and the deuterostomes. How would you justify such placement?

Selected References

See also general references on page 395.

American Society of Zoologists. 1977. Biology of lophophorates. Am. Zool. **17**(1):3–150. *A collection of 13 papers.*

Childress, J. J., H. Felbeck, and G. N. Somero. 1987. Symbiosis in the deep sea. Sci. Am. **256**:114–120 (May). *The amazing story of how the animals around deep-sea vents, including Riftia pachyptila, manage to absorb hydrogen sulfide and transport it to their mutualistic bacteria. For most animals, hydrogen sulfide is highly toxic.*

Crowe, J. H., and A. F. Cooper, Jr. 1971. Cryptobiosis. Sci. Am. **225**:30–36

(Dec.). *Cryptobiotic nematodes, rotifers, and tardigrades can withstand adverse conditions of astonishing rigor, yet perceptible metabolism continues in their state of suspended animation.*

Gould, S. J. 1995. Of tongue worms, velvet worms, and water bears. Natural History **104**(1):6–15. *Intriguing essay on affinities of Pentastomida, Onychophora, and Tardigrada and how they, along with larger phyla, were products of the Cambrian explosion.*

Haugerud, R. E. 1989. Evolution in the pentastomids. Parasitol. Today **5**:126–132. *Much remains to be learned of this puzzling group, but*

there is strong evidence of its crustacean affinities.

Rice, M. E., and M. Todorovic, eds. 1975. Proceedings of the International Symposium on the biology of the Sipuncula and Echiura, 2 vols. Washington, D.C., National Museum of Natural History. *A series of technical articles, but much of interest for further reading on these two phyla.*

Richardson, J. R. 1986. Brachiopods. Sci. Am. **255**:100–106 (Sept.). *Reviews brachiopod biology and adaptations and contends that in the next few million years there may be an increase in the number of species, rather than a further decline.*

Links to the Internet

Visit this textbook's Web site at http://www.mhhe.com/zoology to find live Internet links for each of the references listed below.

1. Pogonophora. Arizona's Tree of Life Web Page. Pictures, characteristics, phylogenetic relationships, references on pogonophorans. References on these interesting tube worms.

2. Web sites on various worms and lesser-known phyla. Arizona's Tree of Life Web Page. Some pictures, characteristics, phylogenetic relationships, but mostly references on these phyla.

3. Tardigrade Appreciation Headquarters. A site devoted to appreciation and information about tardigrades (sometimes called moss piglets), complete with links to other sites on tardigrades.

4. Introduction to the Onychophora. This University of California at Berkeley Museum of Paleontology site contains information on the onychophorans, which may share an ancestor with the arthropods. Links to the fossil record, life history and ecology, systematics, and more on morphology.

5. Brachiopoda. Arizona's Tree of Life Web Page. Pictures, characteristics, phylogenetic relationships, references

on this group of lophophorates. References on brachiopods.

6. Introduction to the Brachiopoda. University of California at Berkeley Museum of Paleontology. Links to the fossil record, life history and ecology, systematics, and more on morphology. Each site typically includes at least one picture of a member of the taxon.

7. Introduction to the Bryozoa. University of California at Berkeley Museum of Paleontology. Links to the fossil record, life history and ecology, systematics, and more on morphology. Each site typically includes at least one picture of a member of the taxon.

Echinoderms, Hemichordates, and Chaetognaths

CHAPTER thirteen

A Design To Puzzle the Zoologist

The distinguished American zoologist Libbie Hyman once described the echinoderms as a "noble group especially designed to puzzle the zoologist." With a combination of characteristics that should delight the most avid reader of science fiction, the echinoderms would seem to confirm Lord Byron's observation that

> Tis strange—but true;
> for truth is always strange;
> Stranger than fiction.

Despite the adaptive value of bilaterality for free-moving animals, and the merits of radial symmetry for sessile animals, echinoderms confounded the rules by becoming free moving but radial. That they evolved from a bilateral ancestor there can be no doubt, for their larvae are bilateral. They undergo a bizarre metamorphosis to a radial adult in which there is a 90° reorientation in body axis, with a new mouth arising on the left side, and a new anus appearing on the right side.

A compartment of the coelom has been transformed in echinoderms into a unique water-vascular system that uses hydraulic pressure to power a multitude of tiny tube feet used in food gathering and locomotion. An endoskeleton of dermal ossicles may fuse together to invest the echinoderm in armor, or it may be reduced in some to microscopic bodies. Many echinoderms have miniature jawlike pincers (pedicellariae) scattered on their body surface, often stalked and some equipped with poison glands, that keep their surface clean by snapping at animals that would settle there.

This constellation of characteristics is unique in the animal kingdom. It has both defined and limited the evolutionary potential of the echinoderms. Despite the vast amount of research that has been devoted to them, we are still far from understanding many aspects of echinoderm biology.

The Echinodermata, along with the chordates, the lophophorates, and the Hemichordata (acorn worms and pterobranchs) are deuterostomes. Typical deuterostome embryogenesis is shown only by some chordates such as amphioxus (p. 276), but these shared characters support monophyly of the Deuterostomia. Nonetheless, their evolutionary history has taken the echinoderms to the point where they are very much unlike any other animal group. The phylum Chaetognatha (arrowworms) traditionally has been included among the deuterostomes, but this arrangement is not supported by recent molecular evidence.[1] Until the issue is clarified, we will cover the chaetognaths in this chapter for convenience.

Phylum Echinodermata

The Echinodermata are marine forms and include the classes Asteroidea (sea stars [or starfishes]), Ophiuroidea (brittle stars), Echinoidea (sea urchins), Holothuroidea (sea cucumbers), and Crinoidea (sea lilies). Echinoderms have a combination of characteristics that are found in no other phylum: (1) an endoskeleton of plates or ossicles, usually spiny, (2) the water-vascular system, (3) the pedicellariae, (4) the dermal branchiae, and (5) secondary radial or biradial symmetry. The water-vascular system and the dermal ossicles have been particularly important in determining the evolutionary potential and limitations of this phylum. Their larvae are bilateral and undergo a metamorphosis to a radial adult.

---

[1]Telford, M. J., and P. W. H. Holland. 1993. Mol. Biol. Evol. **10**:660–676; Wada, H., and N. Satoh. 1994. Proc. Natl. Acad. Sci. **91**:1801–1804.

Ecological Relationships

Echinoderms are all marine; they have no ability to osmoregulate and so are rarely found in waters that are brackish. Virtually all benthic as adults, they are found in all oceans of the world and at all depths, from the intertidal to the abyssal regions.

Some sea stars (Figure 13-1) are particle feeders, but many are predators, feeding particularly on sedentary or sessile prey. Brittle stars (see Figure 13-7) are the most active echinoderms, moving by their arms; and they may be scavengers, browsers, or deposit or filter feeders. Some brittle stars are commensals with sponges. Compared to other echinoderms, sea cucumbers (see Figure 13-14) are greatly extended in the oral-aboral axis and are oriented with that axis more or less parallel to the substrate and lying on one side. Most are suspension or deposit feeders. "Regular" sea urchins (see Figure 13-10), which are radially symmetrical, prefer hard bottoms and feed chiefly on algae or detritus. "Irregular" urchins (sand dollars and heart urchins) (see Figure 13-11), which have become secondarily bilateral, are usually found on sand and feed on small particles. Sea lilies and feather stars (see Figures 13-17 and 13-18) stretch their arms out and up like a flower's petals and feed on plankton and suspended particles.

Class Asteroidea: Sea Stars

Sea stars, often called starfishes, demonstrate the basic features of echinoderm structure and function very well, and they are easily obtainable. We shall consider them first, then comment on the major differences shown by the other groups.

Position in Animal Kingdom

1. Phylum Echinodermata (e-ki′no-der′ma-ta) (Gr. *echinos*, sea urchin, hedgehog, + *derma*, skin, + *ata*, characterized by) belongs to the **Deuterostomia** branch of the animal kingdom, the members of which are enterocoelous coelomates. The other phyla traditionally assigned to this group are Chaetognatha, Hemichordata, and Chordata, but recent evidence questions placement of chaetognaths in the deuterostomes. We are also placing the lophophorate phyla (Phoronida, Ectoprocta, and Brachiopoda) in the Deuterostomia.

2. Primitively, deuterostomes have the following embryological features in common: anus developing from or near the blastopore, and mouth developing elsewhere; coelom budded off from the archenteron (enterocoel); radial and regulative (indeterminate) cleavage; and endomesoderm (mesoderm derived from or with the endoderm) from enterocoelic pouches.

Biological Contributions

1. There is one word that best describes the echinoderms: strange. They have a unique constellation of characteristics found in no other phylum. Among the more striking of the features shown by the echinoderms are as follows:

   a. The system of channels composing the **water-vascular system,** derived from a coelomic compartment.

   b. The **dermal endoskeleton** composed of calcareous ossicles.

   c. The **hemal system,** whose function remains mysterious, also enclosed in a coelomic compartment.

   d. Their **metamorphosis,** which changes a bilateral larva to a radial adult.

Characteristics of Phylum Echinodermata

1. Body not metameric, adult with **radial, pentamerous symmetry** characterized by five or more radiating areas

2. No head or brain; few specialized sensory organs

3. Nervous system with circumoral ring and radial nerves

4. **Endoskeleton** of **dermal calcareous ossicles** with **stereom** structure; covered by an epidermis (ciliated in most); pedicellariae (in some)

5. A **water-vascular system** of coelomic origin that extends from the body surface as a series of tentacle-like projections (podia or tube feet)

6. **Locomotion by tube feet,** which project from the **ambulacral areas,** or by movement of spines, or by movement of arms, which project from central disc of the body

7. Digestive system usually complete; axial or coiled; anus absent in ophiuroids

8. Coelom extensive, forming the perivisceral cavity and the cavity of the water-vascular system; coelom of enterocoelous type

9. So-called **hemal system** present, of uncertain function but playing little, if any, role in circulation of body fluids, and surrounded by extensions of the coelom (perihemal sinuses)

10. Respiration by dermal branchiae, tube feet, respiratory tree (holothuroids), and bursae (ophiuroids)

11. Excretory organs absent

12. Sexes separate (except a few hermaphroditic); fertilization usually external

13. Development through free-swimming, bilateral, larval stages (some with direct development); metamorphosis to radial adult or subadult form

A **B**

C **D**

FIGURE 13-1

Some sea stars (class Asteroidea) from the Pacific. **A,** Cushion star *Pteraster tesselatus* can secrete incredible quantities of mucus as a defense. **B,** *Choriaster granulatus* scavenges dead animals on shallow Pacific reefs. **C,** *Tosia queenslandensis* from the Great Barrier Reef browses encrusting organisms. **D,** *Crossaster papposus*, rose star, feeds on other sea stars.

Sea stars are familiar along shorelines, where sometimes large numbers may aggregate on the rocks. They also live on muddy or sandy bottoms and among coral reefs. They are often brightly colored and range in size from a centimeter in greatest diameter to about a meter across from arm tip to opposite arm tip.

Form and Function

External Features Reflecting their radial pentamerous symmetry, sea stars typically have five arms (rays), but there may be more (Figure 13-1D). The arms merge gradually with the central disc (Figure 13-2A). Beneath the epidermis of sea stars is a mesodermal endoskeleton of small calcareous plates, or **ossicles,** bound together with connective tissue. From these ossicles project the spines and tubercles that are responsible for the spiny surface. The ossicles are penetrated by a meshwork of spaces, usually filled with fibers and dermal cells. This internal meshwork structure is described as **stereom** (see Figure 13-15) and is unique to the echinoderms.

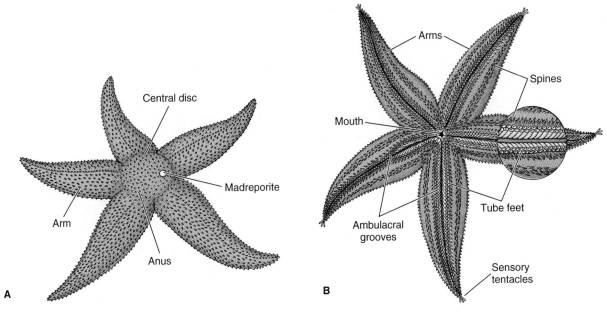

FIGURE 13-2

External anatomy of asteroid. **A,** Aboral view. **B,** Oral view.

Ambulacral (am-bu-la′kral) **grooves** (Figure 13-2B) radiate out along the arms from the centrally located mouth on the under, or oral, side of the animal. The **tube feet (podia)** project from the grooves, which are bordered by movable spines. Viewed from the oral side, the large **radial nerve** can be seen in the center of each ambulacral groove (Figure 13-3C), between the rows of tube feet. The nerve is very superficially located, covered only by thin epidermis. Under the nerve is an extension of the coelom and the radial canal of the water-vascular system (Figure 13-3C). In all other classes of living echinoderms except crinoids, ossicles or other dermal tissue cover over these structures; thus the ambulacral grooves in asteroids and crinoids are **open,** and those of the other groups are **closed.**

The aboral surface is usually rough and spiny, although the spines of many species are flattened, so that the surface appears smooth. Around the bases of the spines in many sea stars are groups of minute pincerlike **pedicellariae** (ped-e-cell-ar′e-ee), bearing tiny jaws manipulated by muscles (Figure 13-4). These help keep the body surface free of debris, protect the papulae, and sometimes aid in food capture. The **papulae** (pap′u-lee) **(dermal branchiae** or **skin gills)** are soft, delicate projections of the coelomic cavity, covered only with epidermis and lined internally with peritoneum; they extend out through spaces between the ossicles (Figure 13-3C) and are concerned with respiration. Also on the aboral side are the inconspicuous **anus** and the circular **madreporite** (Figure 13-3A), a calcareous sieve leading to the water-vascular system.

> *The function of the madreporite is still obscure. One suggestion is that it allows rapid adjustment of hydrostatic pressure within the water-vascular system in response to changes in external hydrostatic pressure resulting from depth changes, as in tidal fluctuations.*

The coelomic compartments of larval echinoderms give rise to several structures in adults, one of which is a spacious body **coelom** filled with fluid. The coelomic fluid circulates around the body cavity and into the papulae, propelled by cilia on the peritoneal lining. Exchange of respiratory gases and excretion of nitrogenous waste, principally ammonia, take place by diffusion through the thin walls of the papulae and tube feet.

Water-Vascular System The water-vascular system is another coelomic compartment and is unique to the echinoderms. It is a system of canals and specialized tube feet that shows exploitation of hydraulic mechanisms to a greater degree than in any other animal group. In sea stars the primary functions of the water-vascular system are locomotion and food gathering, as well as those of respiration and excretion.

Structurally, the water-vascular system opens to the outside through small pores in the madreporite. The madreporite of asteroids is on the aboral surface (Figure 13-2A), and leads into the **stone canal,** which descends toward the **ring canal** around the mouth (Figure 13-3B).

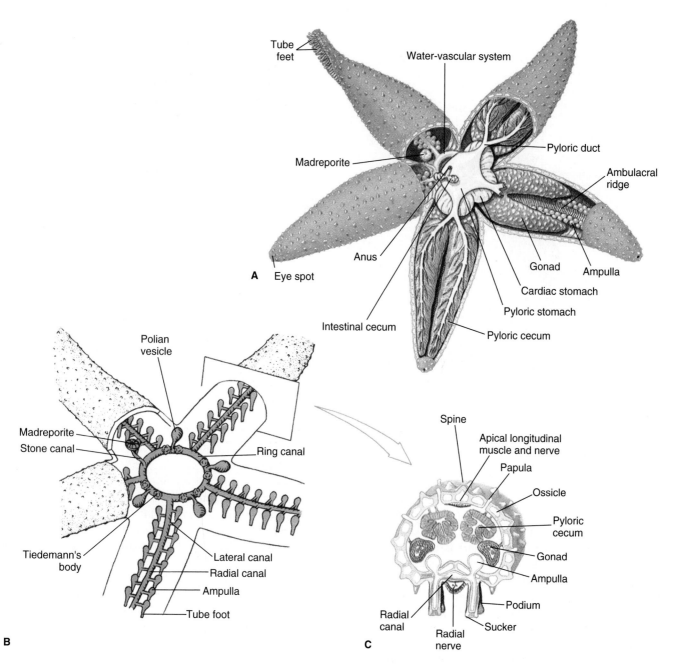

FIGURE 13-3

A, Internal anatomy of a sea star. **B,** Water-vascular system. **C,** Cross section of arm at level of gonads, illustrating open ambulacral groove. Podia (tube feet) penetrate between ossicles. (Polian vesicles are not present in *Asterias.*)

Radial canals diverge from the ring canal, one into the ambulacral groove of each ray. **Polian vesicles** are also attached to the ring canal of most asteroids (but not *Asterias*) and apparently serve as fluid reservoirs for the water-vascular system.

A series of small **lateral canals,** each with a one-way valve, connects the radial canal to the cylindrical podia or tube feet, along the sides of the ambulacral groove in each ray. Each podium is a hollow, muscular tube, the inner end of which is a muscular sac, the **ampulla,** that lies within the body coelom (Figure 13-3), and the outer end of which usually bears a sucker. Some species lack the suckers. The podia pass to the outside between the ossicles in the ambulacral groove.

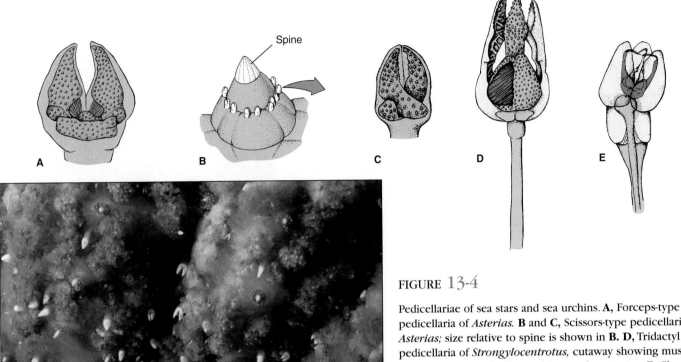

Spine

A B C D E

FIGURE 13-4

Pedicellariae of sea stars and sea urchins. **A,** Forceps-type pedicellaria of *Asterias.* **B** and **C,** Scissors-type pedicellariae of *Asterias;* size relative to spine is shown in **B. D,** Tridactyl pedicellaria of *Strongylocentrotus,* cutaway showing muscle. **E,** Globiferous pedicellaria of *Strongylocentrotus.* **F,** Close-up view of the aboral surface of the sea star *Pycnopodia helianthoides.* Note the large pedicellariae, as well as the groups of small pedicellariae around the spines. Many thin-walled papulae can be seen.

F

Locomotion by means of tube feet illustrates the interesting exploitation of hydraulic mechanisms by echinoderms. The valves in the lateral canals prevent backflow of fluid into the radial canals. The tube foot has in its walls connective tissue that maintains the cylinder at a relatively constant diameter. On contraction of muscles in the ampulla, fluid is forced into the tube foot, extending it. Conversely, contraction of the longitudinal muscles in the tube foot retracts the podium, forcing fluid back into the ampulla. Contraction of muscles in one side of the tube foot bends the organ toward that side. Small muscles at the end of the tube foot can raise the middle of the disclike end, thus creating a suction-cup effect when the end is applied to the substrate. We can estimate that by combining mucous adhesion with suction, a single tube foot can exert a pull equal to 25 to 30 g. Coordinated action of all or many of the tube feet is sufficient to draw the animal up a vertical surface or over rocks. On a soft surface, such as muck or sand, the suckers are ineffective (and numerous sand-dwelling species have no suckers), so the tube feet are employed as legs.

Feeding and Digestive System   The mouth on the oral side leads into a two-part stomach located in the central disc (Figure 13-3). In some species, the large, lower **cardiac stomach** can be everted. The smaller upper **pyloric stomach** connects with **digestive ceca** located in the arms. Digestion is largely extracellular, occurring in the digestive ceca. A short **intestine** leads from the stomach to the inconspicuous anus on the aboral side. Some species lack an intestine and anus.

Many sea stars are carnivorous and feed on molluscs, crustaceans, polychaetes, echinoderms, other invertebrates, and sometimes small fish, but many show particular preferences. Some feed on brittle stars, sea urchins, or sand dollars, swallowing them whole and later regurgitating undigestible ossicles and spines. Some attack other sea stars, and if the predator is small compared to its prey, it may attack and begin eating at the end of one of the prey's arms.

Many asteroids feed heavily on molluscs, and *Asterias* is a significant predator on commercially important clams and oysters. When feeding on a bivalve, a sea star will hump over its prey, attaching its podia to the valves, and then exert a steady pull, using its feet in relays. A force of some 1300 g can be exerted. In half an hour or so the adductor muscles of the bivalve fatigue and relax. With a very small gap available, the star inserts its soft everted stomach into the space between the valves, wraps it around the soft parts of the bivalve, and secretes digestive juices to start digesting them. After feeding, the sea star draws in its stomach by contraction of the stomach muscles and relaxation of body-wall muscles.

Some sea stars feed on small particles, either entirely or in addition to carnivorous feeding. Plankton or other organic particles coming in contact with the animal's oral or aboral surface are carried by the epidermal cilia to the ambulacral grooves and then to the mouth.

Hemal System   Although the hemal system is characteristic of echinoderms, its function remains unclear. It has little or nothing to do with circulation of body fluids, despite its name, which means "blood." It is a system of tissue strands enclosing unlined channels and is itself enclosed in another coelomic compartment, the perihemal channels or sinuses. The main channel of the hemal system connects aboral, gastric, and oral rings that give rise to branches to the gonads, stomach ceca, and arms, respectively.

Nervous and Sensory System   The nervous system in echinoderms comprises three subsystems, each made up of a nerve ring and radial nerves placed at different levels in the disc and arms. An epidermal nerve plexus, or nerve net, connects the systems. Sense organs include ocelli at the arm tips and sensory cells scattered all over the epidermis.

Reproductive System and Regeneration and Autotomy
Most sea stars have separate sexes. A pair of gonads lies in each interradial space (Figure 13-3), and fertilization is external.

Echinoderms can regenerate lost parts. Sea star arms can regenerate readily, even if all are lost. Stars also have the power of **autotomy** (the ability to break off part of their own bodies) and can cast off an injured arm near the base. An arm may take months to regenerate.

If an arm is broken off or removed, and it contains a part of the central disc (about one-fifth), the arm can regenerate a complete new sea star! In former times fishermen used to dispatch sea stars they collected from their oyster beds by chopping them in half with a hatchet—a worse than futile activity. Some sea stars reproduce asexually under normal conditions by cleaving the central disc, each part regenerating the rest of the disc and missing arms (Figure 13-5).

Development   Some species brood their eggs, either under the oral side of the animal or in specialized aboral structures, and development is direct; but in most species the embryonating eggs are free in the water and hatch to free-swimming larvae.

Early embryogenesis shows the typical primitive deuterostome pattern. Gastrulation is by invagination, and the anterior end of the archenteron pinches off to become the coelomic cavity, which expands in a U shape to fill the blastocoel. Each of the legs of the U, at the posterior end, constricts to become a separate vesicle, and these eventually give rise to

FIGURE 13-5

Pacific sea star *Echinaster luzonicus* can reproduce itself by splitting across the disc, then regenerating missing arms. The one shown here has evidently regenerated six arms from the longer one at top left.

the main coelomic compartments of the body (metacoels). The anterior portions of the U (protocoels and mesocoels) give rise to the water-vascular system and the perihemal channels. The free-swimming larva has cilia arranged in bands, and these tracts extend onto the larval arms as development continues. The larva grows three adhesive arms and a sucker at its anterior end and attaches to the substratum. While it is thus attached by this temporary stalk, it undergoes metamorphosis.

> *In echinoderms the metacoel, mesocoel, and protocoel are called the somatocoel, hydrocoel, and axocoel, respectively. During metamorphosis of sea stars, the paired somatocoels become the oral and aboral coelomic cavities, the right axocoel and hydrocoel are lost, and the left axocoel and hydrocoel become the water-vascular system and perihemal channels.*

Metamorphosis involves a dramatic reorganization of a bilateral larva into a radial juvenile. The anteroposterior axis of the larva is lost. *What was the left side becomes the oral surface, and the larval right side becomes the aboral surface* (Figure 13-6). Correspondingly, the larval mouth and anus disappear, and a new mouth and anus form on what were originally the left and right sides, respectively. As internal reorganization proceeds, short, stubby arms and the first podia appear. The animal then detaches from its stalk and begins life as a young sea star.

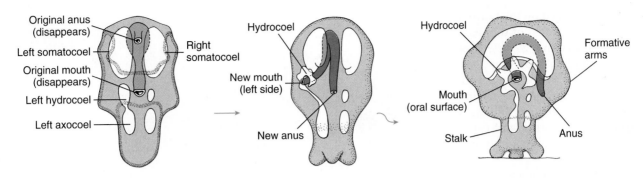

FIGURE 13-6

Asteroid metamorphosis. The left somatocoel becomes the oral coelom, and the right somatocoel becomes the aboral coelom. The left hydrocoel becomes the water-vascular system and the left axocoel the stone canal and perihemal channels. The right axocoel and hydrocoel are lost.

Class Ophiuroidea: Brittle Stars

The brittle stars are the largest of the major groups of echinoderms in numbers of species, and they are probably the most abundant also. They abound in all types of benthic marine habitats, even carpeting the abyssal seafloor in many areas.

Apart from the typical possession of five arms, brittle stars are surprisingly different from asteroids. The arms of brittle stars are slender and sharply set off from the central disc (Figure 13-7). They have no pedicellariae or papulae, and their ambulacral grooves are closed, covered with arm ossicles. The tube feet are without suckers; they aid in feeding but are of limited use in locomotion. In contrast to that in the asteroids, the madreporite of the ophiuroids is located on the oral surface, on one of the oral-shield ossicles (Figure 13-8).

A

FIGURE 13-7

A, Brittle star *Ophiura lutkeni* (class Ophiuroidea). Brittle stars do not use their tube feet for locomotion but can move rapidly (for an echinoderm) by means of their arms. **B,** Basket star *Astrophyton muricatum* (class Ophiuroidea). Basket stars extend their many-branched arms to filter feed, usually at night.

B

Each of the jointed arms consists of a column of articulated ossicles connected by muscles and covered by plates. Locomotion is by arm movement.

Five movable plates surround the mouth, serving as **jaws** (Figure 13-8). There is no anus. The skin is leathery, with dermal plates and spines arranged in characteristic patterns. Surface cilia are mostly lacking.

The visceral organs are all in the central disc, since the arms are too slender to contain them. The **stomach** is saclike and there is no intestine. Indigestible material is cast out of the mouth.

Five pairs of **bursae** (peculiar to ophiuroids) open toward the oral surface by **genital slits** at the bases of the arms. Water circulates in and out of these sacs for exchange of gases. On the coelomic wall of each bursa are small **gonads** that discharge into the bursa their ripe sex cells, which pass through the genital slits into the water for fertilization. Sexes are usually separate; a few ophiuroids are hermaphroditic. The ciliated bands of the larva extend onto delicate, beautiful larval arms, like those of larval echinoids (Figure 13-9C). During metamorphosis to the juvenile, there is no temporarily attached phase, as in asteroids.

Water-vascular, nervous, and hemal systems are similar to those of sea stars.

Brittle stars tend to be secretive, living on hard bottoms where no light penetrates. They are generally negatively phototropic and work themselves into small crevices between rocks, becoming more active at night. They are commonly fully exposed on the bottom in the permanent darkness of the deep sea. Ophiuroids feed on a variety of small particles, either browsing food from the bottom or suspension feeding. Podia are important in transferring food to the mouth. Some brittle stars extend arms into the water and catch suspended particles in mucous strands between the arm spines.

Regeneration and autotomy are even more pronounced in brittle stars than in sea stars. Many seem very fragile, releasing an arm or even part of the disc at the slightest provocation. Some can reproduce asexually by cleaving the disc; each progeny then regenerates the missing parts.

Class Echinoidea: Sea Urchins, Sand Dollars, and Heart Urchins

Echinoids have a compact body enclosed in an endoskeletal **test,** or shell. The dermal ossicles, which have become closely fitting plates, make up the test. Echinoids lack arms, but their tests reflect the typical five-part plan of echinoderms in their five ambulacral areas. Rather than extending from the oral surface to the tips of the arms, as in asteroids, the ambulacral areas follow the contours of the test from the mouth around to the aboral side, ending at the area around the anus **(periproct).** The majority of living species of sea urchins are termed "regular"; they are hemispherical in shape, radially symmetrical, and have medium to long spines (Figure 13-10). Sand dollars and heart urchins (Figure 13-11) are "irregular" because the orders to which they belong have become secondarily bilateral; their spines are usually very short. Regular urchins move by means of their tube feet, with some assistance from their spines, and irregular urchins move chiefly by their spines. Some echinoids are quite colorful.

Echinoids have wide distribution in all seas, from the intertidal regions to the deep oceans. Regular urchins often prefer rocky or hard bottoms, whereas sand dollars and heart urchins like to burrow into a sandy substrate.

The echinoid test is a compact skeleton of 10 double rows of plates that bear movable, stiff spines (Figure 13-12). The five pairs of ambulacral rows have pores (Figure 13-12)

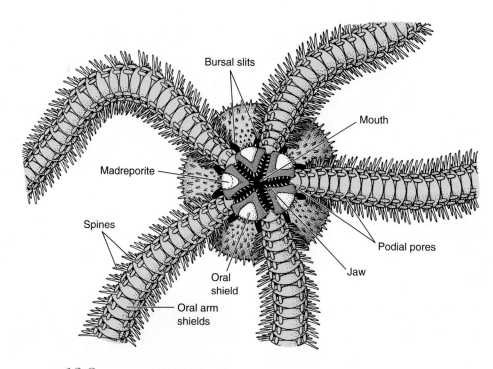

FIGURE 13-8

Oral view of spiny brittle star *Ophiothrix.*

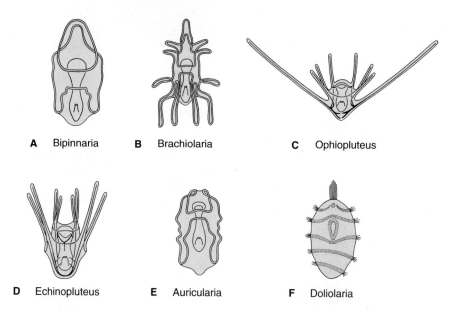

FIGURE 13-9

Larvae of echinoderms. **A,** Bipinnaria of asteroids. **B,** Brachiolaria of asteroids. **C,** Ophiopluteus of ophiuroids. **D,** Echinopluteus of echinoids. **E,** Auricularia of holothuroids. **F,** Doliolaria of crinoids.

through which the long tube feet extend. The spines are moved by small muscles around the bases.

There are several kinds of pedicellariae, the most common of which have three jaws and are mounted on long stalks (Figure 13-4D, E).

Five converging teeth surround the mouth of regular urchins and sand dollars. In some sea urchins branched **gills** (modified podia) encircle the peristome, although these are of little importance in respiratory gas exchange. The anus, **genital openings,** and madreporite are aboral in the periproct region (Figure 13-12). The mouth of sand dollars is located at about the center of the oral side, but the anus has shifted to the margin or even the oral side of the disc, so that an anteroposterior axis and bilateral symmetry can be recognized. Bilateral symmetry is even more accentuated in the heart urchins, with the anus near the posterior end on the oral side and the mouth moved away from the oral pole toward the anterior end (Figure 13-11).

Inside the test (Figure 13-12) is a coiled digestive system and a complex chewing mechanism (in regular urchins and in sand dollars), called **Aristotle's lantern,** to which the teeth are attached (Figure 13-13). A ciliated siphon connects the esophagus to the intestine and enables the water to bypass the stomach to concentrate the food for digestion in the intestine. Sea urchins eat algae and other organic material, and sand dollars collect fine particles on ciliated tracts.

The hemal and nervous systems are basically similar to those of the asteroids. The ambulacral grooves are closed, and the radial canals of the water-vascular system run just beneath the test, one in each of the ambulacral radii (Figure 13-12).

Sexes are separate, and both eggs and sperm are shed into the sea for external fertilization. The larvae may live a planktonic existence for several months and then metamorphose quickly into young urchins. Sea urchins have been used extensively as models in studies of development.

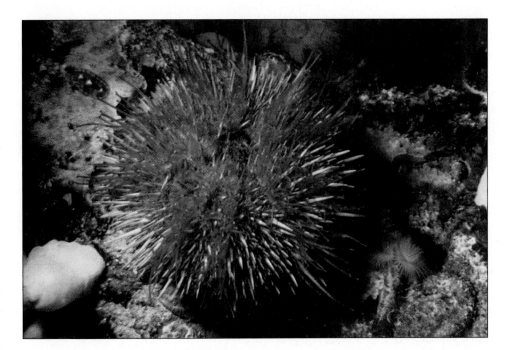

FIGURE **13-10**

Purple sea urchin *Strongylocentrotus purpuratus* is common along the Pacific Coast of North America where there is heavy wave action.

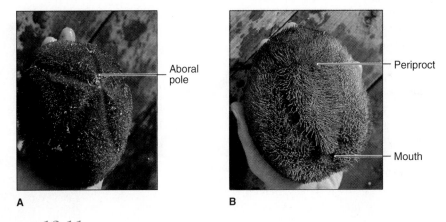

A　　　　　　　　　　　**B**

FIGURE **13-11**

An irregular echinoid *Meoma,* one of the largest heart urchins (test up to 18 cm). *Meoma* occurs in the West Indies and from the Gulf of California to the Galápagos Islands. **A,** Aboral view. **B,** Oral view. Note curved mouth at anterior end and periproct at posterior end.

Class Holothuroidea: Sea Cucumbers

In a phylum characterized by odd animals, class Holothuroidea contains members that both structurally and physiologically are among the strangest. These animals have a remarkable resemblance to the vegetable after which they are named (Figure 13-14). Compared to the other echinoderms, holothurians are greatly elongated in the oral-aboral axis, and the ossicles

(Figure 13-15) are much reduced in most, so that the animals are soft bodied. Some species characteristically crawl on the surface of the sea bottom; others are found beneath rocks, and some are burrowers.

The body wall is usually leathery, with the tiny ossicles embedded in it, although a few species have large ossicles forming a dermal armor. Because of the elongate body form of the sea cucumbers, they characteristically lie on one side. In most species the tube feet are well developed only in the ambulacra normally applied to the substratum. Thus

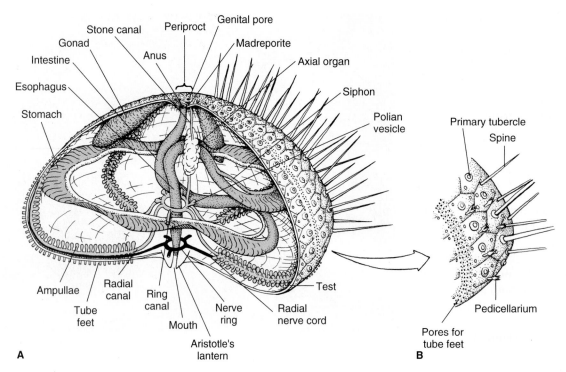

FIGURE 13-12

A, Internal structure of the sea urchin; water-vascular system in orange. **B,** Detail of portion of test.

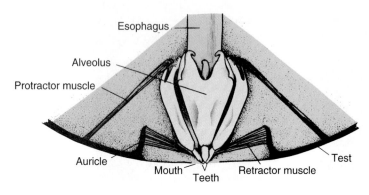

FIGURE 13-13

Aristotle's lantern, the complex mechanism used by the sea urchin for masticating its food. Five pairs of retractor muscles draw the lantern and teeth up into the test; five pairs of protractors push the lantern down and expose the teeth. Other muscles produce a variety of movements. Only major skeletal parts and muscles are shown in this diagram.

FIGURE 13-14

Sea cucumber (class Holothuroidea). Common along the Pacific Coast of North America, *Parastichopus californicus* grows up to 50 cm in length. Its tube feet on the dorsal side are reduced to papillae and warts.

a secondary bilaterality is present, albeit of quite different origin from that of the irregular urchins.

The **oral tentacles** are 10 to 30 retractile, modified tube feet around the mouth. The body wall contains circular and longitudinal muscles along the ambulacra.

The **coelomic cavity** is spacious and filled with fluid. The digestive system empties posteriorly into a muscular **cloaca** (Figure 13-16). A **respiratory tree** composed of two

long, many-branched tubes also empties into the cloaca, which pumps seawater into it. The respiratory tree serves both for respiration and excretion and is not found in any other group of living echinoderms. Gas exchange also occurs through the skin and tube feet.

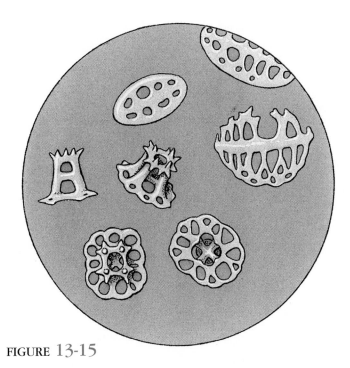

FIGURE **13-15**

Ossicles of sea cucumbers are usually microscopic bodies buried in the leathery dermis. They can be extracted from the tissue with commercial bleach and are important taxonomic characteristics. The ossicles shown here, called tables, buttons, and plates, are from the sea cucumber *Holothuria difficilis*. They illustrate the meshwork (stereom) structure observed in ossicles of all echinoderms at some stage in their development. (×250)

The hemal system is better developed in holothurians than in other echinoderms. The water-vascular system is peculiar in that the madreporite lies free in the coelom.

The sexes are separate, but some holothurians are hermaphroditic. Among the echinoderms, only sea cucumbers have a single gonad, which is considered a primitive character. Fertilization is external.

Sea cucumbers are sluggish, moving partly by means of their ventral tube feet and partly by waves of contraction in the muscular body wall. The more sedentary species trap suspended food particles in the mucus of their outstretched oral tentacles or pick up particles from the surrounding bottom. They then stuff the tentacles into their pharynx, one by one, sucking off the food material. Others crawl along, grazing the bottom with their tentacles.

Class Crinoidea: Sea Lilies and Feather Stars

Crinoids have several primitive characters. As fossil records reveal, crinoids were once far more numerous than now. They differ from other echinoderms by being attached during a substantial part of their lives. Many crinoids are deep-water forms, but feather stars may inhabit shallow waters, especially in the Indo-Pacific and the West Indian–Caribbean regions, where the largest numbers of species are found.

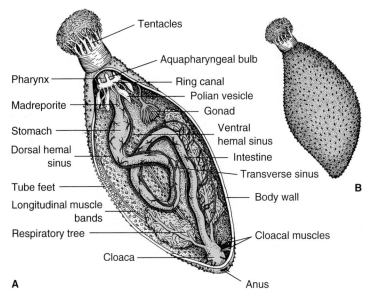

FIGURE **13-16**

Anatomy of the sea cucumber *Sclerodactyla*. **A,** Internal view; hemal system in red. **B,** External view.

FIGURE **13-17**

Comantheria briareus are crinoids found on Pacific coral reefs. They extend their arms into the water to catch food particles both during the day and at night.

The body disc has a leathery skin containing calcareous plates. The epidermis is poorly developed. Five flexible arms branch to form many more arms, each with many lateral **pinnules** arranged like barbs on a feather (Figure 13-17). Sessile forms have a long, jointed **stalk** attached to the aboral side of the body **(calyx)** (Figure 13-18). This stalk is made up of plates, appears jointed, and may bear **cirri.** Madreporite, spines, and pedicellariae are absent.

The upper (oral) surface bears the mouth and the anus. With the aid of tube feet and mucous nets, crinoids feed on small organisms that they catch in the ambulacral grooves. The ambulacral grooves are open and ciliated and serve to carry food to the mouth. Tube feet in the form of tentacles are also found in the grooves. The **water-vascular system** has the echinoderm plan. Sense organs are scanty and primitive.

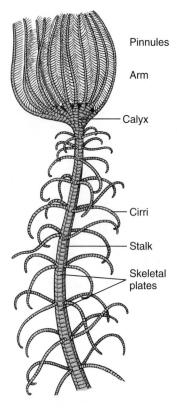

FIGURE 13-18

A stalked crinoid with portion of stalk. Modern crinoid stalks rarely exceed 60 cm, but fossil forms were as much as 20 m long.

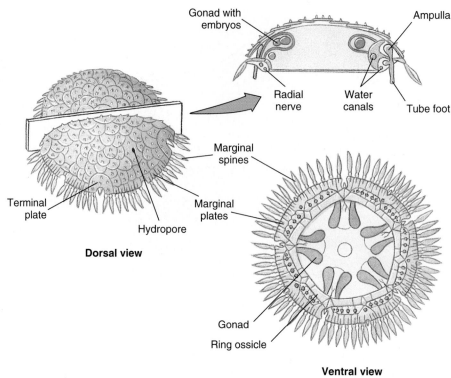

FIGURE 13-19

Xyloplax spp. (class Concentricycloidea) are peculiar little disc-shaped echinoderms. With their podia around the margin, they are the only echinoderms not having podia distributed along ambulacral areas.

> *The flower-shaped body of the sea lily is attached to the substratum by a stalk. During metamorphosis feather stars also become sessile and attached, but after several months they detach and become free moving. Although they may remain attached in the same location for long periods, they are capable of crawling and swimming short distances. They swim by alternate sweeping of their long, feathery arms.*

The sexes are separate, and the gonads are primitive. The larvae swim freely for a time before they become attached and metamorphose. Most living crinoids are from 15 to 30 cm long, but some fossil species had stalks 25 m in length.

Class Concentricycloidea: Sea Daisies

Strange little (less than 1 cm diameter), disc-shaped animals (Figure 13-19) were discovered in water over 1000 m deep off New Zealand. They are the most recently described (1986) class of echinoderms, and only two species are known so far. They have no arms, and their tube feet are around the periphery of the disc, rather than along ambulacral areas. Their water-vascular system includes two concentric ring canals; the outer ring may represent the radial canals because the podia arise from it. A hydropore, homologous to the madreporite, connects the inner ring canal to the aboral surface.

Phylogeny and Adaptive Radiation

Phylogeny

Despite the existence of an extensive fossil record, there have been contesting hypotheses on echinoderm phylogeny. Based on the embryological evidence of the bilateral larvae of echinoderms, there can be little doubt that their ancestors were bilateral and that their coelom had three pairs of spaces (trimeric). Some investigators have held that radial symmetry arose in a free-moving echinoderm ancestor and that sessile groups were derived several times independently from free-moving ancestors. However, this view does not account for the adaptive significance of radial symmetry as an adaptation for sessile existence. The more traditional view is that the first echinoderms were sessile, became radial as an adaptation to that existence, and then gave rise to the free-moving groups. Figure 13-20 is consistent with this hypothesis. It views the evolution of endoskeletal plates with stereom structure and of external ciliary grooves for feeding as early echinoderm (or pre-echinoderm) developments. The extinct carpoids (Homalozoa, Figure 13-20) had stereom ossicles but were not radially symmetrical, and the status of their water-vascular system, if

Classification of Phylum Echinodermata

There are about 6000 living and 20,000 extinct or fossil species of Echinodermata. The traditional classification placed all the free-moving forms that were oriented with oral side down in the subphylum Eleutherozoa, containing most of the living species. The other subphylum, Pelmatozoa, contained mostly forms with stems and oral side up; most of the extinct classes and the living Crinoidea belong to this group. Although alternative schemes have strong supporters, cladistic analysis provides evidence that the two traditional subphyla are monophyletic groups.[2] The following includes only groups with living members.

Subphylum Pelmatozoa (pel-ma′to-zo′a) (Gr. *pelmatos*, a stalk, + *zōon*, animal). Body in form of a cup or calyx, borne on aboral stalk during part or all of life; oral surface directed upward; open ambulacral grooves; madreporite absent; both mouth and anus on oral surface; several fossil classes plus living Crinoidea.

Class Crinoidea (krin-oy′de-a) (Gr. *krinon*, lily, + *eidos*, form, + *-ea*, characterized by): **sea lilies** and **feather stars.** Five arms branching at base and bearing pinnules; ciliated ambulacral grooves on oral surface with tentacle-like tube feet for food gathering; spines, madreporite, and pedicellariae absent. Examples: *Antedon, Nemaster, Comantheria* (Figure 13-17).

Subphylum Eleutherozoa (e-lu′ther-o-zo′a) (Gr. *eleutheros*, free, not bound, + *zōon*, animal). Body form star-shaped, globular, discoidal, or cucumber shaped; oral surface directed toward substratum or oral-aboral axis parallel to substratum; body with or without arms; ambulacral grooves open or closed.

Class Concentricycloidea (kon-sen′tri-sy-kloy′de-a) (L. *cum*, together, + *centrum*, center [having a common center], + Gr., *kyklos*, circle, + *eidos*, form, + *-ea*, characterized by): **sea daisies.** Disc-shaped body, with marginal spines but no arms; concentrically arranged skeletal plates; ring of suckerless podia near body margin; hydropore present; gut present or absent, no anus. Example: *Xyloplax* (Figure 13-19).

Class Asteroidea (as′ter-oy′de-a) (Gr. *aster*, star, + *eidos*, form, + *-ea*, characterized by): **sea stars.** Star shaped, with arms not sharply demarcated from the central disc; ambulacral grooves open, with tube feet on oral side; tube feet often with suckers; anus and madreporite aboral; pedicellariae present. Examples: *Asterias, Pisaster.*

Class Ophiuroidea (o′fe-u-roy′de-a) (Gr. *ophis*, snake, + *oura*, tail, + *eidos*, form, + *-ea*, characterized by): **brittle stars** and **basket stars.** Star shaped, with arms sharply demarcated from central disc; ambulacral grooves closed, covered by ossicles; tube feet without suckers and not used for locomotion; pedicellariae absent. Examples: *Ophiura* (Figure 13-7), *Astrophyton* (Figure 13-7).

Class Echinoidea (ek′i-noy′de-a) (Gr. *echinos*, sea urchin, hedgehog, + *eidos*, form, + *-ea* characterized by): **sea urchins, sea biscuits,** and **sand dollars.** More or less globular or disc-shaped, with no arms; compact skeleton or test with closely fitting plates; movable spines; ambulacral grooves closed; tube feet often with suckers; pedicellariae present. Examples: *Arbacia, Strongylocentrotus* (Figure 13-10), *Lytechinus, Meoma* (Figure 13-11).

Class Holothuroidea (hol′o-thu-roy′de-a) (Gr. *holothourion*, sea cucumber, + *eidos*, form, + *-ea*, characterized by): **sea cucumbers.** Cucumber-shaped, with no arms; spines absent; microscopic ossicles embedded in muscular body wall; anus present; ambulacral grooves closed; tube feet with suckers; circumoral tentacles (modified tube feet); pedicellariae absent; madreporite plate internal. Examples: *Sclerodactyla, Parastichopus* (Figure 13-14), *Cucumaria.*

[2]Brusca, R. C., and G. J. Brusca. 1990. Invertebrates. Sunderland, Massachusetts, Sinauer Associates. Meglitsch, P. A., and F. R. Schram. 1991. Invertebrate zoology, ed. 3. New York, Oxford University Press. Paul, C. R. C., and A. B. Smith. 1984. The early radiation and phylogeny of echinoderms. Biol. Rev. **59:**443–481.

any, is uncertain. Some investigators regard carpoids as a separate subphylum of echinoderms (Homalozoa), and others believe they represent a group of pre-echinoderms that shows affinities to the chordates (Calcichordata, p. 274). The fossil helicoplacoids (Figure 13-20) show evidence of three, true ambulacral grooves, and their mouth was on the side of the body.

Attachment to the substratum by the aboral surface would have led to radial symmetry and the origin of the Pelmatozoa. An ancestor that became free moving and applied its oral surface to the substratum would have given rise to the Eleutherozoa. Phylogeny within the Eleutherozoa is controversial. Most investigators agree that the echinoids and holothuroids are related and form a single clade, but opinions diverge on the relationship of the ophiuroids and asteroids. Figure 13-20 illustrates the view that the ophiuroids arose after the closure of ambulacral grooves, but this scheme treats the evolution of five ambulacral rays (arms) in the ophiuroids and asteroids as independently evolved. Alternatively, if the ophiuroids and asteroids are a single clade, then closed ambulacral grooves must have evolved separately in ophiuroids and in the common ancestor of echinoids and holothuroids.

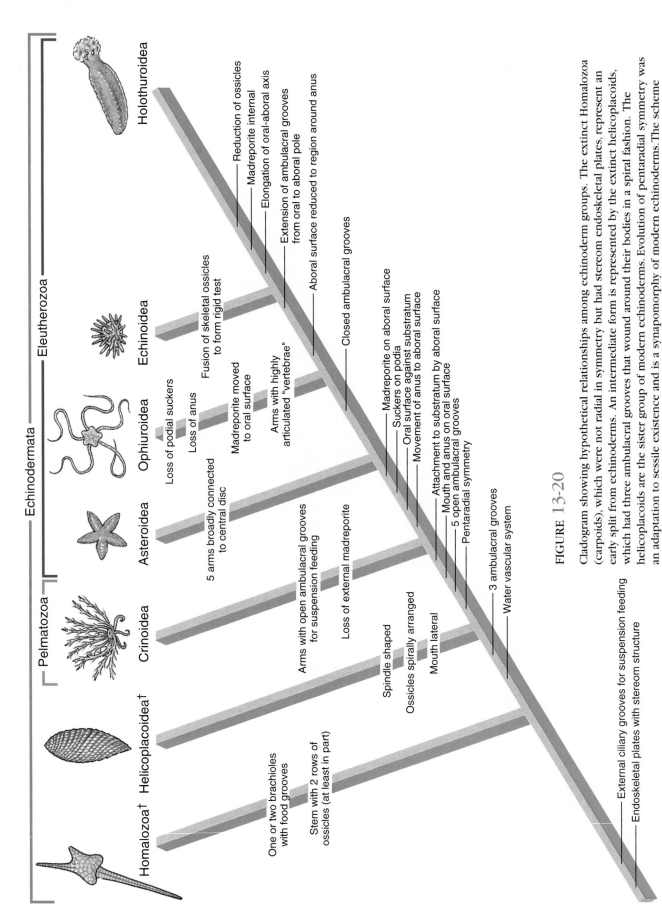

FIGURE 13-20

Cladogram showing hypothetical relationships among echinoderm groups. The extinct Homalozoa (carpoids), which were not radial in symmetry but had stereom endoskeletal plates, represent an early split from echinoderms. An intermediate form is represented by the extinct helicoplacoids, which had three ambulacral grooves that wound around their bodies in a spiral fashion. The helicoplacoids are the sister group of modern echinoderms. Evolution of pentaradial symmetry was an adaptation to sessile existence and is a synapomorphy of modern echinoderms. The scheme depicted here views the ophiuroids as having arisen separately from the asteroids, after the evolution of closed ambulacral grooves, and the possession of five arms would thus have been of separate origin. Alternatively, if the Asteroidea and Ophiuroidea are a monophyletic clade, with five arms being synapomorphic, then closed ambulacral grooves in the ophiuroids would have evolved separately from that character in the echinoids and holothuroids.

The following labels appear along the branches of the cladogram:

- Holothuroidea — Reduction of ossicles; Madreporite internal; Elongation of oral-aboral axis; Extension of ambulacral grooves from oral to aboral pole; Aboral surface reduced to region around anus
- Echinoidea — Fusion of skeletal ossicles to form rigid test; Closed ambulacral grooves
- Ophiuroidea — Loss of podial suckers; Loss of anus; Madreporite moved to oral surface; Arms with highly articulated "vertebrae"
- Asteroidea — 5 arms broadly connected to central disc
- Crinoidea — Arms with open ambulacral grooves for suspension feeding; Loss of external madreporite
- Helicoplacoidea† — Spindle shaped; Ossicles spirally arranged; Mouth lateral
- Homalozoa† — One or two brachioles with food grooves; Stem with 2 rows of ossicles (at least in part)

Shared characters:
- Madreporite on aboral surface
- Suckers on podia
- Oral surface against substratum
- Movement of anus to aboral surface
- Attachment to substratum by aboral surface
- Mouth and anus on oral surface
- 5 open ambulacral grooves
- Pentaradial symmetry
- 3 ambulacral grooves
- Water vascular system
- External ciliary grooves for suspension feeding
- Endoskeletal plates with stereom structure

Higher groupings: Echinodermata, Eleutherozoa, Pelmatozoa

†Extinct groups.

262

Data on the Concentricycloidea are insufficient to place this group on a cladogram.

Adaptive Radiation

The radiation of the echinoderms has been determined by the limitations and potentials of their most important characteristics: radial symmetry, the water-vascular system, and their dermal endoskeleton. If their ancestors had a brain and specialized sense organs, these were lost in the adoption of radial symmetry. Thus, it is not surprising that there are large numbers of creeping, benthic forms with filter-feeding, deposit-feeding, scavenging, and herbivorous habits, comparatively few predators, and very few pelagic species. In this light the relative success of the asteroids as predators is impressive and probably attributable to the extent to which they have exploited the hydraulic mechanism of the tube feet.

Phylum Hemichordata: Acorn Worms

The Hemichordata (hem'i-kor-da'ta) (Gr. *hemi,* half, + *chorda,* string, cord) are marine animals that were formerly considered a subphylum of the chordates, based on their possession of gill slits, a rudimentary notochord, and a dorsal nerve cord. However, zoologists now agree that the so-called hemichordate "notochord" is really an evagination of the mouth cavity and not homologous with the chordate notochord, so the hemichordates are given the rank of a separate phylum.

Hemichordates are wormlike bottom dwellers, living usually in shallow waters. Some are colonial and live in secreted tubes. Many are sedentary or sessile. They are widely distributed, but their secretive habits and fragile bodies make collecting them difficult.

Members of class Enteropneusta (acorn worms) range from 20 mm to 2.5 m in length and 3 to 200 mm in breadth. Members of class Pterobranchia (pterobranchs) are smaller, usually from 5 to 14 mm, not including the stalk. About 70 species of enteropneusts and three small genera of pterobranchs have been described.

Hemichordates have the typical tricoelomate structure of deuterostomes.

Class Enteropneusta

The enteropneusts, or acorn worms (Figure 13-21), are sluggish wormlike animals that live in burrows or under stones, usually in mud or sand flats of intertidal zones.

The mucus-covered body is divided into a tonguelike **proboscis,** a short **collar,** and a long **trunk** (protosome, mesosome, and metasome.)

In the posterior end of the proboscis is a small coelomic sac (protocoel) into which extends the buccal diverticulum, a slender, blindly ending pouch of the gut that reaches forward into the buccal region and was formerly believed to be a noto-

Position in Animal Kingdom

1. Hemichordates belong to the deuterostome branch of the animal kingdom and are enterocoelous coelomates with radial cleavage.
2. Hemichordates show some of both echinoderm and chordate characteristics.
3. A chordate plan of structure is suggested by gill slits and a restricted dorsal tubular nerve cord.
4. Similarity to the echinoderms is shown in larval characteristics.

Biological Contributions

1. A tubular dorsal nerve cord in the collar zone may represent an early stage of the condition in chordates; a diffuse net of nerve cells is similar to the uncentralized, subepithelial plexus of echinoderms.
2. The gill slits in the pharynx, also characteristic of chordates, serve primarily for filter feeding and only secondarily for breathing and are thus comparable to gill slits in the protochordates.

chord. A row of gill pores extends dorsolaterally on each side of the trunk just behind the collar (Figure 13-21). These gill pores open from a series of gill chambers that in turn connect with a series of gill slits in the sides of the pharynx. The primary function of these structures is not respiration, but food gathering (Figure 13-22). Food particles caught in mucus and brought to the mouth by ciliary action on the proboscis and collar are strained out of the branchial water that leaves through the gill slits. The particles are then directed along the ventral part of the pharynx and esophagus to the intestine.

A mid-dorsal vessel carries blood forward to a heart above the buccal diverticulum. The blood then flows into a network of sinuses that may have an excretory function, then posteriorly through a ventral blood vessel and through a network of sinuses in the gut and body wall. Acorn worms have dorsal and ventral nerve cords, and the dorsal cord is hollow in some. Sexes are separate. In some enteropneusts there is a free-swimming larva that closely resembles the larva of sea stars.

Class Pterobranchia

The basic plan of the class Pterobranchia is similar to that of the Enteropneusta, but certain differences are correlated with the sedentary mode of life of pterobranchs. Only two genera are known in any detail. In both genera there are arms with tentacles containing an extension of the coelomic compartment of the mesosome, as in a lophophore (Figure 13-23). One genus has a single pair of gill slits, and the other has none. Both live in tubes from which they project their proboscis and tentacles to feed by mucociliary mechanisms. Some species are dioecious, others are monoecious, and asexual reproduction occurs by budding.

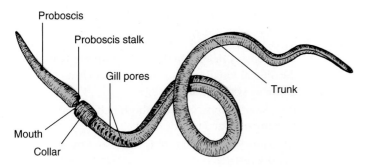

FIGURE 13-21

External lateral view of the acorn worm, *Saccoglossus* (phylum Hemichordata).

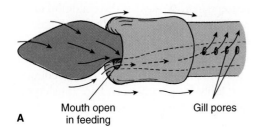

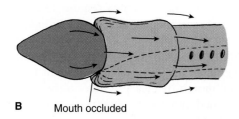

FIGURE 13-22

Food currents of enteropneust hemichordate. **A,** Side view of acorn worm with mouth open, showing direction of currents created by cilia on proboscis and collar. Food particles are directed toward mouth and digestive tract. Rejected particles move toward outside of collar. Water leaves through gill pores. **B,** When mouth is occluded, all particles are rejected and passed onto the collar. Nonburrowing and some burrowing hemichordates use this feeding method.

Phylogeny

Hemichordates share characteristics with both the echinoderms and the chordates. With chordates they share the gill slits, which serve primarily for filter feeding and secondarily for breathing, as they do in some of the protochordates. In addition, a short dorsal, somewhat hollow nerve cord in the collar zone may be homologous to the nerve cord of the chordates. Their early embryogenesis is remarkably like that of echinoderms, and the early larva is almost identical with the bipinnaria larva of asteroids (Figure 13-24). However, the hypothetical relationships shown in Figure 13-25 place the lophophorates as sister groups of the hemichordates and chordates, required by the

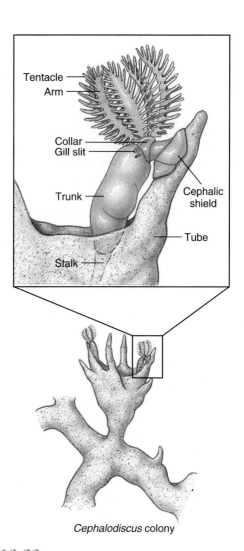

Cephalodiscus colony

FIGURE 13-23

Cephalodiscus, a pterobranch hemichordate. These tiny (5 to 7 mm) forms live in tubes in which they can move freely. Ciliated tentacles and arms direct currents of food and water toward mouth.

proposed synapomorphy for all these groups of a crown of ciliated tentacles containing extensions of the mesocoel.

Phylum Chaetognatha: Arrowworms

The Chaetognatha (ke-tog′na-tha) (Gr. *chaitē,* long flowing hair, + *gnathos,* jaw) is a small group (65 species) of marine animals highly specialized for a planktonic existence. Their relationship to other groups is obscure, but their embryology suggests a relationship to the deuterostomes.

Their small, straight bodies resemble miniature torpedoes, or darts, ranging from 2.5 to 10 cm in length.

Arrowworms usually swim to the surface at night and descend during the day. Much of the time they drift passively, but they can dart forward in swift spurts, using the caudal fin

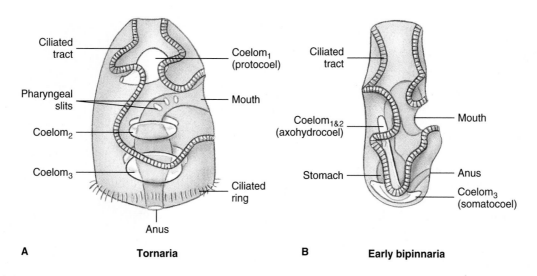

A   **Tornaria**   **B**   **Early bipinnaria**

FIGURE 13-24

Comparison of a hemichordate tornaria (**A**), to an echinoderm bipinnaria (**B**).

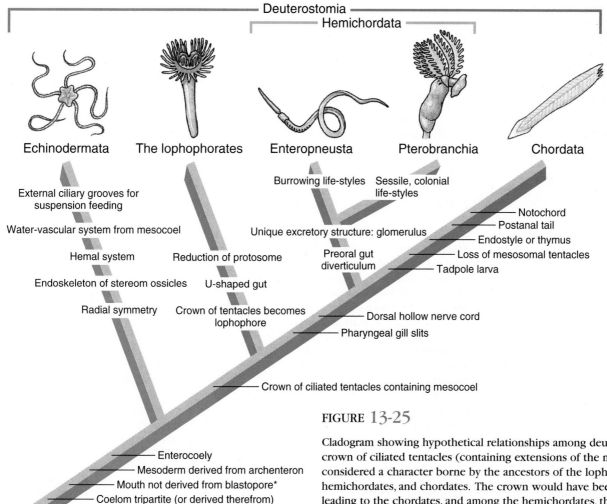

FIGURE 13-25

Cladogram showing hypothetical relationships among deuterostome phyla. The crown of ciliated tentacles (containing extensions of the mesocoel) is here considered a character borne by the ancestors of the lophophorates, hemichordates, and chordates. The crown would have been lost in the line leading to the chordates, and among the hemichordates, the enteropneusts. The pterobranchs retain the primitive character, whereas in the lophophorate phyla it is modified into a lophophore. Uncertainties regarding the relationships of the Chaetognatha preclude placement in this cladogram. The Protostomia serves as outgroup.

(Source: *R. C. Brusca, and G. J. Brusca.* Invertebrates, *1990, Sinauer Associates, Sunderland, MA.*)

*Except in phoronids.

and longitudinal muscles—a fact that no doubt contributes to their success as planktonic predators. Horizontal fins bordering the trunk function in flotation rather than in active swimming.

The body of arrowworms is unsegmented and is made up of the head, trunk, and postanal tail (Figure 13-26). They have teeth and chitinous spines on the head around the mouth. When the animal captures prey, the teeth and raptorial spines spread apart and then snap shut with startling speed. Arrowworms are voracious feeders, living on other planktonic forms, especially copepods, and even small fish (Figure 13-26B).

The body is covered with a thin cuticle. They have a complete digestive system, well-developed coelom, and nervous system with a nerve ring containing several ganglia. Vascular, respiratory, and excretory systems, however, are entirely lacking.

Arrowworms are hermaphroditic with either cross-fertilization or self-fertilization. Juveniles develop directly without metamorphosis. There is no true peritoneum lining the coelom. Cleavage is radial, complete, and equal.

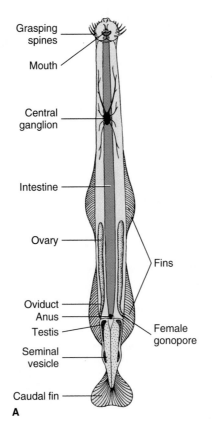

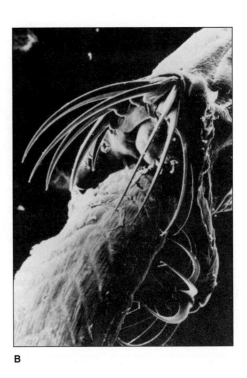

B

FIGURE 13-26

Arrowworms. **A,** Internal structure of *Sagitta*. **B,** Scanning electron micrograph of a juvenile arrowworm, *Flaccisagitta hexaptera* (35 mm length) eating a larval fish.

Summary

The phyla Echinodermata, Chordata, and Hemichordata show the characteristics of the Deuterostomia division of the animal kingdom. Molecular evidence does not support placement of the Chaetognatha among the deuterostomes, but this issue is yet to be clarified

The echinoderms are an important marine group sharply distinguished from other phyla of animals. They have a penta-radial symmetry but were derived from bilateral ancestors.

The sea stars (class Asteroidea) usually have five arms, and the arms merge gradually with a central disc. They have no head and few specialized sensory organs. The mouth is central on the under (oral) side of the body. They have an endoskeleton of dermal ossicles, open ambulacral areas, pedicellariae, and papulae. The water-vascular system is an elaborate hydraulic system derived from one of the coelomic cavities. The madreporite, opening to the outside, connects to the ring canal around the esophagus by way of the stone canal, and radial canals extend from the ring canal along each ambulacral area. Branches from the radial canals lead to the many tube feet, structures that are important in locomotion, food gathering, respiration, and excretion. Many sea stars are predators, whereas others feed on particulate organic matter. Sexes are separate, and the reproductive systems are very simple. The bilateral, free-swimming larva becomes attached, transforms to a radial juvenile, then detaches and becomes a motile sea star.

Brittle stars (class Ophiuroidea) have slender arms that are sharply set off from the central disc; they have no pedicellariae or ampullae, and their ambulacral grooves are closed. Their madreporite is on the oral side. They crawl by means of their arms, and they can move around more rapidly than other echinoderms.

In sea urchins (class Echinoidea), the dermal ossicles have become closely fitting plates, and there are no arms. Some urchins (sand dollars and heart urchins) have evolved a return to bilateral symmetry.

Sea cucumbers (class Holothuroidea) have very small dermal ossicles and a soft body wall. They are greatly elongated in the oral-aboral axis and lie on their side. Sea cucumbers also have undergone some return to bilateral symmetry.

Sea lilies and feather stars (class Crinoidea) are the only group of living echinoderms, other than asteroids, with open ambulacral areas. They are mucociliary particle feeders and lie with their oral side upward.

Hemichordates were formerly considered chordates because their buccal diverticulum was considered a notochord. In common with chordates some of them do have gill slits and a hollow, dorsal nerve cord. Hemichordates show affinities with chordates, echinoderms, and lophophorates, and they are the likely sister group of the chordates.

The arrowworms (phylum Chaetognatha) are a small group but are important as a component of marine plankton.

Review Questions

1. What is the constellation of characteristics possessed by echinoderms that is found in no other phylum?

2. How do we know that echinoderms were derived from an ancestor with bilateral symmetry?

3. What is an ambulacral groove, and what is the difference between open and closed ambulacral grooves? Open ambulacra is considered the primitive condition, and closed ambulacra is considered derived. Can you suggest a reason why this is probably correct?

4. Trace or make a rough copy of Figure 13-3B, without the labels, and then from memory label the parts of the water-vascular system of the sea star.

5. Name the structures involved in following functions in sea stars, brittle stars, sea urchins, sea cucumbers, and crinoids, and briefly describe the action of each: respiration, feeding and digestion, excretion, reproduction.

6. Match the groups below with *all* the correct answers in the lettered column.

    ____ Crinoidea   ____ Asteroidea

    ____ Ophiuroidea   ____ Echinoidea

    ____ Holothuroidea

    a. Closed ambulacral grooves

    b. Oral surface generally upward

    c. With arms

    d. Without arms

    e. Approximately globular or disc shaped

    f. Elongated in oral-aboral axis

    g. With pedicellariae

    h. Madreporite internal

    i. Madreporite on oral plate

7. Define the following: pedicellariae, madreporite, respiratory tree, Aristotle's lantern.

8. Give three examples of how echinoderms are important to humans.

9. Distinguish the Enteropneusta from the Pterobranchia.

10. The Hemichordata were once considered chordates. Why?

11. What is evidence that hemichordates and chordates form a monophyletic group?

12. What is the ecological importance of arrowworms?

Selected References

See also general references on page 395.

Baker, A. N., F. W. E. Row, and H. E. S. Clark. 1986. A new class of Echinodermata from New Zealand. Nature **321**:862–864. *The strange class Concentricycloidea is described.*

Birkeland, C. 1989. The Faustian traits of the crown-of-thorns starfish. Am. Sci. **77**:154–163. *The fast growth in the early years of the life of an* Acanthaster planci *results in loss of body integrity in later life*

Davidson, E. H., B. R. Hough-Evans, and R. J. Britten. 1982. Molecular biology of the sea urchin embryo. Science **217**:17–26 *Many fundamental insights into the process of embryogenesis have been revealed through studies of sea urchins.*

Hughes, T. P. 1994. Catastrophes, phase shifts and large-scale degradation of a Caribbean coral reef. Science **265**: 1547–1551. *Describes the sequence of events, including the die-off of sea urchins, leading to the destruction of the coral reefs around Jamaica.*

Lawrence, J. 1987. A functional biology of echinoderms. Baltimore, The Johns Hopkins University Press. *Treats many aspects of the fascinating biology of these organisms.*

Links to the Internet

Visit this textbook's web site at http://www.mhhe.com/zoology to find live Internet links for each of the references listed below.

Echinoderms

1. Enchinodermata. This site provides links to a variety of information on the echinoderms.

2. Introduction to the Echinoderms. University of California at Berkeley Museum of Paleontology. This site provides information on the echinoderm fossil record, life histories, systematics, and morphology. It also provides a great number of links to sites that focus on each of the echinoderm classes.

3. The CAS Echinoderm Web Page. This site provides information on echinoderm taxonomy and on the echinoderm collection of the California Academy of Science. It also provides links to other echinoderm sites.

4. The Echinoderm Newsletter. This newsletter, prepared by the National Museum of Natural History, provides information on conferences and publications on echinoderms, and gives addresses of biologists studying echinoderms.

5. Echinodermata. Arizona's Tree of Life Web Page. Pictures, characteristics, phylogenetic relationships, references on echinoderms.

6. Phylum Echinodermata. University of Michigan site describing echinoderms. Pictures, general information, and references.

7. Exploring Reef Science. Information on coral-eating starfish on the Great Barrier Reef.

8. Echinoderms. Keys to Marine Invertebrates of the Woods Hole Region. Descriptive information, definition of terminology, and keys to the echinoderms of the Woods Hole Region.

Lesser Deuterostomes

9. BIOSIS: Protochordata. This site provides links to information on the Hemichordata, Urochordata, and Cephalochordata.

10. Introduction to the Hemichordata. This University of California at Berkeley Museum of Paleontology site provides photographs and information on the biology and classification of the hemichordates.

Vertebrate Beginnings: The Chordates

CHAPTER fourteen

It's a Long Way from Amphioxus

Along the more southern coasts of North America, half buried in sand on the seafloor, lives a small fishlike translucent animal quietly filtering organic particles from seawater. Inconspicuous, of no commercial value and largely unknown, this creature is nonetheless one of the famous animals of classical zoology. It is amphioxus, an animal that wonderfully exhibits the four distinctive hallmarks of the phylum Chordata—(1) dorsal, tubular nerve cord overlying (2) a supportive notochord, (3) pharyngeal slits for filter feeding, and (4) a postanal tail for propulsion—all wrapped up in one creature with textbook simplicity. Amphioxus is an animal that might have been designed by a zoologist for the classroom. During the nineteenth century, with interest in vertebrate ancestry running high, amphioxus was considered by many to resemble closely the direct ancestor of the vertebrates. Its exalted position was later acknowledged by Philip Pope in a poem sung to the tune of "Tipperary." It ends with the refrain:

> It's a long way from amphioxus
> It's a long way to us,
> It's a long way from amphioxus
> To the meanest human cuss.
> Well, it's good-bye to fins and gill slits
> And it's welcome lungs and hair,
> It's a long, long way from amphioxus
> But we all came from there.

But amphioxus' place in the sun was not to endure. For one thing, amphioxus lacks one of the most important of vertebrate characteristics, a distinct head with special sense organs and the equipment for shifting to an active predatory mode of life. Absence of a head and several specialized features suggest to zoologists today that amphioxus represents an early departure from the main line of chordate descent. It seems that we are a very long way indeed from amphioxus. Nevertheless, while amphioxus is denied the ancestral vertebrate award, we believe that it more closely resembles the earliest prevertebrate than any other living animal we know.

Animals most familiar to people belong to the phylum Chordata. Humans are members and share the characteristic from which the phylum derives its name—the **notochord** (Gr. *nōton*, back, + L. *chorda*, cord) (Figure 14-1). All members of the phylum possess this structure, either in the larval or embryonic stages or throughout life. The notochord is a rodlike, semirigid body of cells enclosed by a fibrous sheath, which extends, in most cases, the full length of the body between the gut tract and the central nervous system. Its primary purpose is to support and to stiffen the body, that is, to act as a skeletal axis.

The structural plan of chordates retains many of the features of nonchordate invertebrates, such as bilateral symmetry, anteroposterior axis, coelom tube-within-a-tube arrangement, metamerism, and cephalization. However, the exact phylogenetic position of the chordates within the animal kingdom is unclear.

Two possible lines of descent have been proposed. Earlier speculations that focused on the arthropod-annelid-mollusc group (Protostomia branch) of the invertebrates have fallen from favor. Only members of the echinoderm-hemichordate assemblage (Deuterostomia branch) now deserve serious consideration as a chordate sister group. The chordates share with other deuterostomes several important characteristics: radial cleavage (p. 48), an anus derived from the first embryonic opening (blastopore) and a mouth derived from an opening of secondary origin, and a coelom primitively formed by fusion of enterocoelous pouches (except in vertebrates in which the coelom is basically schizocoelous, p.50). These common characteristics indicate a natural unity among the Deuterostomia.

As a whole, a more fundamental unity of plan exists throughout all the organs and systems of the phylum Chordata than in many of the invertebrate phyla. Ecologically the chordates are among the most adaptable of organic forms, able to occupy most kinds of habitat. They illustrate perhaps better than any other animal group the basic evolutionary processes of the origin of new structures, adaptive strategies, and adaptive radiation.

Traditional and Cladistic Classification of the Chordates

The traditional Linnaean classification of the chordates (p. 283) provides a simple and convenient way to indicate the taxa included in each major group. However, in cladistic usage, some of the traditional taxa, such as Agnatha and Reptilia, are no longer recognized. Such taxa do not satisfy the requirement of cladistics that only **monophyletic** groups are valid taxonomic entities, that is, groups that contain all known descendants of a single common ancestor. The reptiles, for example, are considered a **paraphyletic** grouping because this group does not contain all of the descendants of their most recent common ancestor. The common ancestor of reptiles as traditionally recognized is also the ancestor of birds and mammals. Thus, as

Position in the Animal Kingdom

Phylum Chordata (kor-da′ta) (L. *chorda,* cord) belongs to the Deuterostomia branch of the animal kingdom that includes the phyla Echinodermata, Hemichordata, and the three lophophorate phyla—Phoronida, Ectoprocta, and Brachiopoda. These six phyla share many embryological features and are probably descended from an ancient common ancestor. From humble beginnings, the chordates evolved a vertebrate body plan of enormous adaptability that always remains distinctive, while providing almost unlimited scope for specialization in life habit, form, and function.

Biological Contributions

1. The **endoskeleton** of the vertebrates permits continuous growth and the attainment of large body size, and it provides an efficient framework for muscle attachment.
2. The **perforated pharynx** of protochordates that originated as a suspension-feeding device served as the framework for the subsequent evolution of true internal gills with pharyngeal muscular pump, and jaws.
3. The adoption of a **predatory habit** by the early vertebrates and the accompanying evolution of a **highly differentiated brain** and **paired special sense organs** contributed in large measure to the successful adaptive radiation of the vertebrates.
4. The **paired appendages** that appeared in the aquatic vertebrates were successfully adapted later as jointed limbs for efficient locomotion on land or as wings for flight.

shown in the cladogram (Figure 14-3), reptiles, birds, and mammals compose a monophyletic clade called the Amniota, so named because all develop from an egg having special extraembryonic membranes, one of which is the amnion. Therefore according to cladistics, reptiles can be grouped only in a negative manner as amniotes that are not birds or mammals; there are no positive or novel features that unite the reptiles to the exclusion of birds and mammals. Similarly, the agnathans (hagfishes and lampreys) are a paraphyletic grouping because the most recent common ancestor of agnathans is also the ancestor of all the remaining vertebrates (the gnatho-stomes). The reasons why paraphyletic groups are not used in cladistic taxonomy are explained in Chapter 3 (p. 58–59).

The phylogenetic tree of the chordates (Figure 14-2) and the cladogram of the chordates (Figure 14-3) provide different kinds of information. The cladogram shows a nested hierarchy of taxa grouped by their sharing of derived characters. These characters may be morphological, physiological,

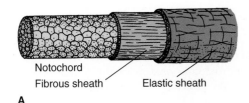

Notochord
Fibrous sheath   Elastic sheath

A

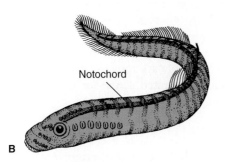

Notochord

B

FIGURE **14-1**

A, Structure of the notochord and its surrounding sheaths. Cells of the notochord proper are thick walled, pressed together closely, and filled with semifluid. Stiffness is caused mainly by turgidity of fluid-filled cells and surrounding connective tissue sheaths. This primitive type of endoskeleton is characteristic of all chordates at some stage of the life cycle. The notochord provides longitudinal stiffening of the main body axis, a base for trunk muscles, and an axis around which the vertebral column develops. **B,** In hagfishes and lampreys it persists throughout life, but in other vertebrates it is largely replaced by the vertebrae. In mammals slight remnants are found in nuclei pulposi of intervertebral discs. The method of notochord formation is different in the various groups of vertebrates. In amphioxus it originates from the endoderm; in birds and mammals it arises as an anterior outgrowth of the embryonic primitive streak.

embryological, behavioral, chromosomal, or molecular in nature. Although the cladogram shows the *relative* time of origin of the novel properties of taxonomic groups and their specific positions in the hierarchical system of evolutionary common descent, it contains no timescale or information on ancestral lineages. By contrast, the branches of a phylogenetic tree are intended to represent real lineages that occurred in the evolutionary past. Geological information regarding ages of lineages is added to the information from the cladogram to generate a phylogenetic tree for the same taxa.

In our treatment of the chordates, we have retained the traditional Linnaean classification (p. 283) because of its conceptual usefulness and because the alternative—thorough revision following cladistic principles—would require extensive change and the virtual abandonment of familiar rankings. However, we have tried to use monophyletic taxa as much as possible, because such usage is consistent with both evolutionary and cladistic taxonomy (see p. 58).

Several of the traditional divisions of the phylum Chordata used in Linnaean classifications are shown in Table 14-1. A fundamental separation is the Protochordata from the Vertebrata. Since the former lack a well-developed head, they

Characteristics of Phylum Chordata

1. Bilateral symmetry; segmented body; three germ layers; well-developed coelom
2. **Notochord** (a skeletal rod) present at some stage in life cycle
3. **Single, dorsal, tubular nerve cord;** anterior end of cord usually enlarged to form a brain
4. **Pharyngeal pouches** present at some stage in life cycle; in aquatic chordates these develop into gill slits
5. **Postanal tail,** usually projecting beyond the anus at some stage but may or may not persist
6. **Segmented muscles** in an unsegmented trunk
7. **Ventral heart,** with dorsal and ventral blood vessels; closed blood system
8. Complete digestive system
9. A cartilaginous or bony **endoskeleton** present in the majority of members (vertebrates)

are also called Acraniata. All vertebrates have a well-developed skull case enclosing the brain and are called Craniata. The vertebrates (craniates) may be variously subdivided into groups based on shared possession of characteristics. Two such subdivisions shown in Table 14-1 are: (1) Agnatha, vertebrates lacking jaws (hagfishes and lampreys), and Gnathostomata, vertebrates having jaws (all other vertebrates) and (2) Amniota, vertebrates whose embryos develop within a fluid-filled sac, the amnion (reptiles, birds, and mammals), and Anamniota, vertebrates lacking this adaptation (fishes and amphibians). The Gnathostomata in turn can be subdivided into Pisces, jawed vertebrates with limbs (if any) in the shape of fins; and the Tetrapoda (Gr. *tetras,* four, + *podos,* foot), jawed vertebrates with two pairs of limbs. Note that several of these groupings are paraphyletic (Protochordata, Acraniata, Agnatha, Anamniota, Pisces) and consequently are not accepted in cladistic classifications. Accepted monophyletic taxa are shown at the top of the cladogram in Figure 14-3 as a nested hierarchy of increasingly more inclusive groupings.

Four Chordate Hallmarks

The four distinctive characteristics that, taken together, set chordates apart from all other phyla are the **notochord; single, dorsal, tubular nerve cord; pharyngeal pouches;** and **postanal tail.** These characteristics are always found at some embryonic stage, although they may be altered or may disappear altogether in later stages of the life cycle.

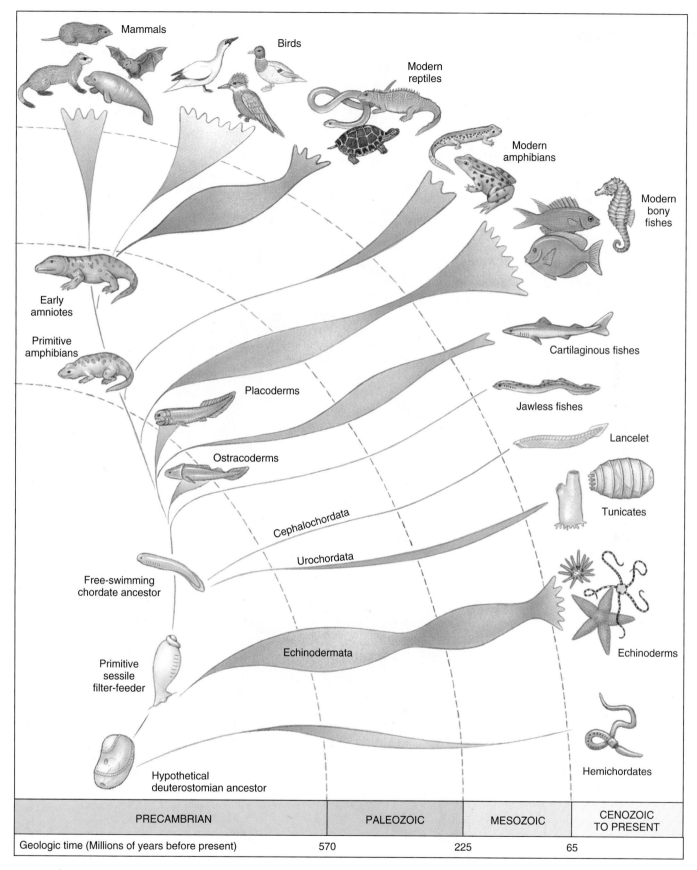

FIGURE 14-2

Phylogenetic tree of the chordates, suggesting probable origin and relationships. Other schemes have been suggested and are possible. The relative abundance in numbers of species of each group through geological time, as indicated by the fossil record, is suggested by the bulging and thinning of that group's line of descent.

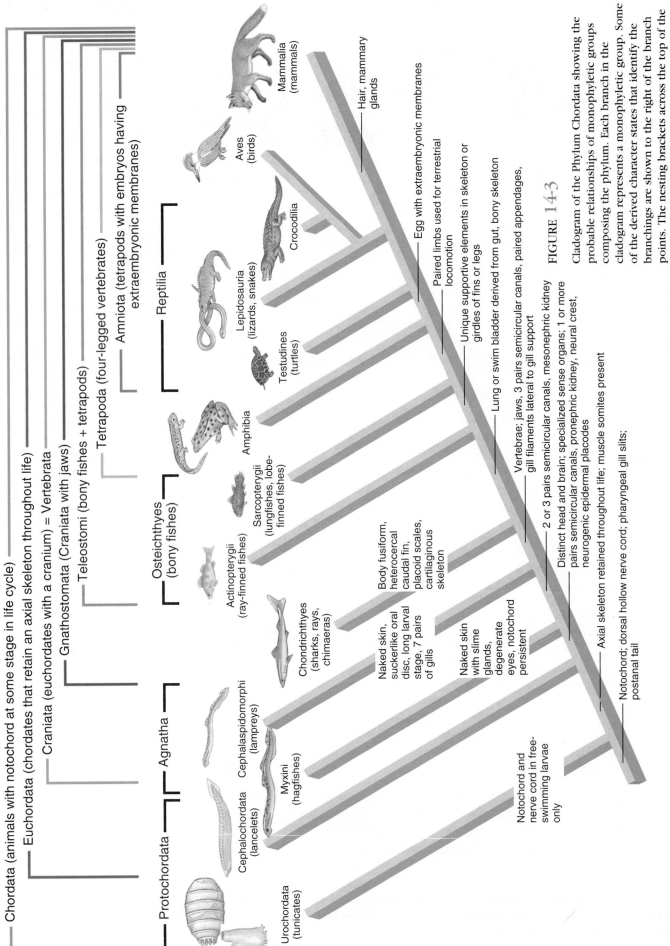

Chordata (animals with notochord at some stage in life cycle)

Euchordata (chordates that retain an axial skeleton throughout life)

Craniata (euchordates with a cranium) = Vertebrata

Gnathostomata (Craniata with jaws)

Teleostomi (bony fishes + tetrapods)

Tetrapoda (four-legged vertebrates)

Amniota (tetrapods with embryos having extraembryonic membranes)

Protochordata

Agnatha

Osteichthyes (bony fishes)

Reptilia

Urochordata (tunicates)

Cephalochordata (lancelets)

Myxini (hagfishes)

Cephalaspidomorphi (lampreys)

Chondrichthyes (sharks, rays, chimaeras)

Actinopterygii (ray-finned fishes)

Sarcopterygii (lungfishes, lobe-finned fishes)

Amphibia

Testudines (turtles)

Lepidosauria (lizards, snakes)

Crocodilia

Aves (birds)

Mammalia (mammals)

Hair, mammary glands

Egg with extraembryonic membranes

Paired limbs used for terrestrial locomotion

Unique supportive elements in skeleton or girdles of fins or legs

Lung or swim bladder derived from gut, bony skeleton

Vertebrae; jaws, 3 pairs semicircular canals, paired appendages, gill filaments lateral to gill support

2 or 3 pairs semicircular canals, mesonephric kidney

Distinct head and brain; specialized sense organs; 1 or more pairs semicircular canals, pronephric kidney, neural crest, neurogenic epidermal placodes

Body fusiform, heterocercal caudal fin, placoid scales, cartilaginous skeleton

Naked skin, suckerlike oral disc, long larval stage, 7 pairs of gills

Naked skin with slime glands, degenerate eyes, notochord persistent

Axial skeleton retained throughout life; muscle somites present

Notochord; dorsal hollow nerve cord; pharyngeal gill slits; postanal tail

Notochord and nerve cord in free-swimming larvae only

FIGURE 14-3

Cladogram of the Phylum Chordata showing the probable relationships of monophyletic groups composing the phylum. Each branch in the cladogram represents a monophyletic group. Some of the derived character states that identify the branchings are shown to the right of the branch points. The nesting brackets across the top of the cladogram identify the monophyletic groupings within the phylum. The term Craniata, although commonly equated with Vertebrata, is preferred by many authorities because it recognizes that the jawless vertebrates (Agnatha) have a cranium but no vertebrae. The lower set of brackets identify the traditional groupings Protochordata, Agnatha, Osteichthyes, and Reptilia. These groups are paraphyletic and not recognized in cladistic usage, but are shown because of their widespread usage.

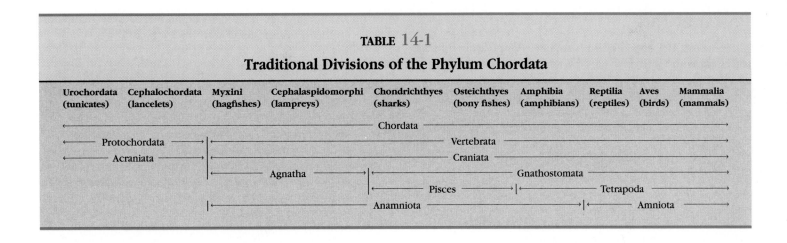

TABLE 14-1

Traditional Divisions of the Phylum Chordata

Urochordata (tunicates)	Cephalochordata (lancelets)	Myxini (hagfishes)	Cephalaspidomorphi (lampreys)	Chondrichthyes (sharks)	Osteichthyes (bony fishes)	Amphibia (amphibians)	Reptilia (reptiles)	Aves (birds)	Mammalia (mammals)
← Chordata →									
← Protochordata →		← Vertebrata →							
← Acraniata →		← Craniata →							
		← Agnatha →		← Gnathostomata →					
				← Pisces →		← Tetrapoda →			
	← Anamniota →							← Amniota →	

Notochord

The notochord is a flexible, rodlike structure, extending the length of the body; it is the first part of the endoskeleton to appear in the embryo. The notochord is an axis for muscle attachment, and because it can bend without shortening, it permits undulatory movements of the body. In most protochordates and in jawless vertebrates, the notochord persists throughout life (Figure 14-1). In all jawed vertebrates a series of cartilaginous or bony vertebrae is formed from the connective tissue sheath around the notochord and replaces the notochord as the chief mechanical axis of the body.

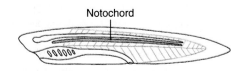

Notochord

Dorsal, Tubular Nerve Cord

In most invertebrate phyla that have a nerve cord, it is ventral to the alimentary canal and is solid, but in the chordates the single cord is dorsal to the alimentary canal and notochord and is a tube (although the hollow center may be nearly obliterated during growth). In vertebrates the anterior end becomes enlarged to form the brain. The hollow cord is produced in the embryo by the infolding of the ectodermal cells on the dorsal side of the body above the notochord. The nerve cord passes through the protective neural arches of the vertebrae, and the anterior brain is surrounded by a bony or cartilaginous cranium.

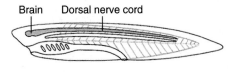

Brain   Dorsal nerve cord

Pharyngeal Pouches and Slits

Pharyngeal slits are perforated slitlike openings that lead from the pharyngeal cavity to the outside. They are formed by the inpocketing of the outside ectoderm (pharyngeal grooves) and the evagination, or outpocketing, of the endodermal lining of the pharynx (pharyngeal pouches). In aquatic chordates, the two pockets break through the pharyngeal cavity where they meet to form the pharyngeal slit. In amniotes these pockets may not break through the pharyngeal cavity and only grooves are formed instead of slits. In tetrapod vertebrates the pharyngeal pouches give rise to several different structures, including Eustachian tube, middle ear cavity, tonsils, and parathyroid glands.

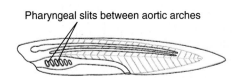

Pharyngeal slits between aortic arches

The perforated pharynx evolved as a filter-feeding apparatus and is used as such in the protochordates. Water with suspended food particles is drawn by ciliary action through the mouth and flows out through the pharyngeal slits, where food is trapped in mucus. In the vertebrates, ciliary action is replaced by a muscular pump that drives water through the pharynx by expanding and contracting the pharyngeal cavity. Also modified are the blood vessels that carry blood through the pharyngeal bars. In protochordates these are simple vessels surrounded by connective tissue. In the early fishes a capillary network was added with only thin, gas-permeable walls separating water outside from blood inside. This improved efficiency of gas transfer. These adaptations led to the development of **internal gills,** completing the conversion of the pharynx from a filter-feeding apparatus in protochordates to a respiratory organ in aquatic vertebrates.

Postanal Tail

The postanal tail, together with somatic musculature and the stiffening notochord, provides the motility that larval tunicates and amphioxus need for their free-swimming existence. As a structure added to the body behind the anus, it clearly has evolved specifically for propulsion in water. Its efficiency is later increased in fishes with the addition of fins. The tail is evident in humans only as a vestige (the coccyx, a series of small vertebrae at the end of the spinal column) but most other mammals have a waggable tail as adults.

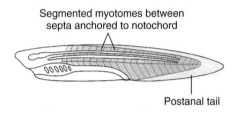

Segmented myotomes between septa anchored to notochord

Postanal tail

Ancestry and Evolution of the Chordates

Since the middle of the nineteenth century when the theory of organic evolution became the focal point for ferreting out relationships among groups of living organisms, zoologists have debated the question of chordate origins. It has been difficult to reconstruct lines of descent because the earliest protochordates were in all probability soft-bodied creatures that stood little chance of being preserved as fossils even under the most ideal conditions. Consequently, such reconstructions largely come from the study of living organisms, especially from an analysis of early developmental stages, which tend to be more insulated from evolutionary change than the differentiated adult forms that they become.

Zoologists at first speculated that the chordates evolved within the protostome lineage (annelids and arthropods) but discarded such ideas when they realized that supposed morphological similarities had no developmental basis. Early in this century when further theorizing became rooted in developmental patterns of animals, it became apparent that the chordates must have originated within the deuterostome branch of the animal kingdom. As explained earlier (p. 49 and Figure 2-14), the Deuterostomia, a grouping that includes the echinoderms, hemichordates, lophophorates, and chordates, has several important embryological features that clearly separate it from the Protostomia and establish its monophyly. Accordingly the deuterostomes are almost certainly a natural grouping of interrelated animals that have their common origin in ancient Precambrian seas. There are several lines of anatomical, developmental, and molecular evidence suggesting that somewhat later, at the base of the Cambrian period some 570 million years ago, the first distinctive chordates arose from a lineage related to echinoderms and hemichordates (Figure 14-2; see also Figure 13-25, p. 265).

> *Most of the early efforts to pin together invertebrate and chordate kinship are now recognized as based on similarities due to analogy rather than homology. Analogous structures are those that perform similar functions but have altogether different origins (such as wings of birds and butterflies). Homologous structures, on the other hand, share a common origin but may look quite different (at least superficially) and perform quite different functions. For example, all vertebrate forelimbs are homologous because they are derived from a pentadactyl limb of the same ancestor, even though they may be modified as differently as the human arm and a bird's wing. Homologous structures share a genetic heritage; analogous structures do not. Obviously, only homologous similarities have any bearing in genealogical connections.*

While modern echinoderms look nothing at all like modern chordates, evolutionary affinity between chordates and echinoderms gains support from fossil evidence. One curious group of fossil echinoderms, the "Calcichordata," have pharyngeal slits and possibly other chordate attributes (Figure 14-4; see also p. 261). These small, nonsymmetrical forms have a head resembling a long-toed medieval boot, a series of branchial slits covered with flaps much like the gill openings of sharks, a postanal tail, and structures that are doubtfully interpreted as notochord and muscle blocks. These creatures apparently used their pharyngeal slits for filter feeding, as do the protochordates today. Although the calcichordates seem to have some of the right chordate characters based on soft anatomy, there is no convincing similarity between the hard skeleton of calcichordates (which was calcium carbonate) and that of vertebrates (which is composed of a complex of calcium and phosphate). Thus, while such fossils bring us closer to an understanding of chordate origins, we are not yet in a position to know the precise characteristics of the long-sought chordate ancestor. We do, however, know a great deal about two living protochordate groups that also descended from it. These we will now consider.

Subphylum Urochordata (Tunicata)

The urochordates ("tail-chordates"), more commonly called tunicates, number some 3000 species. They are found in all seas from near the shoreline to great depths. Most of them are sessile as adults, although some are free living. The name "tunicate" is suggested by the usually tough, nonliving **tunic,** or test, that surrounds the animal (Figure 14-5). As adults, tunicates are highly specialized chordates, for in most species only the larval form, which resembles a microscopic tadpole, bears all the chordate hallmarks. During adult metamorphosis, the notochord (which, in the larva, is restricted to the tail, hence the group name Urochordata) and the tail disappear altogether, while the dorsal nerve cord becomes reduced to a single ganglion.

Urochordata is divided into three classes—**Ascidiacea** (Gr. *askiolion,* little bag, + *acea,* suffix), **Larvacea** (L. *larva,* ghost, + *acea,* suffix), and **Thaliacea** (Gr. *thalia,* luxuriance, + *acea,* suffix). Of these, the members of **Ascidiacea,** commonly known as the ascidians, or sea squirts, are by far the most common and are the best known. Ascidians may be solitary, colonial, or compound. Each of the solitary and colonial forms has its own test, but among the compound forms many individuals may share the same test. In some of these compound ascidians each member has its own incurrent siphon, but the excurrent opening is common to the group.

The typical solitary ascidian (Figure 14-5) is a spherical or cylindrical form that is attached by its base to hard substrates such as rocks, pilings, or the bottoms of ships. Lining the tunic is an inner membrane, the **mantle.** On the outside are two projections: the **incurrent** and **excurrent siphons** (Figure 14-5). Water enters the incurrent siphon and passes into the branchial sac (pharynx) through the mouth. On the midventral side of the branchial sac is a groove, the **endostyle,** which is ciliated and secretes mucus. As the mucous sheet is carried by cilia across the inner surface of the pharynx to the dorsal side, it sieves small food particles from the water passing through the slits in the wall of the branchial sac. Then the mucus with its entrapped food is collected and passed posteriorly to the esophagus. The water, now largely cleared of food particles, is driven by cilia into the atrial cavity and finally out the excurrent siphon. The intestine leads to the anus near the excurrent siphon.

The circulatory system consists of a ventral **heart** near the stomach and two large vessels, one on either side of the heart. An odd feature found in no other chordate is that the heart drives the blood first in one direction for a few beats, then pauses, reverses, and drives the blood in the opposite direction. The excretory system is a type of nephridium near the intestine. The nervous system is restricted to a **nerve ganglion** and a few nerves that lie on the dorsal side of the pharynx. A notochord is lacking. The animals are hermaphroditic. The germ cells are carried out the excurrent siphon into the surrounding water, where cross-fertilization occurs.

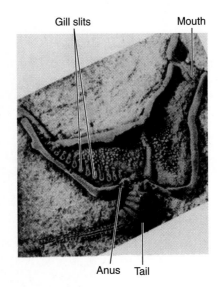

FIGURE 14-4

Fossil of an early echinoderm, a calcichordate, that lived during the Ordovician period (450 million years BP). It shows affinities with both echinoderms and chordates and may belong to a lineage that was ancestral to the chordates.

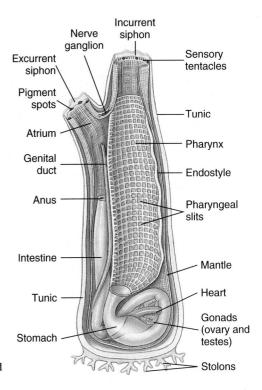

FIGURE 14-5

Structure of a common tunicate, *Ciona* sp.

Of the four chief characteristics of chordates, adult sea squirts have only one, the pharyngeal gill slits. However, the larval form gives away the secret of their true relationship. The tiny tadpole larva (Figure 14-6) is an elongate, transparent form with a head and all four chordate characteristics: a notochord, hollow dorsal nerve cord, propulsive postanal tail, and a large pharynx with endostyle and gill slits. The larva does not feed but swims for several hours before fastening itself vertically by its adhesive papillae to some solid object. It then metamorphoses to become the sessile adult.

The remaining two classes of the Urochordata—**Larvacea** and **Thaliacea**—are mostly small, transparent animals of the open sea (Figure 14-7). Some are small, tadpole-like forms resembling the larval stage of ascidians. Others are spindle shaped or cylindrical forms surrounded by delicate muscle bands. They are mostly carried along by the ocean currents and as such form a part of the plankton. Many are provided with luminous organs and emit a beautiful light at night.

Subphylum Cephalochordata

The cephalochordates are the marine lancelets: slender, laterally compressed, translucent animals about 5 to 7 cm in length (Figure 14-8) that inhabit the sandy bottoms of coastal waters

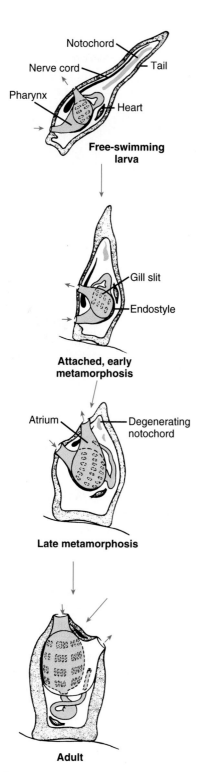

FIGURE 14-6

Metamorphosis of a solitary ascidian from a free-swimming larval stage.

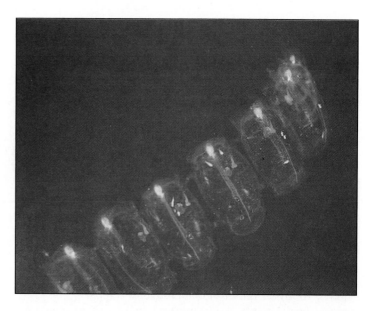

FIGURE 14-7

Colonial thaliacean. The transparent individuals of this delicate, planktonic tunicate are grouped in a chain. Visible within each individual is an orange gonad, an opaque gut, and a long serrated gill bar.

in this diminutive subphylum. Four species of amphioxus (lancelets) are found in North American coastal waters.

Amphioxus is especially interesting because it has the four distinctive characteristics of chordates in simple form. Water enters the mouth, driven by cilia in the buccal cavity, and then passes through numerous pharyngeal slits in the pharynx, where food is trapped in mucus, which is then moved by cilia into the intestine. Here the smallest food particles are separated from the mucus and passed into the midgut caecum (diverticulum), where they are phagocytized and digested intracellularly. As in tunicates, the filtered water passes first into an atrium, and then leaves the body by an atriopore (equivalent to the excurrent siphon of tunicates).

The closed circulatory system is complex for so simple a chordate. The flow pattern is remarkably similar to that of fishes, although there is no heart. Blood is pumped forward in the ventral aorta by peristaltic-like contractions of the vessel wall, and then passes upward through the branchial arteries (aortic arches) in the gill bars to the dorsal aorta. From here the blood is distributed to the body tissues by capillary plexi and then is collected in veins, which return it to the ventral aorta. The blood is colorless, lacking both erythrocytes and hemoglobin.

The nervous system is centered around a hollow nerve cord lying above the notochord. Sense organs are simple, unpaired bipolar receptors located in various parts of the body. The "brain" is a simple vesicle at the anterior end of the nerve cord.

Sexes are separate in amphioxus. The sex cells are set free in the atrium, and then pass out the atriopore to the outside, where fertilization occurs. The larvae hatch soon after the eggs are fertilized and gradually assume the shape and size of adults.

around the world. Lancelets originally bore the generic name *Amphioxus* (Gr. *amphi,* both ends, + *oxys,* sharp), later surrendered by priority to *Branchiostoma* (Gr. *branchia,* gills, + *stoma,* mouth). Amphioxus is still used, however, as a convenient common name for all of the approximately 26 species

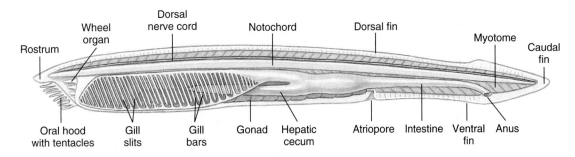

FIGURE 14-8

Amphioxus. This interesting bottom-dwelling cephalochordate illustrates the four distinctive chordate characteristics (notochord, dorsal nerve cord, pharyngeal gill slits, and postanal tail). The vertebrate ancestor probably had a similar body plan.

No other chordate shows the basic diagnostic characteristics of the chordates so well. In addition to the four chordate anatomical hallmarks, amphioxus possesses several structural features that foreshadow the vertebrate plan. Among these are the midgut diverticulum, which secretes digestive enzymes, segmented trunk musculature, and the basic circulatory pattern of more advanced chordates.

Subphylum Vertebrata

The third subphylum of the chordates is the large and diverse Vertebrata, the subject of the next five chapters of this book. This monophyletic group shares the basic chordate characteristics with the other two subphyla, but in addition it reveals a number of novel homologies that the others do not share. The alternative name of the subphylum, Craniata, more accurately describes the group since all have a cranium (bony or cartilaginous braincase) whereas the jawless fishes lack vertebrae.

Adaptations That Have Guided Vertebrate Evolution

From the earliest fishes to the mammals, the evolution of the vertebrates has been guided by the basic adaptations of the living endoskeleton, pharynx and efficient respiration, advanced nervous system, and paired limbs.

Living Endoskeleton

The endoskeleton of vertebrates, as in the echinoderms, is an internal supportive structure and framework for the body. This condition is a departure in animal architecture, since invertebrate skeletons are more commonly exoskeletons. Exoskeletons and endoskeletons have their own particular set of advantages and limitations relating to size. For vertebrates the living endoskeleton possesses an overriding advantage over the secreted, nonliving exoskeleton of arthropods: growing with the body as it does, the endoskeleton permits almost

unlimited body size with much greater economy of building materials. Some vertebrates have become the most massive animals on earth. The endoskeleton forms an excellent jointed scaffolding for muscles, and the muscles in turn protect the skeleton and cushion it from potentially damaging impact.

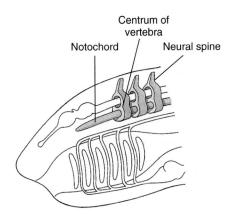

We should note that the vertebrates have not wholly lost the protective function of a firm external covering. The skull and the thoracic rib cage enclose and protect vulnerable organs. Most vertebrates are further protected with a tough integument, often bearing nonliving structures such as scales, hair, and feathers that may provide insulation as well as physical security.

Pharynx and Efficient Respiration

The perforated pharynx (gill slits), present as pharyngeal pouches in all chordates at some stage in their life cycle, evolved as an apparatus for suspension feeding. In primitive chordates, water with suspended food particles is drawn through the mouth by ciliary action and flows out through the gill slits, where food is trapped in mucus. This condition is observed today in amphioxus. As the protovertebrates shifted from suspension feeding to a predatory life habit, the pharynx became modified into a muscular feeding apparatus through which water could be pumped by expanding and contracting

Characteristics of the Subphylum Vertebrata

1. Chief diagnostic features of chordates—**notochord, dorsal nerve cord, pharyngeal pouches,** and **postanal tail**—all present at some stage of the life cycle

2. **Integument** basically of two divisions, an outer **epidermis** of stratified epithelium from the ectoderm and an inner **dermis** of connective tissue derived from the mesoderm; many modifications of skin among the various classes, such as glands, scales, feathers, claws, horns, and hair

3. Distinctive **endoskeleton** consisting of vertebral column (notochord persistent in jawless fishes which lack vertebrae), limb girdles, and two pairs of jointed appendages derived from somatic mesoderm, and a head skeleton (cranium and pharyngeal skeleton) derived largely from neural crest cells.

4. Muscular, perforated pharynx; in fishes pharyngeal slits possess gills and muscular aortic arches; in tetrapods the much reduced pharynx is an embryonic source of glandular tissue

5. **Many muscles** attached to the skeleton to provide for movement

6. Complete digestive system ventral to the spinal column and provided with large digestive glands, liver, and pancreas

7. Circulatory system consisting of a **ventral heart** of two to four chambers; a closed blood vessel system of arteries, veins, and capillaries; blood fluid containing red corpuscles with hemoglobin and white corpuscles; paired aortic arches connecting the ventral and dorsal aortas and branching to the gills in the gill-breathing vertebrates; in the terrestrial types modification of the aortic arch into pulmonary and systemic systems

8. Well-developed **coelom** largely filled with the visceral systems

9. Excretory system consisting of **paired kidneys** (mesonephric or metanephric types in adults) provided with ducts to drain the waste to cloaca or anal region

10. Highly differentiated **brain;** 10 or 12 pairs of **cranial nerves** usually with both motor and sensory functions; a pair of spinal nerves for each primitive myotome; an **autonomic nervous system** in control of involuntary functions of internal organs; **paired special sense organs**

11. **Endocrine system** of ductless glands scattered through the body

12. Nearly always separate sexes; each sex containing paired gonads with ducts that discharge their products either into the cloaca or into special openings near the anus

13. **Body plan** consisting typically of **head, trunk,** and **postanal tail; neck** present in some, especially terrestrial forms; usually two pairs of appendages, although entirely absent in some; coelom divided into a pericardial space and a general body cavity; mammals with a thoracic cavity

---

the pharyngeal cavity. Circulation to the internal gills was improved by the addition of capillary beds (lacking in protochordates) and the development of a ventral heart and muscular aortic arches. All of these changes supported an increased metabolic rate that would have to accompany the switch to an active life of selective predation.

Advanced Nervous System

No single system in the body is more strongly associated with functional and structural advancement than is the nervous system. The prevertebrate nervous system consisted of a brainless nerve cord and rudimentary sense organs, which were mostly chemosensory in function. When the protovertebrates switched to a predatory life-style, new sensory, motor,

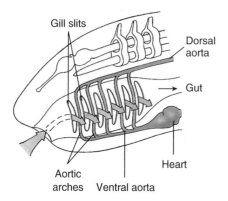

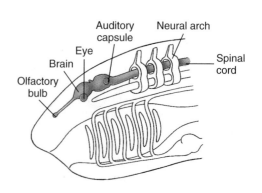

and integrative controls were now essential for the location and capture of larger prey items. In short, the protovertebrates developed a new head, complete with a brain and external paired sense organs especially designed for distance reception. These included paired eyes with lenses and inverted retinas; pressure receptors, such as paired ears designed for equilibrium and later redesigned to include sound reception; and chemical receptors, including taste receptors and exquisitely sensitive olfactory organs.

Paired Limbs

Pectoral and pelvic appendages are present in most vertebrates in the form of paired fins or legs. They originated as swimming stabilizers and later became prominently developed into legs for travel on land. Jointed limbs are especially suited for life on land because they permit finely graded movement against a substrate.

The Search for the Vertebrate Ancestral Stock

Early Chordate Fossils

The earliest Paleozoic vertebrate fossils, the jawless ostracoderm fishes to be considered later in this chapter, share many novel features of organ-system development with living vertebrates. These organ systems therefore must have originated either in an early vertebrate or invertebrate chordate lineage. With one exception, hardly any invertebrate chordates are known as fossils. The exception is *Pikaia gracilens,* a ribbon-shaped, somewhat fishlike creature about 5 cm in length discovered in the famous Burgess Shale of British Columbia (Figure 14-9; see also Figure 1-11, p. 14). *Pikaia* is a mid-Cambrian form that precedes the earliest vertebrate fossils by many millions of years. Although this fossil has not yet been described in detail, we know that it possessed both a notochord and the characteristic chevron-shaped (>) muscle bands (myotomes). Without question *Pikaia* is a chordate. It shows a remarkable resemblance to living amphioxus, at least in overall body organization, and may in fact be an early cephalochordate. *Pikaia* is a provocative fossil but, until other Cambrian chordate fossils are discovered, its relationship to the earliest vertebrates remains uncertain. In the absence of additional fossil evidence, most speculations on vertebrate ancestry have focused on the living cephalochordates and tunicates, since it is widely believed that the vertebrates must have emerged from a lineage resembling one of these protochordate groups.

Garstang's Hypothesis of Chordate Larval Evolution

At first glance, tunicates seem unlikely candidates for the sister group of vertebrates. The adult tunicate, which spends its life

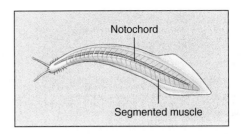

FIGURE 14-9

Pikaia, the earliest known chordate, from the Burgess Shale of British Columbia, Canada.

anchored to some marine surface, lacks a notochord, tubular nerve cord, postanal tail, sense organs, and segmented musculature. Their larvae, however, bear all the right qualifications for chordate membership. Called "tadpole" larvae because of their superficial resemblance to larval frogs, these tiny, site-seeking forms have a notochord, hollow dorsal nerve cord, pharyngeal gill slits, and postanal tail, as well as a brain and sense organs.

At the time of its discovery in 1869, the tadpole larva was considered the descendant of an ancient free-swimming chordate ancestor of the tunicates. The adults then were regarded as degenerate, sessile descendants of the free-swimming form. In 1928, Walter Garstang in England introduced fresh thinking into the vertebrate ancestor debate by turning this sequence around: rather than the ancestral tadpole larva giving rise to a degenerative tunicate sessile adult, he suggested that the sessile adults *were* the ancestral stock. The tadpole larvae then evolved as an adaptation for spreading to new habitats. Next, Garstang suggested that at some point the tadpole failed to metamorphose into an adult but developed gonads and reproduced in the larval stage. With continued larval evolution, a new group of free-swimming animals appeared (Figure 14-10).

Garstang called this process **paedomorphosis** (Gr. *pais,* child, + *morphē,* form), a term that describes the evolutionary retention of juvenile or larval traits in the adult body. Garstang departed from previous thinking by suggesting that evolution may occur in the larval stages of animals—and in this case, lead to the vertebrate lineage. Paedomorphosis is a well-known phenomenon in several different animal groups (paedomorphosis in amphibians is described on p. 317). Furthermore, Garstang's hypothesis agrees with the embryological evidence. Nevertheless, it remains untested and thus speculative.

Position of Amphioxus

For many years zoologists believed that the cephalochordate amphioxus is the closest living relative of vertebrates. No other protochordate shows the basic diagnostic characteristics of the chordates so well. However, amphioxus lacks a brain and all of the specialized sensory equipment that characterize the vertebrates. There are no gills in the pharynx and no mouth or pharyngeal musculature for pumping water through the gill slits; movement of water is entirely by the action of cilia. Despite these specializations and others peculiar to modern cephalochordates, many zoologists believe that amphioxus has retained the primitive pattern of the immediate pre-vertebrate condition. Thus the cephalo-chordates are the probable sister group of the vertebrates (Figure 14-3).

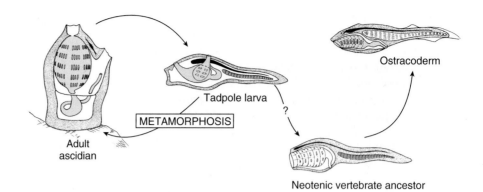

Ostracoderm

Tadpole larva

METAMORPHOSIS

Adult ascidian

Neotenic vertebrate ancestor

FIGURE 14-10

Garstang's hypothesis of larval evolution. Adult tunicates live on the seafloor but reproduce through a free-swimming tadpole larva. More than 500 million years ago, some larvae began to reproduce in the swimming stage. These gave rise to the ostracoderms, the first known vertebrates.

Paedomorphosis, the displacement of ancestral larval or juvenile features into a descendant adult, can be produced by three different evolutionary-developmental processes: neoteny, progenesis, and post-displacement. In neoteny, the growth rate of body form is slowed so that the animal does not attain the ancestral adult form at the time it reaches reproductive maturity. Progenesis is the precocious maturation of gonads in a larval (or juvenile) body that then stops growing and never attains the adult body form. In post-displacement, the onset of a developmental process is delayed relative to reproductive maturation, so that the ancestral adult form is not attained at the time of reproductive maturation. Neoteny, progenesis and post-displacement thus describe different ways in which paedomorphosis can happen. Zoologists use the inclusive term paedomorphosis to describe the results of these evolutionary-developmental processes.

The Earliest Vertebrates: Jawless Ostracoderms

The earliest vertebrate fossils are late Cambrian articulated skeletons from the United States, Bolivia, and Australia. They were small, jawless creatures collectively called ostracoderms (os-trak′o-derm) (Gr. *ostrakon,* shell, + *derma,* skin), which belong to the Agnatha division of the vertebrates. The earliest ostracoderms, called **heterostracans,** lacked paired fins, which later fishes found so important for stability (Figure 14-11). Their swimming movements must have been clumsy, although sufficient to propel them along the ocean bottom where they searched for food. With fixed circular or slitlike mouth openings, they probably filtered small food particles from the water or ocean bottom. However, unlike the ciliary filter-feeding protochordates, the ostracoderms sucked water into the pharynx by muscular pumping, an important innovation that suggests to some authorities that the ostracoderms may have been mobile predators that fed on soft-bodied animals.

The term "ostracoderm" does not describe a natural evolutionary assemblage but rather is a term of convenience for describing several groups of heavily armored, extinct jawless fishes.

During the Devonian period, the ostracoderms underwent a major radiation, resulting in the appearance of several peculiar-looking forms varying in shape and length of the snout, dorsal spines, and dermal plates. One group, the **osteostracans** (Figure 14-11), improved the efficiency of their benthic life by evolving paired pectoral fins. These fins, located just behind the head, provided control over pitch and yaw, which ensured well-directed forward movement. Another group of ostracoderms, the **anaspids,** (Figure 14-11) were more streamlined and more closely resembled modern-day jawless fishes (the lamprey, for example) than any other ostracoderm.

As a group, the ostracoderms were basically fitted for a simple, bottom-feeding life. Yet, despite their anatomical limitations, they enjoyed a respectable radiation in the Silurian and Devonian periods.

For decades, geologists have used strange microscopic, toothlike fossils called **conodonts** (Gr. *kōnos,* cone, + *odontos,* tooth) to date Paleozoic marine sediments without having any idea what kind of creature originally possessed these elements. The discovery in the early 1980s of the fossils of complete conodont animals showed that conodont elements belonged to a small early marine vertebrate (Figure 14-12). It is widely believed that as more is learned about conodont animals they will play an important role in understanding the origin of vertebrates. At present, however, their position in the vertebrate phylogeny is a matter of debate.

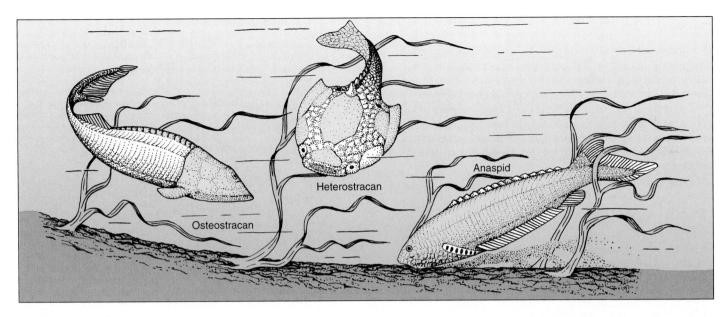

FIGURE 14-11

Three ostracoderms, jawless fishes of Silurian and Devonian times. They are shown as they might have appeared while searching for food on the floor of a Devonian sea. All were probably suspension-feeders, but employed a strong pharyngeal pump to circulate water rather than the much more limiting mode of ciliary feeding used by their protovertebrate ancestors and by amphioxus today. Modern lampreys are believed to be derived from the anaspid group.

Early Jawed Vertebrates

All jawed vertebrates, whether extinct or living, are collectively called **gnathostomes** ("jaw mouth") in contrast to the jawless vertebrates, the **agnathans** ("without jaw"). The living agnathans, the naked hagfishes and lampreys, also are often called cyclostomes ("circle mouth"). The gnathostomes are almost certainly a monophyletic group since the presence of jaws is a derived character state shared by all jawed fishes and tetrapods. The agnathans, however, are defined principally by the absence of a feature—jaws—that characterize the gnathostomes, and the superclass Agnatha therefore may be paraphyletic.

The origin of jaws was one of the most important events in vertebrate evolution. The utility of jaws is obvious: they allow predation on large and active forms of food not available to the jawless vertebrates. There is ample evidence that jaws arose through modifications of the first two of the serially repeated cartilaginous gill arches. The beginnings of this trend can be seen in some of the jawless ostracoderms, where the mouth becomes bordered by strong dermal plates that could be manipulated somewhat like jaws with the gill-arch musculature. Later, the anterior gill arches became

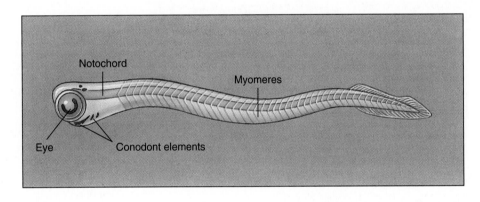

FIGURE 14-12

Restoration of a living conodont animal. The conodont superficially resembled amphioxus, but it possessed a much greater degree of encephalization (large paired eyes, possible auditory capsules) and bone-like mineralized elements—all indicating that the conodont animal was a vertebrate. The conodont elements were probably gill-supporting structures or part of a suspension-feeding apparatus.

hinged and bent forward into the characteristic position of vertebrate jaws (Figure 14-13). Nearly as remarkable as this drastic morphological remodeling is the subsequent evolutionary fate of the many jawbone elements—their transformation into the ear ossicles of the mammalian middle ear.

Among the first jawed vertebrates were the heavily armored **placoderms** (plak′o-derm) (Gr. *plax*, plate, + *derma*, skin); these forms first appear in the fossil record in the early

FIGURE 14-13

How the vertebrates got their jaw. The resemblance between jaws and the gill supports of the primitive fishes such as this carboniferous shark suggests that the upper jaw (palatoquadrate) and lower jaw (Meckel's cartilage) evolved from structures that originally functioned as gill supports. The gill supports immediately behind the jaws are hinged like the jaws and served to link the jaws to the braincase. Relics of this transformation are seen during the development of modern sharks.

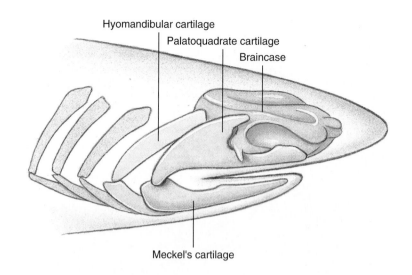

FIGURE 14-14

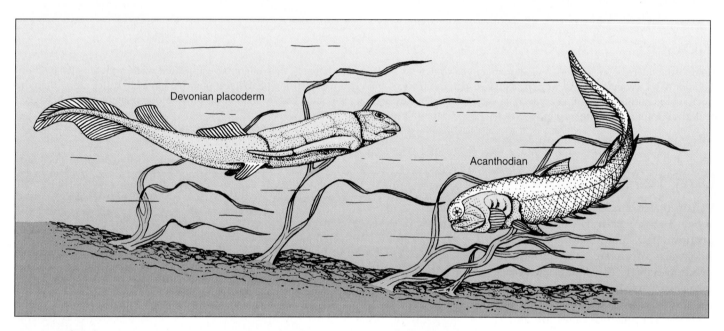

Early jawed fishes of the Devonian period, 400 million years ago. Shown are a placoderm *(left)* and a related acanthodian *(right)*. The jaws and the gill supports from which the jaws evolved develop from neural crest cells, a diagnostic character of vertebrates. Most placoderms were bottom dwellers that fed on detritus although some were active predators. The acanthodians, the earliest-known true jawed fishes, carried less armor than the placoderms. Most were marine but several species entered fresh water.

Devonian period (Figure 14-14). Placoderms evolved a great variety of forms, some very large (one was 10 m in length!) and grotesque in appearance. They were armored fish covered with diamond-shaped scales or with large plates of bone. All became extinct by the end of the Paleozoic era and appear to have left no descendants. However, the **acanthodians** (Figure 14-14), a group of early jawed fishes that were contemporary with the placoderms, may have given rise to the great radiation of bony fishes that dominate the waters of the world today.

Evolution of Modern Fishes and Tetrapods

Reconstruction of the origins of the vast and varied assemblage of modern living vertebrates is based, as we have seen, largely on fossil evidence. Unfortunately the fossil evidence for the earliest vertebrates is often incomplete and tells us much less than we would like to know about subsequent trends in evolution. Affinities become much easier to establish as the

Classification of Phylum Chordata

Phylum Chordata

Group Protochordata (Acrania)

Subphylum Urochordata (u′ro-kor-da′ta) (Gr. *oura,* tail, + L. *chorda,* cord, + *ata,* characterized by) **(Tunicata): tunicates.** Notochord and nerve cord in free-swimming larva only; ascidian adults sessile, encased in tunic.

Subphylum Cephalochordata (sef′a-lo-kor-da′ta) (Gr. *kephalē,* head, + L. *chorda,* cord): **lancelets (amphioxus).** Notochord and nerve cord found along entire length of body and persist throughout life; fishlike in form.

Group Craniata

Subphylum Vertebrata (ver′te-bra′ta) (L. *vertebratus,* backboned). Bony or cartilaginous vertebrae surrounding spinal cord (vertebrae absent in agnathans); notochord only in embryonic stages, persisting in some fishes; also may be divided into two groups (superclasses) according to presence of jaws.

Superclass Agnatha (ag′na-tha) (Gr. *a,* without, + *gnathos,* jaw) **(Cyclostomata): hagfishes, lampreys.** Without true jaws or paired appendages. Probably a paraphyletic group.

Class Myxini (mik-sin′y) (Gr. *myxa,* slime): **hagfishes.** Terminal mouth with four pairs of tentacles; buccal funnel absent; nasal sac with duct to pharynx; 5 to 15 pairs of gill pouches; partially hermaphroditic.

Class Cephalaspidomorphi (sef-a-lass′pe-do-morf′e) (Gr. *kephalē,* head, + *aspidos,* shield, *morphē,* form) **(Petromyzones): lampreys.** Suctorial mouth with horny teeth; nasal sac not connected to mouth; seven pairs of gill pouches.

Superclass Gnathostomata (na′tho-sto′ma-ta) (Gr. *gnathos,* jaw, + *stoma,* mouth): **jawed fishes, all tetrapods.** With jaws and (usually) paired appendages.

Class Chondrichthyes (kon-drik′thee-eez) (Gr. *chondros,* cartilage, + *ichthys,* a fish): **sharks, skates, rays, chimaeras.** Streamlined body with heterocercal tail; cartilaginous skeleton; five to seven gills with separate openings, no operculum, no swim bladder.

Class Osteichthyes (ost′e-ik′thee-eez) (Gr. *osteon,* bone, + *ichthys,* a fish): **bony fishes.** Primitively fusiform body but variously modified; mostly ossified skeleton; single gill opening on each side covered with operculum; usually swim bladder or lung. A paraphyletic group.

Class Amphibia (am-fib′e-a) (Gr. *amphi,* both or double, + *bios,* life): **amphibians.** Ectothermic tetrapods; respiration by lungs, gills, or skin; development through larval stage; skin moist, containing mucous glands, and lacking scales.

Class Reptilia (rep-til′e-a) (L. *repere,* to creep): **reptiles.** Ectothermic tetrapods possessing lungs; embryo develops within shelled egg; no larval stage; skin dry, lacking mucous glands, and covered by epidermal scales. A paraphyletic group.

Class Aves (ay′veez) (L. pl. of *avis,* bird): **birds.** Endothermic vertebrates with front limbs modified for flight; body covered with feathers; scales on feet.

Class Mammalia (ma-may′lee-a) (L. *mamma,* breast): **mammals.** Endothermic vertebrates possessing mammary glands; body more or less covered with hair; well-developed neocerebrum.

fossil record improves. For instance, the descent of birds and mammals from early tetrapod ancestors has been worked out in a highly convincing manner from the relatively abundant fossil record available. By contrast, the ancestry of modern fishes is shrouded in uncertainty.

Despite the difficulty of clarifying early lines of descent for the vertebrates, they are clearly a natural, monophyletic group, distinguished by a large number of shared derived characteristics. They almost certainly have descended from a common ancestor, although we still do not know from which invertebrate group the vertebrate lineage originated. Early in their evolution, the vertebrates divided into the agnathans and the gnathostomes. These two groups differ from each other in many fundamental ways, in addition to the absence of jaws in the former group and their presence in the latter. The appearance of both jaws and paired fins were major innovations in vertebrate evolution, perhaps the most important reasons for the subsequent major radiations of the vertebrates that produced the modern fishes and all of the tetrapods—including you, the reader of this book.

Summary

The phylum Chordata is named for the rod-like notochord that forms a stiffening body axis at some stage in the life cycle of every chordate. All chordates share four distinctive hallmarks that set them apart from all other phyla: notochord, dorsal tubular nerve cord, pharyngeal pouches, and post-anal tail. Two of the three chordate subphyla are invertebrates and lack a well-developed head. They are the Urochordata (tunicates), most of which are sessile as adults, but all of which have a free-swimming larval stage; and the Cephalochordata (lancelets), fishlike forms that include the famous amphioxus.

The chordates have evolutionary affinities to echinoderms, but the exact evolutionary origin of the chordates is not yet, and may never be, known with certainty. Taken as a whole, the chordates have a greater fundamental unity of organ systems and body plan than have many of the invertebrate phyla.

The subphylum Vertebrata includes the backboned members of the animal kingdom (the living jawless vertebrates, the hagfishes and lampreys, actually lack vertebrae but are included with the Vertebrata by tradition because they share numerous homologies with the vertebrates). As a group the vertebrates are characterized by having a well-developed head and by their comparatively large size, high degree of motility, and distinctive body plan, which embodies several distinguishing features that have permitted exceptional adaptive radiation. Most important of these are the living endoskeleton, which allows continuous growth and provides a sturdy framework for efficient muscle attachment and action; a pharynx perforated with gill slits (lost or greatly modified in the reptiles, birds, and mammals) with vastly increased respiratory efficiency; a complex nervous system with clear separation of the brain and spinal cord; and paired limbs.

Review Questions

1. What characteristics are shared by the six deuterostome phyla that indicate a natural grouping of interrelated animals?

2. Explain how the use of a cladistic classification for the vertebrates results in important regroupings of the traditional vertebrate taxa (refer to Figure 14-3). Why are certain traditional groupings such as Reptilia and Agnatha not recognized in cladistic usage?

3. Name four hallmarks shared by all chordates, and explain the function of each.

4. In debating the question of chordate origins, zoologists eventually agreed that the chordates must have evolved within the deuterostome assemblage rather than from a protostome group as earlier argued. What embryological evidences support this view? What characteristics does the fossil echinoderm group Calcichordata possess that suggest it might closely resemble the ancestor of the chordates?

5. Offer a description of an adult tunicate that would identify it as a chordate, yet distinguish it from any other chordate group.

6. Amphioxus long has been of interest to zoologists searching for a vertebrate ancestor. Explain why amphioxus captured such interest and why it no longer is considered to resemble closely the direct ancestor of the vertebrates.

7. Both sea squirts (urochordates) and lancelets (cephalochordates) are suspension-feeding organisms. Describe the suspension-feeding apparatus of a sea squirt and explain in what ways its mode of feeding is similar to, and different from, that of amphioxus.

8. Explain why it is necessary to know the life history of a tunicate to understand why tunicates are chordates.

9. List four adaptations that guided vertebrate evolution, and explain how each has contributed to the success of vertebrates.

10. In 1928 Walter Garstang hypothesized that tunicates resemble the ancestral stock of the vertebrates. Explain this hypothesis.

11. Distinguish between ostracoderms and placoderms. What important evolutionary advances did each contribute to vertebrate evolution? What are conodonts?

12. Explain how we think the vertebrate jaw evolved.

Selected References

See also general references on page 395.

Bone, Q. 1979. The origin of chordates. Oxford Biology Readers, No. 18. New York, Oxford University Press. *Synthesis of hypotheses and disagreements bearing on an unsolved riddle.*

Carroll, R. L. 1988. Vertebrate paleontology and evolution. New York, W.H. Freeman & Company. *Authoritative treatment of the vertebrate fossil record. The first two chapters contain discussions of cladistic classification of the vertebrates, the vertebrate body plan, and the origin of vertebrate characters.*

Gans, C. 1989. Stages in the origin of vertebrates: analysis by means of scenarios. Biol. Rev. **64:**221-268. *Reviews the diagnostic characters of protochordates and ancestral vertebrates and presents a scenario for the protochordate-vertebrate transition.*

Gould, S. J., ed. 1993. The book of life. New York, W.W. Norton & Company. *A sweeping, handsomely illustrated view of (almost entirely) vertebrate life.*

Jeffries, R. P. S. 1986. The ancestry of the vertebrates. Cambridge, Cambridge University Press. *Jeffries argues that the Calcichordata are the direct ancestors of the vertebrates, a view that most zoologists are not willing to accept. Still, this book is an* excellent summary of the deuterostome groups and of the various competing hypotheses of vertebrate ancestry.

Long, J. A. 1995. The rise of fishes: 500 million years of evolution. Baltimore, The Johns Hopkins University Press. *An authoritative, liberally illustrated evolutionary history of fishes.*

Norman, D. 1994. Prehistoric life: the rise of the vertebrates. New York, Macmillan USA. *Although this beautifully illustrated volume focuses on the evolution of land vertebrates, the early chapters deal with the origins of chordates and vertebrates.*

Links to the Internet

Visit this textbook's Web site at *http://www.mhhe.com/zoology* to find live Internet links for each of the references listed below.

1. Chordata. Arizona's Tree of Life Web Page. An introduction, pictures, characteristics, phylogenetic relationships, and references on chordates.

2. BIOSIS: Protochordata. This site provides links to information on the Hemichordata, Urochordata, and Cephalochordata.

3. Introduction to the Urochordata. This University of California at Berkeley Museum of Paleontology site provides photographs and information on the biology and classification of the urochordates.

4. Ascidian News. This site is an online newsletter focusing on the biology of the urochordates. It provides links to other ascidian sites.

5. Introduction to the Cephalochordata University of California at Berkeley Museum of Paleontology. Images, photos, systematics, more information, and links.

Fishes

CHAPTER fifteen

What Is a Fish?

In common (and especially older) usage, the term fish has often been used to describe a mixed assortment of water-dwelling animals. We speak of jellyfish, cuttlefish, starfish, crayfish, and shellfish, knowing full well that when we use the word "fish" in such combinations, we are not referring to a true fish. In earlier times, even biologists did not make such a distinction. Sixteenth-century natural historians classified seals, whales, amphibians, crocodiles, even hippopotamuses, as well as a host of aquatic invertebrates, as fish. Later biologists were more discriminating, eliminating first the invertebrates and then the amphibians, reptiles, and mammals from the narrowing concept of a fish. Today we recognize a fish as a gill-breathing, ectothermic, aquatic vertebrate that possesses fins, and skin that is usually covered with scales. Even this modern concept of the term "fish" is controversial, at least as a taxonomic unit, because fishes do not compose a monophyletic group. The common ancestor of the fishes is also an ancestor to the land vertebrates, which we exclude from the term "fish," unless we use the term in an exceedingly nontraditional way. Because fishes live in a habitat that is basically alien to humans, people have rarely appreciated the remarkable diversity of these vertebrates. Nevertheless, whether appreciated by humans or not, the world's fishes have enjoyed an effusive proliferation that has produced an estimated 24,600 living species—more than all other species of vertebrates combined—with adaptations that have fitted them to almost every conceivable aquatic environment. No other animal group threatens their domination of the seas.

The life of a fish is bound to its body form. Mastery of river, lake, and ocean is revealed in the many ways that fishes have harmonized their design to the physical properties of their aquatic surroundings. Suspended in a medium that is 800 times more dense than air, a trout or pike can remain motionless, varying its neutral buoyancy by adding or removing air from the swim bladder. Or it may dart forward or at angles, using its fins as brakes and tilting rudders. With excellent organs for salt and water exchange, bony fishes can steady and finely tune their body fluid composition in their chosen freshwater or seawater environment. Their gills are the most effective respiratory devices in the animal kingdom for extracting oxygen from a medium that contains less than $\frac{1}{20}$ as much oxygen as air. Fishes have excellent olfactory and visual senses and a unique lateral line system, which with its exquisite sensitivity to water currents and vibrations provides a "distance touch" in water. Thus in mastering the physical problems of their element, early fishes evolved a basic body plan and set of physiological strategies that both shaped and constrained the evolution of their descendants.

> The use of fishes *as the plural form of* fish *may sound odd to most people accustomed to using* fish *in both the singular and the plural. Both plural forms are correct but zoologists use* fishes *to mean more than one kind of fish.*

Ancestry and Relationships of Major Groups of Fishes

The fishes are of ancient ancestry, having descended from an unknown free-swimming protochordate ancestor (hypotheses of chordate and vertebrate origins are discussed in Chapter 14). Whatever their origin, during the Cambrian period, or perhaps even in the Precambrian, the earliest fishlike vertebrates branched into the jawless **agnathans** and the jawed **gnathostomes** (Figure 15-1). All vertebrates have descended from one of these two ancestral groups.

The jawless agnathans, the least derived of the two groups, include the extinct ostracoderms and the living **hagfishes** and **lampreys,** fishes adapted as scavengers or parasites. The agnathans have no vertebrae but are nevertheless included within the subphylum Vertebrata because they have a cranium and many other vertebrate homologies. The ancestry of hagfishes and lampreys is uncertain; they bear little resemblance to the extinct ostracoderms. Although the hagfishes and the more derived lampreys superficially look much alike, they are in fact so different from each other that they have been assigned to separate classes by ichthyologists.

All remaining fishes have paired appendages and jaws and are included, along with the tetrapods (land vertebrates), in the monophyletic lineage of gnathostomes. They appear in the fossil record in the Silurian period with fully formed jaws, and no forms intermediate between agnathans and gnathostomes are known. By the Devonian period, the Age of Fishes, several distinct groups of jawed fishes were well represented. One of these, the placoderms (Figure 14-14, p. 282), became extinct, leaving no descendants, in the Carboniferous period, which followed the Devonian. A second group, the **cartilaginous fishes** of the class Chondrichthyes (sharks, rays, and chimaeras), lost the heavy dermal armor of the early jawed fishes and adopted cartilage rather than bone for the skeleton. Most are active predators with a sharklike body form that has undergone only minor changes over the ages.

Of all the gnathostomes, the **bony fishes** of the class Osteichthyes radiated most extensively and are the dominant fishes today (Figure 15-1). We can recognize two distinct lineages of bony fishes. Of these two, by far the most diverse are the **ray-finned fishes** (subclass Actinopterygii), which radiated to form the modern bony fishes. The other lineage, the **fleshy-finned fishes** (subclass Sarcopterygii), although a relic group today, carry the distinction of including the closest living relatives of the tetrapods. The fleshy-finned fishes are represented today by the **lungfishes** and the **coelacanth**— meager remnants of important stocks that flourished in the Devonian period (Figure 15-1). A classification of the major fish taxa is on p. 307.

Superclass Agnatha: Jawless Fishes

The living jawless fishes are represented by approximately 84 species almost equally divided between two classes: Myxini (hagfishes) and Cephalaspidomorphi (lampreys) (Figures 15-2 and 15-3). Members of both groups lack jaws, internal ossification, scales, and paired fins, and both groups share porelike gill openings and an eel-like body form. In other respects, however, the two groups are morphologically very different. Lampreys bear many derived morphological characters that place them phylogenetically much closer to the jawed bony fishes than to the hagfishes. Because of these differences, hagfishes and lampreys have been assigned to separate vertebrate classes, leaving the grouping "Agnatha" as a paraphyletic assemblage of jawless fishes.

Hagfishes: Class Myxini

The hagfishes are an entirely marine group that feed on dead or dying fishes, annelids, molluscs, and crustaceans. They are neither parasitic like lampreys nor predaceous, but are scavengers. There are only 43 species of hagfishes, of which the best known in North America are the Atlantic hagfish *Myxine glutinosa* (Gr. *myxa,* slime) (Figure 15-3) and the Pacific hagfish *Eptatretus stouti* (NL, *ept*<Gr. *hepta,* seven, + *tretos,* perforated). Although almost completely blind, the hagfish is

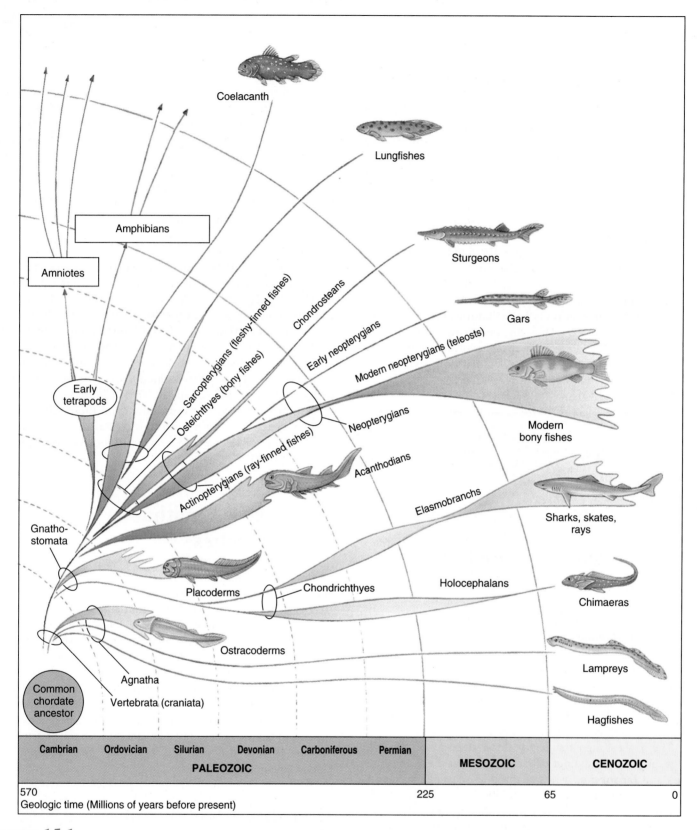

FIGURE 15-1

Graphic representation of the family tree of fishes, showing the evolution of major groups through geological time. Numerous lineages of extinct fishes are not shown. Widened areas in the lines of descent indicate periods of adaptive radiation and the relative number of species in each group. The fleshy-finned fishes (sarcopterygians), for example, flourished in the Devonian period, but declined and are today represented by only four surviving genera (lungfishes and coelacanth). Homologies shared by the sarcopterygians and tetrapods suggest that they are sister groups. The sharks and rays radiated during the Carboniferous period. They came dangerously close to extinction during the Permian period but staged a recovery in the Mesozoic era and are a secure group today. Johnny-come-latelies in fish evolution are the spectacularly diverse modern fishes, or teleosts, which include most living fishes. Note that the class Osteichthyes is a paraphyletic group because it does not include their descendants, the tetrapods; in cladistic usage the Osteichthyes includes the tetrapods (see Figure 15-2).

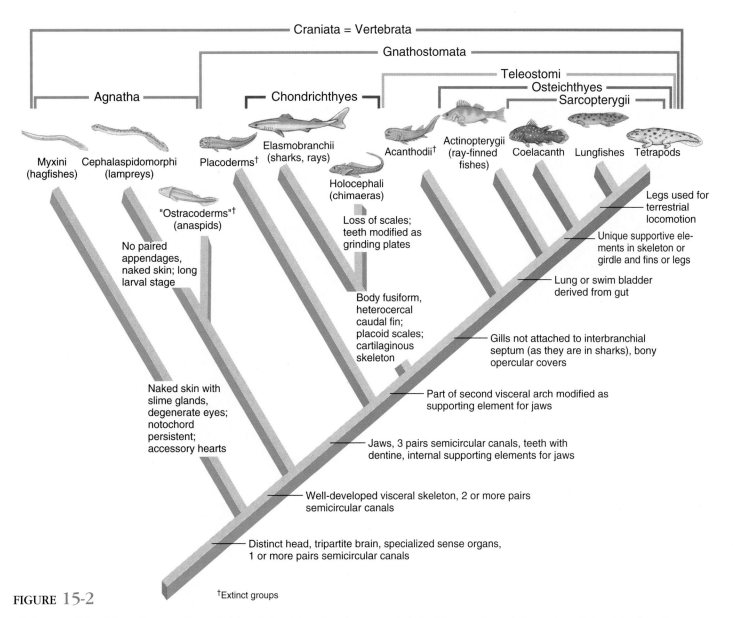

FIGURE 15-2

Cladogram of the fishes, showing the probable relationships of major monophyletic fish taxa. Several alternative relationships have been proposed. Extinct groups are designated by a dagger (†). Some of the shared derived characters that mark the branchings are shown to the right of the branch points. Agnatha is a paraphyletic structural grade not recognized in cladistic classification.

quickly attracted to food, especially dead or dying fish, by its keenly developed senses of smell and touch. After attaching itself to its prey by means of toothed plates, the hagfish thrusts the tongue forward to rasp off pieces of tissue. For extra leverage, the hagfish often ties a knot in its tail, then passes it forward along the body until it is pressed securely against the side of its prey (Figure 15-4).

Unlike any other vertebrate, the body fluids of hagfishes are in osmotic equilibrium with seawater, as in most marine invertebrates. Hagfishes have several other anatomical and physiological peculiarities, including a low-pressure circulatory system served by three accessory hearts in addition to the main heart positioned behind the gills. Hagfishes are also renowned for their ability to generate enormous quantities of slime.

While the unique anatomical and physiological features of the strange hagfishes are of interest to biologists, hagfishes have not endeared themselves to either sports or commercial fishermen. In earlier days of commercial fishing mainly by gill nets and set lines, hagfish often bit into the bodies of captured fish and ate out the contents, leaving behind a useless sack of skin and bones. But as large and efficient otter trawls came into use, hagfishes ceased to be an important pest.

The reproductive biology of hagfishes remains largely a mystery. Both male and female gonads are found in each

Characteristics of the Jawless Fishes

1. Slender, **eel-like** body
2. Median fins but **no** paired appendages
3. **Fibrous** and **cartilaginous skeleton;** notochord persistent; no vertebrae
4. Biting mouth with two rows of eversible teeth in hagfishes; suckerlike oral disc with well-developed teeth in lampreys
5. Heart with one atrium and one ventricle; hagfishes with three accessory hearts; aortic arches in gill region
6. Five to 16 pairs of gills and a single pair of gill apertures in hagfishes; 7 pairs of gills in lampreys
7. **Pronephric kidney** anteriorly and **mesonephric kidney** posteriorly in hagfishes; mesonephric kidney only in lampreys
8. Dorsal nerve cord with differentiated brain; 8 to 10 pairs of cranial nerves
9. Digestive systems **without stomach;** intestine with spiral valve and cilia in lampreys; both lacking in intestine of hagfishes
10. Sense organs of taste, smell, hearing; eyes poorly developed in hagfishes but moderately well developed in lampreys; one pair of **semicircular canals** (hagfishes) or two pairs (lampreys)
11. External fertilization; both ovaries and testes present in an individual but gonads of only one sex functional and no larval stage in hagfishes; separate sexes and long larval stage with radical metamorphosis in lampreys

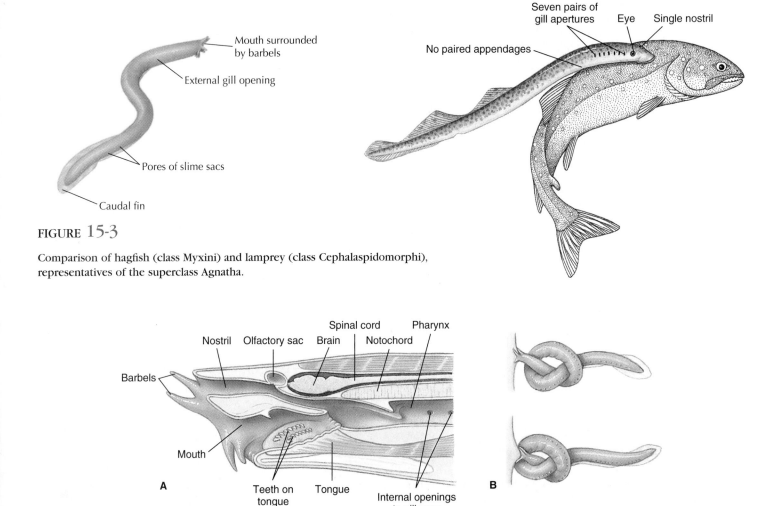

FIGURE 15-3

Comparison of hagfish (class Myxini) and lamprey (class Cephalaspidomorphi), representatives of the superclass Agnatha.

FIGURE 15-4

The Atlantic hagfish *Myxine glutinosa* (Class Myxini). **A,** Sagittal section of the head region showing the rasping tongue (retracted) and internal openings to gill sacs; **B,** Hagfish knotting, showing how it obtains leverage to tear flesh from prey.

animal, but only one gonad becomes functional. The females produce small numbers of surprisingly large, yolky eggs up to 3 cm in diameter. There is no larval stage and growth is direct.

Lampreys: Class Cephalaspidomorphi

Of the 41 described species of lampreys distributed around the world, by far the best known to North Americans is the destructive marine lamprey, *Petromyzon marinus,* of the Great Lakes (Figure 15-3). The name *Petromyzon* (Gr. *petros,* stone, + *myzon,* sucking) refers to the lamprey's habit of grasping a stone with its mouth to hold position in a current. There are 17 species of lampreys in North America of which about half are parasitic; the rest are species that never feed after metamorphosis and die soon after spawning.

In North America all lampreys, marine as well as freshwater forms, spawn in the winter or spring in shallow gravel and sand in freshwater streams. The males begin nest building and are joined later by females. Using their oral discs to lift stones and pebbles and using vigorous body vibrations to sweep away light debris, they form an oval depression. At spawning, with the female attached to a rock to maintain position over the nest, the male attaches to the dorsal side of her head. As the eggs are shed into the nest, they are fertilized by the male. The sticky eggs adhere to pebbles in the nest and soon become covered with sand. The adults die soon after spawning.

The eggs hatch in approximately 2 weeks, releasing small larvae (**ammocoetes**) (Figure 15-5), which stay in the nest until they are approximately 1 cm long; they then burrow into mud or sand and emerge at night to feed on small invertebrates, detritus, and other particulate matter in the water. The larval period lasts from 3 to 17 or more years before the larva rapidly metamorphoses into an adult.

Parasitic lampreys either migrate to the sea, if marine, or remain in fresh water, where they attach themselves by their suckerlike mouth to fish and with their sharp horny teeth rasp through flesh and suck the body fluids (Figure 15-6). To promote the flow of blood, the lamprey injects an anticoagulant into the wound. When gorged, the lamprey releases its hold but leaves the fish with a wound that may prove fatal. The parasitic freshwater adults live a year or more before spawning and then die; the marine forms may live longer.

Nonparasitic lampreys do not feed after emerging as adults, since their alimentary canal degenerates to a nonfunctional strand of tissue. Within a few months and after spawning, they die.

The invasion of the Great Lakes above Lake Ontario by the landlocked sea lamprey, *Petromyzon marinus,* in this cen-

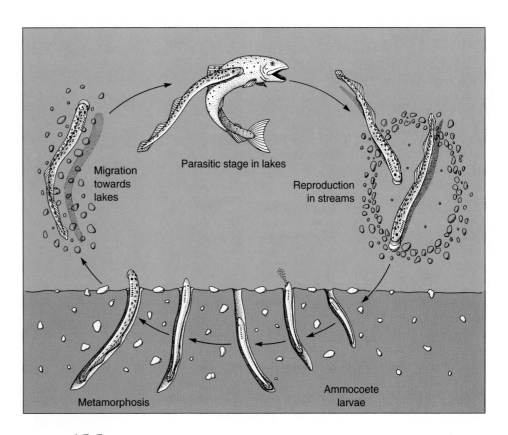

FIGURE 15-5

Life cycle of the "landlocked" form of the sea lamprey *Petromyzon marinus.*

tury had a devastating effect on the fisheries. Lampreys first entered the Great Lakes after the Welland Canal around Niagara Falls, a barrier to further western migration, was deepened between 1913 and 1918. Moving first through Lake Erie to Lakes Huron, Michigan, and Superior, sea lampreys, accompanied by overfishing, caused the total collapse of a multimillion dollar lake trout fishery in the early 1950s. Other less valuable fish species were attacked and destroyed in turn. After reaching a peak abundance in 1951 in Lakes Huron and Michigan and in 1961 in Lake Superior, the sea lampreys began to decline, due in part to depletion of their food and in part to the effectiveness of control measures (mainly chemical larvicides placed in selected spawning streams). Lake trout, aided by a restocking program, are now recovering, but wounding rates are still high in some lakes. Fishery organizations are now experimenting with the release into spawning streams of sterilized male lampreys; when fertile females mate with sterilized males, the female's eggs fail to develop.

Cartilaginous Fishes: Class Chondrichthyes

There are more than 850 living species in the class Chondrichthyes, an ancient, compact, and highly developed group. Although a much smaller and less diverse assemblage than the

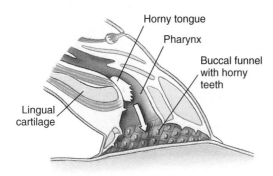

**Attachment to fish with
horny teeth and suction**

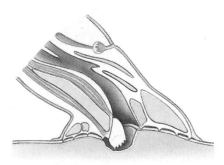

**Tongue protruded for
rasping flesh**

FIGURE 15-6

How the lamprey uses its horny tongue to feed. After firmly
attaching to a fish by its sucker, the protrusable tongue rapidly rasps
an opening through the fish's integument. Body fluid, abraded skin
and muscle are eaten.

Characteristics of the Sharks and Rays (Elasmobranchii)

1. **Body fusiform** (except rays) with a **heterocercal** caudal fin (Figure 15-14)
2. **Mouth ventral** (Figure 15-7); two olfactory sacs that do not connect to the mouth cavity; jaws present
3. Skin with **placoid scales** (Figure 15-16) and mucous glands; teeth of modified placoid scales
4. **Endoskeleton entirely cartilaginous**
5. Digestive system with a J-shaped stomach and **intestine with spiral valve** (Figure 15-8)
6. Circulatory system of several pairs of aortic arches; two-chambered heart
7. Respiration by means of five to seven pairs of gills with **separate and exposed gill slits,** no operculum
8. No swim bladder or lung
9. Mesonephric kidney and rectal gland (Figure 15-8); blood isosmotic or slightly hyperosmotic to seawater; **high concentrations of urea and trimethylamine oxide in blood**
10. Brain of two olfactory lobes, two cerebral hemispheres, two optic lobes, a cerebellum, and a medulla oblongata; 10 pairs of cranial nerves; **three pairs of semicircular canals;** senses of smell, vibration reception (lateral line system), and electroreception well developed
11. Separate sexes; oviparous, ovoviviparous, or viviparous; direct development; **internal fertilization**

bony fishes, their impressive combination of well-developed
sense organs, powerful jaws and swimming musculature, and
predaceous habits ensures them a secure and lasting niche in
the aquatic community. One of their distinctive features is
their cartilaginous skeleton. Although there is some limited
calcification, bone is entirely absent throughout the class—a
curious feature, since the Chondrichthyes are derived from
ancestors having well-developed bone.

Sharks and Rays:
Subclass Elasmobranchii

Sharks, which make up about 45% of the approximately 815
species in the subclass Elasmobranchii, are typically predaceous fishes with five to seven gill slits and gills on each side
and (usually) a spiracle behind each eye. Sharks track their
prey using their lateral line system and large olfactory organs,
since their vision is not well developed. The larger sharks, such
as the massive (but harmless) plankton-feeding whale shark,
may reach 15 m in length, the largest of all fishes. The dogfish
sharks so widely used in zoological laboratories rarely exceed
1 m. More than half of all elasmobranchs are rays, specialized
for a bottom-feeding life-style. Unlike sharks, which swim with
thrusts of the tail, rays propel themselves by wave-like motions
of the "wings," or pectoral fins.

Although to most people sharks have a sinister appearance and a fearsome reputation, they are at the same time
among the most gracefully streamlined of all fishes
(Figure 15-7). Sharks are heavier than water and will sink if
not swimming forward. The asymmetrical **heterocercal
tail,** in which the vertebral column turns upward and
extends into the dorsal lobe of the tail (see Figure 15-14),
provides lift and thrust as it sweeps to and fro in the water,
and the broad head and flat pectoral fins act as planes to
provide head lift.

Sharks are well equipped for their predatory life. The
tough leathery skin is covered with numerous dermal

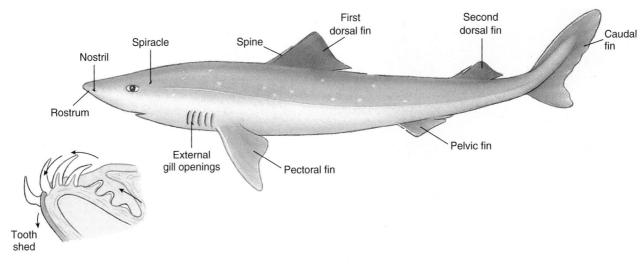

FIGURE 15-7

Dogfish shark, *Squalus acanthias*. Inset: Section of lower jaw shows the formation of new teeth developing inside the jaw. These move forward to replace lost teeth. Rate of replacement varies in different species.

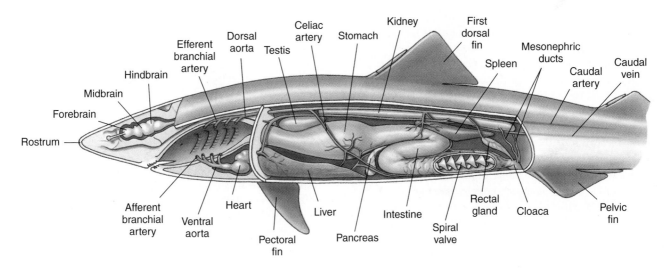

FIGURE 15-8

Internal anatomy of dogfish shark *Squalus acanthias*.

placoid scales (see Figure 15-16) that are modified anteriorly to form replaceable rows of teeth in both jaws (Figure 15-9). Placoid scales in fact consist of dentine enclosed by an enamel-like substance, and they very much resemble the teeth of other vertebrates. Sharks have a keen sense of smell used to guide them to food. Vision is less acute than in most bony fishes, but a well-developed **lateral line system** is used for detecting and locating objects and moving animals (predators, prey, and social partners). It is composed of a canal system extending along the side of the body and over the head (Figure 15-11). Inside are special receptor organs (**neuromasts**) that are extremely sensitive to vibrations and currents in the water. Sharks can also detect and

aim attacks at prey buried in the sand by sensing the bioelectric fields that surround all animals. The receptors, the **ampullary organs of Lorenzini,** are located on the shark's head.

Rays belong to a separate order from the sharks. Rays are distinguished by their dorsoventrally flattened bodies and the much-enlarged pectoral fins that behave as wings in swimming (Figure 15-11). The gill openings are on the underside of the head, and the **spiracles** (on top of the head) are unusually large. Respiratory water enters through these spiracles to prevent clogging the gills, because the mouth is often buried in sand. The teeth are adapted for crushing the prey—mainly molluscs, crustaceans, and an occasional small fish.

FIGURE 15-9

Head of sand tiger shark *Carcharias* sp. Note the series of successional teeth. Also visible in a row below the eye are the ampullae of Lorenzini *(arrow).*

The worldwide shark fishery is experiencing unprecedented pressure, driven by the high price of shark fins for shark-fin soup, an oriental delicacy (which commonly sells for $50.00 per bowl). Coastal shark populations in general have declined so rapidly that "finning" is to be outlawed in the United States; other countries, too, are setting quotas to protect threatened shark populations. Even in the Marine Resources Reserve of the Galápagos Islands, one of the world's exceptional wild places, tens of thousands of sharks have been killed illegally for the Asian shark-fin market. That illegal fishery continues at this writing. Contributing to the threatened collapse of shark fisheries worldwide is the long time required by most sharks to reach sexual maturity; some species take as long as 15 years.

In the order containing the rays (Rajiformes), we commonly refer to members of one family (Rajidae) as skates. Alone among members of the Rajiformes, skates do not bear living young but lay large, yolky eggs enclosed within a horny covering (the "mermaid's purse") that often washes up on beaches. Although the tail is slender, skates have a somewhat more muscular tail than most rays, and they usually have two dorsal fins and sometimes a caudal fin.

In the stingrays, the caudal and dorsal fins have disappeared, and the tail is slender and whiplike. The stingray tail is armed with one or more saw-toothed spines that can inflict dangerous wounds. Electric rays have certain

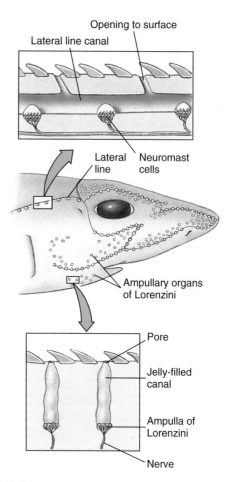

FIGURE 15-10

Sensory canals and receptors in a shark. The ampullae of Lorenzini respond to weak electric fields, and possibly to temperature, water pressure, and salinity. The lateral line sensors, called neuromasts, are sensitive to disturbances in the water, enabling the shark to detect nearby objects by reflected waves in the water.

dorsal muscles modified into powerful electric organs, which can give severe shocks to stun their prey.

Chimaeras: Subclass Holocephali

The approximately 30 species of chimaeras (ky-meer'uz; L. monster), distinguished by such suggestive names as ratfish (Figure 15-12), rabbitfish, spookfish, and ghostfish, are remnants of an aberrant line that diverged from the placoderms at least 350 million years ago (Devonian period of the Paleozoic era). Fossil chimaeras first appeared in the Jurassic period, reached their zenith in the Cretaceous and early Tertiary periods (120 million to 50 million years ago), and have declined ever since. Anatomically they present an odd mixture of sharklike and bony fishlike features. Their food is a mixed diet of seaweed, molluscs, echinoderms, crustaceans, and fishes. Chimaeras are not commercial species

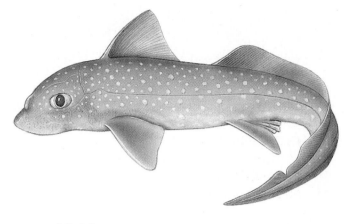

FIGURE 15-12

Chimaera, or ratfish, of North American west coast. This species is one of the most handsome of chimaeras, which tend toward bizarre appearances.

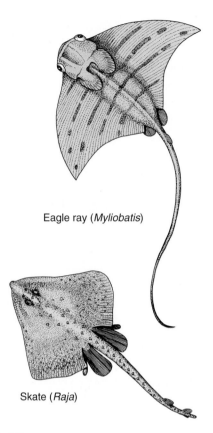

Eagle ray (*Myliobatis*)

Skate (*Raja*)

FIGURE 15-11

Rays are specialized for life on the seafloor. They are flattened dorsoventrally and move by undulations of greatly expanded winglike pectoral fins. One group of rays, the skates, are distinguishd by laying their eggs in a horny capsule, the "mermaid's purse."

and are seldom caught. Despite their grotesque shape, they are beautifully colored with a pearly iridescence.

Bony Fishes: Class Osteichthyes

Origin, Evolution, and Diversity

The bony fishes, the largest and most diverse taxon of all vertebrates, originated in the late Silurian period, approximately 410 million years ago. Details of head structure of earliest complete bony fishes fossils indicate that they probably descended from an ancestor shared with the acanthodians (p. 282). By the middle of the Devonian the bony fishes had developed several key adaptations that contributed to an extensive adaptive radiation. An **operculum** over the gill slits, composed of bony plates attached to the first gill arch, served to increase the efficiency of drawing water across the gill surfaces. These earliest bony fishes also had a pair of lungs, which served as accessory breathing structures. The

fin pattern established at that time persists in bony fishes today: **pectoral and pelvic fins** supported by bony girdles embedded in the body musculature, and median dorsal and anal fins (Figure 15-13). Progressive specialization of jaw structure and feeding mechanisms is another key feature in bony fish evolution. Bony fishes have high levels of activity, supported by efficient gill design for gas exchange, rapid metabolic oxidation of food, and an effective form of undulatory locomotion that persisted in many tetrapods (for example, salamanders, snakes, and many lizards).

By the middle Devonian, the Osteichthyes had divided into two distinct lineages. One lineage, the **ray-finned fishes** (Actinopterygii), includes the modern bony fishes, the largest of all vertebrate radiations. The other lineage is the **fleshy-finned fishes** (Sarcopterygii), a remnant group represented today by the **lungfishes** and the **coelacanth.** Their evolutionary history is of great interest because their descendents include all the land vertebrates (tetrapods) (Figure 15-2).

Ray-Finned Fishes: Subclass Actinopterygii

Ray-finned fishes are an enormous assemblage containing all of our familiar bony fishes—more than 24,600 species. The group had its beginnings in the Devonian freshwater lakes and streams. The ancestral forms were small, bony fishes, heavily armored with ganoid scales (Figure 15-16), and had functional lungs as well as gills.

From these earliest ray-finned fishes, two major groups emerged. Those bearing the most primitive characteristics are the **chondrosteans** (Gr. *chondros,* cartilage, + *osteon,* bone), represented today by the sturgeons, paddlefishes, and bichir *Polypterus* (Gr. *poly,* many, + *pteros,* winged) of African rivers (Figure 15-17). *Polypterus* is an interesting relic with a

Characteristics of Bony Fishes (Osteichthyes)

1. **Skeleton more or less bony,** vertebrae numerous; **tail usually homocercal** (Figure 15-14)
2. Skin with mucous glands and embedded **dermal scales** (Figure 15-15) of three types: **ganoid, cycloid, or ctenoid;** some without scales, no placoid scales (Figure 15-16)
3. Fins both median and paired with **fin rays of cartilage or bone**
4. **Mouth terminal** with many teeth (some toothless); jaws present; olfactory sacs paired and may or may not open into mouth
5. Respiration by gills supported by bony gill arches and covered by a **common operculum**
6. **Swim bladder** often present with or without duct connected to pharynx
7. Circulation consisting of a two-chambered heart, arterial and venous systems, and four pairs of aortic arches
8. Nervous system of brain with small olfactory lobes and cerebrum; large optic lobes and cerebellum; 10 pairs of cranial nerves
9. Sexes separate (some hermaphroditic), gonads paired; fertilization usually external; larval forms may differ greatly from adults

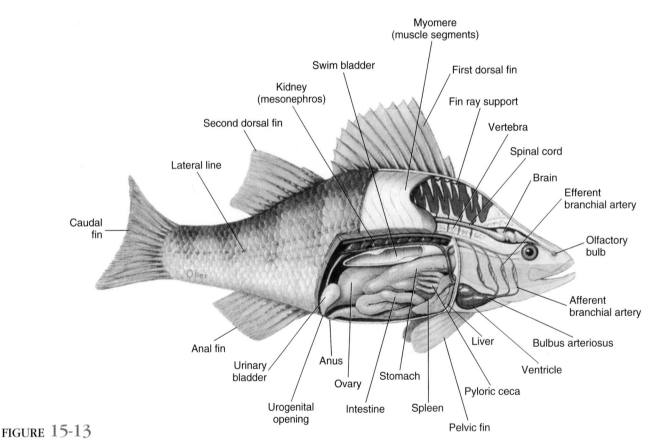

FIGURE 15-13

Internal anatomy of the yellow perch *Perca flavescens,* a freshwater teleost fish.

lunglike swim bladder and many other primitive characteristics; it resembles an ancestral ray-finned fish more than it does any other living descendant. There is no satisfactory explanation for the survival to the present of this fish and the coelacanth *Latimeria* when all of their kin perished millions of years ago.

The second major group to emerge from the early ray-finned stock were **neopterygians** (Gr. *neos,* new, + *pteryx,* fin). The neopterygians appeared in the late Permian and radiated extensively during the Mesozoic era. During the Mesozoic one lineage gave rise to a secondary radiation that led to the modern bony fishes, the teleosts. The two surviving genera of

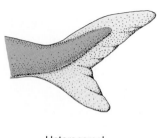

Heterocercal
(shark)

Diphycercal
(lungfish)

Homocercal
(perch)

FIGURE 15-14

Types of caudal fins among fishes.

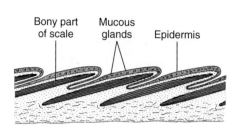

Bony part
of scale

Mucous
glands

Epidermis

FIGURE 15-15

Section through the skin of a bony fish, showing the overlapping scales *(red).* The scales lie in the dermis and are covered by epidermis.

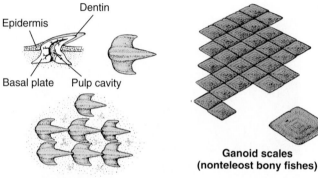

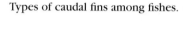

Epidermis

Dentin

Basal plate

Pulp cavity

Placoid scales
(cartilaginous fishes)

Ganoid scales
(nonteleost bony fishes)

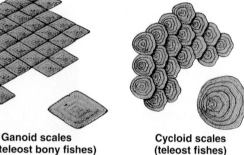

Cycloid scales
(teleost fishes)

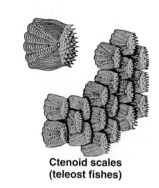

Ctenoid scales
(teleost fishes)

FIGURE 15-16

Types of fish scales. Placoid scales are small, conical toothlike structures characteristic of Chondrichthyes. Diamond-shaped ganoid scales, present in primitive bony fishes such as the gar, are composed of layers of silvery enamel (ganoin) on the upper surface and bone on the lower. Advanced bony fishes have either cycloid or ctenoid scales. These are thin and flexible and are arranged in overlapping rows.

nonteleost neopterygians are the bowfin *Amia* (Gr. tunalike fish) of shallow, weedy waters of the Great Lakes and Mississippi Valley, and the gars *Lepisosteus* (Gr. *lepidos,* scale, + *osteon,* bone) of eastern and southern North America (Figure 15-18). Gars are large predators that belie their lethargic appearance by suddenly dashing forward to grasp their prey with needle-sharp teeth.

The major lineage of neopterygians are the teleosts (Gr. *teleos,* perfect, + *osteon,* bone), the modern bony fishes (Figure 15-13). Diversity appeared early in teleost evolution, foreshadowing the truly incredible variety of body forms among teleosts today. The heavy armorlike scales of the early ray-finned fishes have been replaced in teleosts by light, thin, and flexible **cycloid** and **ctenoid** scales. These look much alike (Figure 15-16) except that ctenoid scales have comblike ridges on the exposed edge that may be an adaptation for reducing frictional drag. Some teleosts, such as some catfishes and sculpins, lack scales altogether. Nearly all teleosts have a **homocercal tail,** with the upper and lower lobes of about equal size (Figure 15-14). The lungs of early forms were trans-

formed in the teleosts to a swim bladder with a buoyancy function. Teleosts have highly maneuverable fins for control of body movement. In small teleosts the fins are often provided with stout, sharp spines, thus making themselves prickly mouthfuls for would-be predators. With these adaptations (and many others), teleosts have become the most diverse of fishes.

The Fleshy-Finned Fishes: Subclass Sarcopterygii

The fleshy-finned fishes are today represented by only seven species: six species (three genera) of lungfishes and a single lobe-finned fish, the coelacanth (seal′a-canth)—survivors of a group once abundant during the Devonian period of the Paleozoic. All of the early sarcopterygians had lungs as well as gills and strong, fleshy, paired lobed fins (pectoral and pelvic) that may have been used like four legs to scuttle along the bottom. They had powerful jaws, a skin covered with heavy, enameled scales, and a **diphycercal** tail (Figure 15-14).

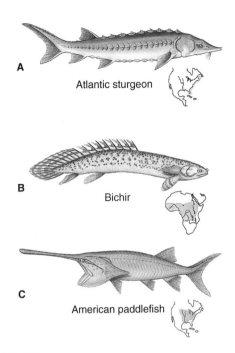

FIGURE 15-17

Chondrostean ray-finned fishes of the subclass Actinopterygii. **A,** Atlantic sturgeon, *Acipenser oxyrhynchus* (now uncommon) of Atlantic coast rivers. **B,** Bichir *Polypterus bichir* of the African Congo. It is a nocturnal predator. **C,** Paddlefish *Polyodon spathula* of the Mississippi River reaches a length of 2 m and a weight of 90 kg.

Of the surviving lungfishes, the least specialized is *Neoceratodus* (Gr. *neos,* new, + *keratos,* horn, + *odes,* form), the living Australian lungfish, which may attain a length of 1.5 m (Figure 15-19). This lungfish is able to survive in stagnant, oxygen-poor water by coming to the surface and gulping air into its single lung, but it cannot live out of water. The South American lungfish, *Lepidosiren* (L. *lepidus,* pretty, + *siren,* Siren, mythical mermaid), and the African lungfish, *Protopterus* (Gr. *protos,* first, + *pteron,* wing), can live out of water for long periods of time. *Protopterus* lives in African streams and rivers that run completely dry during the dry season, with their mud beds baked hard by the hot tropical sun. The fish burrows down at the approach of the dry season and secretes a copious slime that mixes with mud to form a hard cocoon in which it remains dormant until the rains return.

The lobe-finned fishes consist of two groups: the **rhipidistians** which flourished in the late Paleozoic era then became extinct; and the **coelacanths,** a group that also radiated in the Paleozoic and later disappeared except for one remarkable species, the famous coelacanth *Latimeria chalumnae* (Figure 15-20). Since the last coelacanths were believed to have become extinct 70 million years ago, the astonishment of the scientific world can be imagined when the remains of a coelacanth were found on a trawler off the coast of South Africa in 1938. An intensive search was begun in the Comoro Islands area near Madagascar, where, it was learned, native Comoran fishermen occasionally caught them with hand lines

A

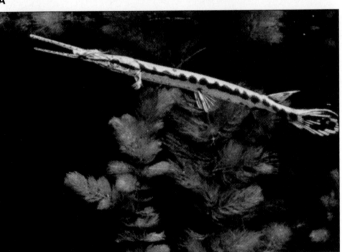

B

FIGURE 15-18

Nonteleost neopterygian fishes. **A,** Bowfin *Amia calva.* **B,** Longnose gar *Lepisosteus osseus.* The bowfin lives in the Great Lakes region and Mississippi basin. Gars are common fishes of eastern and southern North America. They frequent slow-moving streams where they may hang motionless in the water, ready to snatch passing fish.

at great depths. Numerous specimens have now been caught, many in excellent condition, although none has been kept alive beyond a few hours after capture. The "modern" marine coelacanth is a descendant of the Devonian freshwater stock, which reached its evolutionary peak in the Mesozoic era and then disappeared—or so it was believed until 1938.

The fleshy-finned fishes occupy an important position in vertebrate evolution because they include the closest living relatives of the tetrapods. Of the living fleshy-finned fishes, the lungfishes are the sister group of the tetrapods. Because the fleshy-finned fishes as traditionally recognized form a paraphyletic group, cladists include tetrapods as well as fleshy-finned fishes in the Sarcopterygii (Figure 15-2).

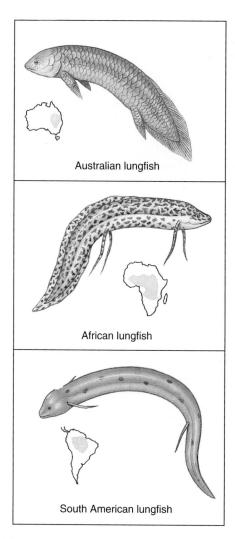

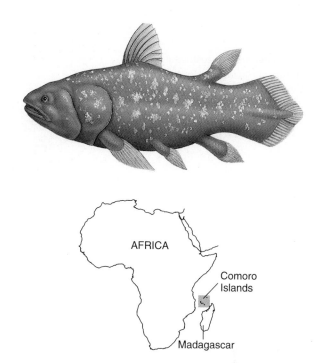

FIGURE 15-20

The coelacanth *Latimeria chalumnae* is a surviving marine relic of a group of lobe-finned fishes that flourished some 350 million years ago.

FIGURE 15-19

Lungfishes are fleshy-finned fishes of the subclass Sarcopterygii. The Australian lungfish *Neoceratodus forsteri* is the least specialized of three lungfish genera. The African lungfish *Protopterus* is best adapted of the three for remaining dormant in mucous-lined cocoons breathing air during prolonged periods of drought.

Structural and Functional Adaptations of Fishes

Locomotion in Water

To the human eye, some fishes appear capable of swimming at extremely high speeds. But our judgment is unconsciously tempered by our own experience that water is a highly resistant medium through which to move. Most fishes, such as a trout or a minnow, can swim maximally about 10 body lengths per second, obviously an impressive performance by human standards. Yet when these speeds are translated into kilometers per hour it means that a 30 cm (1 foot) trout can

swim only about 10.4 km (6.5 miles) per hour. As a general rule, the larger the fish the faster it can swim.

Measuring fish cruising speeds accurately is best done in a "fish wheel," a large ring-shaped channel filled with water that is turned at a speed equal and opposite to that of the fish. Much more difficult to measure are the sudden bursts of speed that most fish can make to capture prey or to avoid being captured. A hooked bluefin tuna was once "clocked" at 66 km per hour (41 mph); swordfish and marlin may be capable of incredible bursts of speed approaching, or even exceeding, 110 km per hour (68 mph). They can sustain such high speeds for no more than 1 to 5 seconds.

The propulsive mechanism of a fish is its trunk and tail musculature. The axial, locomotory musculature is composed of zigzag bands, called **myomeres.** The muscle fibers in each myomere are relatively short and connect the tough connective tissue partitions that separate each myomere from the next. On the surface the myomeres take the shape of a W lying on its side (Figure 15-21) but internally the bands are complexly folded and nested so that the pull of each myomere extends over several vertebrae. This arrangement produces more

FIGURE **15-21**

Trunk musculature of a teleost fish, partly dissected to show internal arrangement of the muscle bands (myomeres). The myomeres are folded into a complex, nested grouping, an arrangement that favors stronger and more controlled swimming.

power and finer control of movement since many myomeres are involved in bending a given segment of the body.

Understanding how fishes swim can be approached by studying the motion of a very flexible fish such as an eel (Figure 15-22). The movement is serpentine, not unlike that of a snake, with waves of contraction moving backward along the body by alternate contraction of the myomeres on either side. The anterior end of the body bends less than the posterior end, so that each undulation increases in amplitude as it travels along the body. While undulations move backward, the bending of the body pushes laterally against the water, producing a **reactive force** that is directed forward, but at an angle. It can be analyzed as having two components: **thrust,** which is used to overcome drag and propels the fish forward, and **lateral force,** which tends to make the fish's head "yaw," or deviate from the course in the same direction as the tail. This side-to-side head movement is very obvious in a swimming eel or shark, but many fishes have a large, rigid head with enough surface resistance to minimize yaw.

The movement of an eel is reasonably efficient at low speed, but its body shape generates too much frictional drag for rapid swimming. Fishes that swim rapidly, such as trout, are less flexible and limit the body undulations mostly to the caudal region (Figure 15-22). Muscle force generated in the large anterior muscle mass is transferred through tendons to the relatively nonmuscular caudal peduncle and tail where thrust is generated. This form of swimming reaches its highest development in the tunas, whose bodies do not flex at all. Virtually all the thrust is derived from powerful beats of the tail fin (Figure 15-23). Many fast oceanic fishes such as marlin, swordfish, amberjacks, and wahoo have swept-back tail fins shaped much like a sickle. Such fins are the aquatic counterpart of the high-aspect ratio wings of the swiftest birds (p. 357).

Swimming is the most economical form of animal locomotion, largely because aquatic animals are almost perfectly supported by their medium and need expend little energy to overcome the force of gravity. If we compare the energy cost per kilogram of body weight of traveling 1 km by different forms of locomotion, we find swimming costs only 0.39 kcal (salmon) as compared with 1.45 kcal for flying (gull) and 5.43 for walking (ground squirrel). However, part of the unfinished business of biology is understanding how fish and aquatic mammals are able

to move through the water while creating almost no turbulence. The secret lies in the way aquatic animals bend their bodies and fins (or flukes) to swim and in the friction-reducing properties of the body surface.

Neutral Buoyancy and the Swim Bladder

All fishes are slightly heavier than water because their skeletons and other tissues contain heavy elements that are present only in trace amounts in natural waters. To keep from sinking, sharks must always keep moving forward in the water. The asymmetrical (heterocercal) tail of a shark provides the necessary tail lift as it sweeps to and fro in the water, and the broad head and flat pectoral fins (Figure 15-8) act as angled planes to provide head lift. Sharks are also aided in buoyancy by having very large livers containing a special fatty hydrocarbon called **squalene** that has a density of only 0.86. The liver thus acts like a large sack of buoyant oil that helps to compensate for the shark's heavy body.

By far the most efficient flotation device is a gas-filled space. Swim bladders are present in most pelagic bony fishes but are absent in tunas, most abyssal fishes, and most bottom dwellers, such as flounders and sculpins. By adjusting the volume of gas in the swim bladder, a fish can achieve neutral buoyancy and remain suspended indefinitely at any depth with no muscular effort. There are severe technical problems, however. If the fish descends to a greater depth, the swim bladder gas is compressed so that the fish becomes heavier and tends to sink. Gas must be added to the bladder to establish a new equilibrium buoyancy. If the fish swims upward, the gas in the bladder expands, making the fish lighter. Unless gas is removed, the fish will rise with ever-increasing speed while the bladder continues to expand.

Fishes adjust gas volume in the swim bladder in two ways. The less specialized fishes (trout, for example) have a **pneumatic duct** that connects the swim bladder to the esophagus; these forms must come to the surface and gulp air to charge the bladder and obviously are restricted to relatively shallow depths. More specialized teleosts have lost the pneumatic duct. In these fishes, the gas must originate in the blood and be secreted into the swim bladder. Gas exchange depends on two highly specialized areas: a

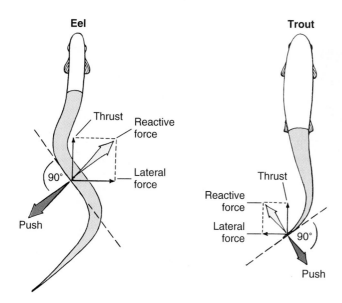

Eel

Thrust
Reactive force
90°
Lateral force
Push

Trout

Thrust
Reactive force
Lateral force
90°
Push

FIGURE **15-22**

Movements of swimming fishes, showing the forces developed by an eel-shaped and spindle-shaped fish.

Source: Vertebrate of Life, *4/e by Pough, et al., 1999. Reprinted by permission of Prentice-Hall, Inc., Upper Saddle River, NJ.*

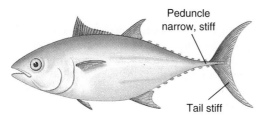

Peduncle narrow, stiff

Tail stiff

FIGURE **15-23**

Bluefin tuna, showing adaptations for fast swimming. Powerful trunk muscles pull on the slender tail stalk. Since the body does not bend, all of the thrust comes from beats of the stiff, sickle-shaped tail.

gas gland that secretes gas into the bladder and a **resorptive area,** or "ovale," that can remove gas from the bladder. The gas gland is supplied by a remarkable network of blood capillaries, called the **rete mirabile** ("marvelous net") that functions as a countercurrent exchange system to trap gases, especially oxygen, and prevent their loss to the circulation (Figure 15-24).

The amazing effectiveness of this device is exemplified by a fish living at a depth of 2400 m (8000 feet). To keep the bladder inflated at that depth, the gas inside (mostly oxygen, but also variable amounts of nitrogen, carbon dioxide, argon, and even some carbon monoxide) must have a pressure exceeding 240 atmospheres, which is much greater than the pressure in a fully charged steel gas cylinder. Yet the oxygen pressure in the fish's blood cannot exceed 0.2 atmosphere—equal to the oxygen pressure at the sea surface.

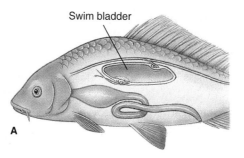

Swim bladder

A

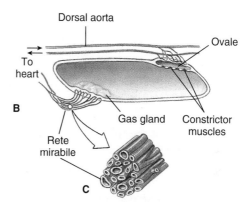

Dorsal aorta

Ovale

To heart

B

Rete mirabile

Gas gland

Constrictor muscles

C

FIGURE **15-24**

Swim bladder of a teleost fish. The swim bladder **(A)** lies in the coelom just beneath the vertebral column. Gas is secreted into the swim bladder by the gas gland **(B).** Gas from the blood is moved into the gas gland by the rete mirabile, a complex array of tightly-packed capillaries that act as a countercurrent multiplier to build up the oxygen concentration. The arrangement of venous and arterial capillaries in the rete is shown in **C.** To release gas during ascent, a muscular valve opens, allowing gas to enter the ovale from which the gas is removed by the circulation.

Respiration

Fish gills are composed of thin filaments, each covered with a thin epidermal membrane that is folded repeatedly into plate-like **lamellae** (Figure 15-25). These are richly supplied with blood vessels. The gills are located inside the pharyngeal cavity and are covered with a movable flap, the **operculum.** This arrangement provides excellent protection to the delicate gill filaments, streamlines the body, and makes possible a pumping system for moving water through the mouth, across the gills, and out the operculum. Instead of opercular flaps as in bony fishes, the elasmobranchs have a series of **gill slits** out of which the water flows. In both elasmobranchs and bony fishes the branchial mechanism is arranged to pump water continuously and smoothly over the gills, although to an observer it appears that fish breathing is pulsatile.

The flow of water is opposite to the direction of blood flow (countercurrent flow), the best arrangement for extracting the greatest possible amount of oxygen from the water.

Some bony fishes can remove as much as 85% of the oxygen from water passing over their gills. Very active fishes, such as herring and mackerel, can obtain sufficient water for their high oxygen demands only by swimming forward continuously to force water into the open mouth and across the gills. This process is called ram ventilation. Such fish will be asphyxiated if placed in an aquarium that restricts free swimming movements, even if the water is saturated with oxygen.

Migration

Eel

For centuries naturalists had been puzzled about the life history of the freshwater eel *Anguilla* (an-gwil′a) (L. eel), a common and commercially important species of coastal streams of the North Atlantic. Eels are **catadromous** (Gr. *kata,* down, + *dromos,* running), meaning that they spend most of their lives in fresh water but migrate to the sea to spawn. Each fall, people saw large numbers of eels swimming down the rivers toward the sea, but no adults ever returned. Each spring countless numbers of young eels, called "elvers" (Figure 15-26), each about the size of a wooden matchstick, appeared in the coastal rivers and began swimming upstream. Beyond the assumption that eels must spawn somewhere at sea, the location of their breeding grounds was completely unknown.

The first clue was provided by two Italian scientists, Grassi and Calandruccio, who in 1896 reported that elvers were not larval eels but rather were relatively advanced juveniles. The true larval eels, they discovered, were tiny, leaf-shaped, completely transparent creatures that bore absolutely no resemblance to an eel. They had been called **leptocephali** (Gr. *leptos,* slender, + *kephalē,* head) by early naturalists, who never suspected their true identity. In 1905 Johann Schmidt, supported by the Danish government, began a systematic study of eel biology that he continued until his death in 1933. With the cooperation of captains of commercial vessels plying the Atlantic, thousands of the leptocephali were caught in different areas of the Atlantic with the plankton nets Schmidt supplied. By noting where larvae in different stages of development were captured, Schmidt and his colleagues eventually reconstructed the spawning migrations.

When the adult eels leave the coastal rivers of Europe and North America, they swim steadily and apparently at

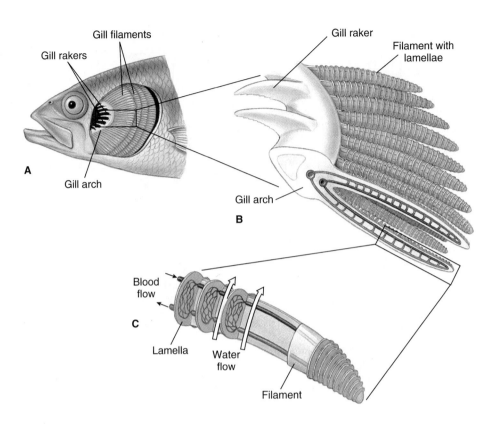

FIGURE 15-25

Gills of fish. Bony, protective flap covering the gills (operculum) has been removed, **A,** to reveal branchial chamber containing the gills. There are four gill arches on each side, each bearing numerous filaments. A portion of gill arch, **B,** shows gill rakers that project forward to strain food and debris, and gill filaments that project to the rear. A single gill filament, **C,** is dissected to show the blood capillaries within the platelike lamellae. Direction of water flow *(large arrows)* is opposite the direction of blood flow.

great depth for one to two months until they reach the Sargasso Sea, a vast area of warm oceanic water southeast of Bermuda (Figure 15-26). Here, at depths of 300 m or more, the eels spawn and die. The minute larvae then begin an incredible journey back to the coastal rivers of Europe. Drifting with the Gulf Stream, those not eaten by numerous predators reach the middle of the Atlantic after two years. By the end of the third year they reach the coastal waters of Europe where the leptocephali metamorphose into elvers, with an unmistakable eel-like body form (Figure 15-26). Here the males and females part company; the males remain in the brackish waters of coastal rivers and estuaries while the females continue up the rivers, often traveling hundreds of miles upstream. After 8 to 15 years of growth, the females, now 1 m or more long, return to the sea to join the smaller males; both return to the ancestral breeding grounds thousands of miles away to complete the life cycle. Since the Sargasso Sea is much closer to the American coastline than it is to Europe, American eel larvae require only about eight months to make the journey.

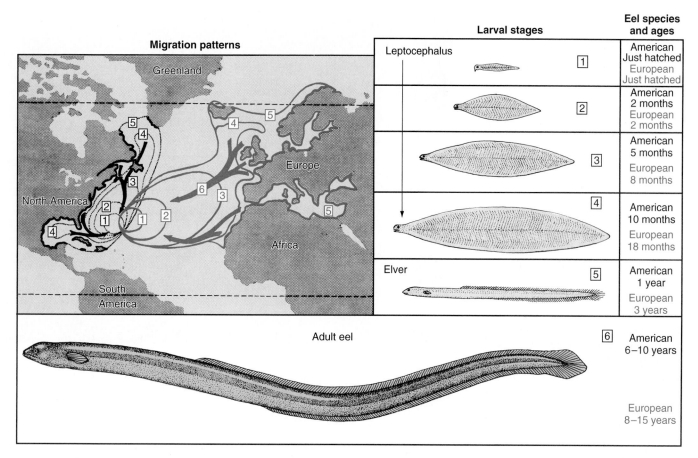

FIGURE 15-26

Life histories of the European eel, *Anguilla anguilla,* and American eel, *Anguilla rostrata. Red,* Migration patterns of European species. *Black,* Migration patterns of American species. Boxed numbers refer to stages of development. Note that the American eel completes its larval metamorphosis and sea journey in one year. It requires nearly three years for the European eel to complete its much longer journey.

Recent enzyme electrophoresis analysis of eel larvae confirmed not only the existence of separate European and American species but also Schmidt's belief that the European and American eels spawn in partially overlapping areas of the Sargasso Sea.

Homing Salmon

The life history of salmon is nearly as remarkable as that of the freshwater eel and certainly has received far more popular attention. Salmon are **anadromous;** that is, they spend their adult lives at sea but return to fresh water to spawn. The Atlantic salmon *(Salmo salar)* and the Pacific salmon (six species of the genus *Oncorhynchus* [on-ko-rink′us]) have this practice, but there are important differences among the seven species. The Atlantic salmon (as well as the closely related steelhead trout) make upstream spawning runs year after year. The six Pacific salmon species (king, sockeye, silver, hump-back, chum, and Japanese masu) each make a single spawning run (Figures 15-27 and 15-28), after which they die.

The virtually infallible homing instinct of the Pacific species is legendary. After migrating downstream as a smolt (a juvenile stage, Figure 15-28), a sockeye salmon ranges many hundreds of miles over the Pacific for nearly four years, grows to 2 to 5 kg in weight, and then returns almost unerringly to spawn in the headwaters of its parent stream. Some straying does occur and is an important means of increasing gene flow and populating new streams.

Experiments by A. D. Hasler and others have shown that homing salmon are guided upstream by the characteristic odor of their parent stream. When the salmon finally reach the spawning beds of their parents (where they themselves were hatched), they spawn and die. The following spring, the newly hatched fry transform into smolts before and during the downstream migration. At this time they are imprinted with the distinctive odor of the stream, which is apparently a mosaic of compounds released by the characteristic vegetation and soil in the watershed of the parent stream. They also seem to imprint on the odors of other streams they pass while migrating downriver and use these odors in reverse sequence as a map during the upriver migration as returning adults.

How do salmon find their way to the mouth of the coastal river from the trackless miles of the open ocean? Salmon move hundreds of miles away from the coast, much too far to be able to detect the odor of their parent stream. Experiments suggest that some migrating fish, like birds, can navigate by orienting to the position of the sun. However, migrant salmon can navigate on cloudy days and at night, indicating that solar navigation, if used at all, cannot be the salmon's only navigational cue. Fish also (again, like birds, see p. 360) appear able to detect the earth's magnetic field and to navigate by orientating to it. Finally, fishery biologists concede that salmon may not require precise navigational abilities at all, but instead may use ocean currents, temperature gradients, and food availability to reach the general coastal area where "their" river is located. From this point, they would navigate by their imprinted odor map, making correct turns at each stream junction until they reach their natal stream.

FIGURE 15-27

Migrating Pacific sockeye salmon.

Reproduction and Growth

In a group as diverse as the fishes, it is no surprise to find extraordinary variations on the basic theme of sexual reproduction. Most fishes favor a simple theme: they are **dioecious**, with **external fertilization** and **external development** of the eggs and embryos. This mode of reproduction is called **oviparous** (meaning "egg-producing"). However, as tropical fish enthusiasts are well aware, the ever-popular guppies and mollies of home aquaria bear their young alive after development in the ovarian cavity of the mother (Figure 15-29). These fish are said to be **ovoviviparous,** meaning "live egg-producing." Some sharks develop a kind of placental attachment through which the young are nourished during gestation. These forms, like placental mammals, are **viviparous** ("alive-producing").

Let us return to the much more common oviparous mode of reproduction. Many marine fishes are extraordinarily profligate egg producers. Males and females aggregate in great schools and, without elaborate courtship behavior, release vast numbers of germ cells into the water to drift with the current. Large female cod may release 4 to 6 million eggs at a single spawning. Less than one in a million will survive the numerous perils of the ocean to reach reproductive maturity.

Salmon runs in the Pacific Northwest have been devasted by a lethal combination of spawning stream degradation by logging, pollution and, especially, by more than 50 hydroelectric dams which obstruct upstream migration of adult salmon and kill downstream migrants as they pass through the dams' power-generating turbines. In addition, the chain of reservoirs behind the dams, which has converted the Columbia and Snake Rivers into a series of lakes, increases mortality of young salmon migrating downstream by slowing their passage to the sea. The result is that the annual run of wild salmon is today only about 3% of the 10 to 16 million fish that ascended the rivers 150 years ago. While recovery plans have been delayed by the power industry, environmental groups argue that in the long run losing the salmon will be more expensive to the regional economy than making the changes now that will allow salmon stocks to recover.

Unlike the minute, buoyant, transparent eggs of pelagic marine teleosts, those of many near-shore and bottom-dwelling (benthic) species are larger, typically yolky, nonbuoyant, and adhesive. Some bury their eggs, many attach them to vegetation, some deposit them in nests, and some even incubate them in their mouths (Figure 15-30). Many benthic spawners guard their eggs. Intruders expecting an easy meal

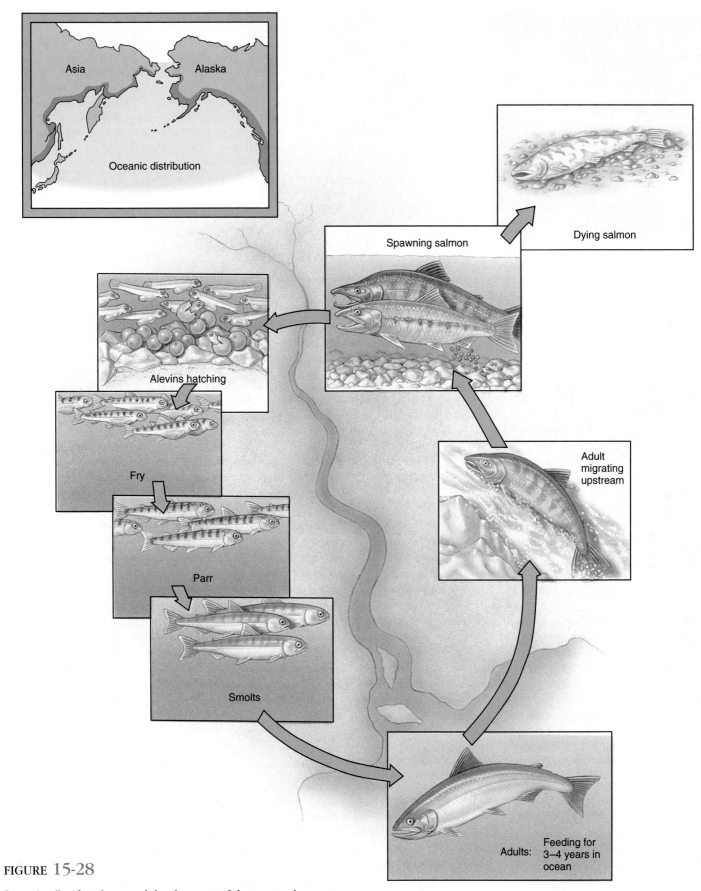

Asia

Alaska

Oceanic distribution

Dying salmon

Spawning salmon

Alevins hatching

Adult migrating upstream

Fry

Parr

Smolts

Adults: Feeding for 3–4 years in ocean

FIGURE 15-28

Spawning Pacific salmon and development of the eggs and young.

FIGURE 15-29

Rainbow surfperch *Hypsurus caryi* giving birth. All of the West Coast surfperches (family Embiotocidae) are ovoviviparous.

FIGURE 15-30

Male banded jawfish *Opistognathus macrognathus* orally brooding its eggs. The male retrieves the female's spawn and incubates the eggs until they hatch. During brief periods when the jawfish is feeding, the eggs are left in the burrow.

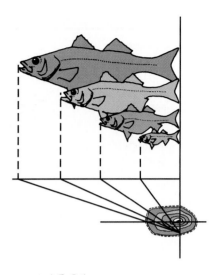

FIGURE 15-31

Scale growth. Fish scales disclose seasonal changes in growth rate. Growth is interrupted during winter, producing year marks (annuli). Each year's increment in scale growth is a ratio to the annual increase in body length. Otoliths (ear stones) and certain bones can also be used in some species to determine age and growth rate.

of eggs may be met with a vivid and often belligerent display by the guard, which is almost always the male.

Freshwater fishes almost invariably produce nonbuoyant eggs. Those, such as perch, that provide no parental care simply scatter their myriads of eggs among weeds or along the bottom. Freshwater fishes that do provide some form of egg care produce fewer, larger eggs that enjoy a better chance for survival.

Elaborate preliminaries to mating are the rule for freshwater fishes. The female Pacific salmon, for example, performs a ritualized mating "dance" with her breeding partner after arriving at the spawning bed in a fast-flowing, gravel-bottomed stream (Figure 15-28). She then turns on her side and scoops out a nest with her tail. As the eggs are laid by the female, they are fertilized by the male (Figure 15-28). After the female covers the eggs with gravel, the exhausted fish dies and drifts downstream.

Soon after the egg of an oviparous species is laid and fertilized, it takes up water and the outer layer hardens. Cleavage follows, and the blastoderm is formed, sitting astride a relatively enormous yolk mass. Soon the yolk mass is enclosed by the developing blastoderm, which then begins to assume a fishlike shape. The fish hatches as a larva carrying a semitransparent sac of yolk which provides its food sup-

ply until the mouth and digestive tract have developed. The larva then begins searching for its own food. After a period of growth the larva undergoes a metamorphosis, especially dramatic in many marine species such as the freshwater eel described previously (Figure 15-26). Body shape is refashioned, fin and color patterns change, and the animal becomes a juvenile bearing the unmistakable definitive body form of its species.

Growth is temperature dependent. Consequently, fish living in temperate regions grow rapidly in summer when temperatures are high and food is abundant but nearly stop growing in winter. Annual rings in the scales reflect this seasonal growth (Figure 15-31), a distinctive record of convenience to fishery biologists who wish to determine a fish's age. Unlike birds and mammals, which stop growing after reaching adult size, most fishes after attaining reproductive maturity continue to grow for as long as they live. This may be a selective advantage for the species, since the larger the fish, the more germ cells it produces and the greater its contribution to future generations.

Classification of Living Fishes

The following Linnaean classification of major fish taxa mostly follows that of Nelson (1994). The probable relationships of these traditional groupings together with the major extinct groups of fishes are shown in a cladogram in Figure 15-2. Other schemes of classification have been proposed. Because of the difficulty of determining relationships among the numerous living and fossil species, we can appreciate why fish classification has undergone, and will continue to undergo, continuous revision.

Phylum Chordata

Subphylum Vertebrata

Superclass Agnatha (ag'na-tha) (Gr. *a*, not, + *gnathos*, jaw) **(Cyclostomata).** No jaws; cartilaginous skeleton; paired fins absent; one or two semicircular canals; notochord persistent. A paraphyletic group that is retained because of traditional usage.

Class Myxini (mik-sy'ny) (Gr. *myxa*, slime): **hagfishes.** Mouth terminal with four pairs of tentacles; nasal sac with duct to pharynx; gill pouches, 5 to 15 pairs; partially hermaphroditic.

Class Cephalaspidomorphi (sef-a-lass'pe-do-morf'e) (Gr. *kephalē*, head, + *aspidos*, shield, + *morphē*, form) **(Petromyzontes): lampreys.** Mouth suctorial with horny teeth; nasal sac not connected to mouth; gill pouches, seven pairs.

Superclass Gnathostomata (na'tho-sto'ma-ta) (Gr. *gnathos*, jaw, + *stoma*, mouth). Jaws present; usually paired limbs; three pairs of semicircular canals; notochord persistent or replaced by vertebral centra. A paraphyletic group.

Class Chondrichthyes (kon-drik'thee-eez) (Gr. *chondros*, cartilage, + *ichthys*, fish): **cartilaginous fishes.** Cartilaginous skeleton; teeth not fused to jaws; no swim bladder; intestine with spiral valve.

Subclass Elasmobranchii (e-laz'mo-bran'kee-i) (Gr. *elasmos*, plated, + *branchia*, gills): **sharks and rays.** Placoid scales or no scales; five to seven gill arches and gills in separate clefts along pharynx. Examples: *Squalus, Raja.*

Subclass Holocephali (hol'o-sef'a-li) (Gr. *holos*, entire, + *kephalē*, head): **chimaeras, or ghostfish.** Gill slits covered with operculum; jaws with tooth plates; single nasal opening; without scales; accessory clasping organs in male; lateral line an open groove.

Class Osteichthyes (os'te-ik'thee-eez) (Gr. *osteon*, bone, + *ichthys*, a fish): **bony fishes.** Body primitively fusiform but variously modified; skeleton mostly ossified; single gill opening on each side covered with operculum; usually swim bladder or lung. A paraphyletic group as traditionally defined; in cladistic usage the Osteichthyes includes the tetrapods to be covered in later chapters.

Subclass Actinopterygii (ak'ti-nop-te-rij'ee-i) (Gr. *aktis*, ray, + *pteryx*, fin, wing): **ray-finned fishes.** Paired fins supported by dermal rays and without basal lobed portions; nasal sacs open only to outside.

Superorder Chondrostei (kon-dros'tee-i) (Gr. *chondros*, cartilage, + *osteon*, bone): **chondrostean ray-finned fishes.** Skeleton mostly cartilaginous; tail heterocercal; notochord persists in adults; intestine with spiral valve. Two living orders containing the bichir *(Polypterus),* sturgeons, and paddlefish.

Superorder Neopterygii (nee-op-te-rij'ee-i) (Gr. *neos*, new, + *pteryx*, fin, wing): **modern bony fishes.** Skeleton mostly bone; body covered with thin scales without bony layer (cycloid or ctenoid) or scaleless; caudal fin mostly homocercal; mouth terminal; notochord a mere vestige; swim bladder mainly a hydrostatic organ and usually not opened to the esophagus. Living neopterygeans are broadly divided into the nonteleosts, two orders represented by the gars *(Lepisosteus)* and the bowfin *(Amia)* ("holosteans" in older classifications); and the teleosts, represented by 38 living orders. There are approximately 23,640 living named species of neopterygians (96% of all living fishes).

Subclass Sarcopterygii (sar-cop-te-rij'ee-i) (Gr. *sarkos*, flesh, + *pteryx*, fin, wing): **fleshy-finned fishes.** Heavy bodied; paired fins with sturdy internal skeleton of basic tetrapod type and musculature; muscular lobes at bases of anal and second dorsal fins; diphycercal tail; intestine with spiral valve. Ten extinct orders and three living orders containing the coelacanth, *Latimeria chalumnae,* and three genera of lungfishes: *Neoceratodus, Lepidosiren,* and *Protopterus.*

Summary

Fishes are poikilothermic, gill-breathing aquatic vertebrates with fins. They include the oldest vertebrate groups, having originated from an unknown chordate ancestor in the Cambrian period or possibly earlier. Four classes of fishes are recognized. The jawless hagfishes (class Myxini) and lampreys (class Cephalaspidomorphi), are ancient groups having an eel-like body form without paired fins; a cartilaginous skeleton (although their ancestors, the ostracoderms, had bony skeletons); a notochord that persists throughout life; and a disclike mouth adapted for sucking or biting. All other vertebrates have jaws, a major development in vertebrate evolution.

Members of the class Chondrichthyes (sharks, rays, and chimaeras) are a compact group having a cartilaginous skeleton (a derived feature), paired fins, excellent sensory equipment, and an active, characteristically predaceous habit. To the fourth class of fishes belong the bony fishes (class Osteichthyes), which may be subdivided into two stems of descent. One stem is a relic group, the fleshy-finned fishes of the subclass Sarcopterygii, represented today by the lungfishes and the coelacanth. The terrestrial vertebrates arose from within one lineage of this group. The second stem is the ray-finned fishes (subclass Actinopterygii), a huge and diverse modern assemblage containing nearly all of the familiar freshwater and marine fishes.

The modern bony fishes (teleost fishes) have radiated into approximately 24,600 species that reveal an enormous diversity of adaptations, body form, behavior, and habitat preference. Most fishes swim by undulatory contractions of the body muscles, which generate thrust (propulsive force) and lateral force. Flexible fishes oscillate the whole body, but in more rapid swimmers the undulations are limited to the caudal region or tail fin alone.

Most pelagic bony fishes achieve neutral buoyancy in water using a gas-filled swim bladder, the most effective gas-secreting device known in the animal kingdom. The gills of fishes, having efficient countercurrent flow between water and blood, facilitate high rates of oxygen exchange.

Many fishes are migratory to some extent, and some, such as freshwater eels and anadromous salmon, make remarkable migrations of great length and precision. Fishes reveal an extraordinary range of sexual reproductive strategies. Most fishes are oviparous, but ovoviviparous and viviparous fishes are not uncommon. The reproductive investment may be in large numbers of germ cells with low survival (many marine fishes) or in fewer germ cells with greater parental care for better survival (freshwater fishes).

Review Questions

1. Provide a brief description of the fishes citing characteristics that would distinguish them from all other animals.

2. What characteristics distinguish the hagfishes and lampreys (superclass Agnatha) from all other fishes?

3. Describe feeding behavior in hagfishes and lampreys. How do they differ?

4. Describe the life cycle of the sea lamprey, *Petromyzon marinus,* and the history of its invasion of the Great Lakes.

5. In what ways are sharks well equipped for the predatory life habit?

6. The lateral line system has been described as a "distant touch" system for sharks. What function does the lateral line system serve? Where are the receptors located?

7. Explain how bony fishes differ from sharks and rays in the following systems or features: skeleton, tail shape, scales, buoyancy, and position of mouth.

8. Match the ray-finned fishes in the right column with the group to which each belongs in the left column:

____ Chondrosteans	a. Perch
____ Nonteleost	b. Sturgeon
neopterygians	c. Gar
____ Teleosts	d. Salmon
	e. Paddlefish
	f. Bowfin

9. Although the chondrosteans are today a relic group, they were one of two major lineages that emerged from early ray-finned fishes of the Devonian period. Give examples of living chondrosteans. What does the term Actinopterygii, the subclass to which the chondrosteans belong, literally mean (refer to the Classification of living fishes on p. 307)?

10. What is the other major lineage of actinopterygians? What are some distinguishing characteristics of modern bony fishes?

11. Only seven species of fleshy-finned fishes are alive today, remnants of a group that flourished in the Devonian period of the Paleozoic. What morphological characteristics distinguish the fleshy-finned fishes? What is the literal meaning of Sarcopterygii, the subclass to which the fleshy-finned fishes belong?

12. Give the geographical locations of the three surviving genera of lungfishes and explain how they differ in their ability to survive out of water. Which of the three is the least specialized?

13. Describe the discovery of the living coelacanth. What is the evolutionary significance of the group to which it belongs?

14. Compare the swimming movements of the eel with that of the trout, and explain why the latter is more efficient for rapid locomotion.

15. Sharks and bony fishes approach or achieve neutral buoyancy in different ways. Describe the methods evolved in each group. Why must a teleost fish adjust the gas volume in its swim bladder when it swims upward or downward? How is gas volume adjusted?

16. What is meant by "countercurrent flow" as it applies to fish gills?

17. Describe the life cycle of the European eel. How does the life cycle of the American eel differ from that of the European?

18. How do adult Pacific salmon find their way back to their parent stream to spawn?

19. What mode of reproduction in fishes is described by each of the following terms: oviparous, ovoviviparous, viviparous?

20. Reproduction in marine pelagic fishes and in freshwater fishes is distinctively different. How and why do they differ?

Selected References

See also general references on page 395.

Bone, Q., and N. B. Marshall. 1982. Biology of fishes. New York, Chapman & Hall. *Concise, well-written, and well-illustrated primer on the functional processes of fishes.*

Conniff, R. 1991. The most disgusting fish in the sea. Audubon **93**(2):100–108 (March). *Recent discoveries shed light on the life history of the enigmatic hagfish that fishermen loathe.*

Horn, M. H., and R. N. Gibson. 1988. Intertidal fishes. Sci. Am. **258**:64–70 (Jan.). *Describes the special adaptations of fishes living in a demanding environment.*

Long, J. A. 1995. The rise of fishes: 500 million years of evolution. Baltimore, The Johns Hopkins University Press. *A lavishly illustrated evolutionary history of fishes.*

Moyle, P. B. 1993. Fish: an enthusiast's guide. Berkeley, University of California Press. *Textbook written in a lively style and stressing function and ecology rather than morphology; abbreviated treatment of the fish groups.*

Nelson, J. S. 1994. Fishes of the world, ed. 3. New York, John Wiley & Sons, Inc. *Authoritative classification of all major groups of fishes.*

Stevens, J. D., ed. 1987. Sharks. New York, Facts on File Publications. *Evolution, biology, and behavior of sharks, handsomely illustrated.*

Thomson, K. S. 1991. Living fossil. The story of the coelacanth. New York, W. W. Norton.

Webb, P. W. 1984. Form and function in fish swimming. Sci. Am. **251**:72–82 (July). *Specializations of fish for swimming and analysis of thrust generation.*

Links to the Internet

Visit this textbook's Web site at http://www.mhhe.com/zoology to find live Internet links for each of the references listed below.

1. Craniata. Arizona's Tree of Life Web Page. An introduction, pictures, characteristics, discussion of the skull, phylogenetic relationships, and references on the craniates.

Jawless Fish

2. Hyperotreti (Hagfishes). Arizona's Tree of Life Web Page. An introduction, pictures, characteristics, discussion of the skull, phylogenetic relationships, and references on hagfishes.

3. Hyperoartia (Lampreys). Arizona's Tree of Life Web Page. An introduction, pictures, characteristics, discussion of the skull, phylogenetic relationships, and references on lampreys.

4. Introduction to the Myxini. University of California at Berkeley Museum of Paleontology. Images, photos, systematics, more information, and links.

Chondrichthyean Fish

5. Class Chondrichthyes. University of Michigan site on actinopterygiian fish. Pictures, much information on the morphology, distribution, and ecology of a large number of sharks. Each fish is linked to Web pages. Images may not be available for display depending on your server.

6. Introduction to the Chondrichthyes. University of California at Berkeley Museum of Paleontology. Images, photos, systematics, more information, and links.

7. The Great White Shark. Images and information. University of California at Berkeley Museum of Paleontology. Images, photos, systematics, more information, and links.

8. Sharks and Their Relatives. Sea World Education Department information on sharks.

9. Mote Marine Laboratory, Center for Shark Research. Shark research, shark myths, information on sharks and cancer, shark attacks, and more.

10. Fiona's Shark Mania. This entertaining site provides a variety of information on sharks.

Osteichthyean Fish

11. The Coelacanth: Living Fossil. This site contains some information, but of greater utility are the many links to other sites with information on the coelacanth.

12. Sarcopterygii. Arizona's Tree of Life Web Page. Pictures, characteristics, phylogenetic relationships, references on sarcopterygian fishes.

13. Class Actinopterygii. University of Michigan site on actinopterygian fish. Pictures, much information on the morphology, distribution, and ecology of a large number of fish. Each fish is linked to additional Web pages. Images may not be available for display depending on your server.

14. Introduction to the Actinopterygii. University of California at Berkeley Museum of Paleontology. Images, photos, systematics, more information, and links.

15. Neopterygii. University of California at Berkeley Museum of Paleontology. Images, photos, systematics, more information, and links.

16. Teleostei. University of California at Berkeley Museum of Paleontology. Images, photos, systematics, more information, and links.

17. Teleostei. Arizona's Tree of Life Web Page. Pictures, characteristics, phylogenetic relationships, references on teleost fishes. A cladogram and references on teleosts. Some links to various groups of teleosts.

18. Marine Fishes at Hawai'i. A pictoral guide to the families of marine fishes found in the waters surrounding the Hawaiian Islands.

19. Zebrafish. Many links to further information on this fish commonly used in research in developmental biology.

General Fish Resources

20. National Marine Fisheries Service. This site contains images of marine fishes, news, information about the Sustained Fisheries Act, a history of fishery science, links to regional NMFS offices, and climatic and oceanographic data.

21. Vertebrate Systematics. University of California at Berkeley Museum of Paleontology. Information on the systematics and natural history of each of the major groups of fishes and links to other fish sites. Click on photographs for more information about each group.

The Early Tetrapods and Modern Amphibians

CHAPTER | sixteen

Vertebrate Landfall

The chorus of frogs beside a pond on a spring evening heralds one of nature's dramatic events. Masses of frog eggs soon hatch into limbless, gill-breathing, fishlike tadpole larvae. Warmed by the late spring sun, they feed and grow. Then, almost imperceptibly, a remarkable transformation takes place. Hindlegs appear and gradually lengthen. The tail shortens. The larval teeth are lost, and the gills are replaced by lungs. Eyelids develop. The forelegs emerge. In a matter of weeks the aquatic tadpole has completed its metamorphosis to an adult frog.

The evolutionary transition from water to land occurred not in weeks but over millions of years. A lengthy series of alterations cumulatively fitted the vertebrate body plan for life on land. The origin of land vertebrates is no less a remarkable feat for this fact—a feat that incidentally would have a poor chance of succeeding today because well-established competitors make it impossible for a poorly adapted transitional form to gain a foothold.

Amphibians are the only living vertebrates that have a transition from water to land in both their ontogeny and phylogeny. Even after some 350 million years of evolution, few amphibians are completely land adapted; most are quasiterrestrial, hovering between aquatic and land environments. This double life is expressed in their name. Even the amphibians that are best adapted for a terrestrial existence cannot stray far from moist conditions. Many, however, have developed ways to keep their eggs out of open water where the larvae would be exposed to enemies.

Adaptation for life on land is a major theme of the remaining vertebrate groups treated in this and the following chapters. These animals form a monophyletic unit known as the **tetrapods.** The amphibians and the amniotes (including reptiles, birds, and mammals) represent the two major extant branches of tetrapod phylogeny. In this chapter, we review what is known about the origins of terrestrial vertebrates and discuss the amphibian lineage in detail. We discuss the major amniote groups in Chapters 17 through 19.

Movement onto Land

The movement from water to land is perhaps the most dramatic event in animal evolution, because it involves the invasion of a habitat that in many respects is more hazardous for life. Life originated in water. Animals are mostly water in composition, and all cellular activities occur in water. Nevertheless, organisms eventually invaded land, carrying their watery composition with them. Vascular plants, pulmonate snails, and tracheate arthropods made the transition much earlier than vertebrates, and winged insects were diversifying at approximately the same time that the earliest terrestrial vertebrates evolved. Although the invasion of land required modification of almost every system in the vertebrate body, aquatic and terrestrial vertebrates retain many basic structural and functional similarities. We see the transition between the aquatic and terrestrial vertebrates most clearly today in the many living amphibians that make this transition during their own life histories.

Beyond the obvious difference in water content, there are several important physical differences that animals must accommodate when moving from water to land. These include (1) oxygen content, (2) density, (3) temperature regulation, and (4) habitat diversity. Oxygen is at least 20 times more abundant in air and it diffuses much more rapidly through air than through water. Consequently, terrestrial animals can obtain oxygen far more easily than aquatic ones once they possess the appropriate adaptations, such as lungs. Air, however, has approximately 1000 times less buoyant density than water and is approximately 50 times less viscous. It therefore provides relatively little support against gravity, requiring the terrestrial animal to develop strong limbs and to remodel the skeleton to achieve adequate structural support. Air fluctuates in temperature more readily than water does, and terrestrial environments therefore experience harsh and unpredictable cycles of freezing, thawing, drying, and flooding. Terrestrial animals require behavioral and physiological strategies to protect themselves from thermal extremes; one such important strategy is the homeothermy (regulated constant body temperature) of birds and mammals.

Despite its hazards, the terrestrial environment offers a great variety of new habitats including coniferous, temperate, and tropical forests, grasslands, deserts, mountains, oceanic islands, and polar regions. The provision of safe shelter for the protection of vulnerable eggs and young may be accomplished much more readily in many of these terrestrial habitats than in aquatic ones.

Early Evolution of Terrestrial Vertebrates

Devonian Origin of the Tetrapods

The Devonian period, beginning some 400 million years ago, was a time of mild temperatures and alternating droughts and floods. During this period some primarily aquatic vertebrates evolved two features that would be important for permitting the subsequent evolution for life on land: lungs and limbs.

The Devonian freshwater environment was unstable. During dry periods, many pools and streams evaporated, water became foul, and the dissolved oxygen disappeared. Only those fishes able to acquire atmospheric oxygen survived such conditions. Gills were unsuitable because in air the filaments collapsed, dried, and quickly lost their function. Virtually all freshwater fishes surviving this period, including the lobe-finned (rhipidistian) fishes and the lungfishes (p. 298), had a kind of lung that developed as an outgrowth of the pharynx. It was relatively simple to enhance the efficiency of the air-filled cavity by improving its vascularity with a rich capillary network, and by supplying it with arterial blood from the last (sixth) pair of aortic arches. Oxygenated blood returned directly to the heart by a pulmonary vein to form a complete pulmonary circuit. Thus the **double circulation** characteristic of all tetrapods originated: a systemic circulation serving the body and a pulmonary circulation supplying the lungs.

Vertebrate limbs also arose during the Devonian period. Although fish fins at first appear very different from the jointed limbs of tetrapods, an examination of the bony elements of the paired fins of the lobe-finned fishes shows that they broadly resemble the equivalent limbs of amphibians. In *Eusthenopteron,* a Devonian lobe-fin, we can recognize an upper arm bone (humerus) and two forearm bones (radius and ulna) as well as other elements that we can homologize with the wrist bones of tetrapods (Figure 16-1). *Eusthenopteron* could walk—more accurately flop—along the bottom mud of pools with its fins, since backward and forward movement of the fins was limited to about 20–25 degrees. *Acanthostega,* one of the earliest known Devonian tetrapods, had well-formed tetrapod legs with clearly formed digits on both fore- and hindlimbs, but the limbs were too weakly constructed to enable the animal to hoist its body off the surface for proper walking on land. *Ichthyostega,* however, with its fully developed shoulder girdle, bulky limb bones, well-developed muscles, and other adaptations for terrestrial life, must have been able to pull itself onto land, although it probably did not walk very well. Thus, the tetrapods evolved their legs underwater and only then, for reasons unknown, began to pull themselves onto land.

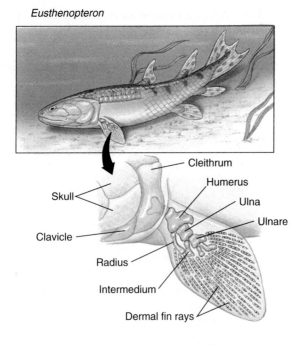

Eusthenopteron

Cleithrum
Humerus
Skull
Ulna
Ulnare
Clavicle
Radius
Intermedium
Dermal fin rays

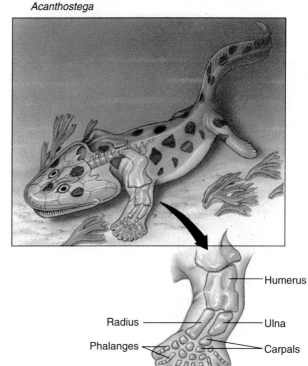

Acanthostega

Humerus
Radius
Ulna
Phalanges
Carpals

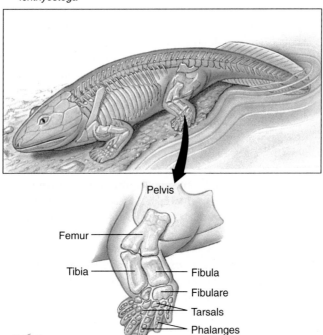

Ichthyostega

Pelvis
Femur
Tibia
Fibula
Fibulare
Tarsals
Phalanges

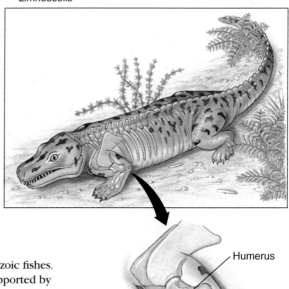

Limnoscelis

Humerus
Radius
Ulna
Phalanges
Carpals

FIGURE 16-1

Evolution of the tetrapod leg. The legs of tetrapods evolved from fins of Paleozoic fishes. *Eusthenopteron,* a late Devonian lobe-finned fish had paired muscular fins supported by bony elements that foreshadowed the bones of tetrapod limbs. The anterior fin contained an upper arm bone (humerus), two forearm bones (radius and ulna), and smaller elements homologous to the wrist bones of tetrapods. As typical of fishes, the pectoral girdle, consisting of the cleithrum, clavicle, and other bones, was firmly attached to the skull. In *Acanthostega,* one of the earliest known Devonian tetrapods (appearing about 360 million years BP), dermal fin rays of the anterior appendage were replaced by eight fully evolved fingers. *Acanthostega* was probably exclusively aquatic because its limbs were too weak for travel on land. *Ichthyostega,* a contemporary of *Acanthostega,* had fully formed tetrapod limbs and must have been able to walk on land. The hindlimb bore seven toes (the number of forelimb digits is unknown). *Limnoscelis,* an anthracosaur amphibian of the Carboniferous (about 300 million years BP) had five digits on both fore- and hindlimbs, the basic pentadactyl model which became the tetrapod standard.

(Sources: R. L. Carroll, Vertebrate Paleontology and Evolution, *1988, W. H. Freeman & Co., NY; M. I. Clack, and J. A. Clack, **347***:66-69, 1990; J. L. Edwards,* American Zoology, ***29***:235-254, 1989; E. Jarvik, Scientific Monthly, **1955***:141-154, March 1955; and C. N. Zimmer,* Discover, **16**(6):118-127, 1995.)

As noted above, evidence points to the lobe-finned fishes as the closest relatives of the tetrapods; in cladistic terms they are the sister group of tetrapods (Figures 16-2 and 16-3). Both the lobe-finned fishes and early tetrapods such as *Acanthostega* and *Ichthyostega* shared several characteristics of skull, teeth, and pectoral girdle. *Ichthyostega* (Gr. *ichthys*, fish, + *stegē*, roof, or covering, in reference to the roof of the skull which was shaped like that of a fish) represents an early offshoot of tetrapod phylogeny that possessed several adaptations, in addition to jointed limbs, that equipped it for life on land (Figure 16-1). These include a stronger backbone and associated muscles to support the body in air, new muscles to elevate the head, strengthened shoulder and hip girdles, a protective rib cage, a more advanced ear structure for detecting airborne sounds, a foreshortening of the skull, and a lengthening of the snout that improved olfactory powers for detecting dilute airborne odors. Yet *Ichthyostega* still resembled aquatic forms in retaining a tail complete with fin rays and in having opercular (gill) bones.

> *The bones of* Ichthyostega, *the most thoroughly studied of all early tetrapods, were first discovered on an east Greenland mountainside in 1897 by Swedish scientists looking for three explorers lost two years earlier during an ill-fated attempt to reach the North Pole by hot-air balloon. Later expeditions by Gunnar Säve-Söderberg uncovered skulls of* Ichthyostega *but Säve-Söderberg died, at age 38, before he was able to make a thorough study of the skulls. After Swedish paleontologists returned to the Greenland site where they found the remainder of* Ichthyostega's *skeleton, Erik Jarvik, one of Säve-Söderberg's assistants, assumed the task of examining the skeleton in detail. This study became his life's work, resulting in the detailed description of* Ichthyostega *available to us today.*

Carboniferous Radiation of the Tetrapods

The capricious Devonian period was followed by the Carboniferous period, characterized by a warm, wet climate during which mosses and large ferns grew in profusion on a swampy landscape. Tetrapods radiated quickly in this environment to produce a great variety of forms, feeding on the abundance of insects, insect larvae, and aquatic invertebrates available. The evolutionary relationships of the early tetrapod groups are still very controversial. We present a tentative cladogram (Figure 16-2) that almost certainly will undergo future revision as new data are collected. Several extinct lineages plus the **Lissamphibia,** which contains the modern amphibians, are placed in a group termed the **temnospondyls** (see Figures 16-2 and 16-3). This group is distinguished by having generally only four digits on the forelimb rather than the five characteristic of most tetrapods.

The lissamphibians diversified during the Carboniferous to produce the ancestors of the three major groups of amphibians alive today, **frogs** (Anura or Salientia), **salamanders** (Caudata or Urodela), and **caecilians** (Apoda or Gymnophiona). The early amphibians improved their adaptations for living in water during this period. Their bodies became flatter for moving about in shallow water. Early salamanders developed weak limbs and the tail became better developed as a swimming organ. Even the anurans (frogs and toads), which are now largely terrestrial as adults, developed specialized hindlimbs with webbed feet better suited for swimming than for movement on land. All amphibians use their porous skin as a primary or accessory breathing organ. This specialization was encouraged by the swampy surroundings of the Carboniferous period but presented serious desiccation problems for life on land.

The Modern Amphibians

The three living amphibian orders comprise more than 3900 species. Most share general adaptations for life on land, including skeletal strengthening and a shifting of special sense priorities from the ancestral lateral line system to the senses of smell and hearing. For this, both the olfactory epithelium and the ear are redesigned to improve sensitivities to airborne odors and sounds.

Nonetheless, most amphibians meet the problems of independent life on land only halfway. In the ancestral life history of amphibians, eggs are aquatic and hatch to produce an aquatic larval form that uses gills for breathing. A metamorphosis follows in which gills are lost and lungs, which are present throughout larval life, are then activated for respiration. Many amphibians retain this general pattern but there are some important exceptions. Some salamanders lack a complete metamorphosis and retain a permanently aquatic, larval morphology throughout life. Others live entirely on land and lack the aquatic larval phase completely. Both of these are evolutionarily derived conditions. Some frogs also have acquired a strictly terrestrial existence by eliminating the aquatic larval stage.

Even the most terrestrial amphibians remain dependent on very moist if not aquatic environments. Their skin is thin, and it requires moisture for protection against desiccation in air. An intact frog loses water nearly as rapidly as a skinless frog. Amphibians also require moderately cool environments. Being ectothermic, their body temperature is determined by and varies with the environment, greatly restricting where they can live. This restriction is especially important for reproduction. Eggs are not well protected from desiccation, and they must be shed directly into the water or onto moist terrestrial surfaces. Completely terrestrial amphibians may lay eggs under logs or rocks, in the moist forest floor, in flooded tree

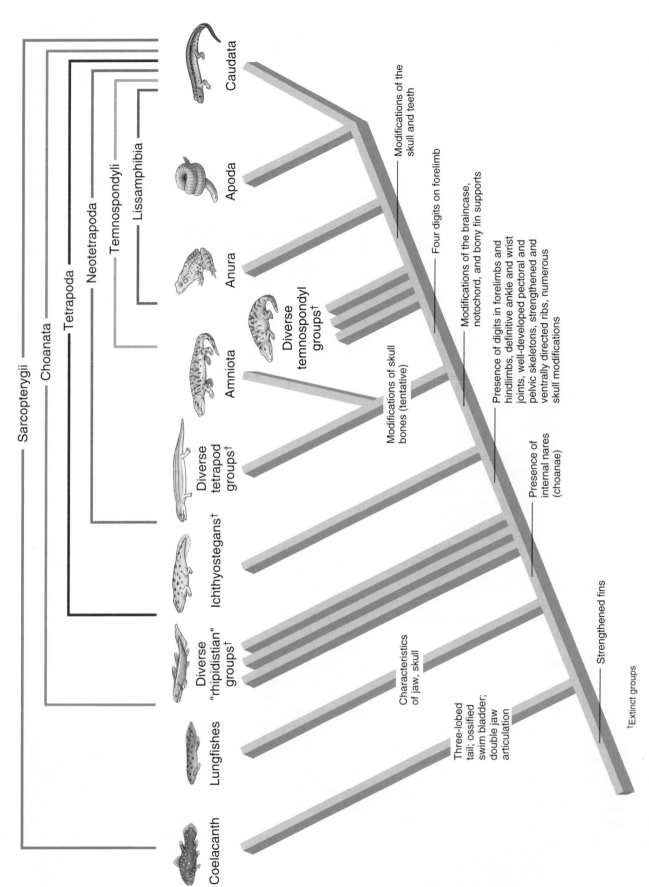

FIGURE 16-2

Tentative cladogram of the Tetrapoda with emphasis on descent of the amphibians. Especially controversial are the relationships of major tetrapod groups (Amniota, Temnospondyli and diverse early tetrapod groups) and outgroups (coelacanth, lungfish, rhipidistian fishes). All aspects of this cladogram are controversial, however, including relationships of the Lissamphibia. The relationships shown for the three groups of Lissamphibia are based on recent molecular evidence. Extinct groups are marked with a dagger symbol (†).

(Source: From E.W. Gaffney in Bulletin of the Carnegie Museum of Natural History, 13:92–105, 1979.)

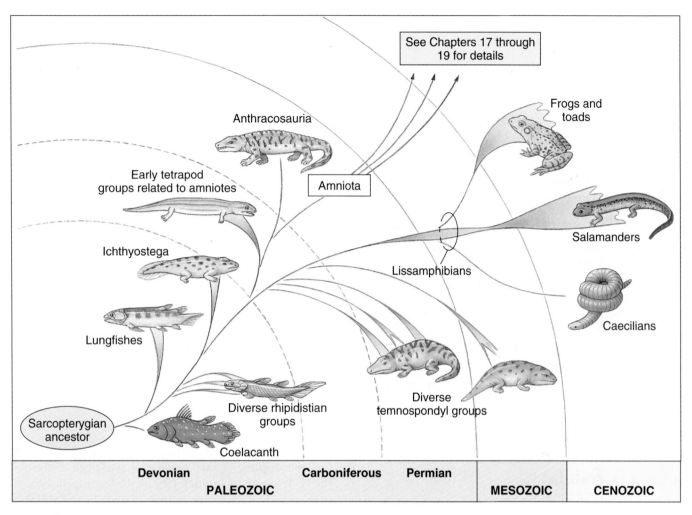

FIGURE 16-3

Early tetrapod evolution and the descent of amphibians. The tetrapods share most recent common ancestry with the extinct Devonian rhipidistian fishes (of *living* groups, the tetrapods are most closely related to the lungfishes). The amphibians share most recent common ancestry with the diverse temnospondyls of the Carboniferous and Permian periods of the Paleozoic, and Triassic period of the Mesozoic.

holes, in pockets on the mother's back (Figure 16-4), or in folds of the body wall. One species of Australian frog even broods its young in its vocal pouch (see Figure 16-4).

We now highlight the special characteristics of the three major groups of amphibians. We will expand the coverage of general amphibian features when discussing the groups in which particular features have been studied most extensively. For most features, this group will be the frogs.

Caecilians: Order Gymnophiona (Apoda)

The order Gymnophiona (jim′no-fy′o-na) (Gr. *gymnos,* naked, + *ophineos,* of a snake) contains approximately 160

species of elongate, limbless, burrowing creatures commonly called **caecilians** (Figure 16-5). They occur in tropical forests of South America (their principal home), Africa, and Southeast Asia. They possess a long, slender body, small scales in the skin of some, many vertebrae, long ribs, no limbs, and a terminal anus. The eyes are small, and most species are totally blind as adults. Their food consists mostly of worms and small invertebrates, which they find underground. Fertilization is internal, and the male is provided with a protrusible copulatory organ. The eggs are usually deposited in moist ground near the water. In some species the eggs are carefully guarded in folds of the body during their development. Viviparity also is common in some caecilians, with the embryos obtaining nourishment by eating the wall of the oviduct.

Characteristics of Modern Amphibians

1. Skeleton mostly bony, with varying numbers of vertebrae; ribs present in some, absent or fused to vertebrae in others

2. Body forms vary greatly from an elongated trunk with distinct head, neck, and tail to a compact, depressed body with fused head and trunk and no intervening neck

3. **Limbs usually four (tetrapod),** although some are legless; forelimbs of some much smaller than hindlimbs, in others all limbs small and inadequate; webbed feet often present; no true nails or claws; **forelimb usually with four digits** but sometimes five and sometimes fewer

4. **Skin smooth and moist with many glands,** some of which may be poison glands; pigment cells (chromatophores) common, of considerable variety; no scales, except concealed dermal ones in some

5. Mouth usually large with small teeth in upper or both jaws; two nostrils open into anterior part of mouth cavity

6. Respiration by lungs (absent in some salamanders), skin, and gills in some, either separately or in combination; external gills in the larval form and may persist throughout life in some

7. **Circulation with three-chambered heart,** two atria and one ventricle, and a **double circulation through the heart;** skin abundantly supplied with blood vessels

8. Ectothermal

9. Excretory system of paired mesonephric kidneys; urea main nitrogenous waste

10. Ten pairs of cranial nerves

11. Separate sexes; fertilization mostly internal in salamanders and caecilians, mostly external in frogs and toads; predominantly oviparous, some ovoviviparous or (rarely) viviparous; metamorphosis usually present; **moderately yolky eggs** (mesolecithal) **with jellylike membrane coverings**

Salamanders: Order Caudata (Urodela)

As its name suggests, the order Caudata (L. *caudatus,* having a tail) are tailed amphibians, some 360 species of salamanders. Salamanders are found in almost all northern temperate regions of the world, and they have great abundance and diversity in North America. Salamanders are found also in the tropical areas of Central and northern South America. Salamanders are typically small; most of the common North American salamanders are less than 15 cm long. Some aquatic forms are considerably longer, and the Japanese giant salamander may exceed 1.5 m in length.

Most salamanders have limbs set at right angles to the body, with forelimbs and hindlimbs of approximately equal size. In some aquatic and burrowing forms, the limbs are rudimentary or absent.

Salamanders are carnivorous both as larvae and adults, preying on worms, small arthropods, and small molluscs. Most eat only things that are moving. Like all amphibians, they are ectotherms and have a low metabolic rate.

Breeding Behavior

Some salamanders are wholly aquatic throughout their life cycle, but most are metamorphic, having aquatic larvae and terrestrial adults that live in moist places under stones and rotten logs. The eggs of most salamanders are fertilized internally, usually after the female picks up a packet of sperm **(spermatophore)** that previously has been deposited by the male on a leaf or stick (Figure 16-6). Aquatic species lay their eggs in clusters or stringy masses in the water. Their eggs hatch to produce an aquatic larva having external gills and a finlike tail. Completely terrestrial species deposit eggs in small, grapelike clusters under logs or in excavations in soft moist earth, and many species remain to guard the eggs (Figure 16-7). These species have **direct development.** They bypass the larval stage and hatch as miniature versions of their parents. The most complex of salamander life cycles is observed in some American newts, whose aquatic larvae metamorphose to form terrestrial juveniles that later metamorphose again to produce secondarily aquatic, breeding adults (Figure 16-8).

Respiration

At various stages of their life history, salamanders may have external gills, lungs, both, or neither of these. They also share the general amphibian condition of having extensive vascular nets in their skin that serve the respiratory exchange of oxygen and carbon dioxide. Salamanders that have an aquatic larval stage hatch with gills, but lose them later if a metamorphosis occurs. Several diverse lineages of salamanders have evolved permanently aquatic forms that fail to undergo a complete metamorphosis and retain their gills and finlike tail throughout life. Lungs, the most widespread respiratory organ

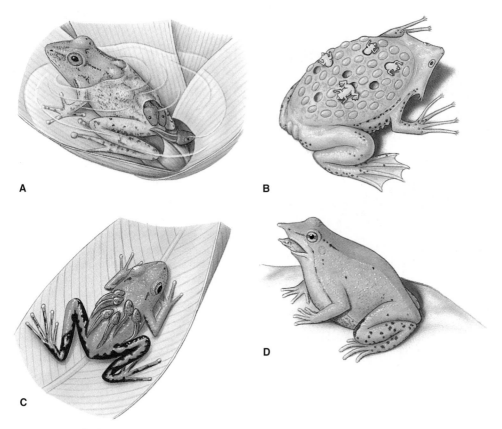

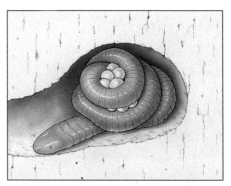

FIGURE 16-5

Female caecilian coiled around eggs in her burrow.

FIGURE 16-4

Reproductive strategies of anurans. **A,** Female South American pygmy marsupial frog *Flectonotus pygmaeus* carries developing larvae in a dorsal pouch. **B,** Female Surinam frog carries eggs embedded in specialized brooding pouches on the dorsum; froglets emerge and swim away when development is complete. **C,** Male poison-dart frog *Phyllobates bicolor* carries tadpoles adhering to its back. **D,** Tadpoles of a male Darwin's frog *Rhinoderma darwinii* develop into froglets in its vocal pouch. When ready to emerge, a froglet crawls into the parent's mouth, which the parent opens to allow the froglet's escape.

the water is so cool and well oxygenated that cutaneous respiration alone was sufficient for life. It is curious that the most completely terrestrial lineage of salamanders evolved in a group that lacks lungs.

Paedomorphosis

Whereas most salamanders complete their development by metamorphosis to the adult body form, some species reach sexual maturity while retaining their gills, aquatic life-style, and other larval characteristics. This condition illustrates **paedomorphosis** (Gr. "child form"), defined as the retention in adult descendants of features that were present only in the pre-adult stages of their ancestors. Some characteristics of the ancestral adult morphology are consequently eliminated. Examples of such nonmetamorphic, permanently-gilled species are mud puppies of the genus *Necturus* (Figure 16-10A), which live on bottoms of ponds and lakes; and the axolotl of Mexico (Figure 16-10B). These species never metamorphose under any conditions.

There are other species of salamanders that reach sexual maturity with larval morphology but, unlike permanent larvae such as *Necturus,* may metamorphose to terrestrial forms under certain environmental conditions. We find good examples in *Ambystoma tigrinum* and related species from North America. Their typical habitat consists of small ponds that can disappear through evaporation in dry weather. When ponds evaporate the aquatic form metamorphoses to a terrestrial form, losing its gills and developing lungs. It then can travel across the land in search of new sources of water in which to live and reproduce.

of terrestrial vertebrates, are present from birth in the salamanders that have them, and become active following metamorphosis. Others, such as amphiumas, while having a completely aquatic life history, nonetheless lose their gills before adulthood and then breathe primarily by lungs. This requires that they periodically raise their nostrils above the surface of the water to get air.

The amphiumas provide an interesting contrast to many species of the large family Plethodontidae (Figures 16-6, 16-7, and 16-9) that are entirely terrestrial but completely lack lungs. The efficiency of cutaneous respiration is increased by the penetration of a capillary network into the epidermis or by the thinning of the epidermis over superficial dermal capillaries. Cutaneous respiration is supplemented by air pumped in and out of the mouth, where respiratory gases are exchanged across the vascularized membranes of the buccal (mouth) cavity (buccopharyngeal breathing). The lungless plethodontids probably originated in streams, where lungs would have been a disadvantage by providing too much buoyancy, and where

FIGURE 16-6

Courtship and sperm transfer in the pygmy salamander, *Desmognathus wrighti*. After judging the female's receptivity by the presence of her chin on his tail base, the male deposits a spermatophore on the ground, then moves forward a few paces. **A,** The white mass of the sperm atop a gelatinous base is visible at the level of the female's forelimb. The male moves ahead, the female following until the spermatophore is at the level of her vent. **B,** The female has recovered the sperm mass in her vent, while the male arches his tail, tilting the female upward and presumably facilitating recovery of the sperm mass.

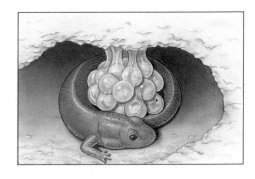

FIGURE 16-7

Female dusky salamander (*Desmognathus* sp.) attending eggs. Many salamanders exercise parental care of the eggs, which includes rotating the eggs and protecting them from fungal infections and predation by various arthropods and other salamanders.

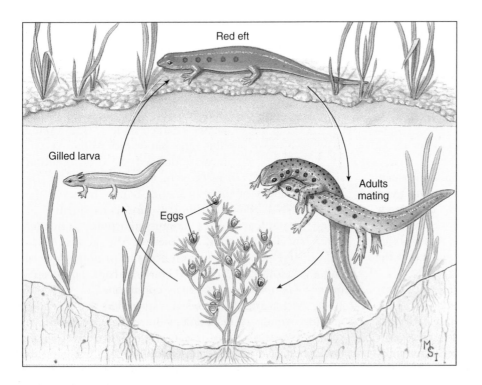

FIGURE 16-8

Life history of the red-spotted newt, *Notophthalmus viridescens* of the family Salamandridae. In many habitats the aquatic larva metamorphoses into a brightly colored "red eft" stage, which remains on land from one to three years before transforming into a secondarily aquatic adult.

FIGURE 16-9

Longtail salamander *Eurycea longicauda*, a common plethodontid salamander.

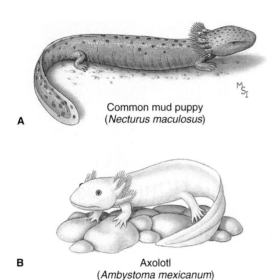

A Common mud puppy
(*Necturus maculosus*)

B Axolotl
(*Ambystoma mexicanum*)

FIGURE 16-10

Paedomorphosis in salamanders. **A,** The mud puppy *Necturus* sp. and **B,** the axolotl *(Ambystoma mexicanum)* are permanently gilled aquatic forms.

In addition to their importance in biomedical research and education, frogs have long served the epicurean frog-leg market. Mainstay of this market are bullfrogs, which are in such heavy demand in Europe (especially France) and the United States—the worldwide harvest is an estimated 200 million bullfrogs (about 10,000 metric tons) annually—that its populations have fallen drastically as the result of excessive exploitation and the draining and pollution of wetlands. Most are Asian bullfrogs imported from India and Bangladesh, some 80 million collected each year from rice fields in Bangladesh alone. With so many insect-eating frogs removed from the ecosystem, rice production is threatened from uncontrolled, flourishing insect populations. In the United States, attempts to raise bullfrogs in farms have not been successful, mainly because bullfrogs are voracious eating machines that normally will accept only living prey, such as insects, crayfish, and other frogs.

Frogs and Toads: Order Anura (Salientia)

The more than 3450 species of frogs and toads that comprise the order Anura (Gr. *an,* without, + *oura,* tail) are for most people the most familiar amphibians. The Anura are an old group, known from the Jurassic period, 150 million years ago. Frogs and toads occupy a great variety of habitats, despite their aquatic mode of reproduction and water-permeable skin, which prevent them from wandering too far from sources of water, and their ectothermy, which bars them from polar and subarctic habitats. The name of the order, Anura, refers to an obvious group characteristic, the absence of tails in adults (although all pass through a tailed larval stage during development). Frogs and toads are specialized for jumping, as suggested by the alternative order name, Salientia, which means leaping.

We see in the appearance and life-style of their larvae further distinctions between the Anura and Caudata. The eggs of most frogs hatch into a tadpole ("polliwog") stage, having a long, finned tail, both internal and external gills, no legs, specialized mouthparts for herbivorous feeding (salamander larvae and some tadpoles are carnivorous), and a highly specialized internal anatomy. They look and act altogether differently from adult frogs. The metamorphosis of a frog tadpole to an adult frog is thus a striking transformation. The permanently gilled larval condition never occurs in frogs and toads as it does in salamanders.

Frogs and toads are divided into 21 families. The best-known frog families in North America are the Ranidae, which contains most of our familiar frogs (Figure 16-11A), and the Hylidae, the tree frogs (Figure 16-11B). True toads, belonging to the family Bufonidae, have short legs, stout bodies, and thick skins usually with prominent warts (Figure 16-12). However, the term "toad" is used rather loosely to refer also to more or less terrestrial members of several other families.

The largest anuran is the West African *Conraua goliath,* which is more than 30 cm long from tip of nose to anus (Figure 16-13). This giant eats animals as big as rats and ducks. The smallest frog recorded is *Phyllobates limbatus,* which is only approximately 1 cm long. This tiny frog, which can be more than covered by a dime, is found in Cuba. The largest American frog is the bullfrog, *Rana catesbeiana* (Figure 16-11A), which reaches a head and body length of 20 cm.

Habitats and Distribution

Probably the most abundant frogs are the approximately 260 species of the genus *Rana* (Gr. frog), found over all the temperate and tropical regions of the world except in New Zealand, the oceanic islands, and southern South America. They are usually found near water, although some, such as the wood frog *R. sylvatica,* spend most of their time on damp forest floors. The larger bullfrogs, *R. catesbeiana,* and green frogs, *R. clamitans,* are nearly always found in or near permanent water or swampy regions. The leopard frogs, *Rana pipiens* and related species, are found in nearly every state and Canadian province and are the most widespread of all North American frogs. The northern leopard frog, *R. pipiens,* is the

A

B

FIGURE 16-11

Two common North American frogs. **A,** Bullfrog, *Rana catesbeiana,* largest American frog and mainstay of the frog-leg epicurean market (family Ranidae). **B,** Green tree frog *Hyla cinerea,* a common inhabitant of swamps of the southeastern United States (family Hylidae). Note adhesive pads on the feet.

FIGURE 16-12

American toad *Bufo americanus* (family Bufonidae). This principally nocturnal yet familiar amphibian feeds on large numbers of insect pests and on snails and earthworms. The warty skin contains numerous glands that produce a surprisingly poisonous milky fluid, providing the toad excellent protection from a variety of potential predators.

species most commonly used in biology laboratories and for classical electrophysiological research.

Within the range of any species, frogs are often restricted to certain localities (for instance, to specific streams or pools) and may be absent or scarce in similar habitats elsewhere. The pickerel frog *(R. palustris)* is especially noteworthy in this respect because it is known to be abundant only in certain localized regions. Recent studies have shown that many populations of frogs worldwide may be suffering declines in numbers and becoming even more patchy than usual in their distributions. In most declining populations the causes are unknown.[1]

Most of the larger frogs are solitary in their habits except during the breeding season. During the breeding period most of them, especially the males, are very noisy. Each male usually takes possession of a particular perch near water, where he may remain for hours or even days, trying to attract a female to that spot. At times frogs are mainly silent, and their presence is not detected until they are disturbed. When they enter the water, they dart about swiftly and reach the bottom of the pool, where they kick up a cloud of muddy water. In swimming, they hold the forelimbs near the body and kick backward with their webbed hindlimbs, which propel them forward. When they come to the surface to breathe, only the head and foreparts are exposed and, since they usually take advantage of any protective vegetation, they are difficult to see.

What is responsible for the widely reported decline in amphibian, especially frog, populations around the world? Puzzling is the evidence that whereas amphibian populations are falling rapidly in many parts of the world, in other areas they are doing well. No single explanation fits all instances of declines. In some instances, changes in sizes of populations are simply random fluctuations caused by periodic droughts and other naturally occurring phenomena. However, several other environmental and human-related factors have been implicated in amphibian declines: habitat destruction and modification; rises in environmental pollutants such as acid rain, fungicides, herbicides, and industrial chemicals; diseases; introduction of nonnative predators and competitors; and depletion of the ozone shield in the stratosphere resulting in severe losses in developing frog embryos from increased ultraviolet radiation. These explanations are radically different from each other. One or more explanations do seem to explain certain population declines; in other instances the reasons for the declines are unknown.

During the winter months most frogs hibernate in the soft mud of the bottoms of pools and streams. Their life processes are at a very low ebb during their hibernation period, and such energy as they need is derived from the glycogen and

---

[1]Sarkar, S. 1996. Ecological theory and anuran declines. *BioScience* 46(3):199–207.

FIGURE 16-13

Conraua (Gigantorana) goliath (family Ranidae) of West Africa, the world's largest frog. This specimen weighed 3.3 kg (approximately 7½ pounds).

fat stored in their bodies during the spring and summer months. The more terrestrial frogs, such as tree frogs, hibernate in the humus of the forest floor. They are tolerant of low temperatures, and many actually survive prolonged freezing of all the extracellular fluid, representing 35% of the body water. Such frost-tolerant frogs prepare for winter by accumulating glucose and glycerol in body fluids, which protects tissues from the normally damaging effects of ice crystal formation.

Adult frogs have numerous enemies, such as snakes, aquatic birds, turtles, raccoons, and humans; fish prey upon tadpoles and only a few survive to maturity. Although usually defenseless, in the tropics and subtropics many frogs and toads are aggressive, jumping and biting at predators. Some defend themselves by feigning death. Most anurans can inflate their lungs so that they are difficult to swallow. When disturbed along the margin of a pond or brook, a frog often remains quite still. When it thinks it is detected it jumps, not always into the water where enemies may be lurking but into grassy cover on the bank. When held in the hand, a frog may cease its struggles for an instant to put its captor off guard and then leap violently, at the same time voiding its urine. Their best protection is their ability to leap and their use of poison glands. Bullfrogs in captivity do not hesitate to snap at tormenters and are capable of inflicting painful bites.

While native American amphibians continue to disappear as wetlands are drained, an exotic frog introduced into southern California has found the climate quite to its liking. The African clawed frog Xenopus laevis *(Figure 16-14) is a voracious, aggressive, primarily aquatic frog that is rapidly displacing native frogs and fish from several waterways and is spreading rapidly. The species was introduced into North America in the 1940s when it was used extensively in human pregnancy tests. When more efficient tests appeared in the 1960s, some hospitals simply dumped surplus frogs into nearby streams, where the prolific breeders have become almost indestructible pests. As is so often the case with alien wildlife introductions, benign intentions frequently lead to serious problems.*

Reproduction

Frogs and toads living in temperate regions of the world breed, feed, and grow only during the warmer seasons of the year, usually in a predictable annual cycle. With warming spring temperatures, in combination with sufficient rainfall, males croak and call vociferously to attract females. After a brief courtship, females enter the water and are clasped by the males in a process called **amplexus** (Figure 16-15). As the female lays the eggs, the male discharges seminal fluid containing sperm over the eggs to fertilize them. After fertilization, the jelly layers absorb water and swell (Figure 16-16). The eggs are laid in large masses, often anchored to vegetation, then abandoned by the parents. The eggs begin development immediately. Within a few days the embryos have developed into tadpoles (Figure 16-16) and hatch, often to face a precarious existence if the eggs were laid in a temporary pond or puddle. In such instances, it becomes a race against time to complete development before the habitat dries up.

At the time of hatching, the tadpole has a distinct head and body with a compressed tail. The mouth is located on the ventral side of the head and is provided with horny jaws for scraping vegetation from objects for food. Behind the mouth is a ventral adhesive disc for clinging to objects. In front of the mouth are two deep pits, which later develop into the nostrils. Swellings are found on each side of the head, and these later become external gills. There are three pairs of external gills, which are transformed into internal gills that become covered with a flap of skin (the operculum) on each side. On the right side the operculum completely fuses with the body wall, but on the left side a small opening, the **spiracle** (L. *spiraculum*, airhole) remains, through which water flows after entering the mouth and passing the internal gills. The hindlegs appear first, whereas the forelimbs are hidden for a time by the folds of the operculum. During metamorphosis the tail is absorbed, the intestine becomes much shorter, the mouth undergoes a transformation into the adult condition, lungs develop, and the gills

FIGURE 16-14

African clawed frog, *Xenopus laevis*. The claws, an unusual feature in frogs, are on the hind feet. This frog has been introduced into California, where it is considered a serious pest.

FIGURE 16-15

A male green frog, *Hyla cinerea*, clasps a larger female during the breeding season in a South Carolina swamp. Clasping (amplexus) is maintained until the female deposits her eggs. Like most tree frogs, these are capable of rapid and marked color changes; the male here, normally green, has darkened during amplexus.

are absorbed. The leopard frog usually completes its metamorphosis within three months; the bullfrog takes much longer to complete the process.

The reproductive mode just described, while typical of most temperate zone anurans, is only one of a great variety of reproductive patterns in tropical anurans. Some of these remarkable strategies are illustrated in Figure 16-4 (p. 317). Some species lay their eggs in foam masses that float on the surface of the water; some deposit their eggs on leaves overhanging ponds and streams into which the emerging tadpoles will drop; some lay their eggs in damp burrows; and others place their eggs in water trapped in tree cavities or in water-filled chambers of some bromeliads (epiphytic plants in the tropical forest canopy). While most frogs abandon their eggs, some, such as the tropical dendrobatids (a family that includes the poison-dart frogs), tend the eggs. When the tadpoles hatch, they squirm up on the parent's back to be carried for varying lengths of time (Figure 16-4C). The marsupial frogs carry the developing eggs in a pouch on the back (Figure 16-4A).

Although most frogs develop through a larval stage (the tadpole), many tropical frogs have evolved direct development. In direct development the tadpole stage is bypassed and the froglet that emerges is a miniture replica of the adult.

Classification of Class Amphibia

Order Gymnophiona (jim′no-fy′o-na) (Gr. *gymnos*, naked, + *ophioneos*, of a snake) **(Apoda): caecilians.** Body elongate; limbs and limb girdle absent; mesodermal scales present in skin of some; tail short or absent; 95 to 285 vertebrae; pantropical, 6 families, 34 genera, approximately 160 species.

Order Caudata (caw-dot′uh) (L. *caudatus*, having a tail) **(Urodela): salamanders.** Body with head, trunk, and tail; no scales; usually two pairs of equal limbs; 10 to 60 vertebrae; predominantly holarctic; 9 living families, 62 genera, approximately 360 species.

Order Anura (uh-nur′uh) (Gr. *an*, without, + *oura*, tail) **(Salientia): frogs, toads.** Head and trunk fused; no tail; no scales; two pairs of limbs; large mouth; lungs; 6 to 10 vertebrae including urostyle (coccyx); cosmopolitan, predominantly tropical; 21 living families; 301 genera; approximately 3450 species.

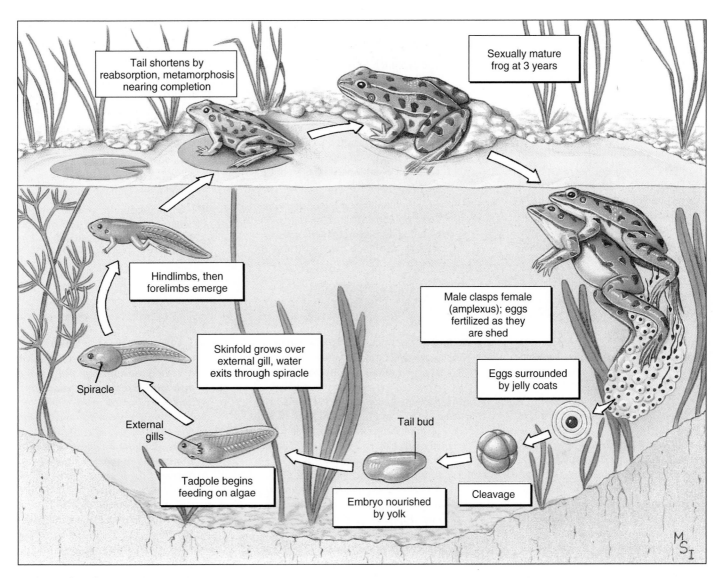

FIGURE 16-16

Life cycle of a leopard frog.

Summary

Amphibians are ectothermic, primitively quadrupedal vertebrates that have glandular skin and that breathe by lungs, gills, or skin. They are the survivors of one of two major branches of tetrapod phylogeny, the other one being represented today by the amniotes. The modern amphibians consist of three major evolutionary groups. The caecilians (order Gymnophiona) are a small tropical group of limbless, elongate forms. The salamanders (order Caudata) are tailed amphibians that have retained the generalized four-legged body plan of their Paleozoic ancestors. The frogs and toads (order Anura) are the largest group of modern amphibians, all of which are specialized for a jumping mode of locomotion.

Most amphibians have a biphasic life cycle that begins with an aquatic larva that later metamorphoses to produce a terrestrial adult that returns to the water to lay eggs. Some frogs, salamanders, and caecilians have evolved direct development that omits the aquatic larval stage and some caecilians have evolved viviparity. Salamanders are unique among amphibians in having evolved several permanently gilled species that retain a larval morphology throughout life, eliminating the terrestrial phase completely. The permanently gilled condition is obligate in some species, but others will metamorphose to a terrestrial form if the pond habitat dries up.

Despite their adaptations for terrestrial life, the adults and eggs of all amphibians require cool, moist environments if not actual pools or streams. The eggs and adult skin have no effective protection against very cold, hot, or dry conditions, greatly restricting the adaptive radiation of amphibians to environments that have moderate temperatures and abundant water.

Review Questions

1. Compared with the aquatic habitat, the terrestrial habitat offers both advantages and problems for an animal making the transition from water to land. Summarize how these differences might have influenced the early evolution of tetrapods.

2. Describe the different modes of respiration used by amphibians. What paradox do the amphiumas and terrestrial plethodontids present regarding the association of lungs with life on land?

3. The evolution of the tetrapod leg was one of the most important advances in vertebrate history. Describe the supposed sequence in its evolution.

4. Compare the general life history patterns of salamanders with those of frogs. Which group shows the greater variety of evolutionary changes of the ancestral biphasic amphibian life cycle?

5. Give the literal meaning of the name Gymnophiona. What animals are included in this amphibian order, what do they look like, and where do they live?

6. What is the literal meaning of the order names Caudata and Anura? What major features distinguish the members of these two orders from each other?

7. Describe the breeding behavior of a typical woodland salamander.

8. How has paedomorphosis been important to the evolution of permanently aquatic salamanders?

9. Briefly describe the reproductive behavior of frogs. In what important ways do frogs and salamanders differ in their reproduction?

Selected References

See also general references on page 395.

Blaustein, A. R., and D. B. Wake. 1995. The puzzle of declining amphibian populations. Sci. Am. **272:**52–57 (Apr.). *Amphibian populations are dwindling in many areas of the world. The causes are multiple but all derive from human activities.*

Conant, R., and J. T. Collins. 1991. A field guide to reptiles and amphibians, ed. 3. The Peterson field guide series. Boston, Houghton Mifflin Company. *Updated version of a popular field guide; color illustrations and distribution maps for all species.*

del Pino, E. M. 1989. Marsupial frogs. Sci. Am. **260:**110–118 (May). *Several species of tropical frogs incubate their eggs on the female's back, often in a special pouch, and emerge as advanced tadpoles or fully formed froglets.*

Duellman, W. E. 1992. Reproductive strategies of frogs. Sci. Am. **267:**80–87 (July). *Many frogs have evolved improbable reproductive strategies that have permitted colonization of land.*

Duellman, W. E., and L. R. Trueb. 1994. Biology of amphibians. Baltimore, Johns Hopkins University Press. *Important comprehensive sourcebook of information on amphibians, extensively referenced and illustrated.*

Halliday, T. R., and K. Adler, eds. 1986. The encyclopedia of reptiles and amphibians. New York, Facts on File, Inc. *Excellent authoritative reference work with high-quality illustrations.*

Hanken, J. 1989. Development and evolution in amphibians. Am. Sci. **77:**336–343 (July–Aug.). *Explains how the diversity in amphibian morphology has been achieved by modifications in development.*

Lewis, S. 1989. Cane toads: an unnatural history. New York, Dolphin/Doubleday. *Based on an amusing and informative film of the same title that describes the introduction of cane toads to Queensland, Australia and the unexpected consequences of their population explosion there. "If Monty Python teamed up with National Geographic, the result would be* Cane Toads.*"*

Moffett, M. W. 1995. Poison-dart frogs: lurid and lethal. National Geographic **187**(5):98–111 (May). *Photographic essay of frogs that can be lethal even to the touch.*

Narins, P. M. 1995. Frog communication. Sci. Am. **273:**78–83 (Aug.). *Frogs employ several strategies to hear and be heard amidst the cacophony of chorusing of many frogs.*

Links to the Internet

Visit this textbook's Web site at http://www.mhhe.com/zoology to find live Internet links for each of the references listed below.

1. Introduction to the Amphibia. This University of California at Berkeley Museum of Paleontology site provides links to information on the fossil record, life history, and systematics of the amphibians.

2. Class Lissamphibia (Amphibia). University of Michigan site on amphibians. Pictures, much information on the morphology, distribution, and ecology of a large number of amphibians. Each species is linked to Web pages. Images and/or audio may not be available for display depending on your server.

3. Savannah River Ecology Laboratory's Herpetology Lab Home Page. The University of Georgia's SREL has been involved in research on reptiles and amphibians since the 1960s. This site has pictures, research summaries, and links to more information on "herps".

4. BIOSIS-Amphibia. Find general information on amphibians as well as specific information on each of the orders of amphibians.

5. North American Amphibian Monitoring Program. Information and links to information on the study and conservation efforts for amphibians. Supported by the USGS and the Patuxent Wildlife Research Center.

Reptiles

CHAPTER | seventeen

Enclosing the Pond

Amphibians, with well-developed legs, redesigned sensory and respiratory systems, and modifications of the post-cranial skeleton for supporting the body in air, have made a notable conquest of land. But, with shell-less eggs and often gill-breathing larvae, their development remains hazardously tied to water. The lineage containing reptiles, birds, and mammals developed an egg that could be laid on land. This shelled egg, perhaps more than any other adaptation, unshackled the early reptiles from the aquatic environment by freeing the developmental process from dependence upon aquatic or very moist terrestrial environments. In fact, the "pond-dwelling" stages were not eliminated but enclosed within a series of extraembryonic membranes that provided complete support for embryonic development. One membrane, the amnion, encloses a fluid-filled cavity, the "pond," within which the developing embryo floats. Another membranous sac, the allantois, serves both as a respiratory surface and as a chamber for the storage of nitrogenous wastes. Enclosing these membranes is a third membrane, the chorion, through which oxygen and carbon dioxide freely pass. Finally, surrounding and protecting everything is a porous, parchmentlike or leathery shell.

With the last ties to aquatic reproduction severed, conquest of land by the vertebrates was ensured. The Paleozoic tetrapods that developed this reproductive pattern were ancestors of a single, monophyletic assemblage called the Amniota, named after the innermost of the three extraembryonic membranes, the amnion. Before the end of the Paleozoic era the amniotes had diverged into multiple lineages that gave rise to all the reptilian groups, the birds, and the mammals.

Members of the paraphyletic class Reptilia (rep-til′e-a) (L. *repto,* to creep) include the first truly terrestrial vertebrates. With nearly 7000 species (approximately 300 species in the United States and Canada) occupying a great variety of aquatic and terrestrial habitats, they are diverse and abundant. Nevertheless, reptiles are perhaps remembered best for what they once were, rather than for what they are now. The Age of Reptiles, which lasted for more than 165 million years, saw the appearance of a great radiation of reptilian lineages into a bewildering array of terrestrial and aquatic forms. Among these were the herbivorous and carnivorous dinosaurs, many of huge stature and awesome appearance, that dominated animal life on land—the ruling reptiles. Then, during a mass extinction at the end of Mesozoic era, they suddenly declined. Among the few reptilian lineages to emerge from the Mesozoic extinction are today's reptiles. One of these, the tuatara *(Sphenodon)* of New Zealand, is the sole survivor of a group that otherwise disappeared 100 million years ago. But others, especially the lizards and snakes, have radiated since the Mesozoic extinction into diverse and abundant groups (Figure 17-1). Understanding the 300-million-year-old history of reptilian life on earth has been complicated by widespread convergent and parallel evolution among the many lineages and by large gaps in the fossil record.

Origin and Adaptive Radiation of Reptiles

As mentioned in the prologue to this chapter, the amniotes are a monophyletic group that evolved in the late Paleozoic. Most paleontologists agree that the amniotes arose from a group of amphibian-like tetrapods, the anthracosaurs, during the early Carboniferous period of the Paleozoic. By the late Carboniferous (approximately 300 million years ago), the amniotes had separated into three lineages. The first lineage, the **anapsids** (Gr. *an,* without, + *apsis,* arch), was characterized by a skull having no temporal opening behind the orbits, that is, the skull behind the orbits was completely roofed with dermal bone (Figure 17-2). This group is represented today only by the turtles. Their morphology is an odd mix of ancestral and derived characters that has scarcely changed at all since the turtles first appeared in the fossil record in the Triassic some 200 million years ago.

The second lineage, the **diapsids** (Gr. *di,* double, + *apsis,* arch), gave rise to all other reptilian groups and to the birds (Figure 17-1). The diapsid skull was characterized by the presence of two temporal openings: one pair located low on the cheeks, and a second pair positioned above the lower pair and separated from them by a bony arch (Figure 17-2). Three subgroups of diapsids appeared. The **lepidosaurs** include the extinct marine ichthyosaurs and all of the modern reptiles with the exception of the turtles and crocodilians. The more derived **archosaurs** comprised the dinosaurs and their relatives, and the living crocodilians and birds. A third, smaller lineage of diapsids, the **sauropterygians** included several extinct aquatic groups, the most conspicuous of which were the large, long-necked plesiosaurs.

The third lineage was the **synapsids** (Gr. *syn,* together, + *apsis,* arch), the mammal-like reptiles. The synapsid skull had a single pair of temporal openings located low on the cheeks and bordered by a bony arch (Figure 17-2). The synapsids were the first group of amniotes to diversify, giving rise first to the pelycosaurs, later to the therapsids, and finally to mammals (Figure 17-1).

Changes in Traditional Classification of Reptiles

With increasing use of cladistic methodology in zoology, and its insistence on hierarchical arrangement of monophyletic groups (see p. 57 and following), important changes have been made in the traditional classification of reptiles. The class Reptilia is no longer recognized by cladists as a valid taxon because it is not monophyletic. As customarily defined, the class Reptilia excludes the birds which descend from the most recent common ancestor of the reptiles. Consequently, the reptiles are a paraphyletic group because they do not include all descendants of their most recent common ancestor. Reptiles can be identified only as amniotes that are not birds or mammals. This is clearly shown in the phylogenetic tree of the amniotes (Figure 17-1).

An example of this problem is the shared ancestry of birds and crocodilians. Based solely on shared derived characteristics, crocodilians and birds are sister groups; that is they are more recently descended from a common ancestor than either is from any other living reptilian lineage. In other words, birds and crocodilians belong to a monophyletic group apart from other reptiles and, according to the rules of cladism, should be assigned to a clade that separates them from the remaining reptiles. This clade is in fact recognized; it is the Archosauria (Figures 17-1 and 17-2), a grouping that also includes the extinct dinosaurs. Therefore birds should be classified as reptiles. The archosaurs plus their sister group, the lepidosaurs (tuataras, lizards, and snakes) comprise a monophyletic group that some taxonomists call the Reptilia. The term Reptilia is thereby redefined to include birds in contrast to its traditional usage. However, evolutionary taxonomists argue that birds represent a novel adaptive zone and grade of organization whereas crocodilians remain within the traditionally recognized reptilian adaptive zone and grade. In this view, the morphological and ecological novelty of birds has been recognized by maintaining the traditional classification that places crocodilians in the class Reptilia and birds in the class Aves. Such conflicts of opinion between proponents of the two major competing schools of taxonomy (cladistics and evolutionary taxonomy) have had the healthy effect of forcing zoologists to reevaluate their views of amniote genealogy and how vertebrate classifications should represent genealogy and degree of divergence. In our treatment we retain the traditional class Reptilia because this is still standard taxonomic practice, but we emphasize that this taxonomy is likely to be discontinued.

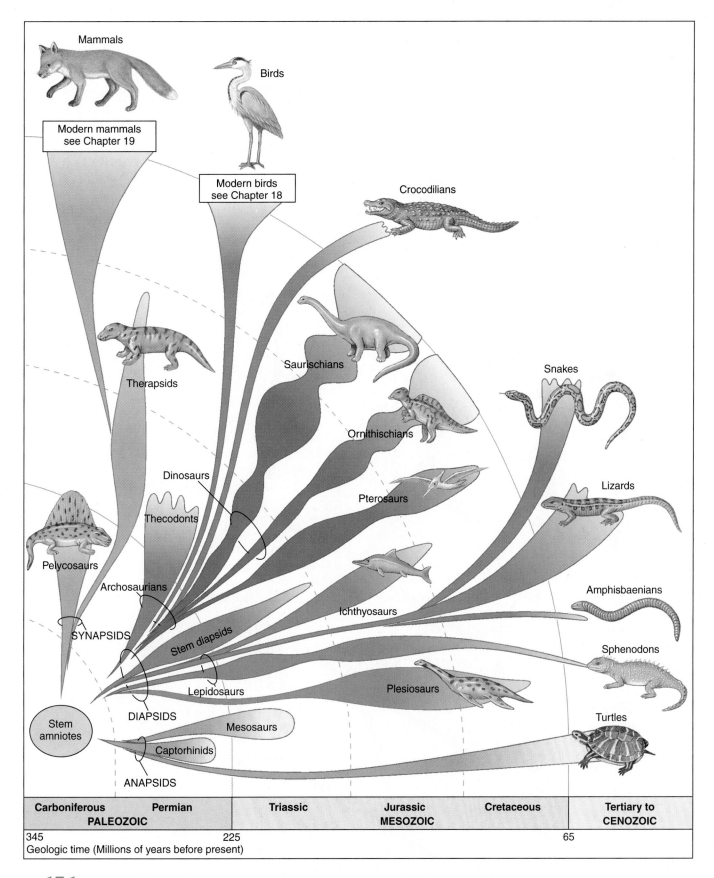

FIGURE 17-1

Evolution of the amniotes. The evolutionary origin of amniotes occurred by the evolution of an amniotic egg that made reproduction on land possible, although this egg may well have developed before the earliest amniotes had ventured far on land. The amniote assemblage, which includes the reptiles, birds, and mammals, evolved from a lineage of small, lizardlike forms that retained the skull pattern of the early tetrapods. First to diverge from the primitive stock were the mammal-like reptiles, characterized by a skull pattern termed the synapsid condition. All other amniotes, including the birds and all living reptiles except the turtles, have a skull pattern known as diapsid. The turtles have a skull pattern known as anapsid. The great Mesozoic radiation of reptiles may have resulted partly by the increased variety of ecological habitats into which the amniotes could move.

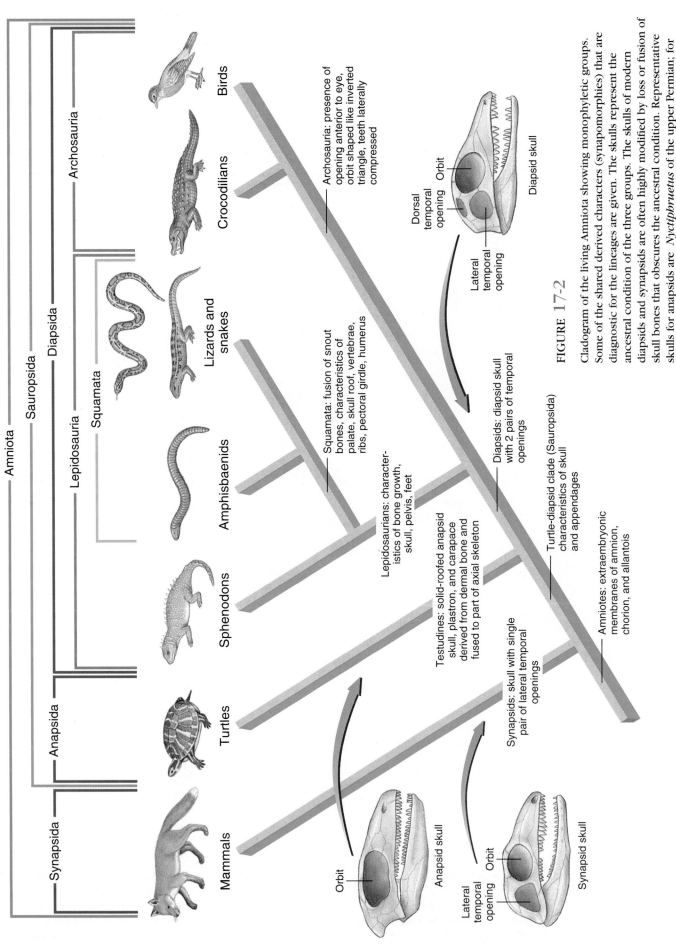

FIGURE 17-2

Cladogram of the living Amniota showing monophyletic groups. Some of the shared derived characters (synapomorphies) that are diagnostic for the lineages are given. The skulls represent the ancestral condition of the three groups. The skulls of modern diapsids and synapsids are often highly modified by loss or fusion of skull bones that obscures the ancestral condition. Representative skulls for anapsids are *Nyctiphruetus* of the upper Permian; for diapsids, *Youngina* of the upper Permian; for synapsids, *Aerosaurus*, a pelycosaur of the lower Permian. The relationships expressed in this cladogram are tentative and controversial, especially that between birds and mammals. Contrary to the view shown here, in which mammals are the outgroup, some authorities support a sister-group relationship between birds and mammals based on several kinds of molecular and physiological evidences.

(Source: F.H. Pough, J.B. Heiser, and W.N. McFarland, 1996, Vertebrate life, ed. 4. Upper Saddle River, NJ, Prentice Hall.)

Labels within the figure:

- Amniota
- Synapsida
- Sauropsida
- Anapsida
- Diapsida
- Archosauria
- Lepidosauria
- Squamata
- Mammals
- Turtles
- Sphenodons
- Amphisbaenids
- Lizards and snakes
- Crocodilians
- Birds

- Archosauria: presence of opening anterior to eye, orbit shaped like inverted triangle, teeth laterally compressed
- Squamata: fusion of snout bones, characteristics of palate, skull roof, vertebrae, ribs, pectoral girdle, humerus
- Lepidosaurians: characteristics of bone growth, skull, pelvis, feet
- Diapsids: diapsid skull with 2 pairs of temporal openings
- Testudines: solid-roofed anapsid skull, plastron, and carapace derived from dermal bone and fused to part of axial skeleton
- Turtle-diapsid clade (Sauropsida) characteristics of skull and appendages
- Synapsids: skull with single pair of lateral temporal openings
- Amniotes: extraembryonic membranes of amnion, chorion, and allantois

Skull labels:
- Dorsal temporal opening
- Orbit
- Lateral temporal opening
- Diapsid skull
- Orbit
- Anapsid skull
- Lateral temporal opening
- Orbit
- Synapsid skull

Characteristics of Class Reptilia

1. Body varied in shape, compact in some, elongated in others; **body covered with an exoskeleton of horny epidermal scales** with the addition sometimes of bony dermal plates; **integument with few glands**

2. **Limbs paired, usually with five toes,** and adapted for climbing, running, or paddling; absent in snakes and some lizards

3. Skeleton well ossified; ribs with sternum (sternum absent in snakes) forming a complete thoracic basket; **skull with one occipital condyle**

4. Respiration by lungs; **no gills;** cloaca used for respiration by some; branchial arches in embryonic life

5. Three-chambered heart; **crocodilians with four-chambered heart;** usually one pair of aortic arches; systemic and pulmonary circuits functionally separated

6. Ectothermic; many thermoregulate behaviorally

7. **Metanephric kidney (paired); uric acid main nitrogenous waste**

8. Nervous system with the optic lobes on the dorsal side of brain; **12 pairs of cranial nerves** in addition to nervus terminalis

9. Sexes separate; **fertilization internal**

10. **Eggs covered with calcareous or leathery shells; extraembryonic membranes (amnion, chorion, and allantois)** present during embryonic life; **no larval stages**

Characteristics of Reptiles That Distinguish Them from Amphibians

1. **Reptiles have tough, dry, scaly skin offering protection against desiccation and physical injury.** The skin consists of a thin epidermis, shed periodically, and a much thicker, well-developed **dermis** (Figure 17-3). The dermis is provided with **chromatophores,** the color-bearing cells that give many lizards and snakes their colorful hues. It is also the layer that, unfortunately for their bearers, is converted into alligator and snakeskin leather, so esteemed for expensive pocketbooks and shoes. The characteristic **scales** of reptiles are formed largely of keratin. Scales are derived mostly from the epidermis; they are not homologous to fish scales, which are bony, dermal structures. In some reptiles, such as alligators, the scales remain throughout life, growing gradually to replace wear. In others, such as snakes and lizards, new scales grow beneath the old, which are shed at intervals. Turtles add new layers of keratin under the old layers of the platelike scutes, which are modified scales. In snakes the old skin (epidermis and scales) is turned inside out when discarded; lizards split out of the old skin leaving it mostly intact and right side out, or it may slough off in pieces.

2. **The shelled (amniotic) egg of reptiles contains food and protective membranes for supporting embryonic development on land.** Reptiles lay their eggs in sheltered locations on land. The young hatch as lung-breathing juveniles rather than as aquatic larvae. The appearance of the shelled egg (Figure 17-4) widened

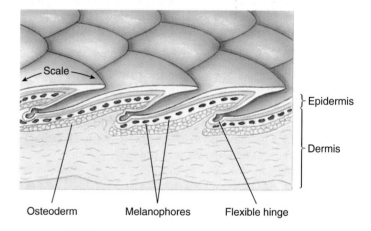

FIGURE 17-3

Section of the skin of a reptile showing the overlapping epidermal scales.

the division between the evolving amphibians and reptiles and, probably more than any other adaptation, contributed to the evolutionary establishment of reptiles.

3. **Reptilian jaws are efficiently designed for applying crushing or gripping force to prey.** The jaws of fish and amphibians are designed for quick jaw closure, but once the prey is seized, little static force can be applied. In reptiles jaw muscles became larger, longer, and arranged for much better mechanical advantage.

4. **Reptiles have some form of copulatory organ, permitting internal fertilization.** Internal fertilization is obviously a requirement for a shelled egg, because the sperm must reach the egg before the

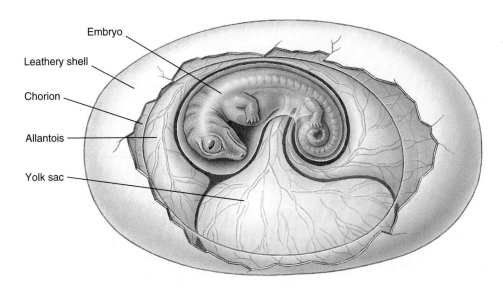

Embryo

Leathery shell

Chorion

Allantois

Yolk sac

FIGURE 17-4

Amniotic egg. The embryo develops within the amnion and is cushioned by amniotic fluid. Food is provided by yolk from the yolk sac and metabolic wastes are deposited within the allantois. As development proceeds, the allantois fuses with the chorion, a membrane lying against the inner surface of the shell; both membranes are supplied with blood vessels that assist in the exchange of oxygen and carbon dioxide across the porous shell. Because this kind of egg is an enclosed, self-contained system, it is often called a "cleidoic" egg (Gr. *kleidoun,* to lock in).

egg is enclosed. Sperm from the paired testes are carried by the vasa deferentia to the copulatory organ, which is an evagination of the cloacal wall. The female system consists of paired ovaries and oviducts. The glandular walls of the oviducts secrete albumin (source of amino acids, minerals, and water for the embryo) and shells for the large eggs.

5. **Reptiles have a more efficient circulatory system and higher blood pressure than amphibians.** In all reptiles the right atrium, which receives unoxygenated blood from the body, is completely partitioned from the left atrium, which receives oxygenated blood from the lungs. Crocodilians have two completely separated ventricles as well (Figure 17-5); in other reptiles the ventricle is incompletely separated. Even in reptiles with incomplete separation of the ventricles, flow patterns within the heart prevent admixture of pulmonary (oxygenated) and systemic (unoxygenated) blood; all reptiles therefore have two functionally separate circulations.

6. **Reptilian lungs are better developed than those of amphibians.** Reptiles depend almost exclusively on lungs for gas exchange, supplemented by respiration through the pharyngeal membranes in some aquatic turtles. Unlike amphibians, which *force* air into the lungs with mouth muscles, reptiles *suck* air into the lungs by enlarging the pleural cavity, either by expanding the rib cage (snakes and lizards) or by movement of internal organs (turtles and crocodilians). Reptiles have no muscular diaphragm, a structure found only in mammals. Cutaneous respiration (gas exchange

across the skin), so important to amphibians, has been completely abandoned by reptiles.

7. **Reptiles have evolved efficient strategies for water conservation.** All amniotes have a metanephric kidney which is drained by its own passageway, the ureter. However, the nephrons of the reptilian metanephros lack the specialized intermediate section of the tubule, the loop of Henle, that enables the kidney to concentrate solutes in the urine. To remove salts from the blood, many reptiles have salt glands located near the nose or eyes (in the tongue of saltwater crocodiles) which secrete a salty fluid that is strongly hyperosmotic to the body fluids. Nitrogenous wastes are excreted by the kidney as uric acid, rather than urea or ammonia. Uric acid has a low solubility and precipitates out of solution readily, allowing water to be conserved; the urine of many reptiles is a semisolid suspension.

8. **All reptiles, except the limbless members, have better body support than the amphibians and more efficiently designed limbs for travel on land.** Nevertheless, most modern reptiles walk with their legs splayed outward and their belly close to the ground. Most dinosaurs, however, (and some modern lizards) walked on upright legs held beneath the body, the best arrangement for rapid movement and for the support of body weight. Many dinosaurs walked on powerful hindlimbs alone.

9. **The reptilian nervous system is considerably more complex than the amphibian system.** Although the reptile's brain is small, the cerebrum is

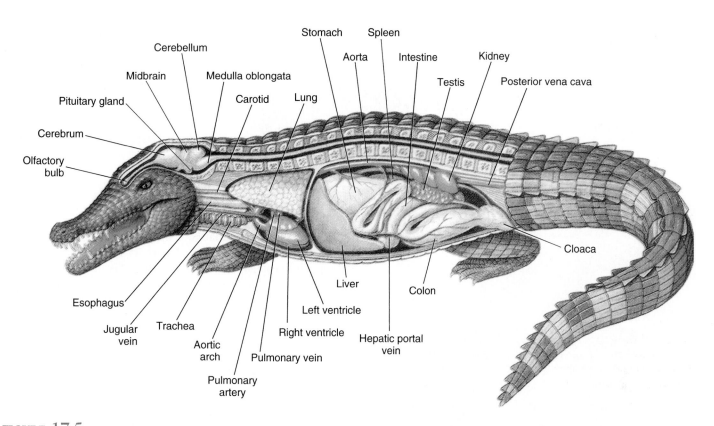

FIGURE 17-5

Internal structure of a male crocodile.

larger relative to the rest of the brain. Connections to the central nervous system are more advanced, permitting complex kinds of behavior unknown in amphibians. With the exception of hearing, sense organs in general are well developed. Jacobson's organ, a specialized olfactory chamber present in many tetrapods, is highly developed in lizards and snakes. Odors are carried to Jacobson's organ by the tongue.

Characteristics and Natural History of Reptilian Orders

Anapsid Reptiles: Subclass Anapsida

Turtles: Order Testudines

Turtles descended from one of the earliest anapsid lineages, probably a group known as the procolophonids of the late Permian, but turtles themselves do not appear in the fossil record until the Upper Triassic, some 200 million years ago. From the Triassic, turtles plodded on to the present with very little change in their early morphology. They are enclosed in shells consisting of a dorsal **carapace** (Fr., from Sp. *carapacho*, covering) and ventral **plastron** (Fr., breastplate). Clumsy and

unlikely as they appear to be within their protective shells, they are nonetheless a varied and ecologically diverse group that seems able to adjust to human presence. The shell is so much a part of the animal that it is fused to thoracic vertebrae and ribs (Figure 17-6). Like a medieval coat of armor, the shell offers protection for the head and appendages, which, in most turtles, can be retracted into it. But because the ribs are fused to the shell, the turtle cannot expand its chest to breathe. Instead, turtles employ certain abdominal and pectoral muscles as a "diaphragm." Air is drawn inward by contracting limb flank muscles to make the body cavity larger. Exhalation is also active: the shoulder girdle is drawn back into the shell, thus compressing the viscera and forcing air out of the lungs.

> The terms "turtle," "tortoise," and "terrapin" are applied variously to different members of the turtle order. In North American usage, they are all correctly called turtles. The term "tortoise" is frequently given to land turtles, especially the large forms. British usage of the terms is different: "tortoise" is the inclusive term, whereas "turtle" is applied only to the aquatic members.

Lacking teeth, the turtle jaw is provided with tough, horny plates for gripping food (Figure 17-7). Sound perception

FIGURE **17-6**

Skeleton and shell of a turtle, showing fusion of vertebrae and ribs with the carapace. The long and flexible neck allows the turtle to withdraw its head into its shell for protection.

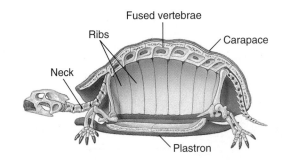

Fused vertebrae

Ribs

Carapace

Neck

Plastron

FIGURE **17-7**

Snapping turtle, *Chelydra serpentina,* showing the absence of teeth. Instead, the jaw edges are covered with a horny plate.

is poor in turtles, and most turtles are mute (the biblical "voice of the turtle" refers to the turtledove, a bird). Compensating for poor hearing is a good sense of smell and color vision. Turtles are oviparous, and fertilization is internal. All turtles, even the marine forms, bury their shelled, amniotic eggs in the ground. An odd feature of turtle reproduction is that in some turtle families, as in all crocodilians and some lizards, the nest temperature determines the sex of the hatchlings. In turtles, low temperatures during incubation produce males and high temperatures produce females.

The great marine turtles, buoyed by their aquatic environment, may reach 2 m in length and 725 kg in weight. One is the leatherback. The green turtle, so named because of its greenish body fat, may exceed 360 kg, although most individuals of this economically valuable and heavily exploited species seldom live long enough to reach anything approaching this size. Some land tortoises may weigh several hundred kilograms, such as the giant tortoises of the Galápagos Islands (Figure 17-8) that so intrigued Darwin during his visit there in 1835. Most tortoises are rather slow moving; one hour of determined trudging carries a large Galápagos tortoise approximately 300 m. A low metabolism probably explains in part the longevity of turtles, for some are believed to live more than 150 years.

Diapsid Reptiles: Subclass Diapsida

The diapsid reptiles, that is, reptiles having a skull with two pairs of temporal openings (Figure 17-2), are classified into three lineages (superorders; see the Classification of Living Reptiles on p. 341). The two with living representatives are the superorder Lepidosauria, containing the lizards, snakes, worm lizards, and *Sphenodon;* and the superorder Archosauria, containing the crocodilians.

Lizards, Snakes, and Worm Lizards: Order Squamata

The squamates are the most recent and diverse products of diapsid evolution, making up approximately 95% of all known living reptiles. Lizards appeared in the fossil record as early as the Permian, but they did not begin their radiation until the Cretaceous period of the Mesozoic when the dinosaurs were at the climax of their radiation. Snakes appeared during the late Cretaceous period, probably from a group of lizards whose descendants include the Gila monster and monitor lizards. Two specializations in particular characterize snakes: extreme elongation of the body and accompanying displacement and rearrangement of internal organs; and specializations for eating large prey.

The diapsid skulls of the squamates are modified from the ancestral diapsid condition by the loss of dermal bone ventral and posterior to the lower temporal opening. This modification has allowed the evolution in most lizards of a **kinetic skull** having movable joints (Figure 17-9). The quadrate, which in other reptiles is fused to the skull, has a joint at its dorsal end, as well as its usual articulation with the lower jaw. In addition, there are joints in the palate and across the roof of the skull that allow the snout to be tilted upward. The specialized mobility of the skull enables lizards to seize and manipulate their prey. It also increases the effective closing force of the jaw musculature. The skull of snakes is even more kinetic than that of lizards. Such exceptional skull mobility is considered a major factor in the diversification of lizards and snakes.

Lizards: Suborder Sauria Lizards are an extremely diverse group, including terrestrial, burrowing, aquatic, arboreal and

FIGURE **17-8**

Mating Galápagos tortoises. The male has a concave plastron that fits over the highly convex carapace of the female, helping to provide stability during mating. Males utter a roaring sound during mating, the only time they are known to emit vocalizations.

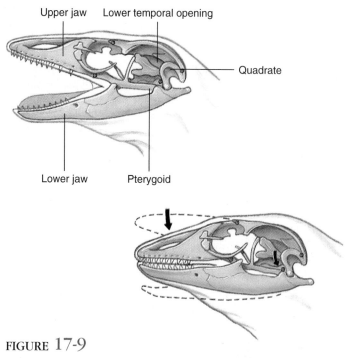

FIGURE **17-9**

Kinetic diapsid skull of a modern lizard (monitor lizard, *Varanus* sp.) showing the joints (indicated by dots) that allow the snout and upper jaw to move on the rest of the skull. The quadrate can move at its dorsal end and ventrally at both the lower jaw and the pterygoid. The front part of the braincase is also flexible, allowing the snout to be raised. Note that the lower temporal opening is very large with no lower border; this modification of the diapsid condition, common in modern lizards, provides space for expansion of large jaw muscles. The upper temporal opening lies dorsal and medial to the postorbital-squamosal arch and is not visible in this drawing.

aerial members. Among the more familiar groups in this varied suborder are the **geckos** (Figure 17-10), small, agile, mostly nocturnal forms with adhesive toe pads that enable them to walk upside down and on vertical surfaces; the **iguanas,** often brightly colored New World lizards with ornamental crests, frills, and throat fans, and a group that includes the remarkable marine iguana of the Galápagos Islands (Figure 17-11); **skinks,** with elongate bodies and reduced limbs; and **chameleons,** a group of arboreal lizards, mostly of Africa and Madagascar. The chameleons are entertaining creatures that catch insects with a sticky-tipped tongue that can be flicked accurately and rapidly to a distance greater than the length of their body (Figure 17-12). The great majority of lizards have four limbs and relatively short bodies, but in many the limbs are degenerate, and a few such as the glass lizards (Figure 17-13) are completely limbless.

Unlike turtles, snakes, and crocodilians, which have distinctive body forms and ways of life, lizards have radiated extensively into a variety of habitats and reveal an array of functional and behavioral specializations. Most lizards have movable eyelids, whereas a snake's eyes are permanently covered with a transparent cap. Lizards have keen vision for daylight (retinas rich in both cones and rods), although one group, the nocturnal geckos, has pure rod retinas for night vision. Most lizards have an external ear that snakes lack. However, as with other reptiles, hearing does not play an important role in the lives of most lizards. Geckos are exceptions because the males are strongly vocal (to announce terri-

tory and discourage the approach of other males), and they must, of course, hear their own vocalizations.

Many lizards have successfully invaded the world's hot and arid regions, aided by characteristics that make desert life possible. Because their skin lacks glands, water loss by this avenue is much reduced. They produce a semisolid urine with a high content of crystalline uric acid, a feature well suited for conserving water also found in other groups that live successfully in arid habitats (birds, insects, and pulmonate snails). Some, such as the Gila monster of the southwestern United States deserts, store fat in their tails, which they draw on during drought to provide both energy and metabolic water (Figure 17-14). Many lizards keep their body temperature relatively constant by behavioral thermoregulation.

Worm Lizards: Suborder Amphisbaenia The somewhat inappropriate common name "worm lizards" describes a group of highly specialized, burrowing forms that are neither worms nor true lizards but certainly are related to the latter. They

The Mesozoic World of Dinosaurs

When, in 1841, the English anatomist Richard Owen coined the term *dinosaur* ("terrible lizard") to describe fossil Mesozoic reptiles of gigantic size, only three poorly known dinosaur genera were distinguished. But with new and marvelous fossil discoveries quickly following, by 1887 zoologists were able to distinguish two groups of dinosaurs based on differences in the structure of the pelvic girdles. The Saurischia ("lizard-hipped") had a simple, three-pronged pelvis with the hip bones arranged much as they are in other reptiles. The large bladelike ilium is attached to the backbone by stout ribs. The pubis and ischium extend ventrally and posteriorly, respectively, and all three bones meet at the hip socket, a deep opening on the side of the pelvis. The Ornithischia ("bird-hipped") had a somewhat more complex pelvis. The ilium and ischium were arranged similarly in ornithischians and saurischians, but the ornithischian pubis was a narrow, rod-shaped bone with anteriorly and posteriorly directed processes lying alongside the ischium. Oddly, while the ornithischian pelvis, as the name suggests, was similar to that of birds, birds are of the saurischian lineage.

Dinosaurs and their living relatives, the birds, are archosaurs ("ruling lizards"), a group that includes thecodonts (early archosaurs restricted to the Triassic), crocodiles, and pterosaurs (refer to the classification of the reptiles on p. 341). As traditionally recognized, the dinosaurs are a paraphyletic group because they do not include birds which are descended from the most recent common ancestor of dinosaurs.

From among the various archosaurian radiations of the Triassic there emerged a thecodont lineage with limbs drawn under the body to provide an upright posture. This lineage gave rise to the earliest dinosaurs of the Late Triassic. In *Herrerasaurus*, a bipedal dinosaur from Argentina, we see one of the most distinctive characteristics of dinosaurs: walking upright on pillar-like legs, rather than on legs splayed outward as with modern amphibians and reptiles. This arrangement allowed the legs to support the great weight of the body while providing an efficient and rapid stride.

Although their ancestry is unclear, two groups of saurischian dinosaurs have been proposed based on differences in feeding habits and locomotion: the carnivorous and bipedal theropods, and the herbivorous and quadrupedal sauropods (sauropodomorphs). *Coelophysis* was an early theropod with a body form typical of all theropods: powerful hindlegs with three-toed feet; long, heavy counterbalancing tail; slender, grasping forelimbs; flexible neck; and a large head with jaws armed with dagger-like teeth. Large predators such as *Allosaurus,* common during the Jurassic, were replaced by even more massively built carnivores of the Cretaceous, such as *Tyrannosaurus,* which reached a length of 14.5 m (47 ft), stood nearly 6 m high, and weighed more than 7200 kg (8 tons). Not all predatory saurischians were massive; several were swift and nimble, such as *Velociraptor* ("speedy predator") of the Upper Cretaceous.

Herbivorous saurischians, the quadrupedal sauropods, appeared in the Late Triassic. Although early sauropods were small- and medium-sized dinosaurs, those of the Jurassic and Cretaceous attained gigantic proportions, the largest terrestrial vertebrates ever to have lived. *Brachiosaurus* reached 25 m (82 ft) in length and may have weighed in excess of 30,000 kg (33 tons). Even larger sauropods have been discovered; *Supersaurus* was 43 m (140 ft) long. With long necks and long front legs, the sauropods were the first vertebrates adapted to feed on trees. They reached their greatest diversity in the Jurassic and began to decline in overall abundance and diversity during the Cretaceous.

The second group of dinosaurs, the Ornithischia, were all herbivorous. Although more varied, even grotesque, in appearance than saurischians, the ornithischians are united by several derived skeletal features that indicate common ancestry. The huge back-plated *Stegosaurus* of the Jurassic is a well known example of armored ornithischians which comprised two of the five major groups of ornithischians. Even more shielded with bony plates than the stegosaurs were the heavily built ankylosaurs, "armored tanks" of the dinosaur world. As the Jurassic gave way to the Cretaceous, several groups of unarmored ornithischians appeared, although many bore impressive horns. The steady increase in ornithiscian diversity in the Cretaceous paralleled a concurrent gradual decline in giant sauropods which had flourished in the Jurassic. *Triceratops* is representative of horned dinosaurs that were common in the Upper Cretaceous. Even more prominent in the Upper Cretaceous were the duck-billed dinosaurs (hadrosaurs) which are believed to have lived in large herds. Many hadrosaurs had skulls elaborated with crests that probably functioned as vocal resonators to produce species-specific calls.

Sixty-five million years ago, the last of the Mesozoic dinosaurs became extinct, leaving birds as the only surviving lineage of archosaurs. There is increasingly convincing evidence that the demise of dinosaurs coincided with the impact on earth of a large asteroid that produced devastating worldwide environmental upheaval. We continue to be fascinated by the awe-inspiring, often staggeringly large creatures that dominated the Mesozoic era for 165 million years—an incomprehensibly long period of time. Today, inspired by clues from fossils and footprints from a lost world, scientists continue to piece together the puzzle of how the various dinosaur groups arose, behaved, and diversified.

SAURISCHIANS

ORNITHISCHIANS

65 MYBP*

Titanosaurus
12 m (40 ft)

Hadrosaur (duck-billed dinosaur)
10 m (33 ft)

136 MYBP*

Velociraptor
1.8 m (6 ft)

Triceratops
9 m (30 ft)

CRETACEOUS

Brachiosaurus
25 m (82 ft)

Stegosaurus
9 m (30 ft)

JURASSIC

Ilium

Ischium

Pubis

190 MYBP*

Allosaurus
11 m (35 ft)

Coelophysis
3 m (10 ft)

Ilium

Pubis

Ischium

TRIASSIC

Herrerasaurus 4 m (13 ft)
One of the oldest known
dinosaurs. Has characteristics
of both saurischians and
ornithischians.

275 MYBP*   *Millions of years before present

FIGURE **17-10**

Tokay, *Gekko gecko,* of Southeast Asia has a true voice and is named after the strident repeated *to-kay, to-kay* call.

FIGURE **17-11**

A large male marine iguana, *Amblyrhynchus cristatus,* of the Galápagos Islands, feeding underwater on algae. This is the only marine lizard in the world. It has special salt-removing glands in the eye orbits and long claws that enable it to cling to the bottom while feeding on small red and green algae, its principal diet. It may dive to depths exceeding 10 m (33 feet) and remain submerged more than 30 minutes.

FIGURE **17-12**

A chameleon snares a dragonfly. After cautiously edging close to its target, the chameleon suddenly lunges forward, anchoring its tail and feet to the branch. A split second later, it launches its sticky-tipped, foot-long tongue to trap the prey. The eyes of this common European chameleon *(Chamaeleo chamaeleon)* are swiveled forward to provide binocular vision and excellent depth perception.

have elongate, cylindrical bodies of nearly uniform diameter, and most lack any trace of external limbs (Figure 17-15). With soft skin divided into numerous rings, and eyes and ears hidden under skin, the amphisbaenians superficially resemble earthworms—a kind of structural convergence that often occurs when two very distantly related groups come to occupy similar habitats. The amphisbaenians have an extensive distribution in South America and tropical Africa.

Snakes: Suborder Serpentes Snakes are entirely limbless and lack both pectoral and pelvic girdles (the latter persists as a vestige in pythons and boas). The numerous vertebrae of snakes, shorter and wider than those of legged vertebrates, permit quick lateral undulations through grass and over rough terrain. The ribs increase rigidity of the vertebral column, providing more resistance to lateral stresses. The elevation of the neural spine gives the numerous muscles more leverage.

FIGURE **17-13**

A glass lizard, *Ophisaurus* sp., of the southeastern United States. This legless lizard feels stiff and brittle to the touch and has an extremely long, fragile tail that readily fractures when the animal is struck or seized. Most specimens, such as this one, have only a partly regenerated tip to replace a much longer tail previously lost. Glass lizards can be readily distinguished from snakes by the deep, flexible groove running along each side of the body. They feed on worms, insects, spiders, birds' eggs, and small reptiles.

FIGURE **17-14**

Gila monster, *Heloderma suspectum,* of southwestern United States desert regions and the related Mexican bearded lizard are the only venomous lizards known. These brightly colored, clumsy-looking lizards feed principally on birds' eggs, nesting birds, mammals, and insects. Unlike venomous snakes, the Gila monster secretes venom from glands in its lower jaw. The chewing bite is painful to humans but seldom fatal.

In addition to the highly kinetic skull that enables snakes to swallow prey several times their own diameter (Figure 17-16), snakes differ from lizards in having no movable eyelids (snakes' eyes are permanently covered with upper and lower transparent eyelids fused together) and no external ears. Most snakes have relatively poor vision, the tree-living snakes of the tropical forest being a conspicuous exception (Figure 17-17). In fact, some arboreal snakes possess excellent binocular vision, which they use to track prey through the branches where scent trails would be difficult to follow. Snakes are totally deaf, although they are sensitive to low-frequency vibrations conducted through the ground.

Nevertheless, most snakes employ the chemical senses rather than vision or vibration detection to hunt their prey. In addition to the usual olfactory areas in the nose, which are not well developed, snakes have a pair of pit-like **Jacobson's organs** in the roof of the mouth. These organs are lined with an olfactory epithelium and are richly innervated. The **forked tongue,** flicked through the air, picks up scent particles (Figure 17-18); the tongue is then withdrawn and sampled molecules are delivered to Jacobson's organs. Information is transmitted to the brain, where scents are identified.

Snakes of the subfamily Crotalinae within the family Viperidae are called **pit vipers** because they possess special heat-sensitive **pit organs** on their heads, located between the nostrils and the eyes (Figures 17-18, 17-19, and 17-20). All of the best-known North American venomous snakes are pit vipers, such as the several species of rattlesnakes, water moccasins, and copperheads. The pits are supplied with a

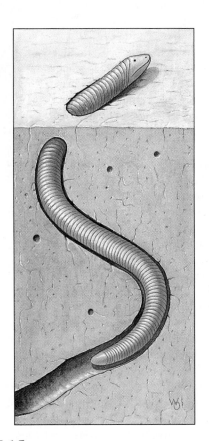

FIGURE **17-15**

A worm lizard of the suborder Amphisbaenia. Worm lizards are burrowing forms with a solidly constructed skull used as a digging tool. The species pictured, *Amphisbaena alba,* is widely distributed in South America.

FIGURE **17-16**

Black rat snake, *Elaphe obsoleta obsoleta,* swallowing a chipmunk.

FIGURE **17-17**

Parrot snake, *Leptophis ahaetulla.* The slender body of this Central American tree snake is an adaptation for sliding along branches without weighting them down.

dense packing of free nerve endings from the fifth cranial nerve. They are exceedingly sensitive to radiant energy (long-wave infrared) and can distinguish temperature differences smaller than 0.003° C from a radiating surface. Pit vipers use the pits to track warm-blooded prey and to aim strikes, which they can make as effectively in total darkness as in daylight.

All vipers have a pair of teeth, modified as fangs, on the maxillary bones. The fangs lie in a membranous sheath when the mouth is closed. When a viper strikes, a special muscle and bone lever system erects the fangs as the mouth opens (Figure 17-20). The fangs are driven into the prey by the thrust, and venom is injected into the wound through a canal in the fangs. A viper immediately releases its prey after the bite and follows it until it is paralyzed or dies. Then the snake swallows it whole. Approximately 8000 bites but only 12 deaths from pit vipers are reported each year in the United States.

The tropical and subtropical countries are the homes of most species of snakes, both venomous and nonvenomous varieties. Even in these countries less than one-third of snakes are venomous. Nonvenomous snakes kill their prey by constriction (Figure 17-21) or by biting and swallowing. Their diet tends to be restricted; many feed principally on rodents, whereas others feed on other reptiles, fishes, frogs, and insects. Some African, Indian, and neotropical snakes have become specialized as egg eaters.

FIGURE **17-18**

A timber rattlesnake, *Crotalus horridus,* flicks its tongue to smell its surroundings. Scent particles trapped on the tongue's surface are transferred to Jacobson's organs, olfactory organs in the roof of the mouth. Note the heat-sensitive pit organ between the nostril and eye.

Venomous snakes are usually divided into four groups based on the type of fangs. The vipers (family Viperidae) have tubular fangs at the front of the mouth; this group includes the American pit vipers previously mentioned and the Old World true vipers, which lack facial heat-sensing pits. Among the latter are the common European adder and the African puff adder. A second family of venomous snakes (family Elapidae) has short, permanently erect fangs so that the venom must be injected by repeated bites. In this group are the cobras (Figure 17-22), mambas, coral snakes, and kraits. The highly venomous sea snakes are usually placed in a third family (Hydrophiidae). The very large family Colubridae, which contains most of the familiar nonvenomous snakes, does include at least two ven-

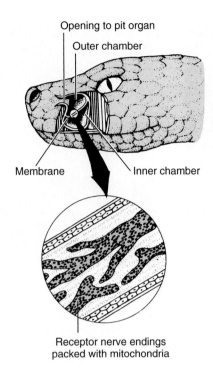

FIGURE 17-19

Pit organ of rattlesnake, a pit viper. Cutaway shows location of a deep membrane that divides the pit into inner and outer chambers. Heat-sensitive nerve endings are concentrated in the membrane.

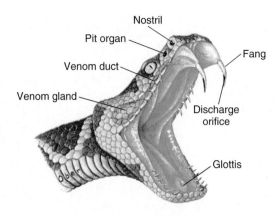

FIGURE 17-20

Head of rattlesnake showing the venom apparatus. The venom gland, a modified salivary gland, is connected by a duct to the hollow fang.

omous (and very dangerous) snakes—the African boomslang and the African twig snake—that have been responsible for many human fatalities. Both are rear-fanged snakes that normally use their venom to quiet struggling prey.

Even the saliva of all harmless snakes possesses limited toxic qualities, and it is logical that there was a natural selection for this toxic tendency as snakes evolved. Snake venoms have traditionally been divided into two types. The **neurotoxic** type acts mainly on the nervous systems, affecting the optic nerves (causing blindness) or the phrenic nerve of the diaphragm (causing paralysis of respiration). The **hemorrhagin** type breaks down red blood corpuscles and blood vessels and produces extensive hemorrhaging of blood into tissue spaces. In fact, most snake venoms are complex mixtures of various fractions that attack different organs in specific ways; they seldom can be assigned categorically to one or the other of the traditional types. Although the sea snakes and the Australian tiger snake have perhaps the most toxic of snake venoms, several larger snakes are more dangerous. The aggressive king cobra, which may exceed 5.5 m in length, is the largest and probably the most dangerous of all venomous snakes. In India, where snakes come in constant contact with people, some 200,000 snakebites cause more than 9000 deaths each year.

Most snakes are **oviparous** (L. *ovum,* egg, + *parere,* to bring forth) species that lay their shelled, elliptical eggs beneath rotten logs, under rocks, or in holes dug in the ground. Most of the remainder, including all the American pit vipers except the tropical bushmaster, are **ovoviviparous** (L. *ovum,* egg, + *vivus,* living, + *parere,* to bring forth), giving birth to well-formed young. Very few snakes are **viviparous** (L. *vivus,* living, + *parere,* to bring forth); in these snakes a primitive placenta forms, permitting the exchange of materials between the embryonic and maternal bloodstreams. Snakes are able to store sperm and can lay several clutches of fertile eggs at long intervals after one mating.

The Tuatara: Order Sphenodonta

The order Sphenodonta is represented by two living species of the genus *Sphenodon* (Gr. *sphenos,* wedge, + *odontos,* tooth) of New Zealand (Figure 17-23). The tuatara is the sole survivor of the sphenodontid lineage that radiated modestly during the early Mesozoic era but declined toward the end of the Mesozoic. The tuatara was once widespread throughout the two main islands of New Zealand but is now restricted to small islets of Cook Strait and off the northeast coast of North Island. On some of these islets, under protection from the New Zealand government, it is prospering.

The tuatara is a lizardlike form 66 cm long or less that lives in burrows often shared with petrels. They are slow-growing animals with a long life; one is recorded to have lived 77 years.

The tuatara has captured the interest of zoologists because of numerous features that are almost identical to those of Mesozoic fossils 200 million years ago. These features include a diapsid skull with two temporal openings bounded by complete arches, and a well-developed median parietal "third eye." In many other respects *Sphenodon* resembles lizards of the early Mesozoic. *Sphenodon* represents one of the slowest rates of evolution known among vertebrates.

FIGURE 17-21

Nonvenomous African house snake, *Boaedon fuluginosus,* constricting a mouse before swallowing it.

FIGURE 17-22

Spectacled, or Indian, cobra, *Naja naja.* Cobras erect the front part of the body when startled and as a threat display. Although the cobra's strike range is limited, all cobras are dangerous because of the extreme toxicity of the venom.

Crocodiles and Alligators: Order Crocodilia

The modern crocodilians and the birds are the only surviving representatives of the archosaurian lineage that gave rise to the great Mesozoic radiation of dinosaurs and their kin. Crocodilians differ little in structural details from crocodilians of the early Mesozoic. Having remained mostly unchanged for nearly 200 million years, crocodilians face an uncertain future in a world dominated by humans.

All crocodilians have an elongate, robust, well-reinforced skull and massive jaw musculature arranged to provide a wide gape and rapid, powerful closure. Teeth are set in sockets, a type of dentition that was typical of all archosaurs as well as the earliest birds. Another adaptation found in no other vertebrate except mammals is a complete secondary palate. This innovation allows crocodilians to breathe when the mouth is filled with water or food (or both).

The estuarine crocodile *(Crocodylus porosus),* found in southern Asia, and the Nile crocodile *(C. niloticus)* (Figure 17-24A) grow to great size (adults

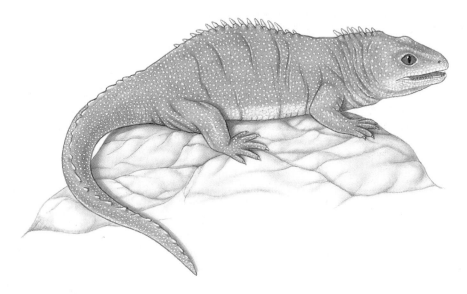

FIGURE 17-23

Tuatara, *Sphenodon* sp., a living representative of the order Sphenodonta. This "living fossil" reptile has, on top of the head, a well-developed parietal "eye" with retina, lens, and nervous connections to the brain. Although covered with scales, this third eye is sensitive to light. The parietal eye may have been an important sense organ in early reptiles. The tuatara is found today only on certain islands off the coastline of New Zealand.

Classification of Living Reptiles

Subclass Anapsida (a-nap′se-duh) (Gr. *an,* without, + *apsis,* arch): **Anapsids.** Amniotes having a skull with no temporal opening.

Order Testudines (tes-tu′din-eez) (L. *testudo,* tortoise) **(Chelonia): turtles.** Body in a bony case of dorsal carapace and ventral plastron; jaws with horny beaks instead of teeth; vertebrae and ribs fused to overlying carapace; tongue not extensible; neck usually retractable; approximately 330 species.

Subclass Diapsida (di-ap′se-duh) (Gr. *di,* double, + *apsis,* arch): **Diapsids.** Amniotes having a skull with two temporal openings.

Superorder Lepidosauria (lep-i-do-sor′ee-uh) (Gr. *lepidos,* scale, + *sauros,* lizard). Diapsid lineage appearing in the Permian; characterized by sprawling posture; no bipedal specializations; diapsid skull often modified by loss of one or both temporal arches.

Order Squamata (squa-ma′ta) (L. *squamatus,* scaly, + *ata,* characterized by): **Snakes, lizards, and amphisbaenians.** Skin of horny epidermal scales or plates, which is shed; quadrate movable; skull kinetic (except worm lizards); vertebrae usually concave in front; paired copulatory organs.

Suborder Lacertilia (lay-sur-till′ee-uh) (L. *lacerta,* lizard) **(Sauria): lizards.** Body slender, usually with four limbs; rami of lower jaw fused; eyelids movable; external ear present; this paraphyletic suborder contains approximately 3300 species.

Suborder Amphisbaenia (am′fis-bee′nee-a) (L. *amphis,* double, + *baina,* to walk): **worm lizards.** Body elongate and of nearly uniform diameter; no legs (except one genus with short front legs); skull bones interlocked for burrowing (not kinetic); limb girdles vestigial; eyes hidden beneath skin; only one lung; approximately 135 species.

Suborder Serpentes (sur-pen′teez) (L. *serpere,* to creep): **snakes.** Body elongate; limbs, ear openings, and middle ear absent; mandibles joined anteriorly by ligaments; eyelids fused into transparent spectacle; tongue forked and protrusible; left lung reduced or absent; approximately 2300 species.

Order Sphenodonta (sfen′o-don′tuh) (Gr. *sphēnos,* wedge, + *odontos,* tooth): **tuatara.** Primitive diapsid skull; vertebrae biconcave; quadrate immovable; median parietal eye present. *Sphenodon* only extant genus (two species).

Superorder Archosauria (ark′o-sor′ee-uh) (Gr. *archōn,* ruling, + *sauros,* lizard). Advanced diapsids, mostly terrestrial, but some specialized for flight; includes birds (covered in the next chapter) and ruling reptiles of the Mesozoic.

Order Crocodilia (croc′o-dil′ee-uh) (L. *crocodilus,* crocodile): **crocodilians.** Skull elongate and massive; nares terminal; secondary palate present; four-chambered heart; vertebrae usually concave in front; forelimbs usually of five digits; hindlimbs of four digits; quadrate immovable; advanced social behavior; 25 species.

Subclass Synapsida (sin-ap′si-duh) (Gr. *syn,* together, + *apsis,* arch). Amniotes having skull with one pair of lateral temporal openings; mammal-like reptiles (extinct); living mammals are descendants of this lineage.

A **B**

FIGURE 17-24

Crocodilians. **A,** Nile crocodile *(Crocodylus niloticus)* basking. The lower jaw tooth fits *outside* the slender upper jaw; alligators lack this feature. **B,** American alligator *(Alligator mississipiensis),* an increasingly noticeable resident of rivers, bayous, and swamps of the southeastern United States.

weighing 1000 kg have been reported) and are swift and aggressive. Crocodiles are known to attack animals as large as cattle, deer, and people. Alligators (Figure 17-24B) are less aggressive than crocodiles and certainly far less dangerous to people. They are unusual among reptiles in being able to make definite vocalizations. The male alligator can give loud bellows in the mating season. In the United States, *Alligator mississipiensis* (Figure 17-24B) is the only species of alligator; *Crocodylus acutus,* restricted to extreme southern Florida, is the only species of crocodile.

Alligators and crocodiles are oviparous. Usually from 20 to 50 eggs are laid in a mass of dead vegetation. The mother hears vocalizations from the hatching young and responds by opening the nest to allow the hatchlings to escape. As with many turtles and some lizards, the incubation temperature of the eggs determines the sex ratio of the offspring. However, unlike turtles (p. 332), low nest temperatures produce only females, while high nest temperatures produce only males.

Summary

The reptiles and other amniotes diverged phylogenetically from a group of labyrinthodont amphibians during the late Paleozoic era, some 300 million years ago. Their success as terrestrial vertebrates is attributed in large part to the evolution of the amniotic egg, which, with its three extraembryonic membranes, provided support for full embryonic development within the protection of a shell. Thus the reptiles could lay their eggs on land. Reptiles are also distinguished from amphibians by their dry, scaly skin that limits water loss; more powerful jaws; internal fertilization; and advanced circulatory, respiratory, excretory, and nervous systems. Like amphibians, reptiles are ectotherms, but most exercise considerable behavioral control over their body temperature.

Before the end of the Paleozoic era, the amniotes began a radiation that separated into three lineages: the anapsids, which gave rise to the turtles; the synapsids, a lineage that led to the modern mammals; and the diapsid lineage, which led to all other reptiles and to the birds. The great burst of reptilian radiation during the Mesozoic era produced a worldwide fauna of great diversity.

The turtles (order Testudines) with their distinctive shells have changed little in design since the Triassic period. Turtles are a small group of long-lived terrestrial, semiaquatic, aquatic, and marine species. They lack teeth. All are oviparous and all, including the marine forms, bury their eggs.

The lizards, snakes, and worm lizards (order Squamata) make up 95% of all living reptiles. Lizards (suborder Lacertilia) are a diversified and successful group adapted for walking, running, climbing, swimming, and burrowing. They are distinguished from snakes by typically having two pairs of legs (some species are legless), united lower jaw halves, movable eyelids, external ears, and absence of fangs. Many lizards are well adapted for survival under hot and arid desert conditions.

Worm lizards (suborder Amphisbaenia) are a small tropical group of legless squamates highly adapted for burrowing.

Snakes (suborder Serpentes), in addition to being entirely limbless, are characterized by their elongate bodies and a highly kinetic skull that permits them to swallow whole prey that may be much larger than the snake's own diameter. Most snakes rely on the chemical senses, especially Jacobson's organs, to hunt prey, rather than on weakly developed visual and auditory senses. The pit vipers have unique infrared-sensing organs for tracking warm-bodied prey. Many snakes are venomous.

The tuatara of New Zealand (order Sphenodonta) is a relict genus and sole survivor of a group that otherwise disappeared 100 million years ago. It bears several features that are almost identical to those of Mesozoic fossil diapsids.

The crocodiles and alligators (order Crocodilia) are the only living reptilian representatives of the archosaurian lineage that gave rise to the extinct dinosaurs and the living birds. Crocodilians have several adaptations for a carnivorous, semiaquatic life, including a massive skull with powerful jaws, and a secondary palate. They have the most complex social behavior of any reptile.

Review Questions

1. What were the three major amniote radiations of the Mesozoic and from which lineage or lineages did the birds and mammals descend? How could you distinguish the skulls characteristic of these different radiations?

2. What changes in egg design allowed the reptiles to lay eggs on land? Why is the egg often called an "amniotic" egg? What are the "amniotes"?

3. Why are the reptiles considered a paraphyletic rather than a monophyletic group? How have cladistic taxonomists revised the content of this taxon to make it monophyletic?

4. Describe ways in which reptiles are more functionally or structurally suited for terrestriality than the amphibians.

5. What are the main characteristics of reptilian skin and how would you distinguish reptilian skin from frog skin?

6. Describe the principal structural features of turtles that would distinguish them from any other reptilian order.

7. How might nest temperature affect egg development in turtles? In crocodilians?

8. What is meant by a "kinetic" skull and what benefit does it confer? How are snakes able to eat such large prey?

9. In what ways are the special senses of snakes similar to those of lizards, and in what ways have they evolved for specialized feeding strategies?

10. How do snakes and crocodilians breathe when their mouths are full of food?

11. What is the function of Jacobson's organ of snakes?

12. What is the function of the "pit" of pit vipers?

13. What is the difference in the structure or location of the fangs of a rattlesnake, a cobra, and an African boomslang?

14. Most snakes are oviparous, but some are ovoviviparous or viviparous. What do these terms mean and what would you have to know to be able to assign a particular snake to one of these reproductive modes?

15. Why is the tuatara (*Sphenodon*) of special interest to biologists? Where would you have to go to see one in its natural habitat?

16. From which diapsid lineage have the crocodilians descended? What other major fossil and living vertebrate groups belong to this same lineage? In what structural and behavioral ways are the crocodilians more advanced than other living reptiles?

Selected References

See also general references on page 395.

Alexander, R. M. 1991. How dinosaurs ran. Sci. Am. **264:**130-136 (April). *By applying the techniques of modern physics and engineering, a zoologist calculates that the large dinosaurs walked slowly but were capable of a quick run; none required the buoyancy of water for support.*

Alvarez, W., and F. Asaro. 1990. An extraterrestrial impact. Sci. Am. **263:**78-84 (Oct.). *This article and an accompanying article by V. E. Courtillot, "A volcanic eruption," present opposing interpretations of the cause of the Cretaceous mass extinction that led to the demise of the dinosaurs.*

Cogger, H. G., and R. G. Zweifel, eds. 1992. Reptiles and amphibians. New York, Smithmark Publishers, Inc. *This comprehensive, up-to-date, and lavishly illustrated volume was written by some of the best-known herpetologists in the field.*

Crews, D. 1994. Animal sexuality. Sci. Am. **270:**108-114 (Jan.) *The reproductive strategies of reptiles, including nongenetic sex determination, provide insights into the origins and functions of sexuality.*

Halliday, T. R., and K. Adler, eds. 1986. The encyclopedia of reptiles and amphibians. New York, Facts on File, Inc. *Comprehensive and beautifully illustrated treatment of the reptilian groups with helpful introductory sections on origins and characteristics.*

Lillywhite, H. B. 1988. Snakes, blood circulation and gravity. Sci. Am. **259:**92-98 (Dec.). *Even long snakes are able to maintain blood circulation when the body is extended vertically (head up posture) through special circulatory reflexes that control blood pressure.*

Lohmann, K. J. 1992. How sea turtles navigate. Sci. Am. **266:**100-106 (Jan.). *Recent evidence suggests that sea turtles use the earth's magnetic field and the direction of ocean waves to navigate back to their natal beaches to nest.*

Norman, D. 1991. Dinosaur! New York, Prentice-Hall. *Highly readable account of the life and evolution of dinosaurs with fine illustrations.*

Zug, G. R. 1993. Herpetology: an introductory biology of amphibians and reptiles. New York, Academic Press, Inc. *Introductory college-level textbook.*

Links to the Internet

Visit this textbook's Web site at http://www.mhhe.com/zoology to find live Internet links for each of the references listed below.

1. Herp Hotlinks. This site provides dozens of links to herpetological Web sites.

2. Class Reptilia. University of Michigan site on amphibians. Pictures, much information on the morphology, distribution, and ecology of a large number of reptiles. Each species is linked to Web pages. Images may not be available for display depending on your server.

3. Reptilia. This site provides links to much information on reptiles.

4. Amphibian and Reptile WWW Sites. This site contains many links to herpetological sites.

5. Classification of Reptiles. This site contains information on living and extinct reptiles.

Snakes

6. Rattlesnakes. This site contains information on rattlesnakes.

Crocodilians

7. Crocodilian Sites on the World Wide Web. A list of links.

8. Crocodilia: Natural History & Conservation: Crocodiles, Caimans.

Turtles

9. National Wildlife Federation, Endangered Sea Turtles Site. General information, links, conservation issues, and more.

Dinosaurs

10. The Dinosauria. University of California at Berkeley Museum of Paleontology site contains information on morphology, systematics, and links to more information about subjects such as hadrosaurs and dinosaur locomotion.

11. Dinosaurs in Cyberspace: Dinolinks. University of California at Berkeley Museum of Paleontology. A list of dinosaur-oriented Web sites, scientific and otherwise. This site has pages and pages of links.

12. Dinobuzz. Current topics concerning dinosaurs from the University of California at Berkeley Museum of Paleontology.

13. National Museum of Natural History/Smithsonian Institution. Interesting misconceptions about dinosaurs.

Birds

CHAPTER eighteen

Long Trip to a Summer Home

Perhaps it was ordained that birds, having mastered flight, would use this power to make the long and arduous seasonal migrations that have captured human wonder and curiosity. For the advantages of migration are many. Moving between southern wintering regions and northern summer breeding regions enables birds to sustain their intense metabolism with abundant and unfailing sources of food. In the far North long summer days and the abundance of insects combine to provide parents with ample food for rearing their young. Predators of birds are not so abundant in the far North, and the brief once-a-year appearance of vulnerable young birds does not encourage the buildup of predator populations. Migration also vastly increases the amount of space available for breeding and reduces aggressive territorial behavior. Finally, migration favors homeostasis—the balancing of physiological processes that maintains internal stability—by allowing birds to avoid climatic extremes.

Still, the wonder of the migratory pageant remains, and there is much yet to learn about its mechanisms. What times migration, and what determines that each bird shall store sufficient fuel for the journey? How did the sometimes difficult migratory routes originate, and what cues do birds use in navigation? And what was the origin of this instinctive force to follow the retreat of winter northward? For it is instinct that drives the migratory waves in spring and fall, instinctive blind obedience that carries most birds successfully to their northern nests, while countless others fail and die, winnowed by the ever-challenging environment.

Of the vertebrates, birds of the class Aves (ay′veez) (L. pl. of *avis,* bird) are the most noticeable, the most melodious, and many think the most beautiful. With more than 9000 species distributed over nearly the entire earth, birds far outnumber all other vertebrates except the fishes. Birds are found in forests and deserts, in mountains and prairies, and on all oceans. Four species are known to have visited the North Pole, and one, a skua, was seen at the South Pole. Some birds live in total darkness in caves, finding their way about by echolocation, and others dive to depths greater than 45 m to prey on aquatic life.

The single unique feature that distinguishes birds from other animals is their feathers. If an animal has feathers, it is a bird; if it lacks feathers, it is not a bird. No other vertebrate group bears such an easily recognizable and foolproof identification tag.

There is great uniformity of structure among birds. Despite approximately 150 million years of evolution, during which they proliferated and adapted themselves to specialized ways of life, we have no difficulty recognizing a bird as a bird. In addition to feathers, all birds have forelimbs modified into wings (although they may not be used for flight); all have hindlimbs adapted for walking, swimming, or perching; all have horny beaks; and all lay eggs. The reason for this great structural and functional uniformity is that birds evolved into flying machines. This fact greatly restricts diversity, so much more evident in other vertebrate classes. For example, birds do not begin to approach the diversity seen in their endothermic evolutionary peers, the mammals, a group that includes forms as dissimilar as whale, porcupine, bat, and giraffe.

A bird's entire anatomy is designed around flight and its perfection. An airborne life for a large vertebrate is a highly demanding evolutionary challenge. A bird must, of course, have wings for support and propulsion. Bones must be light and hollow yet serve as a rigid airframe. The respiratory system must be highly efficient to meet the intense metabolic demands of flight and serve also as a thermoregulatory device to maintain a constant body temperature. A bird must have a rapid and efficient digestive system to process an energy-rich diet; it must have a high metabolic rate; and it must have a high-pressure circulatory system. Above all, birds must have a finely tuned nervous system and acute senses, especially superb vision, to handle the complex demands of headfirst, high-velocity flight.

Characteristics of Class Aves

1. Body usually spindle shaped, with four divisions: head, neck, trunk, and tail; **neck disproportionately long** for balancing and food gathering
2. Limbs paired with the **forelimbs usually modified for flying;** posterior pair variously adapted for perching, walking, and swimming; foot with four toes (two or three toes in some)
3. Epidermal **covering of feathers** and **leg scales;** thin integument of epidermis and dermis; no sweat glands; oil or preen gland at base of tail; **pinna of ear rudimentary**
4. **Fully ossified skeleton with air cavities;** skull bones fused with **one occipital condyle;** each jaw covered with a horny sheath, forming a **beak; no teeth;** ribs with strengthening processes; **tail not elongate;** sternum well developed with keel or reduced with no keel; **single bone in middle ear**
5. Nervous system well developed, with brain and 12 pairs of cranial nerves
6. Circulatory system of **four-chambered heart,** with the **right aortic arch persisting;** reduced renal portal system; nucleated red blood cells
7. Endothermic
8. Respiration by slightly expansible lungs, with thin **air sacs** among the visceral organs and skeleton; **syrinx (voice box)** near junction of trachea and bronchi
9. Excretory system of metanephric kidney; ureters open into cloaca; **no bladder;** semisolid urine; uric acid main nitrogenous waste
10. Sexes separate; testes paired, with the vas deferens opening into the cloaca; **females with left ovary and oviduct only;** copulatory organ in ducks, geese, ratites, and a few others
11. Fertilization internal; **amniotic eggs with much yolk and hard calcareous shells;** embryonic membranes in egg during development; **incubation external;** young active at hatching **(precocial)** or helpless and naked **(altricial);** sex determination by females (females heterogametic)

Origin and Relationships

Approximately 147 million years ago, a flying animal drowned and settled to the bottom of a shallow marine lagoon in what is now Bavaria, Germany. It was rapidly covered with a fine silt and eventually fossilized. There it remained until discovered in 1861 by a workman splitting slate in a limestone quarry. The fossil was approximately the size of a crow, with a skull not unlike that of modern birds except that the beaklike jaws bore small bony teeth set in sockets like those of reptiles (Figure 18-1). The skeleton was decidedly reptilian with a long bony tail, clawed fingers, and abdominal ribs. It might have been classified as a reptile except that it carried the unmistakable imprint of **feathers,** those marvels of biological engineering that only birds possess. *Archaeopteryx lithographica* (ar-kee-op′ter-iks lith-o-graf′e-ka, Gr., meaning "ancient wing inscribed in stone"), as

A   **B**

FIGURE **18-1**

Archaeopteryx, a 147-million-year-old relative of modern birds. **A,** Cast of the second and most nearly perfect fossil of *Archaeopteryx,* which was discovered in a Bavarian stone quarry. Six specimens of *Archaeopteryx* have been discovered, the most recent one in 1987. **B,** Reconstruction of *Archaeopteryx.*

the fossil was named, was an especially fortunate discovery because it proved beyond reasonable doubt the phylogenetic relatedness of birds and reptiles.

The controversy over the dinosaur origin of birds was refueled in late 1996 with announcement of the discovery in China of fossil birds of the Late Jurassic and Early Cretaceous that its discoverers believe were too highly derived to have descended from dinosaurs. The authors assert that Archaeopteryx, *rather than being the ancestor of modern birds, represented a dead-end lineage that became extinct. Modern birds, argue the authors, descended separately from archosaurian ancestors that predated dinosaurs. However, this scenario is strongly rejected by critics who note the uncertain dating of the Chinese discoveries and who point to another fossil discovery from China that, paradoxically, appears to strengthen the dinosaur origin of birds. For the moment,* Archaeopteryx *remains on its perch as a descendant of dinosaurs that gave rise to birds.*

Zoologists had long recognized the similarity of birds and reptiles because of their many shared morphological, developmental, and physiological homologies. The distinguished English zoologist Thomas Henry Huxley was so impressed with these affinities that he called birds "glorified reptiles" and classified them with a group of dinosaurs called theropods that displayed several bird-like characteristics (Figures 18-2 and 18-3). Theropod dinosaurs share many derived characters with birds, the most obvious of which is the elongate, mobile, S-shaped neck. As shown in the cladogram (Figure 18-3), theropods belong to a lineage of diapsid reptiles, the archosaurians, that includes crocodilians and pterosaurs, as well as the dinosaurs. There is now overwhelming evidence that Huxley was correct: birds' closest phylogenetic affinity is to the theropod dinosaurs. The only anatomical feature required to link bird ancestry with the theropod dinosaurs was feathers, and this was provided by the discovery of *Archaeopteryx.* However, recent fossil discoveries have complicated the picture of bird origins and renewed the debate over which amniote lineage was ancestral to birds.

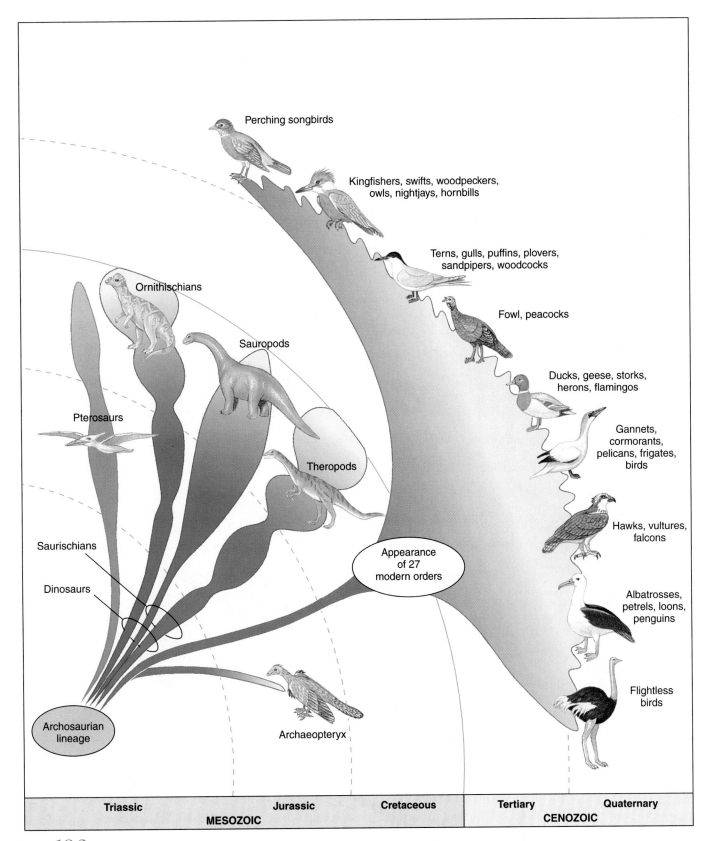

FIGURE 18-2

Evolution of modern birds. Of 27 living bird orders, 9 of the largest are shown. The earliest known bird, *Archaeopteryx,* lived in the Upper Jurassic, about 147 million years ago. *Archaeopteryx* uniquely shares many specialized aspects of its skeleton with the smaller theropod dinosaurs and is considered to have evolved within the theropod lineage. Evolution of modern bird orders occurred rapidly during the Cretaceous and early Tertiary periods.

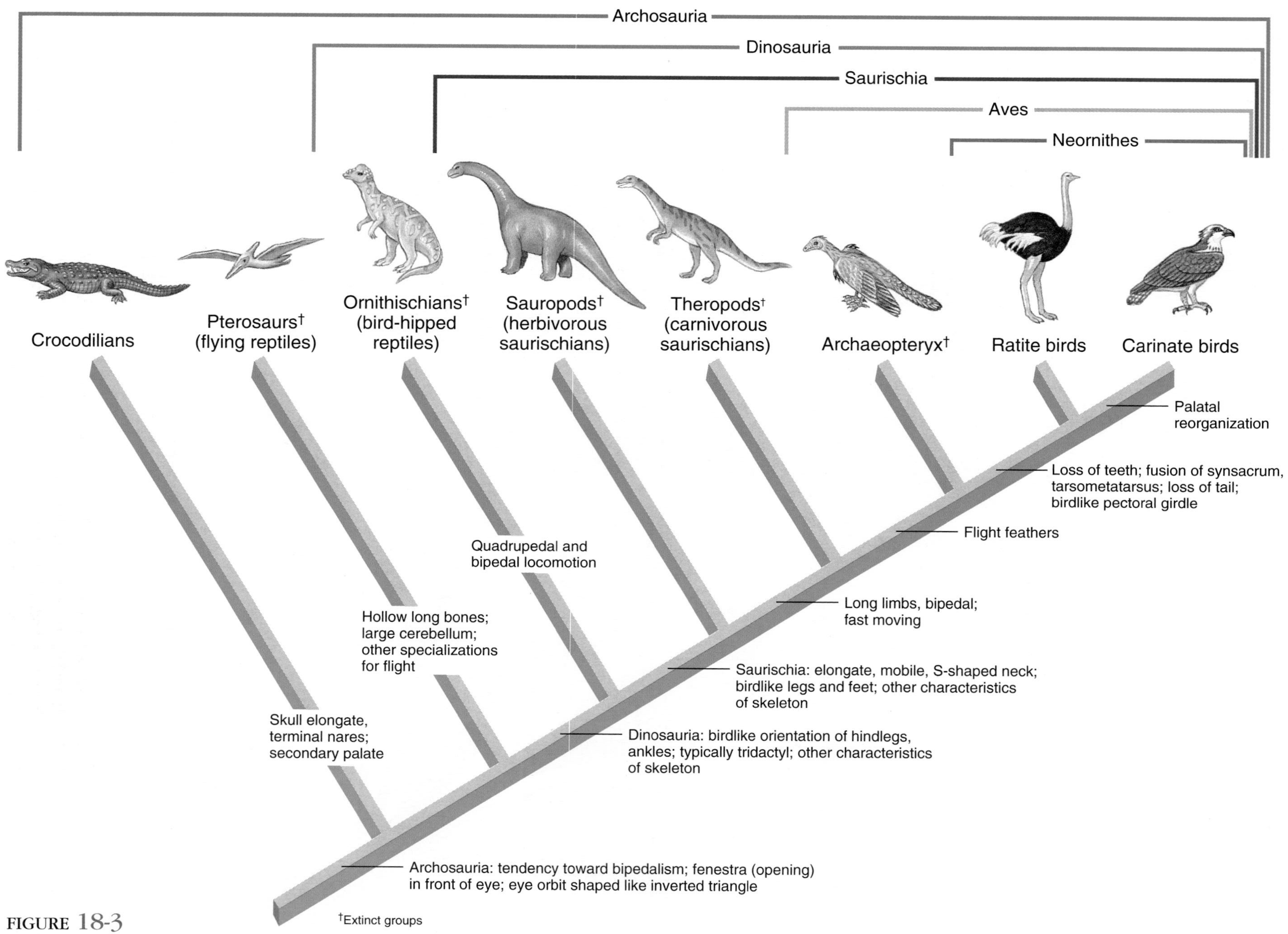
Archosauria
Dinosauria
Saurischia
Aves
Neornithes
Crocodilians
Pterosaurs† (flying reptiles)
Ornithischians† (bird-hipped reptiles)
Sauropods† (herbivorous saurischians)
Theropods† (carnivorous saurischians)
Archaeopteryx†
Ratite birds
Carinate birds
Palatal reorganization
Loss of teeth; fusion of synsacrum, tarsometatarsus; loss of tail; birdlike pectoral girdle
Flight feathers
Quadrupedal and bipedal locomotion
Long limbs, bipedal; fast moving
Hollow long bones; large cerebellum; other specializations for flight
Saurischia: elongate, mobile, S-shaped neck; birdlike legs and feet; other characteristics of skeleton
Skull elongate, terminal nares; secondary palate
Dinosauria: birdlike orientation of hindlegs, ankles; typically tridactyl; other characteristics of skeleton
Archosauria: tendency toward bipedalism; fenestra (opening) in front of eye; eye orbit shaped like inverted triangle
†Extinct groups

FIGURE 18-3

Cladogram of the Archosauria, showing the relationships of several archosaurian groups to modern birds. Shown are a few of the shared derived characters, mostly those related to flight, that were used to construct the genealogy. The outgroup is Lepidosauria (see Figure 17-2, p. 328).

(Sources: J. Gauthier, "Saurischian monophyly and the origin of birds" In K. Padian, The origin of birds and the evolution of flight. *No. 18, 1986, Memoirs California Academy of Science; and J. M. V. Rayner, "Vertebrate flight and the origins of flying vertebrates" in K. C. Allen, and D. E. G. Briggs, 1989,* Evolution and the fossil record, *Smithsonian Institute Press, Washington, D.C.)*

Living birds (Neornithes) are divided into two groups: (1) **ratite** (rat′ite) (L. *ratitus,* marked like a raft, from *ratis,* raft), the large flightless ostrichlike birds and the kiwis, which have a flat sternum with poorly developed pectoral muscles, and (2) **carinate** (L. *carina,* keel), the flying birds that have a keeled sternum on which powerful flight muscles insert. This division originated from the conviction that the flightless birds (ostrich, emu, kiwi, rhea) represented a separate line of descent that never attained flight—a view now rejected. The ostrichlike ratites clearly have descended from flying ancestors. Furthermore, not all carinate, or keeled, birds can fly and many of them even lack keels. Flightlessness has appeared independently among many groups of birds; the fossil record reveals flightless wrens, pigeons, parrots, cranes, ducks, auks, and even a flightless owl. Penguins are flightless although they use their wings to "fly" through the water (Figure 3-6 p. 59). Flightlessness has evolved almost always on islands where few terrestrial predators are found. The flightless birds living on continents today are the large ratites (ostrich, rhea, cassowary, emu), which can run fast enough to escape predators. The ostrich can run 70 km (42 miles) per hour, and claims of speeds of 96 km (60 miles) per hour have been made. The evolution of flightless birds is discussed on p. 18 and in Figure 1-17.

The bodies of flightless birds are dramatically redesigned to remove all of the restrictions of flight. The keel of the sternum is lost, and heavy flight muscles (as much as 17% of the body weight of flying birds), as well as other specialized flight apparatus, disappear. Since body weight is no longer a restriction, flightless birds tend to become large. Several extinct flightless birds were enormous: the giant moas of New Zealand weighed more than 225 kg (500 pounds) and the elephantbird of Madagascar, the largest bird that ever lived, probably weighed nearly 450 kg (about 1000 pounds) and stood nearly 2 m tall.

Adaptations of Bird Structure and Function for Flight

Just as an airplane must be designed and built according to rigid aerodynamic specifications if it is to fly, so too must birds meet stringent structural requirements if they are to stay airborne. All the special adaptations found in flying birds contribute to two things: more power and less weight. Flight by humans became possible when they developed an internal combustion engine and learned how to reduce the weight-to-power ratio to a critical point. Birds accomplished flight millions of years ago. But birds must do much more than fly. They must feed themselves and convert food into high-energy fuel; they must escape predators; they must be able to repair their own injuries; they must be able to air-condition themselves when overheated and heat themselves when too cool; and, most important of all, they must reproduce themselves.

Feathers

A feather is very lightweight, yet it possesses remarkable toughness and tensile strength. The most typical of bird feathers are the **contour feathers,** vaned feathers that cover and streamline the bird's body. A contour feather consists of a hollow **quill,** or calamus, emerging from a skin follicle, and a **shaft,** or rachis, which is a continuation of the quill and bears numerous **barbs** (Figure 18-4). The barbs are arranged in closely parallel fashion and spread diagonally outward from both sides of the central shaft to form a flat, expansive, webbed surface, the **vane.** There may be several hundred barbs in the vane.

If a feather is examined with a microscope, each barb appears to be a miniature replica of the feather with numerous parallel filaments called **barbules** set in each side of the barb and spreading laterally from it. There may be 600 barbules on each side of a barb, adding up to more than 1 million barbules for the feather. The barbules of one barb overlap the barbules of a neighboring barb in a herringbone pattern and are held together with great tenacity by tiny hooks. Should two adjoining barbs become separated—and considerable force is needed to pull the vane apart—they are instantly zipped together again by drawing the feather through the fingertips. The bird, of course, does this with its bill, and much of a bird's time is occupied with preening to keep its feathers in perfect condition.

Like the reptilian scale to which it is homologous, a feather develops from an epidermal elevation overlying a nourishing dermal core. However, rather than flattening like a scale, the feather bud rolls into a cylinder and sinks into the follicle from which it is growing. During growth, pigments (lipochromes and melanin) are added to the epidermal cells. As the feather enlarges and nears the end of its growth, the soft rachis and barbs are transformed into hard structures by the deposition of keratin. The protective sheath splits apart, allowing the end of the feather to protrude and the barbs to unfold.

When fully grown, a feather, like mammalian hair, is a dead structure. The shedding, or molting, of feathers is a highly orderly process. Except in penguins, which molt all at once, feathers are discarded gradually to avoid the appearance of bare spots. Flight and tail feathers are lost in exact pairs, one from each side, so that balance is maintained. Replacements emerge before the next pair is lost, and most birds can continue to fly unimpaired during the molting period; however, many water birds (ducks, geese, loons, and others) lose all their primary feathers at once and are grounded during the molt. Many prepare for molting by moving to isolated bodies of water where they can find food and more easily escape enemies. Nearly all birds molt at least once a year, usually in late summer after the nesting season.

FIGURE **18-4**

Contour feather. Inset enlargement of the vane shows the minute hooks on the barbules that cross-link loosely to form a continuous surface of vane.

Skeleton

One of the major structural requirements for flight is a light, yet sturdy, skeleton (Figure 18-5A). As compared with the earliest known bird, *Archaeopteryx* (Figure 18-5B), the bones of modern birds are phenomenally light, delicate, and laced with air cavities. Such **pneumatized** bones (Figure 18-6) are nevertheless strong. The skeleton of a frigate bird with a 2.1 m (7-foot) wingspan weighs only 114 grams (4 ounces), less than the weight of all its feathers.

As archosaurs, birds evolved from ancestors with diapsid skulls (p. 326). However, the skulls of modern birds are so specialized that it is difficult to see any trace of the original diapsid condition. The bird skull is built lightly and mostly fused into one piece. A pigeon skull weighs only 0.21% of its body weight; by comparison the skull of a rat weighs 1.25% of its body weight. The braincase and orbits are large to accommodate a bulging brain and large eyes needed for quick motor coordination and superior vision.

In *Archaeopteryx,* both jaws contained teeth set in sockets, an archosaurian characteristic. Modern birds are completely toothless, having instead a horny (keratinous) beak molded around the bony jaws. The mandible is a complex of several bones hinged to provide a double-jointed action which permits the mouth to gape widely. Most birds have kinetic skulls (kinetic skulls of lizards are described on p. 332) with a flexible attachment between upper jaw and skull. This attachment allows the upper jaw to move slightly, thus increasing the gape.

The most distinctive feature of the vertebral column is its rigidity. Most of the vertebrae except the **cervicals** (neck vertebrae) are fused together and with the pelvic girdle to form a stiff but light framework to support the legs and provide rigidity for flight. To assist in this rigidity, the ribs are mostly fused with the vertebrae, pectoral girdle, and sternum. Except in flightless birds, the sternum bears a large, thin keel that provides for attachment of the powerful flight muscles. Because *Archaeopteryx* had no sternum (Figure 18-5B), there was no anchorage for the flight muscles equivalent to that of modern birds. This is one of the principal reasons why *Archaeopteryx* could not have done any strenuous wing-beating. *Archaeopteryx* did, however, have a furcula (wishbone) on which enough pectoral muscle could have attached to permit weak flight.

The bones of the forelimbs are highly modified for flight. They are reduced in number, and several are fused together. Despite these alterations, the bird wing is clearly a rearrangement of the basic vertebrate tetrapod limb from which it arose (Figure 16-1, p. 312), and all the elements—upper arm, forearm, wrist, and fingers—are represented in modified form (Figure 18-5A). The birds' legs have undergone less pronounced modification than the wings, since they are still designed principally for walking, as well as for perching, scratching, food gathering, and occasionally for swimming, as were those of their archosaurian ancestors.

Muscular System

The locomotor muscles of the wings are relatively massive to meet the demands of flight. The largest of these is the **pectoralis,** which depresses the wings in flight. Its antagonist is the **supracoracoideus** muscle, which raises the wing (Figure 18-7). Surprisingly, perhaps, this latter muscle is not located on the backbone (anyone who has been served the back of a chicken knows that it offers little meat) but is positioned under the pectoralis on the breast. It is attached by a tendon to the upper side of the humerus of the wing so that it pulls from below by an ingenious "rope-and-pulley" arrangement. Both pectoralis and supracoracoideus are anchored to the keel. With the main muscle mass low in the body, aerodynamic stability is improved.

The main leg muscle mass is located in the thigh, surrounding the femur, and a smaller mass lies over the tibiotarsus (shank or "drumstick"). Thin but strong tendons extend

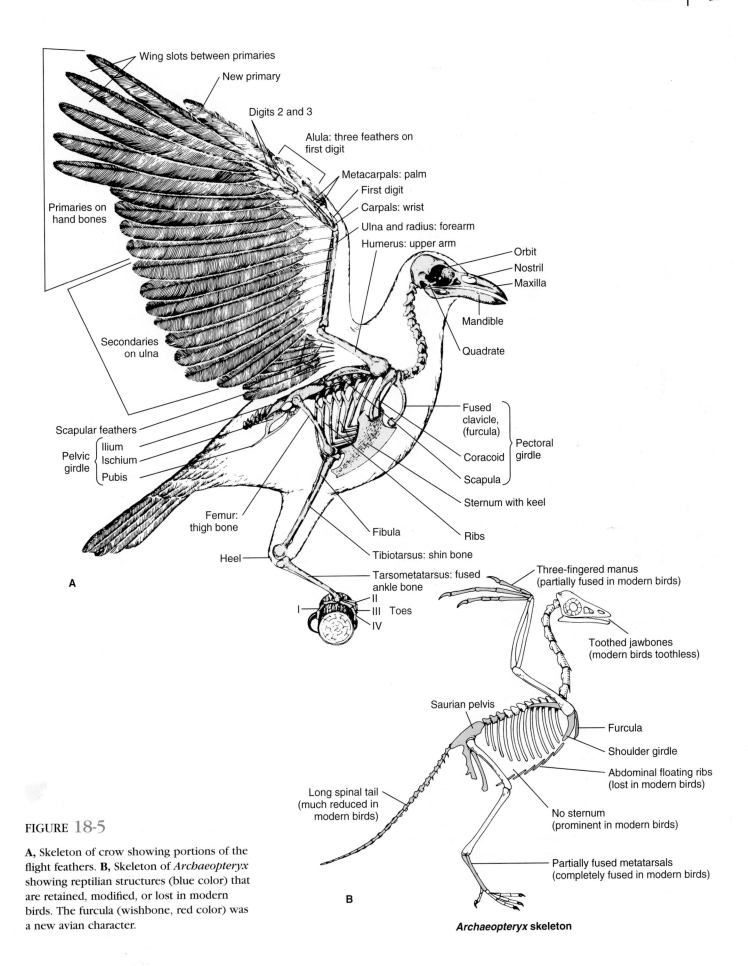

Wing slots between primaries
New primary
Digits 2 and 3
Alula: three feathers on first digit
Metacarpals: palm
First digit
Carpals: wrist
Ulna and radius: forearm
Humerus: upper arm
Orbit
Nostril
Maxilla
Mandible
Quadrate
Primaries on hand bones
Secondaries on ulna
Fused clavicle, (furcula)
Coracoid
Pectoral girdle
Scapula
Scapular feathers
Ilium
Pelvic girdle
Ischium
Pubis
Sternum with keel
Femur: thigh bone
Fibula
Ribs
Heel
Tibiotarsus: shin bone
A
Tarsometatarsus: fused ankle bone
I
II
III   Toes
IV

Three-fingered manus (partially fused in modern birds)
Toothed jawbones (modern birds toothless)
Saurian pelvis
Furcula
Shoulder girdle
Abdominal floating ribs (lost in modern birds)
Long spinal tail (much reduced in modern birds)
No sternum (prominent in modern birds)
Partially fused metatarsals (completely fused in modern birds)
B
***Archaeopteryx* skeleton**

FIGURE 18-5

A, Skeleton of crow showing portions of the flight feathers. **B,** Skeleton of *Archaeopteryx* showing reptilian structures (blue color) that are retained, modified, or lost in modern birds. The furcula (wishbone, red color) was a new avian character.

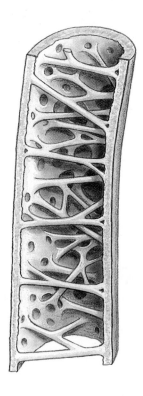

FIGURE 18-6

Hollow wing bone of a songbird showing the stiffening struts and air spaces that replace bone marrow. Such "pneumatized" bones are remarkably light and strong.

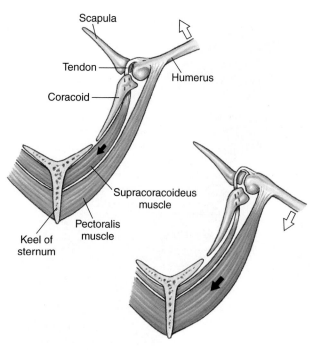

FIGURE 18-7

Flight muscles of a bird are arranged to keep the center of gravity low in the body. Both major flight muscles are anchored on the sternum keel. Contraction of the pectoralis muscle pulls the wing downward. Then, as the pectoralis relaxes, the supracoracoideus muscle contracts and, acting as a pulley system, pulls the wing upward.

downward through sleevelike sheaths to the toes. Consequently the feet are nearly devoid of muscles, explaining the thin, delicate appearance of the bird leg. This arrangement places the main muscle mass near the bird's center of gravity and at the same time allows great agility to the slender, lightweight feet. Because the feet are composed mostly of bone, tendon, and tough, scaly skin, they are highly resistant to damage from freezing. When a bird perches on a branch, an ingenious toe-locking mechanism (Figure 18-8) is activated, which prevents the bird from falling off its perch when asleep. The same mechanism causes the talons of a hawk or owl automatically to sink deeply into its prey as the legs bend under the impact of the strike. The powerful grip of a bird of prey was described by L. Brown.[1]

> When an eagle grips in earnest, one's hand becomes numb, and it is quite impossible to tear it free, or to loosen the grip of the eagle's toes with the other hand. One just has to wait until the bird relents, and while waiting one has ample time to realize that an animal such as a rabbit would be quickly paralyzed, unable to draw breath, and perhaps pierced through and through by the talons in such a clutch.

[1]Brown, L. 1970, Eagles, New York, Arco Publishing.

Digestive System

Birds process an energy-rich diet rapidly and thoroughly with efficient digestive equipment. A shrike can digest a mouse in 3 hours, and berries pass completely through the digestive tract of a thrush in just 30 minutes. Although many animal foods find their way into the diet of birds, insects comprise by far the largest component. Because birds lack teeth, foods that require grinding are reduced in the gizzard (see below). The poorly developed salivary glands mainly secrete mucus for lubricating the food and the slender, horn-covered **tongue.** There are few taste buds, although all birds can taste to some extent. From the short **pharynx** a relatively long, muscular, elastic **esophagus** extends to the **stomach.** In many birds there is an enlargement **(crop)** at the lower end of the esophagus, which serves as a storage chamber.

In pigeons, doves, and some parrots the crop not only stores food but, during the nesting season, produces "milk" by the breakdown of epithelial cells of the crop lining. For the first few days after hatching, the helpless young are fed regurgitated crop milk by both parents. Crop milk is especially rich in fat and protein.

The stomach proper consists of a **proventriculus,** which secretes gastric juice, and the muscular **gizzard,** a region specialized for grinding food. To assist the grinding process, grain-eating birds swallow gritty objects or pebbles,

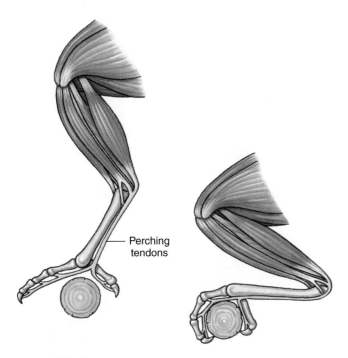

Perching
tendons

FIGURE 18-8

Perching mechanism of a bird. When a bird settles on a branch, tendons automatically tighten, closing the toes around the perch.

which lodge in the gizzard. Certain birds of prey, such as owls, form pellets of indigestible materials, mainly bones and fur, in the proventriculus and eject them through the mouth. At the junction of the intestine with the rectum there are paired **ceca;** these are well developed in herbivorous birds in which they serve as fermentation chambers. The terminal part of the digestive system is the **cloaca,** which also receives the genital ducts and ureters.

The beaks of birds are strongly adapted to specialized food habits—from generalized types, such as the strong, pointed beaks of crows and ravens, to grotesque, highly specialized ones in flamingos, pelicans, and avocets (Figure 18-9). The beak of a woodpecker is a straight, hard chisel-like device. Anchored to a tree trunk with its tail serving as a brace, the woodpecker delivers powerful, rapid blows to excavate nest cavities or expose the burrows of wood-boring insects. It then uses its long, flexible, barbed tongue to seek insects in their galleries. The woodpecker's skull is especially thick to absorb shock.

Circulatory System

The general plan of circulation in birds is not greatly different from that of mammals. The four-chambered heart is large with strong ventricular walls; thus birds share with mammals a complete separation of the respiratory and systemic circulations. The heartbeat is extremely fast, and as in mammals there

is an inverse relationship between heart rate and body weight. For example, a turkey has a heart rate at rest of approximately 93 beats per minute, a chicken has a rate of 250 beats per minute, and a black-capped chickadee has a heart rate of 500 beats per minute when asleep, which may increase to a phenomenal 1000 beats per minute during exercise. Blood pressure in birds is roughly equivalent to that in mammals of similar size. Birds' blood contains **nucleated, biconvex erythrocytes.** (Mammals, the only other endothermic vertebrates, have biconcave erythrocytes without nuclei that are somewhat smaller than those of birds.) The **phagocytes,** or mobile ameboid cells of the blood, are particularly efficient in birds in the repair of wounds and in destroying microbes.

Respiratory System

The respiratory system of birds differs radically from the lungs of reptiles and is marvelously adapted for meeting the high metabolic demands of flight. In birds the finest branches of the bronchi, rather than ending in saclike alveoli as in mammals, are tubelike **parabronchi** through which air flows continuously. Also unique is the extensive system of nine interconnecting **air sacs** that are located in pairs in the thorax and abdomen and even extend by tiny tubes into the centers of the long bones (Figure 18-10A). The air sacs are connected to the lungs in such a way that perhaps 75% of the inspired air bypasses the lungs and flows directly into the air sacs, which serve as reservoirs for fresh air. On expiration, some of this fully oxygenated air is shunted through the lung, while the used air passes directly out (Figure 18-10B). The advantage of such a system is obvious: the lungs receive fresh air during both inspiration and expiration. An almost continuous stream of oxygenated air passes through a system of richly vascularized parabronchi. Although many details of a bird's respiratory system are not fully understood, it is clearly the most efficient respiratory system of any vertebrate.

The remarkable efficiency of the bird respiratory system is emphasized by bar-headed geese that routinely migrate over the Himalayan mountains and have been sighted flying over Mt. Everest (8848 meters or 29,141 feet) under conditions that are severly hypoxic to humans. They reach altitudes of 9000 meters in less than a day, without the acclimatization that is absolutely essential for humans even to approach the upper reaches of Mt. Everest.

Excretory System

The relatively large paired metanephric kidneys are composed of many thousands of **nephrons,** each consisting of a renal corpuscle and a nephric tubule. As in other vertebrates, urine

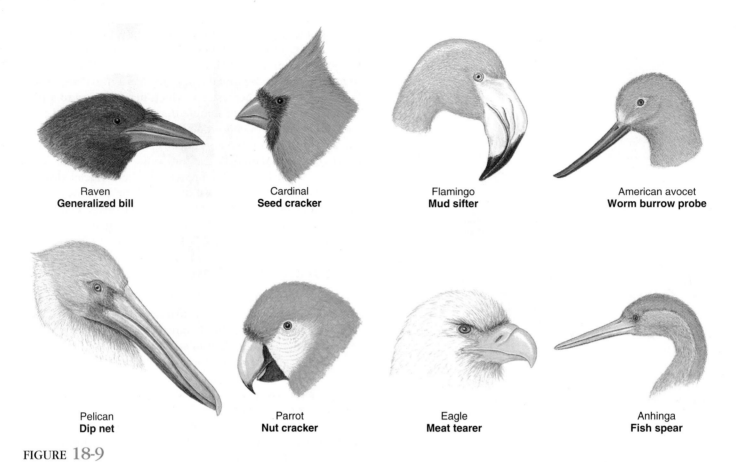

Raven
Generalized bill

Cardinal
Seed cracker

Flamingo
Mud sifter

American avocet
Worm burrow probe

Pelican
Dip net

Parrot
Nut cracker

Eagle
Meat tearer

Anhinga
Fish spear

FIGURE 18-9

Some bills of birds showing variety of adaptations.

is formed by glomerular filtration followed by the selective modification of the filtrate in the tubule.

Birds, like reptiles, excrete their nitrogenous wastes as uric acid rather than urea, an adaptation that originated with the evolution of the shelled (amniotic) egg. In shelled eggs, all excretory products must remain within the eggshell with the growing embryo. If urea were produced, it would quickly accumulate in solution to toxic levels. Uric acid, however, crystallizes out of solution and can be stored harmlessly within the eggshell. Thus, from an embryonic necessity was born an adult virtue. Because of uric acid's low solubility, a bird can excrete 1 g of uric acid in only 1.5 to 3 ml of water, whereas a mammal may require 60 ml of water to excrete 1 g of urea. Uric acid is combined with fecal material in the cloaca. Excess water is reabsorbed in the cloaca, resulting in the formation of a white paste. Thus, despite having kidneys that are less effective in true concentrative ability than mammalian kidneys, birds can form urine containing uric acid nearly 3000 times more concentrated than that in the blood. Even the most effective mammalian kidneys, those of certain desert rodents, can excrete urea only about 25 times the plasma concentration.

Marine birds (also marine turtles) have evolved a unique method for excreting the large loads of salt eaten with their food and in the seawater they drink. Seawater contains approximately 3% salt and is three times saltier than a bird's body fluids. Because the bird kidney cannot concentrate salt in urine above approximately 0.3%, excess salt is removed from the blood by special **salt glands,** one located above each eye (Figure 18-11) . These glands are capable of excreting a highly concentrated solution of sodium chloride—up to twice the concentration of seawater. The salt solution runs out the internal or external nostrils, giving gulls, petrels, and other sea birds a perpetual runny nose.

Nervous and Sensory System

The design of a bird's nervous and sensory system reflects the complex problems of flight and a highly visible existence, in which it must gather food, mate, defend territory, incubate and rear young, and correctly distinguish friend from foe. The brain of a bird has well-developed **cerebral hemispheres, cerebellum,** and **midbrain tectum** (optic lobes). The **cerebral cortex**—chief coordinating center of the mammalian

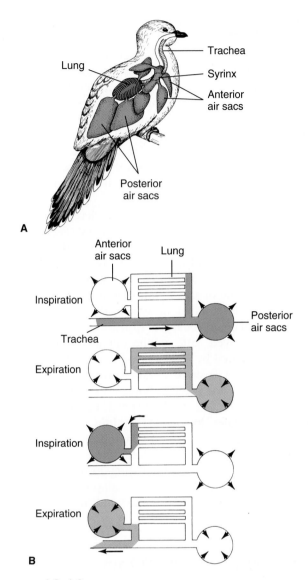

FIGURE 18-10

Respiratory system of a bird. **A,** Lungs and air sacs. One side of the bilateral air sac system is shown. **B,** Movement of a single volume of air through the bird's respiratory system. Two full respiratory cycles are required to move the air through the system.

brain—is thin, unfissured, and poorly developed in birds. But the core of the cerebrum, the **corpus striatum,** has enlarged into the principal integrative center of the brain, controlling such activities as eating, singing, flying, and all complex instinctive reproductive activities. Relatively intelligent birds, such as crows and parrots, have larger cerebral hemispheres than do less intelligent birds, such as chickens and pigeons. The **cerebellum** is a crucial coordinating center where muscle-position sense, equilibrium sense, and visual cues are assembled and used to coordinate movement and balance. The **optic lobes,** laterally bulging structures of the midbrain,

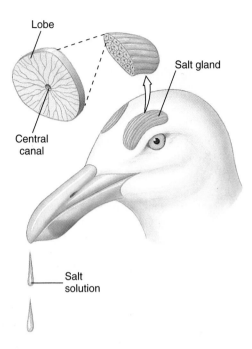

FIGURE 18-11

Salt glands of a marine bird (gull). One salt gland is located above each eye. Each gland consists of several lobes arranged in parallel. One lobe is shown in cross section, much enlarged. Salt is secreted into many radially arranged tubules, then flows into a central canal that leads into the nose.

form a visual apparatus comparable to the visual cortex of mammals.

Except in flightless birds, ducks, and vultures, the senses of smell and taste are poorly developed in birds. This deficiency, however, is more than compensated by good hearing and superb vision, the keenest in the animal kingdom. The organ of hearing, the **cochlea,** is much shorter than the coiled mammalian cochlea, yet birds can hear roughly the same range of sound frequencies as humans. Actually the bird ear far surpasses our capacity to distinguish differences in intensities and to respond to rapid fluctuations in pitch.

The bird eye resembles that of other vertebrates in gross structure but is relatively larger, less spherical, and almost immobile; instead of turning their eyes, birds turn their heads with their long flexible necks to scan the visual field. The light-sensitive **retina** (Figure 18-12) is generously equipped with rods (for dim light vision) and cones (for color vision). Cones predominate in day birds, and rods are more numerous in nocturnal birds. A distinctive feature of the bird eye is the **pecten,** a highly vascularized organ attached to the retina and jutting into the vitreous humor (Figure 18-12). The pecten is thought to provide nutrients to the eye. It may do more, but its function remains largely a mystery.

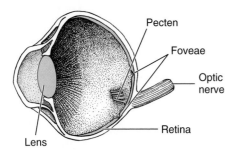

FIGURE **18-12**

Hawk eye has all the structural components of the mammalian eye, plus a peculiar pleated structure, the pecten, believed to provide nourishment to the retina. The extraordinarily keen vision of the hawk is attributed to the extreme density of cone cells in the foveae: 1.5 million per fovea compared to 0.2 million for humans.

Many birds can see into the ultraviolet, enabling them to view environmental features inaccessible to us but accessible to insects (such as flowers with ultraviolet-reflecting "nectar guides" that attract pollinating insects). Several species of ducks, hummingbirds, kingfishers, and passerines (songbirds) can see in the near ultraviolet (UV) down to 370 nm (the human eye filters out ultraviolet light below 400 nm). For what purpose do birds use their UV-sensitivity? Some, such as hummingbirds, may be attracted to nectar-guiding flowers, like insects. But, for the others, the benefit derived from UV-sensitivity is a matter of conjecture.

The **fovea,** or region of keenest vision on the retina, is in a deep pit (in birds of prey and some others), which makes it necessary for the bird to focus exactly on the subject. Many birds moreover have two sensitive spots (foveae) on the retina (Figure 18-12), the central one for sharp monocular views and the posterior one for binocular vision. The visual acuity of a hawk is about eight times that of humans (enabling the hawk to see clearly a crouching rabbit 2 km away), and an owl's ability to see in dim light is more than 10 times that of a human.

Flight

What prompted the evolution of flight in birds, the ability to rise free of earthbound concerns, as almost every human has dreamed of doing? The air was a relatively unexploited habitat stocked with flying insects for food. Flight also offered escape from terrestrial predators and opportunity to travel rapidly and widely to establish new breeding areas and to benefit from year-round favorable climate by migrating north and south with the seasons.

Bird Wing as a Lift Device

The bird wing is an airfoil that is subject to recognized laws of aerodynamics. It is streamlined in cross section, with a slightly concave lower surface **(cambered)** and with small, tight-fitting feathers where the leading edge meets the air (Figure 18-13). Air slips smoothly over the wing, creating lift with minimum drag. Some lift is produced by positive pressure against the undersurface of the wing. But on the upper side, where the airstream must travel farther and faster over a convex surface, negative pressure is created that provides more than two-thirds of the total lift.

The lift-to-drag ratio of an airfoil is determined by the angle of tilt (angle of attack) and the airspeed (Figure 18-13). A wing carrying a given load can pass through the air at high speed and small angle of attack or at low speed and larger angle of attack. As speed decreases, lift can be increased by increasing the angle of attack, but drag forces also increase. Finally a point is reached at which the angle of attack becomes too steep; turbulence appears on the upper surface, lift is destroyed, and stalling occurs. Stalling can be delayed or prevented by placing a **wing slot** along the leading edge so that a layer of rapidly moving air is directed across the upper wing surface. Wing slots were and still are used in aircraft traveling at a low speed. In birds, two kinds of wing slots have developed: (1) the **alula,** or group of small feathers on the thumb (Figures 18-5A and 18-15), which provides a midwing slot, and (2) **slotting between the primary feathers,** which provides a wing-tip slot. In a number of songbirds, these together provide stall-preventing slots for nearly the entire outer (and aerodynamically more important) half of the wing.

Basic Forms of Bird Wings

Bird wings vary in size and form because the successful exploitation of different habitats has imposed special aerodynamic requirements. Four types of bird wings are easily recognized.[2]

Elliptical Wings

Birds such as sparrows, warblers, doves, woodpeckers, and magpies (Figure 18-14A) that must maneuver in forested habitats, have elliptical wings. This type has a **low aspect ratio** (ratio of length to average width). The wings of the highly maneuverable British Spitfire fighter plane of World War II fame conformed closely to the outline of a sparrow's wing. Elliptical wings are slotted between the primary feathers; this slotting helps prevent stalling during sharp turns, low-speed flight, and frequent landing and takeoff. Each separated primary feather behaves as a narrow wing with a high angle of attack, providing high lift at low speed. The high maneuverability of the ellip-

---

[2]Saville, D.B.O. 1957. Adaptive evolution in the avian wing. Evolution **11:**212–224.

tical wing is exemplified by the tiny chickadee, which, if frightened, can change course within 0.03 second.

High-Speed Wings

Birds that feed on the wing, such as swallows, hummingbirds, and swifts, or that make long migrations, such as plovers, sandpipers, terns and gulls, (Figure 18-14B), have wings that sweep back and taper to a slender tip. They are rather flat in section, have a moderately high aspect ratio, and lack the wing-tip slotting characteristic of elliptical wings. Sweepback and wide separation of the wing tips reduce "tip vortex" (Figure 18-13, bottom panel), a drag-creating turbulence that tends to develop at wing tips at faster speeds. This type of wing is aerodynamically efficient for high-speed flight but cannot easily keep a bird airborne at low speeds. The fastest birds, such as sandpipers, clocked at 175 km (109 miles) per hour, belong to this group.

Soaring Wings

The oceanic soaring birds have **high-aspect ratio** wings resembling those of sailplanes. This group includes albatrosses, frigate birds, and gannets (Figure 18-14C). Such long, narrow wings lack wing slots and are adapted for high speed, high lift, and dynamic soaring. They have the highest aerodynamic efficiency of all wings but are less maneuverable than the wide, slotted wings of land soarers. Dynamic soarers have learned how to exploit the highly reliable sea winds, using adjacent air currents of different velocities.

High-Lift Wings

Vultures, hawks, eagles, owls, and ospreys (Figure 18-14D and 18-15)—predators that carry heavy loads—have wings with slotting, alulas, and pronounced camber, all of which promote high lift at low speed. Many of these birds are land soarers, with broad, slotted wings that provide the sensitive response and maneuverability required for static soaring in the capricious air currents over land.

Flapping Flight

This basic form of flight is so complex that complete analysis is still not possible—yet young birds fly almost perfectly on their maiden flight. More than a century ago an English zoologist reared swallow fledglings in a space so confining that they could not fully extend their wings. Yet when released at the age when swallows normally fly, they flew immediately and without practice.

Two forces are required for flapping flight: a vertical *lifting* force to support the bird's weight, and a horizontal *thrusting* force to move the bird forward against the resistive forces of friction. Thrust is provided mainly by the primary feathers at the wing tips, while the secondary feathers of the inner

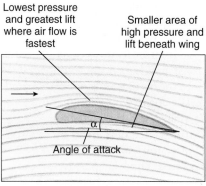

Air flow around wing

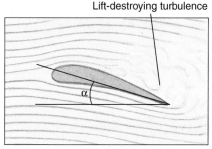

Stalling at low speed

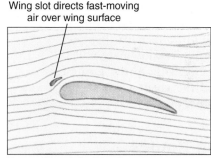

Preventing stall with wing slots

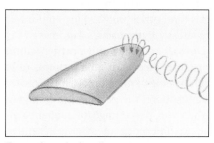

Formation of wing tip vortex

FIGURE 18-13

Air patterns formed by the airfoil, or wing, moving from right to left. At low speed the angle of attack (α) must increase to maintain lift but this increases the threat of stalling. The upper figures show how low-speed stalling can be prevented with wing slots. Wing tip vortex *(bottom),* a turbulence that tends to develop at high speeds, reduces flight efficiency. The effect is reduced in wings that sweep back and taper to a tip.

wing, which do not move so far or so fast, act as an airfoil, providing mainly lift. Greatest power is applied on the downstroke. The primary feathers bend upward and twist to a steep angle of attack, biting into the air like a propeller (Figure 18-16). The entire wing (and the bird's body) is pulled forward. On the upstroke, the primary feathers bend in the opposite direction so that their upper surfaces twist into a positive angle of attack to produce thrust, just as the lower surfaces did on the downstroke. A powered upstroke is essential for hovering flight, as in hummingbirds, and is important for fast, steep takeoffs by small birds with elliptical wings.

Migration and Navigation

We described the advantages of migration in the prologue to this chapter. Not all birds migrate, of course, but the majority of North American and European species do, and the biannual journeys of some are truly extraordinary undertakings. Migration is both the greatest adventure and the greatest risk in the life of a migratory bird.

Migration Routes

Most migratory birds have well-established routes trending north and south. Since most birds (and other animals) live in the Northern Hemisphere, where most of the earth's landmass is concentrated, most birds are south-in-winter and north-in-summer migrants. Of the 4000 or more species of migrant birds (a little less than half the total bird species), most breed in the more northern latitudes of the hemisphere; the percentage of migrants in Canada is far higher than the percentage of migrants in Mexico, for example. Some use different routes in the fall and spring (Figure 18-17). Some, especially certain aquatic species, complete their migratory routes in a very short time. Others, however, make a leisurely trip, often stopping along the way to feed. Some of the warblers are known to take 50 to 60 days to migrate from their winter quarters in Central America to their summer breeding areas in Canada.

Some species are known for their long-distance migrations. The Arctic tern, greatest globe spanner of all, breeds north of the Arctic Circle during the northern summer then migrates to the Antarctic regions for the northern winter. This species is also

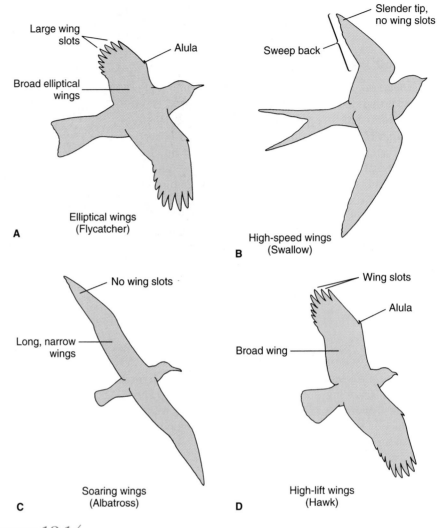

A Elliptical wings (Flycatcher) — Large wing slots, Alula, Broad elliptical wings

B High-speed wings (Swallow) — Slender tip, no wing slots, Sweep back

C Soaring wings (Albatross) — No wing slots, Long, narrow wings

D High-lift wings (Hawk) — Wing slots, Alula, Broad wing

FIGURE 18-14

Four basic forms of bird wings.

known to take a circuitous route in migrations from North America, passing over to the coastlines of Europe and Africa and then to winter quarters, a trip that may exceed 18,000 km (11,200 miles). Other birds that breed in Alaska follow a more direct line down the Pacific coast of North and South America.

Many small songbirds also make great migratory treks (Figure 18-17). Africa is a favorite wintering ground for European birds, and many fly there from Central Asia as well.

Stimulus for Migration

Humans have known for centuries that the onset of the reproductive cycle of birds is closely related to season. Only within the last 60 years, however, has it been proved that the lengthening days of late winter and early spring stimulate the development of the gonads and accumulation of fat—both important internal changes that predispose birds to migrate northward.

FIGURE 18-15

Osprey, *Pandion haliaetus* (order Falconiformes), landing on nest. Note alulas *(arrows)*. Feathers are molted in sequence in exact pairs so that balance is maintained during flight.

FIGURE 18-16

In normal flapping flight of strong fliers like ducks, the wings sweep downward and forward fully extended. Thrust is provided by the primary feathers at the wing tips. To begin the upbeat, the wing is bent, bringing it upward and backward. The wing then extends, ready for the next downbeat.

Increasing day length stimulates the anterior lobe of the pituitary into activity. The release of pituitary gonadotropic hormone in turn sets in motion a complex series of physiological and behavioral changes, resulting in gonadal growth, fat deposition, migration, courtship and mating behavior, and care of the young.

Direction Finding in Migration

Numerous experiments suggest that most birds navigate chiefly by sight. Birds recognize topographical landmarks and follow familiar migratory routes—a behavior assisted by flock migration, during which navigational resources and experience of older birds can be pooled. But in addition to visual navigation, birds make use of a variety of orientation cues at their disposal. Birds have a highly accurate innate sense of time. They also have an innate sense of direction; and recent work adds credence to an old, much debated hypothesis that birds can detect and navigate by the earth's magnetic field. All of these resources are inborn and instinctive, although a bird's navigational abilities may improve with experience.

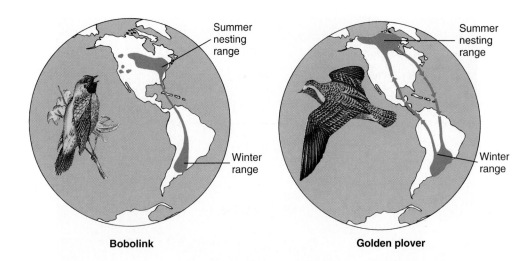

Bobolink   **Golden plover**

FIGURE 18-17

Migrations of the bobolink and golden plover. The bobolink commutes 22,500 km (14,000 miles) each year between nesting sites in North America and its wintering range in Argentina, a phenomenal feat for such a small bird. Although the breeding range has extended to colonies in western areas, these birds take no shortcuts but adhere to the ancestral seaboard route. The golden plover flies a loop migration, striking out across the Atlantic in its southward autumnal migration but returning in the spring by way of Central America and the Mississippi Valley because ecological conditions are more favorable at that time.

In the early 1970s W. T. Keeton showed that the flight bearings of homing pigeons were significantly disturbed by magnets attached to the birds' heads, or by minor fluctuations in the geomagnetic field. But until recently the nature and position of a magnetic receptor in pigeons remained a mystery. Deposits of a magnetic substance called magnetite (Fe$_3$O$_4$) have been discovered in the neck musculature of pigeons and migratory white-crowned sparrows. If this material were coupled to sensitive muscle receptors, as has been proposed, the structure could serve as a magnetic compass that would enable birds to detect and orient their migrations to the earth's magnetic field.

Experiments by German ornithologists G. Kramer and E. Sauer and American ornithologist S. Emlen demonstrated convincingly that birds can navigate by celestial cues: the sun by day and the stars by night. Using special circular cages, Kramer concluded that birds maintain compass direction by referring to the sun, regardless of the time of day (Figure 18-18). This process is called **sun-azimuth orientation** (*azimuth*, compass bearing of the sun). Sauer's and Emlen's ingenious planetarium experiments also strongly suggest that some birds, probably many, are able to detect and navigate by the North Star axis around which the constellations appear to rotate.

Some of the remarkable feats of bird navigation still defy rational explanation. Most birds undoubtedly use a combination of environmental and innate cues to migrate. Migration is a rigorous undertaking. The target is often small, and natural selection relentlessly prunes off individuals making errors in migration, leaving only the best navigators to propagate the species.

Social Behavior and Reproduction

The adage says "birds of a feather flock together," and many birds are indeed highly social creatures. Especially during the breeding season, sea birds gather, often in enormous colonies, to nest and rear young. Land birds, with some conspicuous exceptions, such as starlings and rooks, tend to be less gregarious than sea birds during breeding and to seek isolation for rearing their brood. But these same species that covet separation from their kind during breeding may aggregate for migration or feeding. Togetherness offers advantages: mutual protection from enemies, greater ease in finding mates, less opportunity for individual straying during migration, and mass huddling for protection against low night temperatures during migration. Certain species, such as pelicans (Figure 18-19), may use highly organized cooperative behavior to feed. At no time are the highly organized social interactions of birds more evident than during the breeding season, as they stake out territorial claims, select mates, build nests, incubate and hatch their eggs, and rear their young.

Reproductive System

The testes are tiny bean-shaped bodies during most of the year, and at the breeding season undergo great enlargement, as

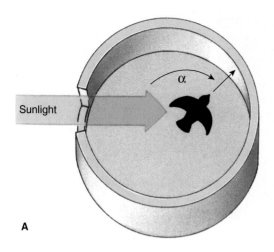

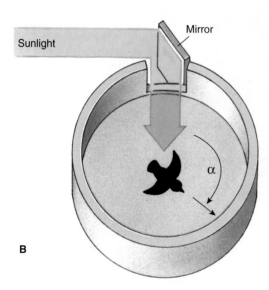

FIGURE 18-18

Gustav Kramer's experiments with sun-compass navigation in starlings. **A,** In a windowed, circular cage, the bird fluttered to align itself in the direction it would normally follow if it were free. **B,** When the true angle of the sun is deflected with a mirror, the bird maintains the same relative position to the sun. This shows that these birds use the sun as a compass. The bird navigates correctly throughout the day, changing its orientation to the sun as the sun moves across the sky.

much as 300 times larger than the nonbreeding size. Before discharge, the millions of sperm are stored in a **seminal vesicle,** which, like the testes, enlarges greatly during the breeding season. Since males of most species lack a penis, copulation is a matter of bringing cloacal surfaces into contact, usually while the male stands on the back of the female (Figure 18-20). Some swifts copulate in flight.

In the female of most birds, only the **left ovary and oviduct** develop (Figure 18-21); those on the right dwindle to vestigial structures (the loss of one ovary is another adaptation of birds for reducing weight). Eggs discharged from the ovary

are picked up by the oviduct, which runs posteriorly to the cloaca. While the eggs are passing down the oviduct, **albumin,** or egg white, from special glands is added to them; farther down the oviduct, the shell membrane, shell, and shell pigments are secreted about the egg.

Fertilization takes place in the upper oviduct several hours before the layers of albumin, shell membranes, and shell are added. Sperm remain alive in the female oviduct for many days after a single mating.

Mating Systems

The two most common types of mating systems in animals are **monogamy,** in which an individual mates with only one partner each breeding season, and **polygamy,** in which an individual mates with two or more partners each breeding period. Monogamy is rare in most animal groups, but in birds it is the general rule: more than 90% are monogamous. In a few bird species such as swans and geese, partners are chosen for life and often remain together throughout the year. Seasonal monogamy is more common, however; the great majority of migrant birds pair up during the breeding season but lead independent lives the rest of the year.

One reason that monogamy is much more common among birds than among mammals is that female birds are not equipped, as mammals are, with a built-in food supply for the young. Accordingly, the ability of the two sexes to provide parental care, especially food for the young, is more equal in birds than in mammals. A female bird will choose a male whose parental investment in their young is apt to be high and avoid a male that has mated with another female. If the male had mated with another female, he could at best divide his time between his two mates and might even devote most of his attention to the alternate mate. Consequently, females enforce monogamy.

Monogamy in birds is also encouraged by the need for the male to secure and defend a territory before he can attract a mate. The male may sing a great deal to announce his presence to females and to discourage rival males from entering his territory. The female wanders from one territory to another, seeking a male with foraging territory that offers the best chances for reproductive success. Usually a male is able to defend an area that provides just enough resources for one nesting female.

The most common form of polygamy in birds, when it occurs, is **polygyny** ("many females"), in which a male mates with more than one female. In many species of grouse, the males gather in a collective display ground, the **lek,** which is divided into individual territories, each vigorously defended by a displaying male (Figure 18-22). There is nothing of value in the lek to the female except the male, and all he can offer are his genes, for only the females care for the young. Usually there are a dominant male and several subordinate males in the lek. Competition among males for females is intense, but females appear to choose the dominant male for mating because, presumably, social rank correlates with genetic quality.

A

B

FIGURE 18-19

Cooperative feeding behavior by the white pelican, *Pelecanus onocrotalus.* **A,** Pelicans form a horseshoe to drive fish together. **B,** Then they plunge simultaneously to scoop fish in their huge bills. These photographs were taken 2 seconds apart.

FIGURE 18-20

Copulation in birds. In most bird species the male lacks a penis. The male copulates by standing on the back of the female, pressing his cloaca against that of the female, and passing sperm to the female.

Nesting and Care of Young

To produce offspring, all birds lay eggs that must be incubated by one or both parents. The eggs of most songbirds (order Passeriformes) require approximately 14 days for hatching; those of ducks and geese require at least twice that long. Most of the duties of incubation fall on the female, although in many instances both parents share the task, and occasionally only the male incubates the eggs.

Most birds build some form of nest in which to rear their young. Some birds simply lay their eggs on the bare ground or rocks, making no pretense of nest building. Others build elaborate nests such as the pendant nests constructed by orioles, the delicate lichen-covered mud nests of hummingbirds (Figure 18-23) and flycatchers, the chimney-shaped mud nests of cliff swallows, the floating nests of rednecked grebes, and the huge brush-pile nests of Australian brush turkeys. Most birds take considerable pains to conceal their nests from enemies. Woodpeckers, chickadees, bluebirds, and many others place their nests in tree hollows or other

cavities; kingfishers excavate tunnels in the banks of streams for their nests; and birds of prey build high in lofty trees or on inaccessible cliffs. Nest parasites such as the brown-headed cowbird and the European cuckoo build no nests at all but simply lay their eggs in the nests of birds smaller than themselves. When the eggs hatch, the foster parents care for the cowbird young which outcompete the host's own hatchlings.

Newly hatched birds are of two types: **precocial** and **altricial.** Precocial young, such as quail, fowl, ducks, and most water birds, are covered with down when hatched and can run or swim as soon as their plumage is dry (Figure 18-24). The altricial ones, on the other hand, are naked and helpless at birth and remain in the nest for a week or more. The young of both types require care from parents for some time after hatching. They must be fed, guarded, and protected against rain and sun. The parents of altricial species must carry food to their young almost constantly, for most young birds will eat more than their weight each day. This enormous food consumption explains the rapid growth of the young and their quick exit from the nest. The food of the young, depending on the species, includes worms, insects, seeds, and fruit.

Nesting success is very low with many birds, especially in altricial species. One investigation several years ago of 170 altricial bird nests reported that only 21% produced at least one young. The annual censusing of birds shows that nesting success is even lower today. Of the many causes of nesting failures, predation by raccoons, skunks, opossums, blue jays, crows, and others, especially in suburban and rural woodlots,

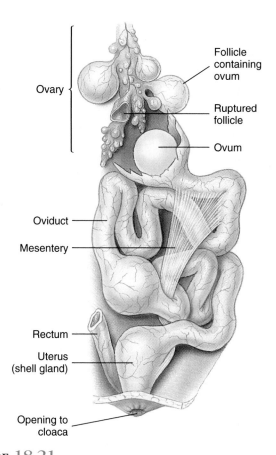

Ovary

Follicle containing ovum

Ruptured follicle

Ovum

Oviduct

Mesentery

Rectum

Uterus (shell gland)

Opening to cloaca

FIGURE 18-21

Reproductive system of a female bird.

and nest parasitism by the brown-headed cowbird are the most important factors.

Bird Populations

Bird populations, like those of other animal groups, vary in size from year to year. Snowy owls, for example, are subject to population cycles that closely follow cycles in their food supply, mainly rodents. Voles, mice, and lemmings in the north have a fairly regular 4-year cycle of abundance (p. 384); at population peaks, predator populations of foxes, weasels, and buzzards, as well as snowy owls, increase because there is abundant food for rearing their young. After a crash in the rodent population, snowy owls move south, seeking alternative food supplies. They occasionally appear in large numbers in southern Canada and the northern United States, where their total absence of fear of humans makes them easy targets for thoughtless hunters.

Occasionally the activities of people bring about spectacular changes in bird distribution. Both starlings (Figure 18-25) and house sparrows have been accidentally or deliberately introduced into numerous countries, to become the two most abundant bird species on earth, with the exception of domestic fowl.

FIGURE 18-22

Dominant male sage grouse, *Centrocercus urophasianus,* surrounded by several hens that have been attracted by his "booming" display.

FIGURE 18-23

Anna's hummingbird, *Calypte anna,* feeding its young in its nest of plant down and spider webs and decorated on the outside with lichens. The female builds the nest, incubates the two pea-sized eggs, and rears the young with no assistance from the male. Anna's hummingbird is a common resident of California. It is the only hummer to overwinter in the United States.

Humans also are responsible for the extinction of many bird species. More than 80 species of birds have, since 1695, followed the last dodo to extinction. Many were victims of changes in their habitat or competition with better-adapted species. But several have been hunted to extinction, among them the passenger pigeon, which only a century ago darkened the skies over North America in incredible numbers estimated in the billions (Figure 18-26).

Altricial
One-day-old meadowlark

Precocial
One-day-old ruffed grouse

FIGURE **18-24**

Comparison of 1-day-old altricial and precocial young. The altricial meadowlark *(left)* is born nearly naked, blind, and helpless. The precocial ruffed grouse *(right)* is covered with down, alert, strong legged, and able to feed itself.

A

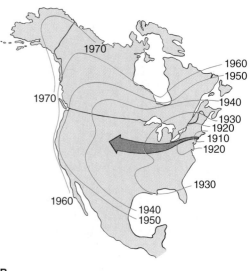

B

FIGURE **18-25**

A, European starling, *Sturnus vulgaris.* Starlings are omnivorous, eating mostly insects in spring and summer and shifting to wild fruits in the fall. **B,** Colonization of North America by European starlings after the introduction of 120 birds into Central Park in New York City in 1890. There are now perhaps 100 million starlings in the United States alone, testimony to the great reproductive potential of birds.

Today, game bird hunting is a well-managed renewable resource in the United States and Canada, and while hunters kill millions of game birds each year, none of the 74 bird species legally hunted are endangered. Hunting interests, by acquiring large areas of wetlands for migratory bird refuges and sanctuaries, have contributed to the recovery of both game and nongame birds.

Lead poisoning of waterfowl is a side effect of hunting. Before long-delayed federal regulations went into effect in 1991, requiring the use of nonlead shot for all inland and coastal waterfowl hunting, shotguns scattered more than 3000 tons of lead each year in the United States alone. When waterfowl eat the pellets (which they mistake for seeds), the pellets are ground and eroded in their gizzards, facilitating the absorption of lead into the blood. Lead poisoning paralyzes or weakens birds, leading to death by starvation. Today, birds are still dying from ingesting lead shot that has accumulated over the years.

FIGURE **18-26**

Sport-shooting of passenger pigeons in Louisiana during the nineteenth century. Relentless sport and market hunting before the establishment of state and federal hunting regulations, eventually dropped the population too low to sustain colonial breeding. The last passenger pigeon died in captivity in 1914.

Classification of Class Aves[3]

The class Aves contains more than 9600 species distributed among some 27 orders of living birds and a few fossil orders. Very few birds remain to be discovered. Of the 27 orders, four (or five depending on the classification system) are **ratite,** or flightless, birds (ostriches, rheas, cassowaries and emus, and kiwis), although flightlessness is not restricted to these groups. The remaining 23 orders are **carinate** birds (birds with a keeled sternum).

Class Aves (L. *avis,* bird)

Subclass Archaeornithes (Gr. *archaios,* ancient, + *ornis,* bird). Birds of the late Jurassic and early Cretaceous bearing many primitive characteristics. *Archaeopteryx.*

Subclass Neornithes (Gr. *neos,* new, + *ornis,* bird). Extinct and living birds with well-developed sternum and usually with keel; tail reduced; metacarpals and some carpals fused together. Cretaceous to Recent.

Superorder Paleognathae (Gr. *palaios,* ancient, + *gnathos,* jaw). Modern birds with primitive archosaurian palate. Ratites (with unkeeled sternum) and tinamous (with keeled sternum).

Order Struthioniformes (stroo'thi-on-i-for'meez) (L. *struthio,* ostrich, + *forma,* form): **ostrich.** One species, the flightless ostrich of Africa (Figure 18-27) is the largest of living birds.

Order Rheiformes (re'i-for'meez) (Gr. mythology, *Rhea,* mother of Zeus; + form): **rheas.** Two species of flightless birds of South America; often called the American ostriches.

Order Casuariiformes (kazh'u-ar'ee-i-for'meez) (N.L. *Casuarius,* type genus, + form): **cassowaries, emus.** Four species of flightless birds found in Australia and New Guinea.

Order Apterygiformes (ap'te-rij'i-for'meez) (Gr. *a,* not, + *pteryx,* wing, + form): **kiwis.** Three species of kiwis, flightless birds of New Zealand.

Order Tinamiformes (tin-am'i-for'meez) (N.L. *Tinamus,* type genus, + form): **tinamous.** Ground-dwelling, grouselike birds of Central and South America. About 60 species.

Superorder Neognathae (Gr. *neos,* new, + *gnathos,* jaw). Modern birds with flexible palate.

Order Sphenisciformes (sfe-nis'i-for'meez) (Gr. *sphēniskos,* dim. of *sphen,* wedge, from the shortness of the wings, + form): **penguins.** Web-footed marine swimmers of the southern seas. About 17 species.

Order Gaviiformes (gay'vee-i-for'meez) (L. *gavia,* bird, probably sea mew, + form): **loons.** Four species of loons, divers with short legs and heavy bodies.

Order Podicipediformes (pod'i-si-ped'i-for'meez) (L. *podex,* rump; *pes, pedis,* foot): **grebes.** Short-legged divers; 18 species, worldwide distribution.

Order Procellariiformes (pro-sel-lar'ee-i-for'meez) (L. *procella,* tempest, + form): **albatrosses, petrels, fulmars, shearwaters.** Marine birds with hooked beak and tubular nostrils. About 100 species, worldwide distribution.

Order Pelecaniformes (pele-can-i-form'eez) (Gr. *pelekan,* pelican, + form): **pelicans, cormorants, gannets, boobies, anhingus, frigatebirds,** and **tropicbirds.** Colonial fish-eaters with throat pouch. About 55 species, worldwide distribution, especially in the tropics.

Order Ciconiiformes (si-ko'nee-i-for'meez) (L. *ciconia,* stork, + form): **herons, bitterns, storks, ibises, spoonbills, flamingos** (Figure 18-28). Long-necked, long-legged, mostly colonial waders. About 90 species, worldwide distribution.

Order Anseriformes (an'ser-i-for'meez) (L. *anser,* goose + form): **swans, geese, ducks.** Birds having broad bills with marginal filtering ridges. About 150 species, worldwide distribution.

Order Falconiformes (fal'ko-ni-for'meez) (L. *falco,* falcon, + form): **eagles, hawks, vultures, falcons, condors, buzzards.** Diurnal birds of prey. About 270 species, worldwide distribution.

Order Galliformes (gal'li-for'meez) (L. *gallus,* cock, + form): **quail, grouse, pheasants, ptarmigan, turkeys, domestic fowl.** Chickenlike ground-nesting herbivores with strong beaks and heavy feet. About 250 species, worldwide distribution.

Order Gruiformes (groo′i-for′meez) (L. *grus,* crane, + form): **cranes, rails, coots, gallinules.** Prairie and marsh breeders. About 215 species, worldwide distribution.

Order Charadriiformes (ka-rad′ree-i-for′meez) (N.L. *Charadrius,* genus of plovers, + form): **gulls** (Figure 18-29), **oyster catchers, plovers, sandpipers, terns, woodcocks, turnstones, lapwings, snipe, avocets, phalaropes, skuas, skimmers, auks, puffins.** Shorebirds. About 330 species, worldwide distribution.

Order Columbiformes (ko-lum′bi-for′meez) (L. *columba,* dove, + form): **pigeons, doves.** Birds with short necks, short legs, and a short, slender bill. About 290 species, worldwide distribution.

Order Psittaciformes (sit′ta-si-for′meez) (L. *psittacus,* parrot, form): **parrots, parakeets.** Birds with hinged and movable upper beak, fleshy tongue. About 320 species, pantropical distribution.

Order Cuculiformes (ku-koo′li-for′meez) (L. *cuculus,* cuckoo, + form): **cuckoos, roadrunners.** About 150 species, worldwide distribution.

Order Strigiformes (strij′i-for′meez) (L. *strix,* screech owl, + form): **owls.** Nocturnal predators with large eyes, powerful beaks and feet, and silent flight. About 135 species, worldwide distribution.

Order Caprimulgiformes (kap′ri-mul′ji-for′meez) (L. *caprimulgus,* goatsucker, + form): **goatsuckers, nighthawks, whippoorwills.** Night and twilight feeders with small, weak legs and wide mouths fringed with bristles. About 95 species, worldwide distribution.

Order Apodiformes (up-pod′i-for′meez) (Gr. *apous,* footless, + form): **swifts, hummingbirds.** Small birds with short legs and rapid wingbeat. About 400 species, worldwide distribution.

Order Coliiformes (ka-lee′i-for′meez) (Gr. *kolios,* green woodpecker, + form): **mousebirds.** Six species of small southern African birds of uncertain relationship.

Order Trogoniformes (tro-gon′i-for′meez) (Gr. *trōgon,* gnawing, + form): **trogons.** Richly colored, long-tailed birds. About 35 species, pantropical distribution.

Order Coraciiformes (ka-ray′see-i-for′meez or kor′uh-sigh′uh-for′meez) (N.L. *coracii* from Gr. *korakias,* a kind of raven, + form): **kingfishers, hornbills,** and others. Birds with strong bills that nest in cavities. About 200 species, worldwide distribution.

Order Piciformes (pis′i-for′meez) (L. *picus,* woodpecker, + form): **woodpeckers, toucans, puffbirds, honeyguides.** Birds with highly specialized bills and having two toes extending forward and two backward. About 380 species, worldwide distribution.

Order Passeriformes (pas′er-i-for′meez) (L. *passer,* sparrow, + form): **perching songbirds** (Figure 18-30). The largest order of birds, containing 56 families and 60% of all birds. Most have a highly developed syrinx. Their feet are adapted for perching on thin stems and twigs. The young are altricial. More than 5000 species, worldwide distribution.

---

[3]The traditional bird classification given here, called a morphological taxonomy, is based on the careful comparison of shared derived anatomical characters within and between bird groups. A new and still controversial biochemical classification based on degrees of similarity between DNAs of living birds from all over the world is believed by its proponents to represent true phylogenetic relationships much better than the traditional morphological classification. The biochemical taxonomy has produced several astonishing realignments. Most prominent of these is the sweeping revision of the order Ciconiiformes which, as revised, includes penguins, loons, grebes, albatrosses, and birds of prey, all previously placed in separate orders. DNA hybridization studies establish the close relatedness of these groups, whose true genetic affinities are masked by divergent evolution. Biochemical taxonomy, now under review by the American Ornithological Union, is certain to produce significant revision of the traditional taxonomy which has been the standard for more than a century. Proctor and Lynch (1993) compare the biochemical classification reported by Sibley and Ahlquist (1990) with the traditional morphological classification.

FIGURE 18-27

Ostrich *Struthio camelus* of Africa, the largest of all living birds. Order Struthioniformes.

FIGURE 18-30

Ground finch *Geospiza fuliginosa,* one of the famous Darwin's finches of the Galápagos Islands. Order Passeriformes.

FIGURE 18-28

Greater flamingos *Phoenicopterus ruber* on an alkaline lake in East Africa. Order Ciconiiformes.

Of particular concern is the recent sharp decline of songbirds in the United States and southern Canada. Amateur birdwatchers and ornithologists have recorded that many songbird species that were abundant as recently as 40 years ago are now suddenly scarce. There are several reasons for the decline. Intensification of agriculture, permitted by the use of herbicides, pesticides, and fertilizers, has deprived ground-nesting birds of fields that were left fallow before the use of these agents. The excessive fragmentation of forests throughout much of the United States has increased exposure of nests of forest-dwelling species to nest predators such as blue jays, raccoons, and opossums, and to nest parasites such as the brown-headed cowbird. House cats also kill millions of small birds every year. From a study of radio-collared farm cats in Wisconsin, researchers estimated that in that state alone, cats may kill 19 million songbirds in a single year.

FIGURE 18-29

Laughing gulls *Larus atricilla* in flight. Order Charadriiformes.

The rapid loss of tropical forests—approximately 170,000 square kilometers each year, an area about the size of the state of Washington[4]—is depriving some 250 species of songbird migrants of their wintering homes. Of all the long-term threats facing songbird populations, tropical deforestation is the most serious and most intractable to change. If the rate of deforestation accelerates in the next few decades as expected, the world's tropical forests will have disappeared by 2040 (Terborgh, 1992).

Some birds, such as robins, house sparrows, and starlings, can accommodate these changes, and may even thrive on them. But for most birds the changes are adverse. Terborgh (1992) warns that unless we take leadership in managing our natural resources wisely we soon could be facing the silent spring that Rachel Carson envisioned 35 years ago.

[4]Brown, L.R. 1993. State of the world 1993. New York, W.W. Norton & Company.

Summary

The more than 9600 species of living birds are egg-laying, endothermic vertebrates covered with feathers and having the forelimbs modified as wings. Birds are closest phylogenetically to the theropods, a group of Mesozoic dinosaurs with several birdlike characteristics. The oldest known fossil bird, *Archaeopteryx* from the Jurassic period of the Mesozoic era, had numerous reptilian characteristics and was almost identical to certain theropod dinosaurs except that it had feathers. It is probably not in the direct lineage leading to modern birds but can be considered a sister group to modern birds.

The adaptations of birds for flight are of two basic kinds: those reducing body weight and those promoting more power for flight. Feathers, the hallmark of birds, are complex derivatives of reptilian scales and combine lightness with strength, water repellency, and high insulative value. Body weight is further reduced by elimination of some bones, fusion of others (to provide rigidity for flight), and the presence in many bones of hollow, air-filled spaces. The light, horny bill, replacing the heavy jaws and teeth of reptiles, serves as both hand and mouth for all birds and is variously adapted for different feeding habits.

Adaptations that provide power for flight include a high metabolic rate and body temperature coupled with an energy-rich diet; a highly efficient respiratory system consisting of a system of air sacs arranged to pass air through the lungs during both inspiration and expiration; powerful flight and leg muscles arranged to place muscle weight near the bird's center of gravity; and an efficient, high-pressure circulation.

Birds have keen eyesight, good hearing, poorly developed sense of smell, and superb coordination for flight. The metanephric kidneys produce uric acid as the principal nitrogenous waste.

Birds fly by applying the same aerodynamic principles as an airplane and using similar equipment: wings for lift and support, a tail for steering and landing control, and wing slots for control at low flight speed. Flightlessness in birds is unusual but has evolved independently in several bird orders, usually on islands where terrestrial predators are absent; all are derived from flying ancestors.

Bird migration refers to regular movements between summer nesting places and wintering regions. Spring migration to the north, where more food is available for nestlings, enhances reproductive success. Many cues are used for finding direction during migration, including innate sense of direction and ability to navigate by the sun, the stars, or the earth's magnetic field.

The highly developed social behavior of birds is manifested in vivid courtship displays, mate selection, territorial behavior, and incubation of eggs and care of the young.

Review Questions

1. Explain the significance of the discovery of *Archaeopteryx*. Why did this fossil prove beyond reasonable doubt that birds share an ancestor with some reptilian groups?
2. Birds are broadly divided into two groups: ratite and carinate. Explain what these terms mean and briefly discuss the appearance of flightlessness in birds.
3. The special adaptations of birds all contribute to two essentials for flight: more power and less weight. Explain how each of the following contributes to one or both of these two essentials: feathers, skeleton, muscle distribution, digestive system, circulatory system, respiratory system, excretory system, reproductive system.
4. How do marine birds rid themselves of excess salt?
5. In what ways are the bird's ears and eyes specialized for the demands of flight?
6. Explain how the bird wing is designed to provide lift. What design features help to prevent stalling at low flight speeds?
7. Describe the four basic forms of bird wings. How does wing shape correlate with bird size and nature of flight (whether powered or soaring)?
8. What are the advantages of seasonal migration for birds?
9. Describe the different navigational resources birds may use in long-distance migration.
10. What are some of the advantages of social aggregation among birds?
11. More than 90% of all bird species are monogamous. Explain why monogamy is so much more common among birds than among mammals.
12. Briefly describe an example of polygyny among birds.
13. Define the terms precocial and altricial as they relate to birds.
14. Offer some examples of how human activities have affected bird populations.

Selected References

See also general references on page 395.

Brooke, M., and T. Birkhead, eds. 1991. The Cambridge encyclopedia of ornithology. New York, Cambridge University Press. *Comprehensive, richly illustrated treatment that includes a survey of all modern bird orders.*

Burton, R. 1985. Bird behavior. New York, Alfred A. Knopf, Inc. *Well-written and well-illustrated summary of bird behavior.*

Elphick, J. ed. 1995. The atlas of bird migration: tracing the great journeys of the world's birds. New York, Random House. *Lavishly illustrated collection of maps of birds' breeding and wintering areas, migration routes, and many facts about each bird's migration journey.*

Emlen, S. T. 1975. The stellar-orientation system of a migratory bird. Sci. Am.

233:102–111 (Aug.). *Describes fascinating research with indigo buntings, revealing their ability to navigate by the center of celestial rotation at night.*

Feduccia, A. 1996. The origin and evolution of birds. New Haven, Yale University Press. *An updated successor to the author's* The Age of Birds *(1980) but more comprehensive; rich source of information on the evolutionary relationships of birds.*

Norbert, U. M. 1990. Vertebrate flight. New York, Springer-Verlag. *Detailed review of the mechanics, physiology, morphology, ecology, and evolution of flight. Covers bats as well as birds.*

Proctor, N. S., and P. J. Lynch. 1993. Manual of ornithology: avian structure and function. New Haven, Connecticut, Yale University Press.

Sibley, C. G., and J. E. Ahlquist. 1990. Phylogeny and classification of birds: a study in molecular evolution. New Haven, Yale University Press. *A comprehensive application of DNA annealing experiments to the problem of resolving avian phylogeny.*

Terborgh, J. 1992. Why American songbirds are vanishing. Sci. Am. **266:**98–104 (May). *The number of songbirds in the United States has been dropping*

sharply. *The author suggests the reasons why.*

Waldvogel, J. A. 1990. The bird's eye view. Am. Sci. **78:**342–353 (July–Aug.). *Birds possess visual abilities unmatched by humans. So how can we know what they really see?*

Wellnhofer, P. 1990. *Archaeopteryx.* Sci. Am. **262:**70–77 (May). *Description of perhaps the most important fossil ever discovered.*

Welty, J. C., and L. Baptista. 1988. The life of birds, ed. 4. Philadelphia, Saunders College Publishing. *Among the best of the ornithology texts; lucid style and well illustrated.*

Links to the Internet

Visit this textbook's Web site at http://www.mhhe.com/zoology to find live Internet links for each of the references listed below.

1. Class Aves. University of Michigan site on amphibians. Pictures, much information on the morphology, distribution, and ecology of a large number of birds, with links to nearly all orders. Each taxon is linked to Web pages. Images may not be available for display depending on your server.

2. Birding on the Web. This site provides links, answers to frequently asked questions, announcements, links to bird chatlines, information on bird classification, and other information.

3. BIRDNET: Bird Accounts. An ornithological information source, supported by the Ornithological Council, provides information on bird orders and species lists for most orders.

4. Peterson Online: Birds. This site presents information that helps both

the novice or the skilled birder to identify birds. It includes games in bird identificaiton.

5. The Raptor Center. The Raptor Center provides medical treatment for injured birds of prey and is dedicated to the preservation of raptors. This site describes the mission of the center, answers frequently asked questions, includes a glossary of terms, and provides information for those interested in the work of the Raptor Center.

6. Patuxent Wildlife Research Center. A plethora of links to individual bird species, their life histories, morphology, pictures, and videos.

7. The Birds of North America. The Academy of Natural Sciences supports this site, where you can look up natural history information on almost all breeding birds in North America.

8. Sound and Vision: Online Bird Identification Guide. Information on many bird species may be found at this site, including photographs and recordings.

9. Cornell Lab of Ornithology. This site provides a large amount of information for birders.

10. Vertebrate Flight Exhibit. This University of California at Berkeley Museum of Paleontology site has an introduction to flight, as well as the physics, evolution, and origin of vertebrate flight.

11. Names of North American Birds. A description of the correct format for writing common names of birds, then a list of over 1900 bird species found in North America, Mexico, and Hawaii by common and scientific name.

12. Threatened Birds of the United States. A list of extinct, threatened, vulnerable, or rare birds of the United States, categorized by order.

Mammals

CHAPTER | nineteen

The Tell-Tale Hair

I If Fuzzy Wuzzy, the bear that had no hair (according to the children's rhyme), was truly hairless, he could not have been a mammal or a bear. For hair is as much an unmistakable characteristic of mammals as feathers are of birds. If an animal has hair it is a mammal; if it lacks hair it must be something else. It is true that many aquatic mammals are nearly hairless (whales, for example) but hair can usually be found (with a bit of searching) at least in vestigial form somewhere on the body of the adult. Unlike feathers, which evolved from converted reptilian scales, mammalian hair is a completely new epidermal structure. Mammals use their hair for protection from the elements, for protective coloration and concealment, for waterproofing and buoyancy, and for behavioral signaling; they have turned hairs into sensitive vibrissae on their snouts and into prickly quills. Perhaps most important of all, mammals use their hair for thermal insulation, which allows them to enjoy the great advantages of homeothermy. Warm-blooded animals in most climates and at sunless times benefit from this natural and controllable protective insulation.

Hair, of course, is only one of several features that together characterize a mammal and help us to understand the mammalian evolutionary achievement. Among these are a highly developed placenta for feeding the embryo; mammary glands for nourishing the newborn; and a surpassingly advanced nervous system that far exceeds in performance that of any other animal group. It is doubtful, however, that even with this winning combination of adaptations, the mammals could have triumphed as they have without their hair.

Mammals, with their highly developed nervous system and numerous ingenious adaptations, occupy almost every environment on earth that supports life. Although not a large group (about 4450 species as compared with more than 9000 species of birds, approximately 24,600 species of fishes, and 800,000 species of insects), the class Mammalia (mam-may′lee-a) (L. *mamma*, breast) is overall the most biologically differentiated group in the animal kingdom. Many potentialities that dwell more or less latently in other vertebrates are highly developed in mammals. Mammals are exceedingly diverse in size, shape, form, and function. They range in size from the recently discovered Kitti's hognosed bat, weighing only 1.5 g, to the whales, some of which exceed 100 tons.

Yet, despite their adaptability and in some instances because of it, mammals have been influenced by the presence of humans more than any other group of animals. We have domesticated numerous mammals for food and clothing, as beasts of burden, and as pets. We use millions of mammals each year in biomedical research. We have introduced alien mammals into new habitats, occasionally with benign results but more frequently with unexpected disaster. Although history provides us with numerous warnings, we continue to overcrop valuable wild stocks of mammals. The whale industry has threatened itself with total collapse by exterminating its own resource—a classic example of self-destruction in the modern world, in which competing segments of an industry are intent only on reaping all they can today as though tomorrow's supply were of no concern whatever. In some cases destruction of a valuable mammalian resource has been deliberate, such as the officially sanctioned (and tragically successful) policy during the Indian wars of exterminating the bison to drive the Plains Indians into starvation. Although commercial hunting has declined, the ever-increasing human population with the accompanying destruction of wild habitats has harassed and disfigured the mammalian fauna.

We are becoming increasingly aware that our presence on this planet as the most powerful product of organic evolution makes us responsible for the character of our natural environment. Since our welfare has been and continues to be closely related to that of the other mammals, it is clearly in our interest to preserve the natural environment of which all mammals, ourselves included, are a part. We need to remember that nature can do without us but we cannot exist without nature.

Origin and Evolution of Mammals

The evolutionary descent of mammals from their earliest amniote ancestors is perhaps the most fully documented transition in vertebrate history. From the fossil record, we can trace the derivation over 150 million years of endothermic, furry mammals from their small, ectothermic, hairless ancestors. The structure of the skull roof permits us to identify three major groups of amniotes that diverged in the Carboniferous period of the Paleozoic era, the **synapsids, anapsids,** and **diapsids** (p. 326). The synapsid group, which includes the mammals and their ancestors, has a pair of openings in the skull roof for the attachment of jaw muscles (Figure 19-2). Synapsids were the first amniote group to radiate widely into terrestrial habitats. The anapsid group is characterized by solid skulls and includes the turtles and their ancestors (see p. 331). The diapsids have two pairs of openings in the skull roof (Figure 17-2, p. 328) and this group contains the dinosaurs, lizards, snakes, crocodilians, birds, and their ancestors.

The earliest synapsids radiated extensively into diverse herbivorous and carnivorous forms that are often collectively called **pelycosaurs** (Figures 19-1 and 19-2). The pelycosaurs share a general outward resemblance to lizards, but this resemblance is misleading. The pelycosaurs are not closely related to lizards, which are diapsids, nor are they a monophyletic group. From one group of early carnivorous synapsids arose the **therapsids** (Figure 19-2), the only synapsid group to survive beyond the Paleozoic. In the therapsids we see for the first time an efficient erect gait with upright limbs positioned beneath the body. Since stability was reduced by raising the animal from the ground, the muscular coordination center of the brain, the cerebellum, took on an expanded role. The therapsids radiated into numerous herbivorous and carnivorous forms but most disappeared during the great extinction at the end of the Permian.

Only the last therapsid subgroup to evolve, the **cynodonts,** survived to enter the Mesozoic. The cynodonts evolved several novel features including a high metabolic rate, which supported a more active life; increased jaw musculature, permitting a stronger bite; several skeletal changes, supporting greater agility; and a secondary bony palate (Figure 19-3), enabling the animal to breathe while holding prey or chewing food. The secondary palate would be important to subsequent mammalian evolution by permitting the young to breathe while suckling. Toward the end of the Triassic certain cynodont groups arose that closely resembled mammals, sharing with them several derived features of the skull and teeth.

Fishes, amphibians, most reptiles, and birds have a **primary palate,** *which is the roof of the mouth cavity formed by the ventral skull bones. In these vertebrates, there is no separation of nasal passages from the mouth cavity. In mammals and crocodilians the nasal passages are completely separated from the mouth by the development of a secondary bony roof, the* **secondary palate.** *Mammals extend the separation of oral and nasal cavities even farther backward by adding to this "hard palate" a fleshy soft palate; these structures are shown in Figure 19-3.*

The earliest mammals of the late Triassic were small mouse- or shrew-sized animals with enlarged cranium, jaws redesigned for a shearing action, and a new type of dentition

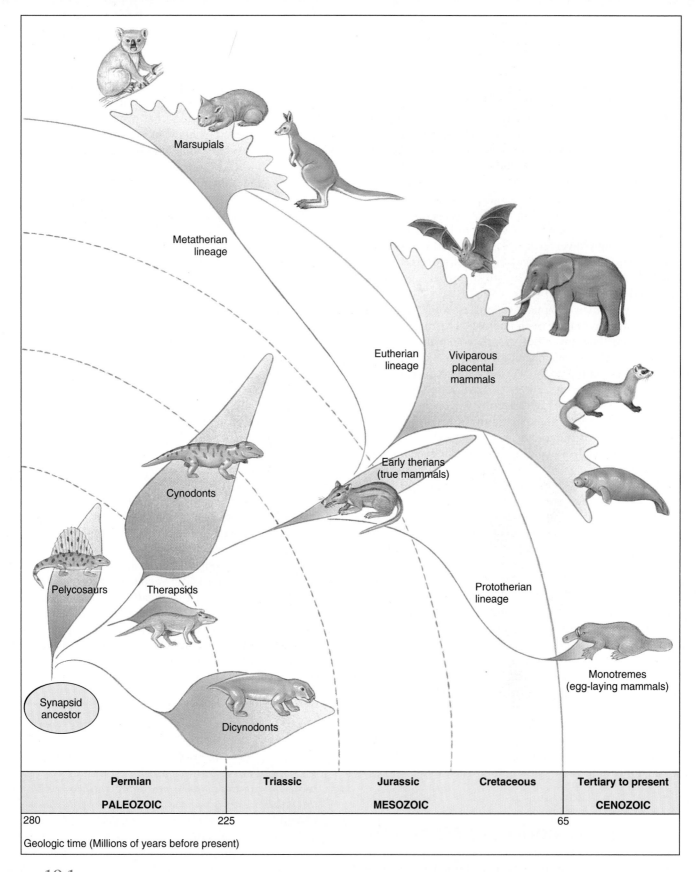

FIGURE 19-1

Evolution of the major groups of synapsids. The synapsid lineage, characterized by lateral temporal openings in the skull, began with the pelycosaurs, early mammal-like amniotes of the Permian. The pelycosaurs radiated extensively and evolved changes in the jaws, teeth, and body form that presaged several mammalian characteristics. These trends continued in their successors, the therapsids, especially in the cynodonts. One lineage of cynodonts gave rise in the Triassic to the therians, the true mammals. Fossil evidence, as currently interpreted, indicates that all three groups of living mammals—monotremes, marsupials, and placentals—are derived from the same lineage. The great radiation of modern placental orders occurred during the Cretaceous and Tertiary periods.

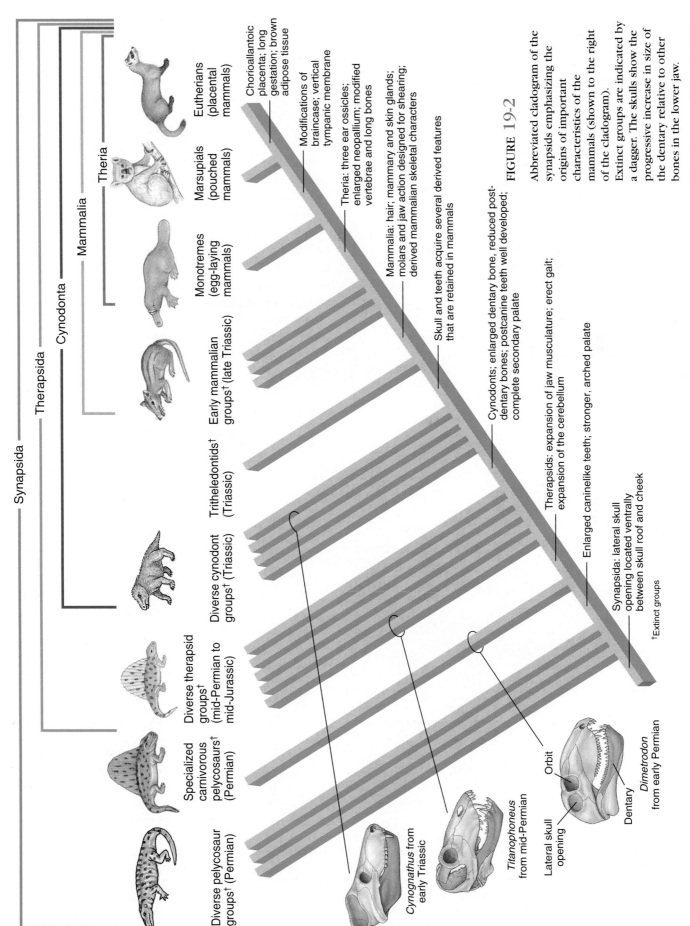

Eutherians (placental mammals)

Chorioallantoic placenta; long gestation; brown adipose tissue

Modifications of braincase; vertical tympanic membrane

Marsupials (pouched mammals)

Theria

Theria: three ear ossicles; enlarged neopallium; modified vertebrae and long bones

Monotremes (egg-laying mammals)

Mammalia

Mammalia: hair; mammary and skin glands; molars and jaw action designed for shearing; derived mammalian skeletal characters

Early mammalian groups† (late Triassic)

Skull and teeth acquire several derived features that are retained in mammals

Tritheledontids† (Triassic)

Cynodonta

Cynodonts; enlarged dentary bone, reduced post-dentary bones; postcanine teeth well developed; complete secondary palate

Diverse cynodont groups† (Triassic)

Therapsids: expansion of jaw musculature; erect gait; expansion of the cerebellum

Diverse therapsid groups† (mid-Permian to mid-Jurassic)

Therapsida

Enlarged caninelike teeth; stronger, arched palate

Specialized carnivorous pelycosaurs† (Permian)

Synapsida: lateral skull opening located ventrally between skull roof and cheek

Diverse pelycosaur groups† (Permian)

Synapsida

†Extinct groups

Cynognathus from early Triassic

Titanophoneus from mid-Permian

Lateral skull opening

Orbit

Dentary

Dimetrodon from early Permian

FIGURE 19-2

Abbreviated cladogram of the synapsids emphasizing the origins of important characteristics of the mammals (shown to the right of the cladogram). Extinct groups are indicated by a dagger. The skulls show the progressive increase in size of the dentary relative to other bones in the lower jaw.

(Sources: T. S. Kemp, Mammal-like reptiles and the origin of mammals, 1982, Academic Press, NY; K. Gauthier, et al., "Amniote phylogeny and the importance of fossils" in Cladistics, 4:105–209, 1998; R. L. Carroll, Vertebrate paleontology and evolution, 1988, W. H. Freeman & Co., NY; F. H. Pough, et al., Vertebrate life, 3d ed., 1989, Macmillan Co., NY; and T. Rowe, "Phylogenetic systematics and the early history of mammals" in F. S. Szalay, et al., Mammal phylogeny, vol 1, 1993, Springer-Verlag, NY.)

Characteristics of Mammals

1. **Body covered with hair,** but reduced in some
2. **Integument** with **sweat, scent, sebaceous,** and **mammary glands**
3. Skull with **two occipital condyles** and **secondary bony palate;** middle ear with **three ossicles** (malleus, incus, stapes); **seven cervical vertebrae** (except some xenarthrans [edentates] and the manatee); **pelvic bones fused**
4. Mouth with **diphyodont teeth** (milk, or deciduous, teeth replaced by a permanent set of teeth); teeth heterodont in most (varying in structure and function); lower jaw a **single enlarged bone (dentary)**
5. Movable eyelids and **fleshy external ears (pinnae)**
6. Four limbs (reduced or absent in some) adapted for many forms of locomotion
7. Circulatory system of a four-chambered heart, **persistent left aorta,** and **nonnucleated, biconcave red blood corpuscles**
8. Respiratory system of lungs with alveoli, and voice box (larynx); **secondary palate** (anterior bony palate and posterior continuation of soft tissue, the soft palate) separates air and food passages (Figure 19-3); **muscular diaphragm** for air exchange separates thoracic and abdominal cavities
9. Excretory system of metanephros kidneys and ureters that usually open into a bladder
10. Brain highly developed, especially **neocerebrum;** 12 pairs of cranial nerves
11. Endothermic and homeothermic
12. Separate sexes
13. Internal fertilization; **embryos develop in a uterus** with **placental attachment** (placenta rudimentary in marsupials and absent in monotremes); **fetal membranes (amnion, chorion, allantois)**
14. Young nourished by **milk from mammary glands**

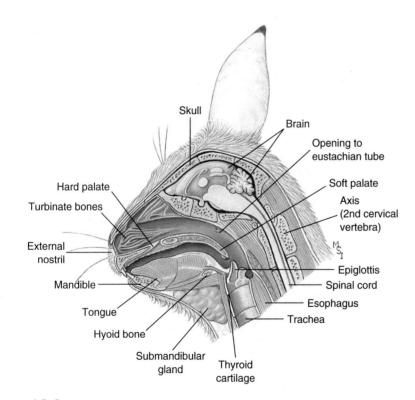

FIGURE 19-3

Sagittal section through the head of a rabbit. The hard and soft palates together form the secondary palate, a roof that separates mouth and nasal cavities, and a characteristic of all mammals and some reptiles.

in which the teeth were replaced only once (deciduous and permanent teeth). This event contrasts with the primitive amniote pattern of continual tooth replacement throughout life. The earliest mammals were almost certainly endothermic, although their body temperature would have been rather lower than modern placental mammals. Hair was essential for insulation, and the presence of hair implies that sebaceous and sweat glands must also have evolved at this time to lubricate the hair and promote heat loss. The fossil record is silent on the appearance of mammary glands, but they must have evolved before the end of the Triassic.

Oddly, the early mammals of the mid-Triassic, having developed nearly all of the novel attributes of modern mammals, had to wait for another 150 million years before they could achieve their great diversity. In the meantime the dinosaurs became diverse and abundant, while all nonmammalian synapsid groups became extinct. But mammals survived, first as shrewlike, probably nocturnal, creatures. Then, in the Tertiary, especially during the Eocene epoch that began about 54 million years ago, the modern mammals began to expand rapidly. The great Cenozoic radiation of mammals is partly attributed to the numerous environments vacated by the many amniote groups that became extinct at the end of the Cretaceous. The mammalian radiation was almost certainly promoted by the fact that mammals were agile, endothermic, intelligent, adaptable, and gave birth to living young, which they protected and nourished from their own milk supply, thus dispensing with vulnerable eggs laid in nests.

The class Mammalia includes 21 orders: one order containing the monotremes, one order containing the marsupials, and 19 orders of placentals. A complete classification is on pp. 389–390.

Structural and Functional Adaptations of Mammals

Integument and Its Derivatives

The mammalian skin and its modifications especially distinguish mammals as a group. As the interface between the animal and its environment, the skin is strongly molded by the animal's way of life. In general the skin is thicker in mammals than in other classes of vertebrates, although as in all vertebrates it is made up of **epidermis** and **dermis.** Among the mammals the dermis becomes much thicker than the epidermis. The epidermis is relatively thin where it is well protected by hair, but in places that are subject to much contact and use, such as the palms or soles, its outer layers become thick and cornified with keratin.

Hair

Hair is especially characteristic of mammals, although humans are not very hairy creatures, and in whales hair is reduced to only a few sensory bristles on the snout. A hair grows out of a hair follicle that, although an epidermal structure, is sunk into the dermis of the skin (Figure 19-4). The hair grows continuously by rapid proliferation of cells in the follicle. As the hair shaft is pushed upward, new cells are carried away from their source of nourishment and die, turning into the same dense type of fibrous protein, called **keratin,** that constitutes nails, claws, hooves, and feathers.

Mammals characteristically have two kinds of hair forming the **pelage** (fur coat): (1) dense and soft **underhair** for insulation and (2) coarse and longer **guard hair** for protection against wear and to provide coloration. The underhair traps a layer of insulating air. In aquatic animals, such as the fur seal, otter, and beaver, it is so dense that it is almost impossible to wet. In water the guard hairs become wet and mat down, forming a protective blanket over the underhair (Figure 19-5).

When a hair reaches a certain length, it stops growing. Normally it remains in the follicle until a new growth starts, whereupon it falls out. In most mammals, there are periodic molts of the entire coat. In humans, hair is shed and replaced throughout life (although balding males confirm that replacement is not assured!).

A hair is more than a strand of keratin. It consists of three layers: the medulla or pith in the center of the hair, the cortex with pigment granules next to the medulla, and the outer cuticle composed of imbricated scales. The hair of different mammals shows a considerable range of structure. It may be deficient in cortex, such as the brittle hair of deer, or it may be deficient in medulla, such as the hollow, air-filled hairs of the wolverine. The hairs of rabbits and some others are scaled to interlock when pressed together. Curly hair, such as that of sheep, grows from curved follicles.

In the simplest cases, such as foxes and seals, the coat is shed once each year during the summer months. Most mammals have two annual molts, one in the spring and one in the fall. The summer coat is always much thinner than the winter coat and in some mammals it may be a different color. Several northern mustelid carnivores, such as weasels, have white winter coats and brown-colored summer coats. It was once believed that the white inner pelage of arctic animals conserves body heat by reducing radiation loss; in fact, dark and white pelages radiate heat equally well. The winter white of arctic animals is simply camouflage in a land of snow. The varying hare of North America has three annual molts: the white winter coat is replaced by a brownish gray summer coat, and this is replaced in autumn by a grayer coat, which is soon shed to reveal the winter white coat beneath (Figure 19-6).

Outside the Arctic, most mammals wear somber colors that are protective. Often the species is marked with "salt-and-pepper" coloration or a disruptive pattern that helps make it inconspicuous in its natural surroundings. Examples are the spots of leopards and fawns and the stripes of tigers. Other mammals, such as skunks, advertise their presence with conspicuous warning coloration.

The hair of mammals has become modified to serve many purposes. The bristles of hogs, the spines of porcupines and their kin, and the vibrissae on the snouts of most mammals are examples. **Vibrissae,** commonly called "whiskers," are really sensory hairs that provide a tactile sense to many mammals. The slightest movement of a vibrissa generates impulses in sensory nerve endings that travel to special sensory areas in the brain. Vibrissae are especially long in nocturnal and burrowing animals.

Porcupines, hedgehogs, echidnas, and a few other mammals have developed an effective and dangerous spiny armor. When cornered, the common North American porcupine turns its back toward the attacker and lashes out with the barbed tail. The lightly-attached quills break off at their bases

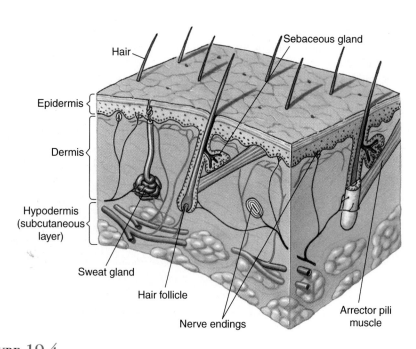

FIGURE 19-4

Structure of human skin (epidermis and dermis) and hypodermis, showing hair and glands.

FIGURE 19-5

American beaver, *Castor canadensis,* about to cut down an aspen tree. This second largest rodent (the South American capybara is larger) has a heavy waterproof pelage consisting of long, tough guard hairs overlying the thick, silky underhair so valued in the fur trade. Order Rodentia, family Castoridae.

when they enter the skin and, aided by backward-pointing hooks on the tips, work deeply into tissues. Dogs are frequent victims (Figure 19-7) but fishers, wolverines, and bobcats are able to flip the porcupine onto its back to expose vulnerable underparts.

Horns and Antlers

Three kinds of horns or hornlike substances are found in mammals. **True horns,** found in **ruminants** (for example, cud-chewers such as sheep and cattle), are sheaths of keratinized epidermis that embrace a hollow core of bone arising from the skull. Horns are not normally shed, usually are not branched (although they may be greatly curved), and are found in both sexes (except pronghorn antelope in which they occur only in the male).

Antlers of the deer family are solid bone when mature. During their annual growth, antlers develop beneath a covering of highly vascular soft skin called **"velvet"** (Figure 19-8). When growth of the antlers is complete just before the breeding season, blood vessels constrict and the stag tears off the velvet by rubbing the antlers against trees. Antlers are dropped after the breeding season. New buds appear a few months later to herald the next set of antlers. For several years each new pair of antlers is larger and more elaborate than the previous set. The annual growth of antlers places a strain on the mineral metabolism, since during the growing season a large moose or elk must accumulate 50 or more pounds of calcium salts from its vegetable diet.

The **rhinoceros horn** is the third kind of horn. Hairlike keratinized filaments that arise from dermal papillae are cemented together to form a single horn.

The escalating trade in rhinoceros products—especially rhinoceros horn—during the last three decades, is pushing Asian and African rhinos to the brink of extinction. Rhinoceros horn is valued in China as an agent for reducing fever, and for treating heart, liver, and skin diseases; and in North India as an aphrodisiac. Such supposed medicinal values are totally without pharmacological basis. The principal use of rhinoceros horns, however, is to fashion handles for daggers in the Middle East. Because of their phallic shape, rhinoceros horn daggers are traditional gifts at puberty rites. Between 1969 and 1977, horns from 8000 slaughtered rhinos were imported into North Yemen alone.

Glands

Of all vertebrates, mammals have the greatest variety of integumentary glands. Most fall into one of four classes: sweat, scent, sebaceous, and mammary. All are derivatives of the epidermis.

FIGURE 19-6

Snowshoe, or varying, hare, *Lepus americanus* in **A,** brown summer coat and, **B,** white winter coat. In winter, extra hair growth on the hind feet broadens the animal's support in snow. Snowshoe hares are common residents of the taiga (northern coniferous forests) and are an important food for lynxes, foxes, and other carnivores. Population fluctuations of hares and their predators are closely related. Order Lagomorpha.

FIGURE 19-7

Dogs are frequent victims of the porcupine's impressive armor. Unless removed (usually by a veterinarian) the quills will continue to work their way deeper in the flesh causing great distress and may lead to the victim's death.

Sweat glands are simple, tubular, highly coiled glands that occur over much of the body in most mammals. They are not present in other vertebrates. Two kinds of sweat glands may be distinguished: eccrine and apocrine (Figure 19-4). **Eccrine glands** secrete a watery sweat that, when evaporated on the skin's surface, draws heat from the skin and cools it. Eccrine glands occur in hairless regions, especially the foot pads, in most mammals, although in horses and most primates they are scattered all over the body. Eccrine glands are much reduced or absent in rodents, rabbits, whales, and others. **Apocrine glands,** the second type of sweat glands, are larger than eccrine glands and have longer and more winding ducts. Their secretory coil is in the dermis and extends deep into the hypodermis. They always open into the follicle of a hair or where a hair has been. Apocrine glands develop approximately at sexual puberty and are restricted (in the human species) to the axillae (armpits), mons pubis, breasts, external auditory canals, prepuce, scrotum, and a few other places. Their secretion is not watery, like ordinary sweat of eccrine glands, but is a milky, whitish or yellow secretion that dries on the skin to form a plasticlike film. Apocrine glands are not involved in heat regulation, but their activity is correlated with certain aspects of the sex cycle, among other possible functions.

Scent glands are present in nearly all mammals. Their location and functions vary greatly. They are used in communication with members of the same species, to mark territorial boundaries, for warning, or for defense. Scent-producing glands are located in orbital, metatarsal, and interdigital regions (deer); behind the eyes and on the cheek (pica and woodchuck); penis (muskrats, beavers, and many canines); base of the tail (wolves and foxes); back of the head (dromedary); and anal region (skunks, minks, and weasels). The latter, the most odoriferous of all glands, open by ducts into the anus; their secretions can be discharged forcefully for several feet. During the mating season many mammals give off strong scents for attracting the opposite sex. Humans also are endowed with scent glands. But civilization has taught us to dislike our own scent, a concern that has stimulated a lucrative deodorant industry to produce an endless output of soaps and odor-masking compounds.

Sebaceous glands are intimately associated with hair follicles, although some are free and open directly onto the surface. The cellular lining of the gland itself is discharged in the secretory process and must be renewed for further secretion. These gland cells become distended with a fatty accumulation, then die, and are expelled as a greasy mixture called **sebum** into the hair follicle. Called a "polite fat" because it does not turn rancid, it serves as a dressing to keep the skin and hair pliable and glossy. Most mammals have sebaceous glands all over the body; in humans they are most numerous in the scalp and on the face.

Mammary glands, which provide the name for mammals, are probably modified apocrine glands. Whatever their evolutionary origin, they occur on all female mammals and in a rudimentary form on all male mammals. They develop by the thickening of the epidermis to form a milk line along each side of the abdomen in the embryo. On certain parts of these lines the mammae appear while the intervening parts of the ridge disappear. In the human female the mammary glands begin to increase in size at puberty because of fat accumulation and reach their maximum development in approximately the twentieth year. The breasts (or mammae) undergo additional

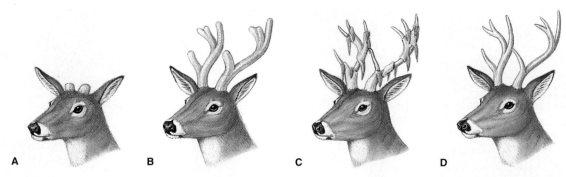

FIGURE 19-8

Annual growth of buck deer antlers. **A,** Antlers begin growth in late spring, stimulated by pituitary gonadotropins. **B,** The bone grows very rapidly until halted by a rapid rise in testosterone production by the testes. **C,** The skin (velvet) dies and sloughs off. **D,** Testosterone levels peak during the fall breeding season. The antlers are shed in January as testosterone levels subside.

development during pregnancy. In other mammals the mammae are swollen only periodically when they are distended with milk during pregnancy and subsequent nursing of the young.

Food and Feeding

Mammals have exploited an enormous variety of food sources; some mammals require highly specialized diets, whereas others are opportunistic feeders that thrive on diversified diets. For all mammals, food habits and physical structure are inextricably linked. A mammal's adaptations for attack and defense and its specializations for finding, capturing, chewing, swallowing, and digesting food all determine a mammal's shape and habits.

Teeth, perhaps more than any other single physical characteristic, reveal the life habit of a mammal (Figure 19-9). All mammals have teeth, except monotremes, anteaters, and certain whales, and their modifications are correlated with what the mammal eats.

As the mammals evolved during the Mesozoic, major changes occurred in the teeth and jaws. Unlike the uniform **homodont** dentition of the reptiles, mammalian teeth became differentiated to perform specialized functions such as cutting, seizing, gnawing, tearing, grinding, or chewing. Teeth differentiated in this manner are called **heterodont.** Typically, the mammalian dentition is differentiated into four types: **incisors,** with simple crowns and sharp edges, used mainly for snipping or biting; **canines,** with long conical crowns, specialized for piercing; **premolars,** with compressed crowns and one or two cusps, suited for shearing and slicing; and **molars,** with large bodies and variable cusp arrangement, used for crushing and grinding. The primitive tooth formula, which expresses the number of each tooth type in one-half of the upper and lower jaw, was I 3/3, C 1/1, PM 4/4, M 3/3. Members of the order Insectivora (e.g., shrews), some omnivores, and carnivores come closest to this primitive pattern (Figure 19-9).

Unlike reptiles, mammals do not continuously replace their teeth throughout their lives. Most mammals grow just two sets of teeth: a temporary set, called **deciduous,** or **milk,** teeth, which is replaced by a permanent set when the skull has grown large enough to accommodate a full set. Only the incisors, canines, and premolars are deciduous; the molars are never replaced and the single permanent set must last a lifetime.

Feeding Specializations

The feeding apparatus of a mammal—the teeth and jaws, tongue, and alimentary canal—are adapted to its particular feeding habits. Mammals are customarily divided among four basic categories—insectivores, carnivores, omnivores, and herbivores—but many other feeding specializations have evolved in mammals, as in other living organisms, and the feeding habits of many mammals defy exact classification. The principal feeding specializations of mammals are shown in Figure 19-9.

Insectivores are small mammals, usually opportunistic feeders, that feed on a variety of small invertebrates, such as worms and grubs, as well as insects. Examples are shrews, moles, anteaters, and most bats. The insectivorous category is not a sharply distinguished one because carnivores and omnivores often include insects in their diets.

Herbivorous mammals that feed on grasses and other vegetation form two main groups: the **browsers** and **grazers,** such as the ungulates (hooved mammals including horses, deer, antelope, cattle, sheep, and goats), and the **gnawers,** such as the rodents, and rabbits and hares. In herbivores, the canines are reduced in size or absent, whereas the molars, which are adapted for grinding, are broad and usually high-crowned. Rodents have chisel-sharp incisors that grow throughout life and must be worn away to keep pace with their continual growth (Figure 19-9).

Herbivorous mammals have a number of interesting adaptations for dealing with their fibrous diet of plant food.

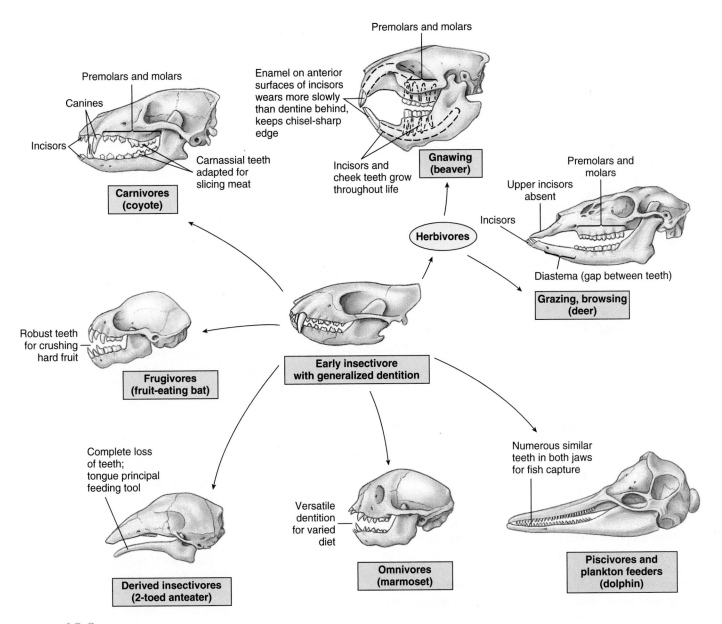

FIGURE 19-9

Feeding specializations of major trophic groups of eutherian mammals. The early eutherians were insectivores; all other types are descended from them.

Cellulose, the structural carbohydrate of plants, is composed of long chains of glucose units linked by a type of chemical bond that few enzymes can attack. No vertebrates synthesize cellulose-splitting enzymes. Instead, herbivorous vertebrates harbor anaerobic bacteria and protozoa in large fermentation chambers in the gut. These microorganisms break down and metabolize the cellulose, releasing a variety of fatty acids, sugars, and starches that the host animal can absorb and use.

In some herbivores, such as horses and zebras, rabbits and hares, elephants, and many rodents, the gut has a spacious sidepocket, or diverticulum, called a **cecum,** which serves as a fermentation chamber and absorptive area. Hares, rabbits,

and some rodents often eat their fecal pellets (**coprophagy**), giving the food a second pass through the fermenting action of the intestinal bacteria.

The **ruminants** (cattle, bison, buffalo, goats, antelopes, sheep, deer, giraffes, and okapis) have a huge **four-chambered stomach** (Figure 19-10). When a ruminant feeds, grass passes down the esophagus to the **rumen,** where it is broken down by bacteria and protozoa and then formed into small balls of cud. At its leisure the ruminant returns the cud to its mouth where the cud is deliberately chewed at length to crush the fiber. Swallowed again, the food returns to the rumen where the cellulolytic bacteria and protozoa continue fermentation.

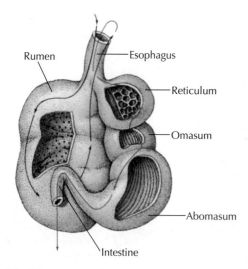

FIGURE 19-10

Ruminant's stomach. Food passes first to the rumen (sometimes through the reticulum) and then is returned to the mouth for chewing (chewing the "cud," or rumination) (*black arrow*). After reswallowing, food returns to the rumen or passes directly to reticulum, omasum, and abomasum for final digestion (*red arrow*).

The pulp passes to the **reticulum,** then to the **omasum,** where water, soluble food, and microbial products are absorbed. The remainder proceeds to the **abomasum** ("true" acid stomach), where proteolytic enzymes are secreted and normal digestion takes place.

Herbivores in general have large, long digestive tracts and must eat a considerable amount of plant food to survive. A large African elephant weighing 6 tons must consume 135 to 150 kg (300 to 400 pounds) of rough fodder each day to obtain sufficient nourishment for life.

Carnivorous mammals feed mainly on herbivores. This group includes foxes, dogs, weasels, wolverines, fishers, cats, lions, and tigers. Carnivores are well-equipped with biting and piercing teeth and powerful clawed limbs for killing their prey. Since their protein diet is much more easily digested than is the woody food of herbivores, their digestive tract is shorter and the cecum small or absent. Carnivores organize their feeding into discrete meals rather than feeding continuously (as do most herbivores) and therefore have much more leisure time for play and exploration (Figure 19-11).

Note that the terms "insectivores" and "carnivores" have two different uses in mammals: to describe diet and to denote specific taxonomic orders of mammals. For example, not all carnivores belong to the order Carnivora (many marsupials, pinnipeds, cetaceans, and all insectivores are carnivorous) and not all members of the order Carnivora are carnivorous. Many are opportunistic feeders and some, such as the panda, are strict vegetarians.

Omnivorous mammals live on both plants and animals for food. Examples are pigs, raccoons, rats, bears, and most primates (including humans). Many carnivorous forms also eat fruits, berries, and grasses when hard pressed. The fox, which usually feeds on mice, small rodents, and birds, eats frozen apples, beechnuts, and corn when its normal food sources are scarce.

For most mammals, searching for food and eating occupy most of their active life. Seasonal changes in food supplies are considerable in temperate zones. Living may be easy in the summer when food is abundant, but in winter many carnivores must range far and wide to eke out a narrow existence. Some mammals migrate to regions where food is more abundant, while others hibernate and sleep the winter months away. Many mammals, such as squirrels, chipmunks, gophers, and certain mice, build up stores of food during periods of plenty for use during the winter (Figure 19-12).

Migration

Migration is a much more difficult undertaking for mammals than for birds. Not surprisingly, few mammals make regular seasonal migrations, preferring instead to center their activities in a defined and limited home range. Nevertheless, there are some striking examples of mammalian migrations. More migrators are found in North America than on any other continent.

An example is the barren-ground caribou of Canada and Alaska, which undertakes direct and purposeful mass migrations spanning 160 to 1100 km (100 to 700 miles) twice annually (Figure 19-13). From winter ranges in the boreal forests (taiga), they migrate rapidly in late winter and spring to calving ranges on the barren grounds (tundra). The calves are born in mid-June. As summer progresses, they are increasingly harassed by warble and nostril flies that bore into their flesh, by mosquitos that drink their blood (estimated at a liter per caribou each week during the height of the mosquito season), and by wolves that prey on the calves. They move southward in July and August, feeding little along the way. In September they reach the forest and feed there almost continuously on low ground vegetation. Mating (rut) occurs in October.

Caribou have suffered a drastic decline in numbers since the nineteenth century when there were several million of them. By 1958 less than 200,000 remained in Canada. The decline has been attributed to several factors, including habitat alteration from exploration and development in the North, but especially to excessive hunting. For example the Western Arctic herd in Alaska exceeded 250,000 caribou in 1970. Following five years of heavy unregulated hunting, a 1976 census revealed only about 65,000 animals left. After restricting hunting, the herd had increased to 140,000 by 1980 and was expected to reach its original population of 250,000 in the 1990s. However, the proposed scheme to open the Arctic National Wildlife Refuge to petroleum development threatens this recovery.

FIGURE 19-11

Lionesses, *Panthera leo,* eating a wildebeest. Lions stalk prey and then charge suddenly to surprise the victim. They lack stamina for a long chase. Lions gorge themselves with the kill, then sleep and rest for periods as long as one week before eating again. Order Carnivora, family Felidae.

FIGURE 19-12

Eastern chipmunk, *Tamias striatus,* with cheek pouches stuffed with seeds to be carried to a hidden cache. It will try to store at least a half-bushel of food for the winter. It hibernates but awakens periodically to eat some of its cached food. Order Rodentia, family Sciuridae.

The plains bison, before its deliberate near extinction by humans, made huge circular migrations to separate summer and winter ranges.

The longest mammalian migrations are made by the oceanic seals and whales. One of the most remarkable migrations is that of fur seals, which breed on the Pribilof Islands approximately 300 km (185 miles) off the coast of Alaska and north of the Aleutian Islands. From wintering grounds off southern California the females journey as much as 2800 km (1740 miles) across open ocean, arriving in the spring at the Pribilofs where they congregate in enormous numbers (Figure 19-14). The young are born within a few hours or days after arrival of the cows. Then the bulls, having already arrived and established territories, collect harems of cows, which they guard with vigilance. After the calves have been nursed for approximately three months, cows and juveniles leave for their long migration southward. The bulls do not follow but remain in the Gulf of Alaska during the winter.

Although we might expect bats, the only winged mammals, to use their gift of flight to migrate, few of them do. Most spend the winter in hibernation. The four species of American bats that do migrate spend their summers in the northern or western states and their winters in the southern United States or Mexico.

Flight and Echolocation

Mammals have not exploited the skies to the same extent that they have the terrestrial and aquatic environments. How-ever, many mammals scamper about in trees with amazing agility; some can glide from tree to tree, and one group, the bats, is capable of full flight. Gliding and flying evolved independently in several groups of mammals, including marsupials, rodents, flying lemurs, and bats. Anyone who has watched a gibbon perform in a zoo realizes there is something akin to flight in this primate, too. Among the arboreal squirrels, all of which are nimble acrobats, by far the most efficient is the flying squirrel (Figure 19-15). These forms actually glide rather than fly, using the gliding skin that extends from the sides of the body.

Bats, the only group of flying mammals, are nocturnal and thus hold a niche unoccupied by most birds. Their achievement is attributed to two things: flight and the capacity to navigate by echolocation. Together these adaptations enable bats to fly and avoid obstacles in absolute darkness, to locate and catch insects with precision, and to find their way deep into caves (a habitat largely ignored by both mammals and birds) where they sleep away the daytime hours.

When in flight, bats emit short pulses 5 to 10 milliseconds in duration in a narrow directed beam from the mouth or nose (Figure 19-16). Each pulse is frequency modulated; that is, it is highest at the beginning, up to 100,000 hertz (Hz, cycles per second), and sweeps down to perhaps 30,000 Hz at the end. Sounds of this frequency are ultrasonic to the human ear, which has an upper limit of about 20,000 Hz. When a bat is searching for prey, it produces about 10 pulses per second. If a prey is detected, the rate increases rapidly up to 200 pulses per second in the final phase of approach and capture. The pulses are spaced so that the echo of each is received before

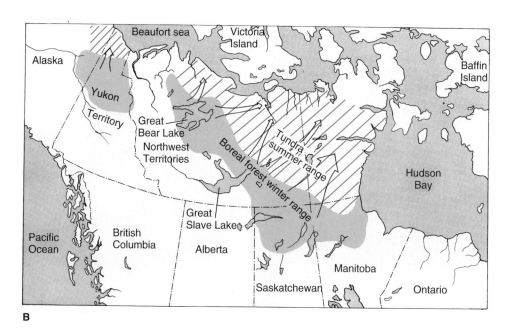

FIGURE 19-13

Barren-ground caribou, *Rangifer tarandus,* of Canada and Alaska. **A,** Adult male caribou in autumn pelage and antlers in velvet. **B,** Summer and winter ranges of some major caribou herds in Canada and Alaska (other herds not shown occur on Baffin Island and in western and central Alaska). The principal spring migration routes are indicated by arrows; routes vary considerably from year to year. The same species is known as reindeer in Europe. Order Artiodactyla, family Cervidae.

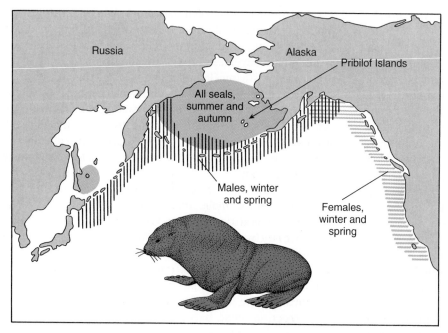

FIGURE 19-14

Annual migrations of the fur seal, showing the separate wintering grounds of males and females. Both males and females of the large Pribilof population migrate in early summer to the Pribilof Islands, where the females give birth to their pups and then mate with the males. Order Pinnipedia, family Otariidae.

the next pulse is emitted, an adaptation that prevents jamming. Since the transmission-to-reception time decreases as the bat approaches an object, it can increase pulse frequency to obtain more information about the object. Pulse length is also shortened as the bat nears the object. It is interesting that some prey of bats, certain nocturnal moths for example, have evolved ultrasonic detectors used to detect and avoid approaching bats.

The external ears of bats are large, like hearing trumpets, and shaped variously in different species. Less is known about the inner ear of bats, but it obviously is capable of receiving the ultrasonic sounds emitted. Biologists believe that bat navigation is so refined that a bat builds up a mental image of its surroundings from echo scanning that approaches the resolution of a visual image from eyes of diurnal animals.

For reasons not fully understood, all bats are nocturnal, even the fruit-eating bats that use vision and olfaction instead of echolocation to find their food. The tropics and subtropics have many kinds of bats, including the famed vampire

FIGURE 19-15

Flying squirrel, *Glaucomys sabrinus*, coming in for a landing. Area of undersurface is nearly trebled when gliding skin is spread. Glides of 40 to 50 m are possible. Good maneuverability during flight is achieved by adjusting the position of the gliding skin with special muscles. Flying squirrels are nocturnal and have superb night vision. Order Rodentia, family Sciuridae.

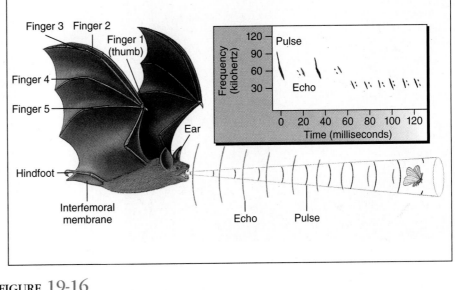

FIGURE 19-16

Echolocation of an insect by the little brown bat *Myotis lucifugus*. Frequency modulated pulses are directed in a narrow beam from the bat's mouth. As the bat nears its prey, it emits shorter, lower signals at a faster rate. Order Chiroptera.

bat. Vampire bats are provided with razor-sharp incisors used to shave away the epidermis of their prey, exposing underlying capillaries. After infusing an anticoagulant to keep the blood flowing, the bat laps up its meal and stores it in a specially modified stomach.

Reproduction

Most mammals have definite mating seasons, usually in the winter or spring and timed to coincide with the most favorable time of the year for rearing the young after birth. Many male mammals are capable of fertile copulation at any time, but the female mating function is restricted to a time during a periodic cycle, known as the **estrous cycle.** The female receives the male only during a relatively brief period known as **estrus,** or heat (Figure 19-17).

There are three different patterns of reproduction in mammals. One pattern is represented by the egg-laying (oviparous) mammals, the **monotremes.** The duck-billed platypus has one breeding season each year. The ovulated eggs, usually two, are fertilized in the oviduct. As they continue down the oviduct, various glands add albumin and then a thin, leathery shell to each egg. When laid, the eggs are about the size of a robin's egg. The platypus lays its eggs in a burrow nest where they are incubated for about 12 days. After hatching, the young suck milk from the openings of the mother's mammary glands for a pro-

longed period. Thus in monotremes there is no gestation (period of pregnancy) and the developing embryo draws on nutrients stored in the egg, much as do the embryos of reptiles and birds. But in common with all other mammals, monotremes rear their young on milk.

The **marsupials** are pouched, viviparous mammals that exhibit a second pattern of reproduction. Although only the eutherians (p. 389) are called "placental mammals," the marsupials do have a primitive type of yolk sac placenta. The embryo (blastocyst) of a marsupial is at first encapsulated by shell membranes and floats free for several days in the uterine fluid. After "hatching" from the shell membranes, the embryo does not implant, or "take root" in the uterus as in eutherians, but it does erode a shallow depression in the uterine wall in which it lies and absorbs nutrient secretions from the mucosa by way of the vascularized yolk sac. Gestation (the intrauterine period of development) is brief in marsupials, and all marsupials give birth to tiny young that are effectively still embryos, both anatomically and physiologically (Figure 19-18). However, early birth is followed by a prolonged interval of lactation and parental care (Figure 19-19).

The third pattern of reproduction is that of the viviparous **placental mammals,** the eutherians. In placentals, the reproductive investment is in prolonged gestation, unlike marsupials in which the reproductive investment is in prolonged lactation (Figure 19-19). The embryo remains in the mother's uterus, nourished by food supplied through a chorioallantoic type of placenta, an intimate connection between mother and young. The length of gestation is longer in placentals than marsupials, and in large mammals it is much longer. For example,

FIGURE 19-17

African lions *Panthera leo* mating. Lions breed at any season, although predominantly in spring and summer. During the short period a female is receptive, she may mate repeatedly. Three or four cubs are born after gestation of 100 days. Once the mother introduces the cubs into the pride, they are treated with affection by both adult males and females. Cubs go through an 18- to 24-month apprenticeship learning how to hunt and then are frequently driven from the pride to manage themselves. Order Carnivora, family Felidae.

mice have a gestation period of 21 days; rabbits and hares, 30 to 36 days; cats and dogs, 60 days; cattle, 280 days; and elephants, 22 months. But there are important exceptions (nature seldom offers perfect correlations). Baleen whales, the largest mammals, carry their young for only 12 months, while bats, no larger than mice, have gestation periods of 4 to 5 months. The condition of the young at birth also varies. An antelope bears its young well furred, eyes open, and able to run about. Newborn mice, however, are blind, naked, and helpless. We all know how long it takes a human baby to gain its footing. Human growth is in fact slower than that of any other mammal, and this is one of the distinctive attributes that sets us apart from other mammals.

A curious phenomenon that lengthens the gestation period of many mammals is delayed implantation. The blastocyst remains dormant while its implantation in the uterine wall is postponed for periods of a few weeks to several months. For many mammals (for example, bears, seals, weasels, badgers, bats, and many deer) delayed implantation is a device for extending gestation so that the young are born at the time of year that is best for their survival.

FIGURE 19-18

Opossums, *Didelphis marsupialis,* 15 days old, fastened to teats in mother's pouch. When born after a gestation period of only 12 days, they are the size of honeybees. They remain attached to the nipples for 50 to 60 days. Order Marsupialia, family Didelphidae.

Mammalian Populations

A population of animals includes all the members of a species that share a particular space and potentially interbreed. All mammals (like other organisms) live in ecological communities, each composed of numerous populations of different animal and plant species. Each species is affected by the activities of other species and by other changes, especially climatic, that occur. Thus populations are always changing in size. Populations of small mammals are lowest before the breeding season and greatest just after the addition of the new members. Beyond these expected changes in population size, mammalian populations may fluctuate from other causes.

Irregular fluctuations are commonly produced by variations in climate, such as unusually cold, hot, or dry weather, or by natural catastrophes, such as fires, hailstorms, and hurricanes. These are **density-independent** causes because they affect a population whether it is crowded or dispersed. However, the most spectacular fluctuations are **density dependent;** that is, they correlate with population crowding.

Cycles of abundance are common among many rodent species. One of the best known examples is the mass migrations of the Scandinavian and arctic North American lemmings following population peaks. Lemmings (Figure 19-20) breed all year, although more in the summer than in the winter. The gestation period is only 21 days; young born at the beginning of the summer are weaned in 14 days and are capable of reproducing by the end of the summer. At the peak of their population density, having devastated the vegetation by tunneling and grazing, lemmings begin long, mass migrations to find new undamaged habitats for food and space. They swim

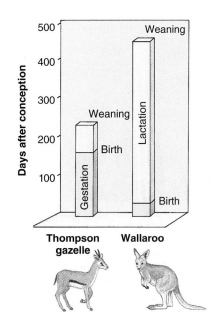

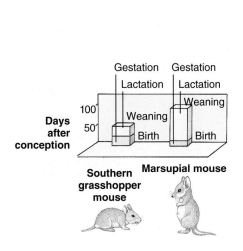

FIGURE 19-20

Collared lemming, *Dicrostonyx* sp., a small rodent of the far north. Populations of lemmings fluctuate widely. Order Rodentia, family Muridae.

FIGURE 19-19

Comparison of gestation and lactation periods between matched pairs of ecologically similar species of marsupial and placental mammals. The graph shows that marsupials have shorter intervals of gestation and much longer intervals of lactation than in similar species of placentals.

across streams and small lakes as they go but cannot distinguish these from large lakes, rivers, and the sea, in which they drown. Since lemmings are the main diet of many carnivorous mammals and birds, any change in lemming population density affects all their predators as well.

The renowned fecundity of meadow mice, and the effect of removing the natural predators from rodent populations, is felicitously expressed in this excerpt from Thornton Burgess's "Portrait of a Meadow Mouse."

He's fecund to the nth degree
In fact this really seems to be
His one and only honest claim
To anything approaching fame.
In just twelve months, should all survive,
A million mice would be alive—
His progeny. And this, 'tis clear,
Is quite a record for a year.
Quite unsuspected, night and day
They eat the grass that would be hay.
On any meadow, in a year,
The loss is several tons, I fear.
Yet man, with prejudice for guide,
The checks that nature doth provide
Destroys. The meadow mouse survives
And on stupidity he thrives.

In his book The Arctic *(1974. Montreal, Infacor, Ltd.), Canadian naturalist Fred Bruemmer describes the growth of lemming populations in arctic Canada: "After a population crash one sees few signs of lemmings; there may be only one to every 10 acres. The next year, they are evidently numerous; their runways snake beneath the tundra vegetation, and frequent piles of rice-sized droppings indicate the lemmings fare well. The third year one sees them everywhere. The fourth year, usually the peak year of their cycle, the populations explode. Now more than 150 lemmings may inhabit each acre of land and they honeycomb it with as many as 4000 burrows. Males meet frequently and fight instantly. Males pursue females and mate after a brief but ardent courtship. Everywhere one hears the squeak and chitter of the excited, irritable, crowded animals. At such times they may spill over the land in manic migrations."*

Varying hares (snowshoe rabbit) of North America show 10-year cycles in abundance. The well-known fecundity of rabbits enables them to produce litters of three or four young as many as five times per year. The density may increase to 4000 hares competing for food in each square mile of northern forest. Predators (owls, minks, foxes, and especially lynxes) also increase (Figure 19-21). Then the population

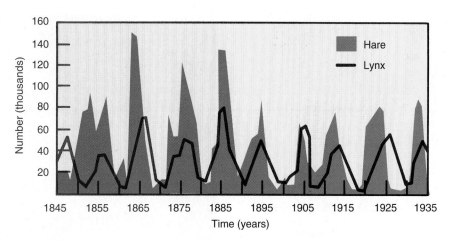

FIGURE **19-21**

Changes in population size of varying hare and lynx in Canada as indicated by pelts received by the Hudson's Bay Company. The abundance of lynx (predator) follows that of the hare (prey).

FIGURE **19-22**

A prosimian, the Mindanao tarsier, *Tarsius syrichta carbonarius* of Mindanao Island in the Philippines.

crashes precipitously for reasons that have long been a puzzle to scientists. Rabbits die in great numbers, not from lack of food or from an epidemic disease (as was once believed) but evidently from some density-dependent psychogenic cause. As crowding increases, hares become more aggressive, show signs of fear and defense, and stop breeding. The entire population reveals symptoms of pituitary-adrenal gland exhaustion, an endocrine imbalance called "shock disease," which results in death. These dramatic crashes are not well understood. Whatever the causes, population crashes that follow superabundance, although harsh, permit the vegetation to recover, thus providing the survivors with a much better chance for successful breeding.

Human Evolution

Darwin devoted an entire book, *The Descent of Man and Selection in Relation to Sex,* largely to human evolution. The idea that humans shared common descent with apes and other animals was repugnant to the Victorian world, which responded with predictable outrage (Figure 1-16, p. 18). When Darwin's views were first debated, few human fossils had been unearthed, but the current accumulation of fossil evidence has strongly vindicated Darwin's belief that humans descended from primate ancestors. All primates share certain significant characteristics: grasping fingers on all four limbs, flat fingernails instead of claws, and forward-pointing eyes with binocular vision and excellent depth perception. The following synopsis highlights the current hypotheses of the evolutionary descent of the primates.

The earliest primate was probably a small, nocturnal animal similar in appearance to tree shrews. This ancestral primate stock split into two major lineages, one of which gave rise to the **prosimians,** which include lemurs, tarsiers (Figure 19-22), and lorises, and the other to the **simians,** which include the monkeys (Figure 19-23) and apes (Figure 19-24). Prosimians and many simians are arboreal (tree-dwellers), which is probably the ancestral life-style for both groups. Arboreality probably selected for increased intelligence. Flexible limbs are essential for active animals moving through trees. Grasping hands and feet, in contrast to the clawed feet of squirrels and other rodents, enable the primates to grip limbs, hang from branches, seize food and manipulate it, and, most significantly, use tools. Highly developed sense organs, especially good vision, and proper coordination of limb and finger muscles are essential for an active arboreal life. Of course, sense organs are no better than the brain processing the sensory information. Precise timing, judgment of distance, and alertness require a large cerebral cortex.

The earliest simian fossils appeared in Africa some 40 million years ago. Many of these primates became day-active rather than nocturnal, making vision the dominant special sense, now enhanced by color vision. We recognize three major simian groups whose precise phylogenetic relationships are unknown. These are (1) the New World monkeys of South America (ceboids), including the howler monkey (Figure 19-23A), spider monkey, and the tamarins; (2) the Old World monkeys (cercopithecoids), including baboons (Figure 19-23B), mandrills, and colobus monkeys; and (3) the anthropoid apes (Figure 19-24). In addition to their geographic separation, Old World monkeys differ from New World monkeys in lacking a grasping tail while having close-set nostrils, better opposable, grasping thumbs, and more advanced teeth. Apes first appear in 25-million-year-old fossils. At this time the woodland savannas were arising in Africa, Europe, and North America. Perhaps motivated by the greater abundance of food on the ground, these apes left the trees and became largely terrestrial. Because of the benefits of standing upright (better view of predators, freeing of hands for using tools and gathering food), emerging hominids gradually evolved upright posture.

FIGURE 19-23

Monkeys. **A,** Red-howler monkeys, an example of the New World monkeys. **B,** The olive baboon, an example of the Old World monkeys.

FIGURE 19-24

The gorilla, an example of the anthropoid apes.

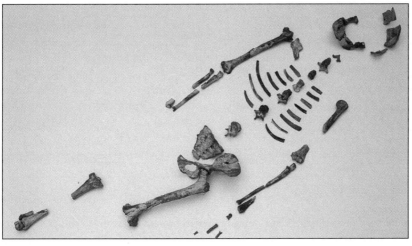

FIGURE 19-25

Lucy *(Australopithecus afarensis),* the most nearly complete skeleton of an early hominid ever found. Lucy is dated at 2.9 million years old. A nearly complete skull of *A. afarensis* was discovered in 1994.

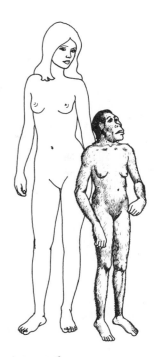

FIGURE 19-26

A reconstruction of the appearance of Lucy *(right)* compared with a modern human *(left).*

Evidence of the earliest hominids of this period is sparse. Not until about 4.4-million years ago, after a lengthy fossil gap, do the first "near humans" appear in the fossil record. The best documented of early hominid fossils is *Australopithecus afarensis,* a short, bipedal hominid with a face and brain size resembling those of a chimpanzee. Numerous fossils of this species have now been unearthed, the most celebrated of which was the 40% complete skeleton of a female discovered in 1974 by Donald Johanson and named "Lucy" (Figures 19-25 and 19-26). Many paleoanthropologists believe that *Australopithecus afarensis* represents the ancestral stock of all human and humanlike forms that followed (Figure 19-27).

Until very recently, *A. afarensis,* dated to nearly 4 million years ago, was the oldest known hominid. However, in 1994 a 4.4-million-year-old hominid fossil named *Ardipithecus ramidus* was uncovered in Africa. *Ardipithecus ramidus,* a mosaic of primitive ape-like traits and derived hominid traits, appears to be ancestral to the australopithecine species. In 1995 a new species called *Australopithecus anamensis* was discovered, dated to 4.2 million years ago. Some paleoanthropologists

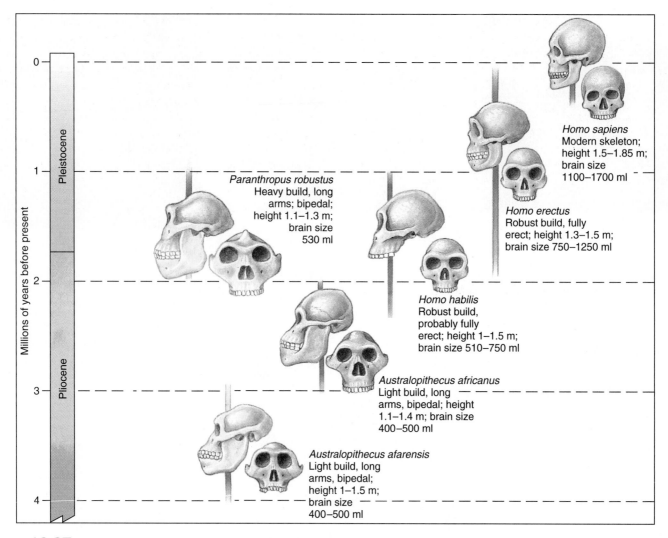

FIGURE **19-27**

Hominid skulls, showing several of the best-known hominid lines preceding modern humans *(Homo sapiens)*. The time span of existence for each species, as indicated by the fossil record, is suggested by the vertical red lines.

believe that this fossil may be intermediate in a lineage leading from *A. ramidus* to *A. afarensis;* others argue for caution and await new fossil discoveries.

Between 3 and 4 million years ago two quite separate hominid lines emerged that coexisted for at least 2 million years. One was the bipedal *Australopithecus africanus,* the "southern African ape," with a brain size only about one-third as large as that of modern humans. A different line of australopithecines was large and robust *(Paranthropus robustus,* Figure 19-27) and probably approached the size of a gorilla. Until they became extinct between 1.75 and 1 million years ago, the australopithecines shared the countryside with an advanced, fully erect hominid, *Homo habilis,* the first true human (Figure 19-27). *Homo habilis,* meaning "able man," was more lightly built but larger brained than the australopithecines and unquestionably used stone and bone tools. This species appeared about 2 million years ago and survived for perhaps half a million years.

About 1.5 million years ago *Homo erectus* appeared, probably as a descendant of *Homo habilis. Homo erectus* was a large hominid standing 150 to 170 cm (5 to 5½ feet) tall, with a low but distinct forehead, strong brow ridges, and a brain capacity of around 1000 cc (about intermediate between the brain capacity of *Homo habilis* and modern humans) (Figure 19-27). *Homo erectus* was a social species living in tribes of 20 to 50, had a successful and complex culture, and became widespread throughout the tropical and temperate Old World.

After the disappearance of *Homo erectus* about 300,000 years ago, subsequent human evolution and the establishment of *Homo sapiens* ("wise man") threaded a complex course. From among the many early subcultures of *Homo sapiens,* the Neanderthals emerged about 130,000 years ago. With a brain capacity well within the range of modern humans, the Neanderthals were proficient hunters and tool users. They dominated the Old World in the late Pleistocene epoch. About 30,000 years ago the Neanderthals were replaced and quite possibly exterminated by modern humans, tall people with a culture very different from that of the Neanderthals. Implement crafting developed rapidly, and human culture became enriched with aesthetics, artistry, and sophisticated language.

Biologically, *Homo sapiens* is a product of the same processes that have directed the evolution of every organism from the time of life's origin. Mutation, isolation, genetic drift, and natural selection have operated for us as they have for other animals. Yet we have what no other animal has, a non-genetic cultural evolution that provides a constant feedback between past and future experience. Our symbolic languages, capacities for conceptual thought, knowledge of our history, and abilities to manipulate our environment emerge from this nongenetic cultural endowment. Finally, we owe much of our cultural and intellectual achievements to our arboreal ancestry which bequeathed us with binocular vision, superb visuotactile discrimination, and manipulative skills in the use of our hands. If the horse (with one toe instead of five fingers) had human mental capacity, could it have accomplished what humans have?

Classification of Living Mammalian Orders[1]

All modern mammals are placed in two subclasses, the Prototheria, containing the monotremes, and the Theria, containing the marsupials and the placentals. Of the 19 recognized placental orders, seven of the small orders are omitted from the following classification.

Class Mammalia

Subclass Prototheria (pro′to-thir′ee-a) (Gr. *prōtos,* first, + *thēr,* wild animal). Cretaceous and early Cenozoic mammals. Extinct except for the egg-laying monotremes.

Infraclass Ornithodelphia (or′ni-tho-del′fee-a) (Gr. *ornis,* bird, + *delphys,* womb). Monotreme mammals.

Order Monotremata (mon′o-tre′ma-tah) (Gr. *monos,* single, + *trēma,* hole): **egg-laying (oviparous) mammals: duck-billed platypus, spiny anteater.** Three species in this order from Australia, Tasmania, and New Guinea; most noted member of order is the duck-billed platypus *(Ornithorhynchus anatinus);* spiny anteater, or echidna *(Tachyglossus),* has a long, narrow snout adapted for feeding on ants, its chief food.

Subclass Theria (thir′ee-a) (Gr. *thēr,* wild animal). Marsupial and placental mammals.

Infraclass Metatheria (met′a-thir′e-a) (Gr. *meta,* after, + *thēr,* wild animal). Marsupial mammals.

Order Marsupialia (mar-su′pe-ay′le-a) (Gr. *marsypion,* little pouch): **viviparous pouched mammals: opossums, kangaroos, koalas, Tasmanian wolves, wombats, bandicoots, numbats, and others.** Mammals characterized by an abdominal pouch, the **marsupium,** in which they rear their young; young nourished via a yolk-sac placenta; mostly Australian with representatives in the Americas; 260 species.

Infraclass Eutheria (yu-thir′e-a) (Gr. *eu,* true, + *thēr,* wild animal). The viviparous placental mammals.

Order Insectivora (in-sec-tiv′o-ra) (L. *insectum,* an insect, + *vorare,* to devour): **insect-eating mammals: shrews** (Figure 19-28), **hedgehogs, tenrecs, moles.** Small, sharp-snouted animals with primitive characters that feed principally on insects; 390 species.

Order Chiroptera (ky-rop′ter-a) (Gr. *cheir,* hand, + *pteron,* wing): **bats.** Flying mammals with forelimbs modified into wings; use of echolocation by most bats; most nocturnal; second largest mammalian order, exceeded in species numbers only by the order Rodentia; 986 species.

Order Primates (pry-may′teez) (L. *prima,* first): **prosimians, monkeys, apes, humans.** First in the animal kingdom in brain development with especially large cerebral hemispheres; mostly arboreal, apparently derived from insectivores with retention of many primitive characteristics; five digits (usually provided with flat nails) on both forelimbs and hindlimbs; group singularly lacking in claws, scales, horns, and hooves; two suborders[2]; 233 species.

Suborder Prosimii (pro-sim′ee-i) (Gr. *pro,* before, + *simia,* ape): **lemurs, bush babies, tarsiers, lorises, pottos.** Arboreal, mostly nocturnal, primates restricted to the tropics of the Old World.

Suborder Anthropoidea (an′thro-poy′de-a) (Gr. *anthropos,* man): **monkeys, gibbons, apes, humans.**

Order Xenarthra (ze-nar′thra) (Gr. *xenos,* intrusive, + *arthron,* joint) (formerly Edentata

Continued on next page

[L. *edentatus,* toothless]): **anteaters, armadillos, sloths.** Either toothless (anteaters) or with simple peglike teeth (sloths and armadillos); restricted to South and Central America with the nine-banded armadillo in the southern United States; 30 species.

Order Lagomorpha (lag′o-mor′fa) (Gr. *lagos,* hare; + *morphē,* form): **rabbits, hares, pikas** (Figure 19-29). Dentition resembling that of rodents but with four upper incisors rather than two as in rodents; 69 species.

Order Rodentia (ro-den′che-a) (L. *rodere,* to gnaw): **gnawing mammals: squirrels** (Figure 19-30), **rats, woodchucks.** Most numerous of all mammals both in numbers and species; dentition with two upper and two lower chisel-like incisors that grow continually and are adapted for gnawing; 1814 species.

Order Cetacea (see-tay′she-a) (L. *cetus,* whale): **whales** (Figure 19-31), **dolphins, porpoises.** Anterior limbs of cetaceans modified into broad flippers; posterior limbs absent; nostrils represented by a single or double blowhole on top of the head; teeth, when present, all alike and lacking enamel; hair limited to a few hairs on muzzle, no skin glands except the mammary and those of eye; no external ear; 79 species.

Order Carnivora (car-niv′o-ra) (L. *caro,* flesh, + *vorare,* to devour): **flesh-eating mammals: dogs, wolves, cats, bears** (Figure 19-32), **weasels.** All with predatory habits; teeth especially adapted for tearing flesh; in most, canines used for killing prey; worldwide except in the Australian and Antarctic regions; 240 species.

Order Pinnipedia (pi-ni-peed′e-a) (L. *pinna,* feather, + *ped,* foot): **sea lions, seals, and walruses.** Aquatic carnivores with limbs modified as flippers for swimming; nearly all marine; food consists mostly of fish; 34 species.

Order Proboscidea (pro′ba-sid′e-a) (Gr. *proboskis,* elephant's trunk, from *pro,* before, + *boskein,* to feed): **proboscis mammals: elephants.** Living land animals, have two upper incisors elongated as tusks, and the molar teeth are well developed; two extant species: the Indian elephant, with relatively small ears, and the African elephant, with large ears.

Order Perissodactyla (pe-ris′so-dak′ti-la) (Gr. *perissos,* odd, + *dactylos,* toe): **odd-toed hoofed mammals: horses, asses, zebras, tapirs, rhinoceroses.** Mammals with an odd number (one or three) of toes and with well-developed hooves (Figure 19-33); all herbivorous; both Perissodactyla and Artiodactyla often referred to as ungulates, or hoofed mammals, with teeth adapted for chewing; 17 species.

Order Artiodactyla (ar′te-o-dak′ti-la) (Gr. *artios,* even, + *daktylos,* toe): **even-toed hoofed mammals: swine, camels, deer and their allies, hippopotamuses, antelopes, cattle, sheep, goats.** Each toe sheathed in a cornified hoof; most have two toes, although the hippopotamus and some other have four (Figure 19-33); many, such as the cow, deer, and sheep, with horns or antlers; many are ruminants, that is, herbivores with partitioned stomachs; 211 species.

---

[1]Based on Nowak, R.M. 1991. Walker's Mammals of the world, ed. 5. Baltimore, The Johns Hopkins University Press.
[2]G.G. Simpson's (1945) division of the primates into prosimian and anthropoid suborders is followed here, but it is no longer recognized by many mammalogists who hold conflicting views on classification, especially between the order and family levels.

FIGURE 19-28

The shorttail shrew, *Blarina brevicauda,* eating a grasshopper. This tiny but fierce mammal, with a prodigious appetite for insects, mice, snails, and worms, spends most of its time underground and so is seldom seen by humans. Shrews are believed to resemble the insectivorous ancestors of placental mammals. Order Insectivora, family Soricidae.

FIGURE 19-29

A pika, *Ochotona princeps,* atop a rockslide in Alaska. This little rat-sized mammal does not hibernate but prepares for winter by storing dried grasses beneath boulders. Order Lagomorpha.

FIGURE 19-30

Eastern gray squirrel, *Sciurus carolinensis.* This common resident of Eastern towns and hardwood forests serves as an important reforestation agent by planting numerous nuts that sprout into trees. Order Rodentia, family Sciuridae.

FIGURE 19-31

Humpback whale, *Megaptera novaeangliae,* breaching. Among the most acrobatic of whales, humpbacks appear to breach to stun fish schools or to communicate information to other herd members. Order Cetacea, family Balaenopteridae.

FIGURE 19-32

Grizzly bear, *Ursus borribilis,* of Alaska. Grizzlies, once common in the lower 48 states, are now confined largely to northern wilderness areas. Order Carnivora, family Ursidae.

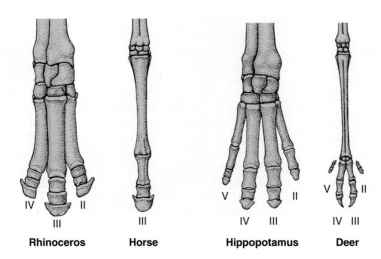

IV   II

III

Rhinoceros

III

Horse

V   II

IV   III

Hippopotamus

V   II

IV   III

Deer

FIGURE 19-33

Odd-toed and even-toed ungulates. The rhinoceros and horse (order Perissodactyla) are odd-toed; the hippopotamus and deer (order Artiodactyla) are even-toed. The lighter, faster mammals run on only one or two toes.

Summary

Mammals are endothermic and homeothermic vertebrates whose bodies are insulated by hair and who nurse their young with milk. The approximately 4450 species of mammals are descended from the synapsid lineage of amniotes that arose in the Carboniferous period of the Paleozoic era. Their evolution can be traced from the pelycosaurs of the Permian period to the therapsids of the late Permian and Triassic periods of the Mesozoic era. One group of the therapsids, the cynodonts, gave rise during the Triassic to the therians, the true mammals. Mammalian evolution was accompanied by the appearance of many important derived characteristics, among these the enlarged brain with greater sensory integration, high metabolic rate, endothermy, and many changes in the skeleton that supported a more active life. Mammals diversified rapidly during the Tertiary period of the Cenozoic era.

Mammals are named for the glandular milk-secreting organs of the female (rudimentary in the male), a unique adaptation which, combined with prolonged parental care, buffers the infants from the demands of foraging for themselves and eases the transition to adulthood. Hair, the integumentary outgrowth that covers most mammals, serves variously for mechanical protection, thermal insulation, protective coloration, and waterproofing. Mammalian skin is rich in glands: sweat glands that function in evaporative cooling, scent glands used in social interactions, and sebaceous glands that secrete lubricating skin oil. All placental mammals have deciduous teeth that are replaced by permanent teeth (diphyodont dentition). The four groups of teeth—incisors, canines, premolars, and molars—may be highly modified in different mammals for specialized feeding tasks, or they may be absent.

The food habits of mammals strongly influence their body form and physiology. Insectivores feed mainly on insects and other small invertebrates. Herbivorous mammals have special adaptations for harboring the intestinal bacteria that break down cellulose of plant materials, and they have developed adaptations for detecting and escaping predators. Carnivorous mammals feed mainly on herbivores, have a simple digestive tract, and have developed adaptations for a predatory life. Omnivores feed on both plant and animal foods.

Some marine, terrestrial, and aerial mammals migrate; some migrations, such as those of fur seals and caribou, are extensive. Migrations are usually made toward favorable climatic and optimal food and calving conditions, or to bring the sexes together for mating.

Mammals with true flight, the bats, are nocturnal and thus avoid direct competition with birds. Most employ ultrasonic echolocation to navigate and feed in darkness.

The living mammals with the most primitive characters are the egg-laying monotremes of the Australian region. After hatching, the young are nourished with the mother's milk. All other mammals are viviparous. Embryos of marsupials have brief gestation periods, are born underdeveloped, and complete their early growth in the mother's pouch, nourished by milk. The remaining 19 of the 21 orders of mammals are eutherians, mammals that develop an advanced placental attachment between mother and embryos through which the embryos are nourished for a prolonged period.

Mammal populations fluctuate from both density-dependent and density-independent causes and some mammals, particularly rodents, may experience extreme cycles of abundance in population density. The unqualified success of mammals as a group cannot be attributed to greater organ system perfection, but rather to their impressive overall adaptability—the capacity to fit more perfectly in total organization to environmental conditions and thus exploit virtually every habitat on earth.

Darwinian evolutionary principles give us great insight into our own origins. Humans are primates, a mammalian group that descended from a shrewlike ancestor. The common ancestor of all modern primates was arboreal and had grasping fingers and forward-facing eyes capable of binocular vision. Primates radiated over the last 80 million years to form two major lines of descent: the prosimians (lemurs, lorises, and tarsiers) and the simians (monkeys, apes, and hominids). The earliest hominids appeared about 4.4 million years ago. The bipedal australopithecines that appeared about 4 million years ago gave rise to, and coexisted with, the species *Homo habilis*, the first user of stone tools. *Homo erectus* appeared about 1.5 million years ago and was eventually replaced by *Homo sapiens* some 300,000 years ago.

Review Questions

1. Describe the evolution of mammals, tracing the synapsid lineage from early amniote ancestors to true mammals. How would you distinguish the skull of a synapsid from that of diapsid?
2. Describe some of the structural and functional adaptations that appeared in the early amniotes that foreshadowed the mammalian body plan. Which mammalian attributes do you think were especially important to the successful radiation of mammals?
3. Hair is believed to have evolved in the therapsids as an adaptation for insulation, but modern mammals have adapted hair for several other purposes. Describe these.
4. What is distinctive about each of the following: ruminant horns, deer antlers, and the rhinoceros horn? Briefly describe the growth cycle of antlers.
5. Describe the location and principal function(s) of each of the following skin glands: sweat glands (of two kinds, eccrine and apocrine), scent glands, sebaceous glands, and mammary glands.
6. Define the terms "diphyodont" and "heterodont" and explain how both terms apply to mammalian dentition.
7. Describe the food habits of each of the following groups: insectivores, herbivores, carnivores, and omnivores. Can you give the common names of some mammals belonging to each group?
8. Most herbivorous mammals depend on cellulose as their main energy source, yet no mammal synthesizes cellulose-splitting enzymes. How are the digestive tracts of mammals specialized for digestion of cellulose?
9. Describe the annual migrations of barren-ground caribou and fur seals.
10. Explain what is distinctive about the life habit and mode of navigation in bats.
11. Describe and distinguish the patterns of reproduction in monotremes, marsupials, and placental mammals. What aspects of mammalian reproduction are present in *all* mammals but in no other class of vertebrates?
12. What is the difference between density-dependent and density-independent causes of fluctuations in the size of mammalian populations?
13. Describe the hare-lynx population cycle, considered a classic example of a prey-predator relationship (Figure 19-21). From your examination of the cycle, can you formulate a hypothesis to explain the oscillations?
14. What anatomical characteristics set the primates apart from other mammals?
15. What role does the fossil named "Lucy" play in the reconstruction of human evolutionary history?
16. In what ways do the genera *Australopithecus* and *Homo,* which coexisted for at least 2 million years, differ?
17. When approximately did the different species of *Homo* appear and how did they differ socially?
18. What major attributes make the human position in animal evolution unique?

Selected References

See also general references on page 395.

Eisenberg, J. F. 1981. The mammalian radiations: an analysis of trends in evolution, adaptation, and behavior. Chicago, University of Chicago Press. *Wide-ranging, authoritative synthesis of mammalian evolution and behavior.*

Grzimek's encyclopedia of mammals. 1990. vol. 1–5. New York, McGraw-Hill Publishing Company. *Valuable source of information on all mammalian orders.*

Jones, S., R. D. Martin, and D. Pilbeam. 1992. Cambridge encyclopedia of human evolution. Cambridge, England, Cambridge University Press. *Comprehensive and informative encyclopedia written for the nonspecialist. Highly readable and highly recommended.*

Macdonald, D., ed. 1984. The encyclopedia of mammals. New York, Facts on File Publications. *Coverage of all mammalian orders and families, enhanced with fine photographs and color artwork.*

Nowak, R. M. 1991. Walker's mammals of the world, ed. 5. Baltimore, The Johns Hopkins University Press. *The definitive illustrated reference work on mammals, with descriptions of all extant and recently extinct species.*

Preston-Mafham, R., and K. Preston-Mafham. 1992. Primates of the world. New York, Facts on File Publications. *A small "primer" with high quality photographs and serviceable descriptions.*

Rice, J. A., ed. 1994. The marvelous mammalian parade. Natural History 103(4):39–91. *A special multi-authored section on mammalian evolution.*

Rismiller, P. D., and R. S. Seymour. 1991. The echidna. Sci. Am. **294**:96–103 (Feb.). *Recent studies of this fascinating monotreme have revealed many secrets of its natural history and reproduction.*

Savage, R. J. G., and M. R. Long. 1986. Mammal evolution: an illustrated guide. New York, Facts on File Publications. *Profusely illustrated survey of fossil mammals.*

Stringer, C. B. 1990. The emergence of modern humans. Sci. Am. **263**:98–104 (Dec.). *A review of the geographical origins of modern humans.*

Suga, N. 1990. Biosonar and neural computation in bats. Sci. Am. **262**:60–68 (June). *How the bat nervous system processes echolocation signals.*

Links to the Internet

Visit this textbook's Web site at http://www.mhhe.com/zoology to find live Internet links for each of the references listed below.

1. Class Mammalia. University of Michigan site on mammals. Pictures, much information on the morphology, distribution, and ecology of a large number of mammals, with links to nearly all orders. Each individual family is also linked to Web pages. Images may not be available for display depending on your server.

2. Animal Resources. This site provides many links to information on a variety of animals. Many mammals are featured.

3. Introduction to Synapsida. This University of California at Berkeley Museum of Paleontlogy site provides much information regarding the ancient synapsid ancestors of the mammals.

4. Phylogeny of the Mammals. University of California at Berkeley Museum of Paleontology. Click on an icon to find information on the classification and biology of any of the mammalian subclasses and orders.

5. The Philadelphia Zoo—Mammals! This site provides general information on mammals and links to specific information on zoo mammals.

6. Mammal Species of the World. This site is a database of mammalian taxonomy.

7. IMMA. This site, supported by the International Marine Mammal Association, provides much information on marine fisheries, marine mammals in captivity, and conservation of marine mammals.

8. Yerkes Regional Primate Research Center. This site provides much information on the primate research being conducted at the Yerkes Center.

9. Big Cats Online. This site contains information on big cat species.

10. Primate Center at Duke University. Information on primate research, captive breeding, biodiversity, and endangered primates, particularly prosimians.

11. Gorillas Online. Links to pictures, information, FAQs, and more links about gorillas.

12. Bat Conservation International. Information and links to much information about bats.

General References

The references below pertain to groups covered in more than one chapter. They include a number of very valuable field manuals that aid in identification, as well as general texts.

Barrington, E. J. W. 1979. Invertebrate structure and function, ed. 2. New York, John Wiley & Sons, Inc. *Excellent account of function in major invertebrate groups.*

Brusca, R. C., and G. J. Brusca. 1990. Invertebrates. Sunderland, Massachusetts, Sinauer Associates, Inc. *Invertebrate text organized around the bauplan ("body plan") concept—structural range, architectural limits, and functional aspects of a design—for each phylum. Includes cladistic analysis of phylogeny for most groups.*

Conway Morris, S., J. D. George, K. Gibson, and H. M. Platt (eds.). 1985. The origins and relationships of lower invertebrates. Oxford, Clarendon Press. *Technical discussions of phylogenetic relationships among lower invertebrates; essential for serious students of this topic.*

Fotheringham, N. 1980. Beachcomber's guide to Gulf Coast marine life. Houston, Texas, Gulf Publishing Company. Coverage arranged by habitats. *No keys, but common forms that occur near shore can be identified.*

Gosner, K. L. 1979. A field guide to the Atlantic seashore: invertebrates and seaweeds of the Atlantic coast from the Bay of Fundy to Cape Hatteras. The Peterson Field Guide Series. Boston, Houghton Mifflin Company. *A helpful aid for students of invertebrates found along the northeastern coast of the United States.*

Humann, P. 1992. Reef creature identification. Florida, Caribbean, Bahamas. Jacksonville, Florida, New World Publications. *Excellent field guide to aid identification of Atlantic reef invertebrates except corals.*

Hyman, L. H. 1940-1967. The invertebrates, 6 vols. New York, McGraw-Hill Book Company. *Informative discussions on the phylogenies of most invertebrates are treated in this outstanding series of monographs. Volume I contains a discussion of the colonial hypothesis of the origin of metazoa, and volume 2 contains a discussion of the origin of bilateral animals, body cavities, and metamerism.*

Kaplan, E. H. 1988. A field guide to southeastern and Caribbean seashores: Cape Hatteras to the Gulf Coast, Florida, and the Caribbean. A Peterson Field Guide Series. Boston, Houghton Mifflin Company. *More than just a field guide, this comprehensive book is filled with information on the biology of seashore animals; complements Gosner's field guide which covers animals norlh of Cape Hatteras.*

Kozloff, E. N. 1987. Marine invertebrates of the Pacific Northwest. Seattle, University of Washington Press. *Contains keys for many marine groups.*

Kozloff, E. N. 1990. Invertebrates. Philadelphia, Saunders College Publishing. *A good invertebrate text, less exhaustive than Ruppert and Barnes (1994) or Brusca and Brusca (1990). Cladistic analysis not emphasized.*

Lane, R. P., and H. W. Crosskey. 1992. Medically important insects and arachnids. London, Chapman and Hall. *The most up-to-date medical entomology text available.*

Meglitsch, P. A., and F. R. Schram. 1991. Invertebrate zoology, ed. 3. New York, Oxford University Press. *Schram's thorough revision of the older Meglitsch text. Includes cladistic treatments.*

Morris, R. H., D. P. Abbott, and E. C. Haderlie. 1980. Intertidal invertebrates of California. Stanford, Stanford University Press. *An essential reference on the most important invertebrates of the intertidal zone in California. Contains 900 color photographs.*

Nielson, C. 1995. Animal evolution: interrelationships of the living phyla. New York, Oxford University Press. *Cladistic analysis of morphology is used to develop sister-group relationships of the living Metazoa. An advanced but essential reference.*

Pennak, R. W. 1989. Freshwater invertebrates of the United States, ed. 3. New York, John Wiley & Sons, Inc. *Contains keys for identification of freshwater invertebrates with brief accounts of each group. Indispensable for freshwater biologists.*

Ricketts, E. F., J. Calvin, and J. W. Hedgpeth (revised by D. W. Phillips). 1985. Between Pacific tides, ed. 5. Stanford, Stanford University Press. *A revision of a classic work in marine biology. It stresses the habits and habitats of the Pacific coast invertebrates, and the illustrations are revealing. It includes an excellent, annotated systematic index and bibliography.*

Roberts, L. S., and J. Janovy, Jr. 1996. Foundations of parasitology, ed. 5. Dubuque, Wm. C. Brown Publishers. *Highly readable and up-to-date account of parasitic protistans, worms and arthropods.*

Ruppert, E. E., and R. D. Barnes. 1994. Invertebrate zoology, ed. 6. Philadelphia, Saunders College Publishing. *Authoritative, detailed coverage of invertebrate phyla.*

Smith, D. L. 1977. A guide to marine coastal plankton and marine invertebrate larvae. Dubuque, Iowa, Kendall/Hunt Publishing Company. *Valuable manual for identification of marine plankton, which is usually not covered in most field guides.*

Willmer, P. 1990. Invertebrate relationships. Patterns in animal evolution. Cambridge, Cambridge University Press. *Articulate statement of invertebrate phylogeny from a non-cladist. Good account of the polyphyletic hypothesis for the origin of arthropods.*

Glossary

This glossary lists definitions, pronunciations, and derivations of the most important recurrent technical terms, units, and names (excluding taxa) used in the text.

A

abomasum (ab′ō-mā′səm) (L. *ab*, from, + *omasum*, paunch). Fourth and last chamber of the stomach of ruminant mammals.

aboral (ab-o′ rəl) (L. *ab*, from, + *os*, mouth) A region of an animal opposite the mouth.

acanthodians (a′kan-thō′dē-əns) (Gr. *akantha*, prickly, thorny). A group of the earliest known true jawed fishes from Lower Silurian to Lower Permian.

aciculum (ə-sik′ū-ləm) (L. *acicula*, dm. of *acus*, a point) Supporting structure in notopodium and neuropodium of some polychaetes.

acoelomate (a-sēl′ə-māt′) (Gr. *a*, not, + *koilōma*, cavity) Without a coelom, as in flatworms and proboscis worms.

acontium (ə-kän′chē-əm), pl. **acontia** (Gr. *akontion*, dart) Threadlike structure bearing nematocysts located on mesentery of sea anemone.

adaptation (L. *adaptatus*, fitted) An anatomical structure, physiological process, or behavioral trait that evolved by natural selection and improves an organism's ability to survive and leave descendants.

adaptive radiation Evolutionary diversification that produces numerous ecologically disparate lineages from a single ancestral one, especially when this diversification occurs within a short interval of geological time.

adaptive value Degree to which a characteristic helps an organism to survive and reproduce; lends greater fitness in environment; selective advantage.

adaptive zone A characteristic reaction and mutual relationship between environment and organism ("way of life") demonstrated by a group of evolutionarily related organisms.

adductor (ə-duk′tər) (L. *ad*, to, + *ducere*, to lead) A muscle that draws a part toward a median axis, or a muscle that draws the two valves of a mollusc shell together.

adipose (ad′ə-pōs) (L. *adeps*, fat) Fatty tissue; fatty.

aerobic (a-rō′bik) (Gr. *aēr*, air, + *bios*, life) Oxygen-dependent form of respiration.

bat/āpe/ärmadillo/herring/fēmale/finch/līce/ crocodile/crōw/duck/ūnicorn/tüna/ə indicates unaccented vowel sound "uh" as in mammal, fishes, cardinal, heron, vulture/stress as in bi-ol′o-gy, bi′o-log′i-cal

afferent (af′ə-rənt) (L. *ad*, to, + *ferre*, to bear) Adjective meaning leading or bearing toward some organ, for example, nerves conducting impulses toward the brain or blood vessels carrying blood toward an organ; opposed to efferent.

agnathan (ag′nə-thən) (Gr. *a*, without, + *gnathos*, jaw) A jawless fish of the superclass Agnatha of the phylum Chordata.

alate (ā′lāt) (L. *alatus*, wing) Winged.

albumin (al-bū′mən) (L. *albumen*, white of egg) Any of a large class of simple proteins that are important constituents of vertebrate blood plasma and tissue fluids and also present in milk, whites of eggs, and other animal substances.

allele (ə-lēl′) (Gr. *allēlōn*, of one another) Alternative forms of genes coding for the same trait; situated at the same locus in homologous chromosomes.

allelic frequency The proportion of all copies of a particular gene in a population **gene pool** represented by a particular **allele.**

allopatric (Gr. *allos*, other, + *patra*, native land) In separate and mutually exclusive geographical regions.

allopatric speciation The hypothesis that new species are formed by dividing an ancestral species into geographically isolated subpopulations that evolve **reproductive barriers** between them through independent evolutionary divergence from their common ancestor.

altricial (al-tri′shəl) (L. *altrices*, nourishers) Referring to young animals (especially birds) having the young hatched in an immature dependent condition.

alula (al′yə-lə) (L. dim. of *ala*, wing) The first digit or thumb of a bird's wing, much reduced in size.

ambulacra (am′byə-lak′rə) (L. *ambulare*, to walk) In echinoderms, radiating grooves where podia of water-vascular system characteristically project to outside.

ameboid (ə-mē′boid) (Gr. *amoibē*, change, + *oid*, like) Ameba-like in putting forth pseudopodia.

amictic (ə-mik′tic) (Gr. *a*, without, + *miktos*, mixed or blended) Pertaining to female rotifers, which produce only diploid eggs that cannot be fertilized, or to the eggs produced by such females. Compare with mictic.

ammocoetes (am-ə-sēd′ēz) (Gr. *ammos*, sand, + *koitē*, bed) Larvae of any of various species of lampreys.

amnion (am′nē-än) (Gr. *amnion*, membrane around the fetus) The innermost of the extraembryonic membranes forming a fluid-filled sac around the embryo in amniotes.

amniote (am′nē-ōt) Having an amnion; as a noun, an animal that develops an amnion in embryonic life, that is, reptiles, birds, and mammals; adj. **amniotic.**

amplexus (am-plek′səs) (L. embrace) The copulatory embrace of frogs or toads.

ampulla (am-pūl′ə) (L. flask) Membranous vesicle; dilation at one end of each semicircular canal containing sensory epithelium; muscular vesicle above the tube foot in water-vascular system of echinoderms.

anadromous (an-ad′rə-məs) (Gr. *anadromos*, running upward) Refers to fishes that migrate up streams from the sea to spawn.

anaerobic (an′ə-rō′bik) (Gr. *an*, not, + *aēr*, air, + *bios*, life) Not dependent on oxygen for respiration.

analogy (L. *analogous*, ratio) Similarity of function but not of origin.

anapsids (ə-nap′səds) (Gr. *an*, without, + *apsis*, arch) Amniotes in which the skull lacks temporal openings, with turtles the only living representatives.

ancestral character state The condition of a taxonomic character inferred to have been present in the most recent common ancestor of a taxonomic group being studied cladistically.

androgenic gland (an′drō-jen′ək) (Gr. *anēr*, male, + *gennaein*, to produce) Gland in Crustacea that causes development of male characteristics.

annulus (an′yəl-əs) (L. ring) Any ringlike structure, such as superficial rings on leeches.

antenna (L. sail yard) A sensory appendage on the head of arthropods, or the second pair of the two such pairs of structures in crustaceans.

antennal gland Excretory gland of Crustacea located in the antennal metamere.

anterior (L. comparative of *ante*, before) The head of an organism or (as the adjective) toward that end.

anthropoid (an′thrə-poyd) (Gr. *anthrōpos*, man, + *eidos*, form) Resembling humans; especially the great apes of the family Pongidae.

aperture (ap′ər-chər) (L. *apertura* from *aperire*, to uncover) An opening; the opening into the first whorl of a gastropod shell.

apical (ā′pə-kl) (L. *apex*, tip) Pertaining to the tip or apex.

apical complex A certain combination of organelles found in the protozoan phylum Apicomplexa.

apocrine (ap′ə-krən) (Gr. *apo*, away, + *krinein*, to separate) Applies to a type of mammalian sweat gland that produces a viscous secretion by breaking off a part of the cytoplasm of secreting cells.

apopyle (ap′-ə-pīl) (Gr. *apo*, away from, + *pylē*, gate) In sponges, opening of the radial canal into the spongocoel.

arboreal (är-bōr′ē-al) (L. *arbor*, tree) Living in trees.

archaeocyte (ar′kē-ə-sīt) (Gr. *archaios*, beginning, + *kytos*, hollow vessel) Ameboid cells of varied function in sponges.

archenteron (ärk-en′tə-rän) (Gr. *archē*, beginning, + *enteron*, gut) The main cavity of an embryo in the gastrula stage; it is lined with endoderm and represents the future digestive cavity.

archosaur (är′kə-sor) (Gr. *archōn*, ruling, + *sauros*, lizard) Advanced diapsid vertebrates, a group that includes the living crocodiles and the extinct pterosaurs and dinosaurs.

Aristotle's lantern Masticating apparatus of some sea urchins.

artiodactyl (är′ti-o-dak′təl) (Gr. *artios*, even, + *daktylos*, toe) One of an order of mammals with two or four digits on each foot.

asconoid (Gr. *askos*, bladder) Simplest form of sponges, with canals leading directly from the outside to the interior.

asexual Without distinct sexual organs; not involving formation of gametes.

atoke (ā′tōk) (Gr. *a*, without, + *tokos*, offspring) Anterior, nonreproductive part of a marine polychaete, as distinct from the posterior, reproductive part (epitoke) during the breeding season.

atrium (ā′trē-əm) (L. *atrium*, vestibule) One of the chambers of the heart; also, the tympanic cavity of the ear; also, the large cavity containing the pharynx in tunicates and cephalochordates.

auricle (aw′ri-kl) (L. *auricula*, dim. of *auris*, ear) One of the less muscular chambers of the heart; atrium; the external ear, or pinna; any earlike lobe or process.

auricularia (ə-rik′u-lar′ē-ə) (L. *auricula*, a small ear). A type of larva found in Holothuroidea.

autogamy (aw-täg′ə-me) (Gr. *autos*, self, + *gamos*, marriage) Condition in which the gametic nuclei produced by meiosis fuse within the same organism that produced them to restore the diploid number.

autotomy (aw-tät′ə-me) (Gr. *autos*, self, + *tomos*, a cutting) The breaking off of a part of the body by the organism itself.

autotroph (aw′tō-trōf) (Gr. *autos*, self, + *trophos*, feeder) An organism that makes its organic nutrients from inorganic raw materials.

autotrophic nutrition (Gr. *autos*, self, + *trophia*, denoting nutrition) Nutrition characterized by the ability to use simple inorganic substances for the synthesis of more complex organic compounds, as in green plants and some bacteria.

axial (L. *axis*, axle) Relating to the axis, or stem; on or along the axis.

axocoel (aks′ə-sēl) (Gr. *axon*, axle, + *koilos*, hollow) Anterior coelomic compartment in echinoderms; corresponds to protocoel.

axolotl (ak′sə-lot′l) (Nahuatl *atl*, water, + *xolotl*, doll, servant, spirit) Salamanders of the species *Ambystoma mexicanum*, which do not metamorphose, and retain aquatic larval characteristics throughout adulthood.

axon (ak′sän) (Gr. *axōn*) Elongate extension of a neuron that conducts impulses away from the cell body and toward the synaptic terminals.

axoneme (aks′ə-nēm) (L. *axis*, axle, + Gr. *nēma*, thread) The microtubules in a cilium or flagellum, usually arranged as a circlet of nine pairs enclosing one central pair; also, the microtubules of an axopodium.

axopodium (ak′sə-pō′dē-əm) (Gr. *axon*, an axis, + *podion*, small foot) Long, slender, more or less permanent pseudopodium found in certain sarcodine protozoa. (Also **axopod.**)

B

basal body Also known as kinetosome and blepharoplast, a cylinder of nine triplets of microtubules found basal to a flagellum or cilium; same structure as a centriole.

basis, basipodite (bā′səs, bā-si′pə-dīt) (Gr. *basis*, base, + *pous*, *podos*, foot) The distal or second joint of the protopod of a crustacean appendage.

benthos (ben′thäs) (Gr. depth of the sea) Organisms that live along the bottom of the seas and lakes; adj., **benthic.** Also, the bottom itself.

bilateria (bī′lə-tir′ē-ə) (L. *bi-*, two, + *latus*, side) Bilaterally symmetrical animals.

binary fission A mode of asexual reproduction in which the animal splits into two approximately equal offspring.

binomial nomenclature The Linnean system of naming species in which the first word is the name of the genus (first letter capitalized) and the second word is the specific epithet (uncapitalized), usually an adjective modifying the name of the genus. Both of these words are written in italics.

biogenetic law A statement postulating a characteristic relationship between **ontogeny** and **phylogeny.** Examples include Haeckel's law of **recapitulation** and Von Baer's law that general characteristics (those shared by many species) appear earlier in ontogeny than more restricted ones; neither of these statements is universally true.

bioluminescence Method of light production by living organisms in which usually certain proteins (luciferins), in the presence of oxygen and an enzyme (luciferase), are converted to oxyluciferins with the liberation of light.

bipinnaria (L. *bi*, double, + *pinna*, wing, + *aria*, like or connected with). Free-swimming, ciliated, bilateral larva of the asteroid echinoderms; develops into the brachiolaria larva.

biramous (bī-rām′əs) (L. *bi*, double, + *ramus*, a branch). Adjective describing appendages with two distinct branches, contrasted with uniramous, unbranched.

blastocoel (blas′tə-sēl) (Gr. *blastos*, germ, + *koilos*, hollow) Cavity of the blastula.

blastomere (Gr. *blastos*, germ, + *meros*, part) An early cleavage cell.

blastopore (Gr. *blastos*, germ, + *poros*, passage, pore) External opening of the archenteron in the gastrula.

blastula (Gr. *blastos*, germ, + *ula*, dim) Early embryological state of many animals; consists of a hollow mass of cells.

blepharoplast (blə-fä′rə-plast) (Gr. *blepharon*, eyelid, + *plastos*, formed) See **basal body.**

BP Before the present.

brachial (brak′ē-əl) (L. *brachium*, forearm) Referring to the arm.

branchial (brank′ē-əl) (Gr. *branchia*, gills) Referring to gills.

buccal (buk′əl) (L. *bucca*, cheek) Referring to the mouth cavity.

budding Reproduction in which the offspring arises as an outgrowth from the parent and is initially smaller than the parent. Failure of the offspring to separate from the parent leads to colony formation.

bursa pl. **bursae** (M.L. *bursa*, pouch, purse made of skin) A saclike cavity. In ophiuroid echinoderms, pouches opening at bases of arms and functioning in respiration and reproduction (genitorespiratory bursae).

C

calyx (kā′-liks) (L. bud cup of a flower). Any of various cup-shaped zoological structures.

capitulum (ka-pi′tə-ləm) (L. small head) Term applied to small, headlike structures of various organisms, including projection from body of ticks and mites carrying mouthparts.

carapace (kar′ə-pās) (F. from Sp. *carapacho*, shell) Shieldlike plate covering the cephalothorax of certain crustaceans; dorsal part of the shell of a turtle.

carinate (kar′ə-nāt) (L. *carina*, keel) Having a keel, in particular the flying birds with a keeled sternum for the insertion of flight muscles.

carnivore (kar′nə-vōr) (L. *carnivorus*, flesh eating) One of the flesh-eating mammals of the order Carnivora. Also, any organism that eats animals. Adj., **carnivorous.**

cartilage (L. *cartilago*; akin to L. *cratis*, wickerwork) A translucent elastic tissue that makes up most of the skeleton of embryos, very young vertebrates, and adult cartilaginous fishes, such as sharks and rays; in higher forms much of it is converted into bone.

caste (kast) (L. *castus*, pure, separated) One of the polymorphic forms within an insect society, each caste having its specific duties, as queen, worker, soldier, and so on.

catadromous (kə-tad′rə-məs) (Gr. *kata*, down, + *dromos*, a running) Refers to fishes that migrate from fresh water to the ocean to spawn.

catastrophic species selection Differential survival among species during a time of mass extinction based on character variation that permits some species but not others to withstand severe environmental disturbances, such as those caused by an asteroid impact.

caudal (käd′l) (L. *cauda*, tail) Constituting, belonging to, or relating to a tail.

cecum, caecum (sē′kəm) (L. *caecus*, blind) A blind pouch at the beginning of the large intestine; any similar pouch.

cellulose (sel′ū-lōs) (L. *cella*, small room) Chief polysaccharide constituent of the cell wall of green plants and some fungi; an insoluble carbohydrate $(C_6H_{10}O_5)_n$ that is converted into glucose by hydrolysis.

centriole (sen′trē-ōl) (Gr. *kentron*, center of a circle, + L. *ola*, small) A minute cytoplasmic organelle usually found in the centrosome and considered to be the active division center of the animal cell; organizes spindle fibers during mitosis and meiosis. Same structure as basal body or kinetosome.

cephalization (sef′ə-li-zā-shən) (Gr. *kephalē*, head) The process by which specialization, particularly of the sensory organs and appendages, became localized in the head end of animals.

cephalothorax (sef′ə-lä-thō′raks) (Gr. *kephalē*, head, + thorax) A body division found in many Arachnida and higher Crustacea in which the head is fused with some or all of the thoracic segments.

cerata (sə-ra′tə) (Gr. *keras*, a horn, bow) Dorsal processes on some nudibranch for gaseous exchange.

cercaria (ser-kar′ē-ə) (Gr. *kerkos*, tail, + L. *aria*, like or connected with) Tadpolelike juveniles of trematodes (flukes).

cervical (sər′və-kəl) (L. *cervix*, neck) Relating to a neck.

chelicera (kə-lis′ə-rə), pl. **chelicerae** (Gr. *chēlē*, claw, + *keras*, horn) One of a pair of the most anterior head appendages on the members of the subphylum Chelicerata.

chelipeds (kēl′ə-peds) (Gr. *chēlē*, claw, + L. *pes*, foot) Pincerlike first pair of legs in most decapod crustaceans; specialized for seizing and crushing.

chemoautotroph (ke-mō-aw′tō-trōf) (Gr. *chemeia*, transmutation, + *autos*, self, + *trophos*, feeder) An organism utilizing inorganic compounds as a source of energy.

chitin (kī′tən) (Fr. *chitine*, from Gr. *chitōn*, tunic) A horny substance that forms part of the cuticle of arthropods and is found sparingly in certain other invertebrates; a nitrogenous polysaccharide insoluble in water, alcohol, dilute acids, and digestive juices of most animals.

chloragogen cells (klōr′ə-gog-ən) (Gr. *chloros*, light green, + *agōgos*, a leading, a guide) Modified peritoneal cells, greenish or brownish, clustered around the digestive tract of certain annelids; apparently they aid in elimination of nitrogenous wastes and in food transport.

chlorophyll (klō′rə-fil) (Gr. *chloros*, light green, + *phyllon*, leaf) Green pigment found in plants and in some animals necessary for photosynthesis.

chloroplast (Gr. *chloros*, light green, + *plastos*, molded) A plastid containing chlorophyll and usually other pigments, found in cytoplasm of plant cells.

choanocyte (kō-an′ə-sīt) (Gr. *choanē*, funnel, + *kytos*, hollow vessel) One of the flagellate collar cells that line cavities and canals of sponges.

chorion (kō′rē-on) (Gr. *chorion*, skin) The outer of the double membrane that surrounds the embryo of reptiles, birds, and mammals; in mammals it contributes to the placenta.

chromatophore (krō-mat′ə-fōr) (Gr. *chrōma*, color, + *herein*, to bear) Pigment cell, usually in the dermis, in which usually the pigment can be dispersed or concentrated.

chromosomal theory of inheritance The well-established principle, initially proposed by Sutton and Boveri in 1903–1904, that nuclear chromosomes are the physical bearers of genetic material in eukaryotic organisms. It is the foundation for modern evolutionary genetics.

chrysalis (kris′ə-lis) (L. from Gr. *chrysos*, gold) The pupal stage of a butterfly.

cilium (sil′ē-əm), pl. **cilia** (L. eyelid) A hairlike, vibratile organelle process found on many animal cells. Cilia may be used in moving particles along the cell surface or, in ciliate protozoans, for locomotion.

cirrus (sir′əs) (L. curl) A hairlike tuft on an insect appendage; locomotor organelle of fused cilia; male copulatory organ of some invertebrates.

clade (klād) (Gr. *klados*, branch) A taxon or other group consisting of a single species and all of its descendants, forming a distinct branch on a phylogenetic tree.

cladistics (klad-is′təks) (Gr. *klados*, branch, sprout) A system of arranging taxa by analysis of primitive and derived characteristics so that the arrangement will reflect phylogenetic relationships.

cladogram (klād′ə-gram) (Gr. *klados*, branch, + *gramma*, letter). A branching diagram showing the pattern of sharing of evolutionarily derived characters among species or higher taxa.

clitellum (klī-tel′əm) (L. *clitellae*, pack-saddle) Thickened saddlelike portion of certain midbody segments of many oligochaetes and leeches.

cloaca (klō-ā′kə) (L. sewer) Posterior chamber of digestive tract in many vertebrates, receiving feces and urogenital products. In certain invertebrates, a terminal portion of digestive tract that serves also as respiratory, excretory, or reproductive duct.

cnidocil (nī′dō-sil) (Gr. *knidē*, nettle, + L. *cilium*, hair) Triggerlike spine on nematocyst.

cnidocyte (nī′dō-sīt) (Gr. *knidē*, nettle, + *kytos*, hollow vessel) Modified interstitial cell that holds the nematocyst; during development of the nematocyst, the cnidocyte is a cnidoblast.

cochlea (kōk′lē-ə) (L. snail, from Gr. *kochlos*, a shellfish) A tubular cavity of the inner ear containing the essential organs of hearing; occurs in crocodiles, birds, and mammals; spirally coiled in mammals.

cocoon (kə-kun′) (Fr. *cocon*, shell) Protective covering of a resting or developmental stage, sometimes used to refer to both the covering and its contents; for example, the cocoon of a moth or the protective covering for the developing embryos in some annelids.

coelenteron (sē-len′tər-on) (Gr. *koilos*, hollow, + *enteron*, intestine) Internal cavity of a cnidarian; gastrovascular cavity; archenteron.

coelom (sē′lōm) (Gr. *koilōma*, cavity) The body cavity in triploblastic animals, lined with mesodermal peritoneum.

coelomoduct (sē-lō′mə-dukt) (Gr. *koilos*, hollow, + *ductus*, a leading) A duct that carries gametes or excretory products (or both) from the coelom to the exterior.

collagen (käl′ə-jən) (Gr. *kolla*, glue, + *genos*, descent) A tough, fibrous protein occurring in vertebrates as the chief constituent of collagenous connective tissue; also occurs in invertebrates, for example, the cuticle of nematodes.

collencyte (käl′ən-sīt) (Gr. *kolla*, glue, + *kytos*, hollow vessel) A type of cell in sponges that secretes fibrillar collagen.

colloblast (käl′ə-blast) (Gr. *kolla*, glue, + *blastos*, germ) A glue-secreting cell on the tentacles of ctenophores.

comb plate One of the plates of fused cilia that are arranged in rows for ctenophore locomotion.

commensalism (kə-men′səl-iz′əm) (L. *com*, together with, + *mensa*, table) A relationship in which one individual lives close to or on another and benefits, and the host is unaffected; often symbiotic.

common descent Darwin's theory that all forms of life are derived from a shared ancestral population through a branching of evolutionary lineages.

comparative biochemistry Studies of the structures of biological macromolecules, especially proteins and nucleic acids, and their variation within and among related species to reveal homologies of macromolecular structure.

comparative cytology Studies of the structures of chromosomes within and among related species to reveal homologies of chromosomal structure.

comparative method Use of patterns of similarity and dissimilarity among species or populations to infer their phylogenetic relationships; use of phylogeny to examine evolutionary processes and history.

comparative morphology Studies of organismal form and its variation within and among related species to reveal homologies of organismal characters.

conjugation (kon′jū-gā′shən) (L. *conjugare*, to yoke together) Temporary union of two ciliate

bat/āpe/ärmadillo/herring/fēmale/finch/līce/crocodile/crōw/duck/ūnicorn/tüna/ə indicates unaccented vowel sound "uh" as in mammal, fishes, cardinal, heron, vulture/stress as in bi-ol′o-gy, bi′o-log′i-cal

protozoa while they are exchanging chromatin material and undergoing nuclear phenomena resulting in binary fission. Also, formation of cytoplasmic bridges between bacteria for transfer of plasmids.

conodont kō′ə-dänt (Gr. *kōnos*, cone, + *odontos*, tooth) Toothlike element from a Paleozoic animal now believed to have been an early marine vertebrate.

conspecific (L. *com*, together, + *species*) A member of the same species.

contractile vacuole A clear fluid-filled cell vacuole in protozoa and a few lower metazoa; takes up water and releases it to the outside in a cyclical manner, for osmoregulation and some excretion.

control That part of a scientific experiment to which the experimental variable is not applied but that is similar to the experimental group in all other respects.

coprophagy (kə-prä′fə-jē) (Gr. *kopros*, dung, + *phagein*, to eat) Feeding on dung or excrement as a normal behavior among animals; reinjestion of feces.

copulation (Fr. from L. *copulare*, to couple) Sexual union to facilitate the reception of sperm by the female.

corneum (kor′nē-əm) (L. *corneus*, horny) Epithelial layer of dead, keratinized cells. Stratum corneum.

corona (kə-rō′nə) (L. crown) Head or upper portion of a structure; ciliated disc on anterior end of rotifers.

corpora allata (kor′pə-rə əl-la′tə) (L. *corpus*, body, + *allatum*, aided) Endocrine glands in insects that produce juvenile hormone.

cortex kor′teks) (L. bark) The outer layer of a structure.

coxa, coxopodite (kox′ə, kəx-ä′pə-dīt) (L. *coxa*, hip, + Gr. *pous, podos*, foot) The proximal joint of an insect or arachnid leg; in crustaceans, the proximal joint of the protopod.

Cretaceous extinction A **mass extinction** that occurred 65 million years ago in which 76% of existing species, including all dinosaurs, became extinct, marking the end of the Mesozoic era.

cryptobiotic (Gr. *kryptos*, hidden, + *bīoticus*, pertaining to life) Living in concealment; refers to insects and other animals that live in secluded situations, such as underground or in wood; also tardigrades and some nematodes, rotifers, and others that survive harsh environmental conditions by assuming for a time a state of very low metabolism.

ctenoid scales (ten′oyd) (Gr. *kteis, ktenos*, comb) Thin, overlapping dermal scales of the more advanced fishes; exposed posterior margins have fine, toothlike spines.

cuticle (kū′ti-kəl) (L. *cutis*, skin) A protective, noncellular, organic layer secreted by the external epithelium (hypodermis) of many invertebrates. In higher animals the term refers to the epidermis or outer skin.

cycloid scales (sī′kloyd) (Gr. *kyklos*, circle) Thin, overlaping dermal scales of the more primitive fishes; posterior margins are smooth.

cynodonts (sin′ə-dänts) (Gr. *kynodōn*, canine tooth) A group of mammal-like carnivorous synapsids of the Upper Permian and Triassic.

cyst (sist) (Gr. *kystis*, a bladder, pouch) A resistant, quiescent stage of an organism, usually with a secreted wall.

cysticercus (sis′tə-ser′kəs) (Gr. *kystis*, bladder, + *kerkos*, tail) A type of juvenile tapeworm in which an invaginated and introverted scolex is contained in a fluid-filled bladder.

cystid (sis′tid) (Gr. *kystis*, bladder) In an ectoproct, the dead secreted outer parts plus the adherent underlying living layers.

cytopharynx (Gr. *kytos*, hollow vessel, + *pharynx*, throat) Short tubular gullet in ciliate protozoa.

cytoplasm (si′tə-plasm) (Gr. *kytos*, hollow vessel, + *plasma*, mold) The living matter of the cell, excluding the nucleus.

cytoproct (sī′tə-prokt) (Gr. *kytos*, hollow vessel, + *prōktos*, anus) Site on a protozoan where indigestible matter is expelled.

cytopyge (sī′tə-pīj) (Gr. *kytos*, hollow vessel, + *pyge*, rump or buttocks). In some protozoa, localized site for expulsion of wastes.

cytostome (sī′tə-stōm) (Gr. *kytos*, hollow vessel, + *stoma*, mouth) The cell mouth in many protozoa.

D

Darwinism Theory of evolution emphasizing common descent of all living organisms, gradual change, multiplication of species and natural selection.

data sing. **datum** (Gr. *dateomai*, to divide, cut in pieces) The results in a scientific experiment, or descriptive observations, upon which a conclusion is based.

deciduous (də-sij′ü-wəs) (L. *deciere*, to fall off) Shed or falling off at the end of a growing period.

deduction (L. *deductus*, led apart, split, separated) Reasoning from the general to the particular, that is, from given premises to their necessary conclusion.

definitive host The host in which sexual reproduction of a symbiont takes place; if no sexual reproduction, then the host in which the symbiont becomes mature and reproduces; contrast intermediate host.

derived character state A condition of a taxonomic character inferred by cladistic analysis to have arisen within a taxon being examined cladistically rather than having been inherited from the most recent common ancestor of all members of the taxon.

dermal (Gr. *derma*, skin) Pertaining to the skin; cutaneous.

dermis The inner, sensitive mesodermal layer of skin; corium.

determinate cleavage The type of cleavage, usually spiral, in which the fate of the blastomeres is determined very early in development; mosaic cleavage.

detritus (də-trī′tus) (L. that which is rubbed or worn away) Any fine particulate debris of organic or inorganic origin.

Deuterostomia (dū′də-rō-stō′mē-ə) (Gr. *deuteros*, second, secondary, + *stoma*, mouth) A group of higher phyla in which cleavage is indeterminate and primitively radial. The endomesoderm is enterocoelous, and the mouth is derived away from the blastopore. Includes Echinodermata, Chordata, and a number of minor phyla. Compare with **Protostomia.**

diapause (dī′ə-pawz) (Gr. *diapausis*, pause) A period of arrested development in the life cycle of insects and certain other animals in which physiological activity is very low and the animal is highly resistant to unfavorable external conditions.

diapsids (dī-ap′səds) (Gr. *di*, two, + *apsis*, arch) Amniotes in which the skull bears two pairs of temporal openings; includes reptiles (except turtles) and birds.

diffusion (L. *diffusion*, dispersion) The movement of particles or molecules from area of high concentration of the particles or molecules to area of lower concentration.

digitigrade (dij′ə-də-grād) (L. *digitus*, finger, toe, + *gradus*, step, degree) Walking on the digits with the posterior part of the foot raised; compare **plantigrade.**

dimorphism (dī-mor′fizm) (Gr. *di*, two, + *morphē*, form) Existence within a species of two distinct forms according to color, sex, size, organ structure, and so on. Occurrence of two kinds of zooids in a colonial organism.

dioecious (dī-ē′shəs) (Gr. *di*, two, + *oikos*, house) Having male and female organs in separate individuals.

diphycercal (dif′ i-ser′kəl) (Gr. *diphyēs*, twofold, + *kerkos*, tail) A tail that tapers to a point, as in lungfishes; vertebral column extends to tip without upturning.

diphyodont (di-fī′ə-dänt) (Gr. *diphyēs*, twofold, + *odous*, tooth) Having deciduous and permanent sets of teeth successively.

diploblastic (di′plə-blas′tək) (Gr. *diploos*, double, + *blastos*, bud) Organism with two germ layers, endoderm and ectoderm.

diploid (dip′loid) (Gr. *diploos*, double, + *eidos*, form) Having the somatic (double, or 2n) numbers of chromosomes or twice the number characteristic of a gamete of a given species.

directional selection Natural selection that favors one extreme value of a continuously varying trait and disfavors other values.

disruptive selection Natural selection that favors simultaneously two different extreme values of a continuously varying trait but disfavors intermediate values.

distal (dis′təl). Farther from the center of the body than a reference point.

dorsal (dor′səl) (L. *dorsum*, back) Toward the back, or upper surface, of an animal.

E

eccrine (ek′rən) (Gr. *ek*, out of, + *krinein*, to separate) Applies to a type of mammalian sweat gland that produces a watery secretion.

ecdysiotropin (ek-dē-zē-ə-trō′pən) (Gr. *ekdysis*, to strip off, escape, + *tropos*, a turn, change) Hormone secreted in brain of insects that stimulates prothoracic gland to secrete molting hormone. Prothoracicotropic hormone; brain hormone.

ecdysis (ek′də-sis) (Gr. *ekdysis*, to strip off, escape) Shedding of outer cuticular layer; molting, as in insects or crustaceans.

ecdysone (ek-dī′sōn) (Gr. *ekdysis*, to strip off). Molting hormone of arthropods, stimulates growth and ecdysis, produced by prothoracic glands in insects and Y organs in crustaceans.

ectoderm (ek′tō-derm) (Gr. *ektos*, outside, + *derma*, skin) Outer layer of cells of an early embryo (gastrula stage); one of the germ layers, also sometimes used to include tissues derived from ectoderm.

ectognathous (ek′tə-nā′thəs) (Gr. *ektos*, outside, without, + *gnathos*, jaw). Derived character of most insects; mandibles and maxillae not in pouches.

ectolecithal (ek′tō-les′ə-thəl) (Gr. *ektos*, ouside, + *lekithos*, yolk). Yolk for nutrition of the embryo contributed by cells that are separate from the egg cell and are combined with the zygote by envelopment within the eggshell.

ectoplasm (ek′tō-plazm) (Gr. *ektos*, outside, + *plasma*, form) The cortex of a cell or that part of cytoplasm just under the cell surface; contrasts with **endoplasm.**

ectothermic (ek′tō-therm′ic) (Gr. *ektos*, outside, + *thermē*, heat) Having a variable body temperature derived from heat acquired from the environment, contrasts with **endothermic.**

efferent (ef′ə-rənt) (L. *ex*, out, + *ferre*, to bear) Leading or conveying away from some organ, for example, nerve impulses conducted away from the brain, or blood conveyed away from an organ; contrasts with **afferent.**

egestion (ē-jes′chən) (L. *egestus*, to discharge) Act of casting out indigestible or waste matter from the body by any normal route.

elephantiasis (el-ə-fən-tī′ə-səs) Disfiguring condition caused by chronic infection with filarial worms *Wuchereria bancrofti* and *Brugia malayi.*

embryogenesis (em′brē-ō-jen′ə-səs) (Gr. *embryon*, embryo, + *genesis*, origin) The origin and development of the embryo; embryogeny.

emigrate (L. *emigrare*, to move out) To move *from* one area to another to take up residence.

encystment Process of cyst formation.

endemic (en-dem′ik) (Gr. *en*, in, + *demos*, populace) Peculiar to a certain region or country; native to a restricted area; not introduced.

endoderm (en′də-dərm) (Gr. *endon*, within, + *derma*, skin) Innermost germ layer of an embryo, forming the primitive gut; also may refer to tissues derived from endoderm.

endognathous (en′də-nā-thəs) (Gr. *endon*, within, + *gnathous*, jaw). Ancestral character in insects, found in orders Diplura, Collembola, and Protura, in which the mandibles and maxillae are located in pouches.

endolecithal (en′də-les′ə-thəl) (Gr. *endon*, within, + *lekithos*, yolk). Yolk for nutrition of the embryo incorporated into the egg cell itself.

endoplasm (en′də-pla-zm) (Gr. *endon*, within, + *plasma*, mold or form) The portion of cytoplasm that immediately surrounds the nucleus.

endopod, endopodite (en′də-päd, en-dop′ə-dīt) (Gr. *endon*, within, + *pous, podos*, foot) Medial branch of a biramous crustacean appendage.

endoskeleton (Gr. *endon*, within, + *skeletos*, hard) A skeleton or supporting framework within the living tissues of an organism; contrasts with **exoskeleton.**

endostyle (en′də-stīl) (Gr. *endon*, within, + *stylos*, a pillar) Ciliated groove(s) in the floor of the pharynx of tunicates, cephalochordates, and larval cyclostomes, used for accumulating and moving food particles to the stomach.

endothermic (en′də-therm′ik) (Gr. *endon*, within, + *thermē*, heat) Having a body temperature determined by heat derived from the animal's own oxidative metabolism; contrasts with **ectothermic.**

enterocoel (en′tər-ō-sēl′) (Gr. *enteron*, gut, + *koilos*, hollow) A type of coelom formed by the outpouching of a mesodermal sac from the endoderm of the primitive gut.

enterocoelomate (en′ter-ō-sēl′ō-māte) (Gr. *enteron*, gut, + *koilōma*, cavity, + Engl. *ate*, state of) An animal having an enterocoel, such as an echinoderm or a vertebrate.

enterocoelous mesoderm formation Embryonic formation of mesoderm by a pouchlike outfolding from the archenteron, which then expands and obliterates the blastocoel, thus forming a large cavity, the coelom, lined with mesoderm.

enteron (en′tə-rän) (Gr. intestine) The digestive cavity.

ephyra (ef′ə-rə) (Gr. *Ephyra*, Greek city) Refers to castlelike appearance. Medusa bud from a scyphozoan polyp.

epidermis (ep′ə-dər′məs) (Gr. *epi*, on, upon, + *derma*, skin) The outer, nonvascular layer of skin of ectodermal origin; in invertebrates, a single layer of ectodermal epithelium.

epipod, epipodite (ep′ə-päd, e-pip′ə-dīt) (Gr. *epi*, on, upon, + *pous, podos*, foot) A lateral process on the protopod of a crustacean appendage, often modified as a gill.

epistome (ep′i-stōm) (Gr. *epi*, on, upon, + *stoma*, mouth) Flap over the mouth in some lophophorates bearing the protocoel.

epithelium (ep′i′thē′lē-um) (Gr. *epi*, on, upon, + *thēlē*, nipple) A cellular tissue covering a free surface or lining a tube or cavity.

epitoke (ep′i′tōk) (Gr. *epitokos*, fruitful) Posterior part of a marine polychaete when swollen with developing gonads during the breeding season; contrast with **atoke.**

erthrocyte (ə-rith′rō-sīt) (Gr. *erythros*, red, + *kytos*, hollow vessel) Red blood cell; has hemoglobin to carry oxygen from lungs or gills to tissues; during formation in mammals, erythrocytes lose their nuclei, those of other vertebrates retain the nuclei.

estrous cycle Periodic episodes of estrus, or "heat," when females of most mammalian species become sexually receptive.

estrus (es′trəs) (L. *oestrus*, gadfly, frenzy) The period of heat, or rut, especially of the female during ovulation of the eggs. Associated with maximum sexual receptivity.

eukaryotic, eucaryotic (ū′ka-rē-ot′ik) (Gr. *eu*, good, true, + *karyon*, nut, kernel) Organisms whose cells characteristically contain a membrane-bound nucleus or nuclei; contrasts with **prokaryotic.**

eutely (u′te-lē) (Gr. *euteia*, thrift) Condition of a body composed of a constant number of cells or nuclei in all adult members of a species, as in rotifers, acanthocephalans, and nematodes.

evagination (ē-vaj-ə-nā′shən) (L. *e*, out, + *vagina*, sheath) An outpocketing from a hollow structure.

evolution (L. *evolvere*, to unfold). Organic evolution encompasses all changes in the characteristics and diversity of life on earth throughout its history.

evolutionary sciences Empirical investigation of ultimate causes in biology using the comparative method.

evolutionary species concept A single lineage of ancestral-descendant populations that maintains its identity from other such lineages and has its own evolutionary tendencies and historical fate; differs from the biological species concept by explicitly including a time dimension and including asexual lineages.

evolutionary taxonomy A system of classification, formalized by George Gaylord Simpson, that groups species into Linnean higher taxa representing a hierarchy of distinct adaptive zones; such taxa may be monophyletic or paraphyletic but not polyphyletic.

exopod, exopodite (ex′ə-päd, ex-äp′ə-dīt) (Gr. *exō*, outside, + *pous, podos*, foot) Lateral branch of a biramous crustacean appendage.

exoskeleton (ek′sō-skel′ə-tən) (Gr. *exō*, outside, + *skeletos*, hard) A supporting structure secreted by ectoderm or epidermis; external, not enveloped by living tissue, as opposed to endoskeleton.

experiment (L. *experiri*, to try) A trial made to support or disprove a hypothesis.

experimental method A general procedure for testing hypotheses by predicting how a biological system will respond to a disturbance, making the disturbance under controlled

bat/āpe/ärmadillo/herring/fēmale/finch/līce/ crocodile/crōw/duck/ūnicorn/tüna/ə indicates unaccented vowel sound "uh" as in mammal, fishes, cardinal, heron, vulture/stress as in bi-ol′o-gy, bi′o-log′i-cal

conditions, and then comparing the observed results with the predicted ones.

experimental sciences Empirical investigation of proximate causes in biology using the **experimental method.**

F

filipodium (fi'li-pō'dē-əm) (L. *filum*, thread, + Gr. *pous, podos*, a foot) A type of pseudopodium that is very slender and may branch but does not rejoin to form a mesh.

filter feeding Any feeding process by which particulate food is filtered from water in which it is suspended.

fission (L. *fissio*, a splitting) Asexual reproduction by a division of the body into two or more parts.

fitness Degree of adjustment and suitability for a particular environment. Genetic fitness is relative contribution of one genetically distinct organism to the next generation; organisms with high genetic fitness are naturally selected and become prevalent in a population.

flagellum (flə-jel'əm) pl. **flagella** (L. a whip) Whiplike organelle of locomotion.

flame bulb Specialized hollow excretory or osmoregulatory structure of one or several small cells containing a tuft of flagella (the "flame") and situated at the end of a minute tubule; connected tubules ultimately open to the outside. See **protonephridium.**

fluke (O.E. *flōc*, flatfish) A member of class Trematoda or class Monogenera. Also, certain of the flatfishes (order Pleuronectiformes).

food vacuole A digestive organelle in the cell.

foraminiferan (for'əm-i-nif'-ər-ən) (L. *foramin*, hole, perforation, + *fero*, to bear). A member of the class Granuloreticulosea (phylum Sarcomastigophora) bearing a test with many openings.

fossil (fos'əl). Any remains or impression of an organism from a past geological age that has been preserved by natural processes, usually by mineralization in the earth's crust.

fossorial (fä-sōr'ē-əl) (L. *fossor*, digger) Adapted for digging.

fouling Contamination of feeding or respiratory areas of an organism by excrement, sediment, or other matter. Also, accumulation of sessile marine organisms on the hull of a boat or ship so as to impede its progress through the water.

fovea (fō'vē-ə) (L. a small pit) A small pit or depression; especially the fovea centralis, a small rodless pit in the retina of some vertebrates, a point of acute vision.

G

gamete (ga'mēt, gə-mēt') (Gr. *gamos*, marriage) A mature haploid sex cell; usually male and female gametes can be distinguished. An egg or a sperm.

gametic meiosis Meiosis that occurs during formation of the gametes, as in humans and other metazoa.

gametocyte (gə-mēt'ə-sīt) (Gr. *gametēs*, spouse, + *kytos*, hollow vessel) The mother cell of a gamete, that is, immature gamete.

ganglion (gang'lē-ən) pl. **ganglia** (Gr. little tumor). An aggregation of nerve tissue containing nerve cells.

ganoid scales (ga'noyd) Gr. *ganos*, brightness) Thick, bony, rhombic scales of some primitive bony fishes; not overlapping.

gastrodermis (gas'tro-dər'mis) (Gr. *gaster*, stomach, + *derma*, skin) Lining of the digestive cavity of cnidarians.

gastrovascular cavity (Gr. *gaster*, stomach, + L. *vasculum*, small vessel) Body cavity in certain lower invertebrates that functions in both digestion and circulation and has a single opening serving as both mouth and anus.

gemmule (je'mūl) (L. *gemma*, bud, + *ula*, dim.) Asexual, cystlike reproductive unit in freshwater sponges; formed in summer or autumn and capable of overwintering.

gene (Gr. *genos*, descent) The part of a chromosome that is the hereditary determiner and is transmitted from one generation to another. Specifically, a gene is a nucleic acid sequence (usually DNA) that encodes a functional polypeptide or RNA sequence.

gene pool A collection of all of the alleles of all of the genes in a population.

genetic drift Change in gene frequencies by chance processes in the evolutionary process of animals. In small populations one allele may drift to fixation, becoming the only representative of that gene locus.

genotype (jēn'ə-tīp) (Gr. *genos*, offspring, + *typos*, form) The genetic constitution, expressed and latent, of an organism; the total set of genes present in the cells of an organism; contrasts with **phenotype.**

genus (je'nus) pl. **genera** (L. race) A group of related species with taxonomic rank between family and species.

germ layer In the animal embryo, one of three basic layers (ectoderm, endoderm, mesoderm) from which the various organs and tissues arise in the multicellular animal.

germovitellarium (jer'mə-vit-əl-ar'ē-əm) (L. *germen*, a bud, offshoot, + *vitellus*, yolk). Closely associated ovary (germarium) and yolk-producing structure (vitellarium) in rotifers.

germ plasm The germ cells of an organism, as distinct from the somatoplasm; the hereditary material (genes) of the germ cells.

gestation (je-stā'shən) (L. *gestare*, to bear) The period in which offspring are carried in the uterus.

glochidium (glō-kid'ē-əm) (Gr. *glochis*, point, + *idion*, dimin. suffix) Bivalved larval stage of freshwater mussels.

glycogen (glī'kə-jən) (Gr. *glykys*, sweet, + *genes*, produced) A polysaccharide constituting the principal form in which carbohydrate is stored in animals; animal starch.

gnathobase (nath'ə-bās') (Gr. *gnathos*, jaw, + base). A median basic process on certain appendages in some arthropods, usually for biting or crushing food.

gnathostomes (nath'ə-stōmz) (Gr. *gnathos*, jaw, + *stoma*, mouth). Vertebrates with jaws.

gonad (gō'nad) (N.L. *gonas*, a primary sex organ) An organ that produces gametes (ovary in the female and testis in the male).

gonangium (gō-nan'jē-əm) (N.L. *gonas*, primary sex organ, + *angeion*, dimin. of vessel) Reproductive zooid of hydroid colony (Cnidaria).

gonoduct (Gr. *gonos*, seed, progeny, + duct) Duct leading from a gonad to the exterior.

gonopore (gän'ə-pōr) (Gr. *gonos*, seed, progeny, + *poros*, an opening) A genital pore found in many invertebrates.

grade (L. *gradus*, step) A level of organismal complexity or adaptive zone characteristic of a group of evolutionarily related organisms.

gradualism (graj'ə-wal-iz'əm) A component of Darwin's evolutionary theory postulating that evolution occurs by the temporal accumulation of small, incremental changes, usually across very long periods of geological time; it opposes claims that evolution can occur by large, discontinuous or macromutational changes.

green gland Excretory gland of certain Crustacea; the antennal gland.

gynecophoric canal (gī'nə-kə-fōr'ik) (Gr. *gynē*, woman, + *pherein*, to carry) Groove in male schistosomes (certain trematodes) that carries the female.

H

habitat (L. *habitare*, to dwell) The place where an organism normally lives or where individuals of a population live.

halter (hal'tər) pl. **halteres** (hal-ti'rēz) (Gr. leap) In Diptera, small club-shaped structure on each side of the metathorax representing the hind wings; believed to be sense organs for balancing; also called balancer.

heterochrony (hed'ə-rō-krōn-ē) (Gr. *heteros*, different, + *chronos*, time). Evolutionary change in the relative time of appearance or rate of development of characteristics from ancestor to descendant.

haploid (Gr. *haploos*, single) The reduced, or n, number of chromosomes, typical of gametes, as opposed to the diploid, or 2n, number found in somatic cells. In certain groups, some mature organisms have a haploid number of chromosomes.

Hardy-Weinberg equilibrium Mathematical demonstration that the Mendelian hereditary process does not change the populational frequencies of alleles or genotypes across generations, and that change in allelic or genotypic frequencies requires factors such as natural selection, genetic drift in finite populations, recurring mutation, migration of individuals among populations, and nonrandom mating.

hemal system (hē'məl) (Gr. *haima*, blood) System of small vessels in echinoderms; function unknown.

hemimetabolous (hē′mē-mə-ta′bə-ləs) (Gr. *hēmi,* half, + *metabolē,* change) Refers to gradual metamorphosis during development of insects, without a pupal stage.

hemocoel (hē′mə-sēl) (Gr. *haima,* blood, + *koilos,* hollow) Main body cavity of arthropods; may be subdivided into sinuses, through which the blood flows.

hemoglobin (Gr. *haima,* blood, + L. *globulus,* globule) An iron-containing respiratory pigment occurring in vertebrate red blood cells and in blood plasma of many invertebrates; a compound of an iron porphyrin heme and a protein globin.

hemolymph (hē′mə-limf) (Gr. *haima,* blood + L. *lympha,* water) Fluid in the coelom or hemocoel of some invertebrates that represents the blood and lymph of vertebrates.

herbivore ([h]erb′ə-vōr′) (L. *herba,* green crop, + *vorare,* to devour) Any organism subsisting on plants. Adj., **herbivorous.**

hermaphrodite (hər-maf′rə-dīt) (Gr. *hermaphroditos,* containing both sexes; from Greek mythology. Hermaphroditos, son of Hermes and Aphrodite) An organism with both male and female functional reproductive organs. **Hermaphroditism** may refer to an aberration in unisexual animals; **monoecism** implies that this is the normal condition for the species.

heterocercal (het′ər-o-sər′kəl) (Gr. *heteros,* different, + *kerkos,* tail) In some fishes, a tail with the upper lobe larger than the lower, and the end of the vertebral column somewhat upturned in the upper lobe, as in sharks.

heterochrony (hed′ə-rō-krōn-ē) (Gr. *heteros,* different, + *chronos,* time) Evolutionary change in the relative time of appearance or rate of development of characteristics from ancestor to descendant.

heterodont (hed′ə-ro-dänt) (Gr. *heteros,* different, + *odous,* tooth) Having teeth differentiated into incisors, canines, and molars for different purposes.

heterotroph (het′ə-rō-trōf) (Gr. *heteros,* different, + *trophos,* feeder) An organism that obtains both organic and inorganic raw materials from the environment in order to live; includes most animals and those plants that do not carry on photosynthesis.

heterozygous Refers to an organism in which homologous chromosomes contain different allelic forms (often dominant and recessive) of a gene; derived from a zygote formed by union of gametes of dissimilar allelic constitution.

hexamerous (hek-sam′ər-əs) (Gr. *hex,* six, + *meros,* part) Six parts, specifically, symmetry based on six or multiples thereof.

hibernation (L. *hibernus,* wintry) Condition, especially of mammals, of passing the winter in a torpid state in which the body temperature drops nearly to freezing and the metabolism drops close to zero.

hierarchical system A scheme arranging organisms into a series of taxa of increasing inclusiveness, as illustrated by Linnean classification.

histology (hi-stäl′-ə-jē) (Gr. *histos,* web, tissue, + *logos,* discourse) The study of the microscopic anatomy of tissues.

holometabolous (hō′lō-mə-ta′bə-ləs) (Gr. *holo,* complete, + metabolē, change) Complete metamorphosis during development.

holophytic nutrition (hōl′ō-fit′ik) (Gr. *holo,* whole, + *phyt,* plant) Occurs in green plants and certain protozoa and involves synthesis of carbohydrates from carbon dioxide and water in the presence of light, chlorophyll, and certain enzymes.

holozoic nutrition (hōl′ō-zō′ik) (Gr. *holo,* whole, + *zoikos,* of animals) Type of nutrition involving ingestion of liquid or solid organic food particles.

homeothermic (hō′mē-ō-thər′mik) (Gr. *homeo,* alike, + *thermē,* heat) Having a nearly uniform body temperature, regulated independent of the environmental temperature; "warm-blooded."

home range The area over which an animal ranges in its activities. Unlike territories, home ranges are not defended.

hominid (häm′ə-nid) (L. *homo, hominis,* man) A member of the family Hominidae, now represented by one living species, *Homo sapiens.*

homocercal (hō′mə-ser′kəl) (Gr. *homos,* same, common, + *kerkos,* tail) A tail with the upper and lower lobes symmetrical and the vertebral column ending near the middle of the base, as in most teleost fishes.

homodont (hō′mō-dänt) (Gr. *homos,* same, + *odous,* tooth) Having all teeth similar in form.

homology (hō-mäl′ə-jē) (Gr. *homologos,* agreeing) Similarity of parts or organs of different organisms caused by evolutionary derivation from a corresponding part or organ in a remote ancestor, and usually having a similar embryonic origin. May also refer to a matching pair of chromosomes. Serial homology is the correspondence in the same individual of repeated structures having the same origin and development, such as the appendages of arthropods. Adj., **homologous.**

homonoid (häm′ə-noyd) Relating to the Hominoidea, a superfamily of primates to which the great apes and humans are assigned.

homoplasy (hō′mə-plā′sē). Phenotypic similarity among characteristics of different species or populations (including molecular, morphological, behavioral, or other features) that does not accurately represent patterns of common evolutionary descent (= nonhomologous similarity); it is produced by evolutionary parallelism, convergence and/or reversal, and is revealed by incongruence among different characters on a cladogram or phylogenetic tree.

hyaline (hī′ə-lən) (Gr. *hyalos,* glass) Adj., glassy, translucent. Noun, a clear, glassy structureless material occurring in, for example, cartilage, vitreous bodies, mucin, and glycogen.

hydatid cyst (hī-da′təd) (Gr. *hydatis,* watery vesicle) A type of cyst formed by juveniles of certain tapeworms (*Echinococcus*) in their vertebrate hosts.

hydranth (hī′dranth) (Gr. *hydōr,* water, + *anthos,* flower) Nutritive zooid of hydroid colony.

hydrocoel (hī′drə-sēl) (Gr. *hydōr,* water, + *koilos,* hollow) Second or middle coelomic compartment in echinoderms; left hydrocoel gives rise to water vascular system.

hydrocoral (Gr. *hydōr,* water, + *korallion,* coral) Certain members of the cnidarian class Hydrozoa that secrete calcium carbonate, resembling true corals.

hydroid The polyp form of a cnidarian as distinguished from the medusa form. Any cnidarian of the class Hydrozoa, order Hydroida.

hydrostatic skeleton A mass of fluid or plastic parenchyma enclosed within a muscular wall to provide the support necessary for antagonistic muscle action; for example, parenchyma in acoelomates and perivisceral fluids in pseudocoelomates serve as hydrostatic skeletons.

hypodermis (hī′pə-dər′mis) (Gr. *hypo,* under, + L. *dermis,* skin) The cellular layer lying beneath and secreting the cuticle of annelids, arthropods, and certain other invertebrates.

hypostome (hī′pə-stōm) (Gr. *hypo,* under, + *stoma,* mouth) Name applied to structure in various invertebrates (such as mites and ticks), located at posterior or ventral area of mouth; elevation supporting mouth of hydrozoan.

hypothesis (Gr. *hypothesis,* foundation, supposition) A statement or proposition that can be tested by experiment.

hypothetico-deductive method The central procedure of scientific inquiry in which a postulate is advanced to explain a natural phenomenon and then is subjected to observational or experimental testing that potentially could reject the postulate.

I

immediate cause See **proximate cause.**

inbreeding The tendency among members of a population to mate preferentially with close relatives.

indeterminate cleavage A type of embryonic development in which the fate of the blastomeres is not determined very early as to tissues or organs, for example, in echinoderms and vertebrates.

indigenous (in-dij′ə-nəs) (L. *indigna,* native) Pertains to organisms that are native to a particular region; not introduced.

induction (L. *inducere, inductum,* to lead) Reasoning from the particular to the general;

that is, deriving a general statement (hypothesis) based on individual observations. In embryology, the alteration of cell fates as the result of interaction with neighboring cells.

infraciliature (in′frə-sil′ē-ə-tər) (L. *infra*, below, + *cilia*, eyelashes) The organelles just below the cilia in ciliate protozoa.

inheritance of acquired characteristics The discredited Lamarckian notion that organisms, by striving to meet the demands of their environments, obtain new adaptations and pass them by heredity to their offspring.

instar (inz′tär) (L. form) Stage in the life of an insect or other arthropod between molts.

integument (in-teg′ū-mənt) (L. *integumentum*, covering) An external covering or enveloping layer.

intermediary meiosis Meiosis that occurs neither during gamete formation nor immediately after zygote formation, resulting in both haploid and diploid generations, such as in foraminiferan protozoa.

intermediate host A host in which some development of a symbiont occurs, but in which maturation and sexual reproduction do not take place (contrasts with **definitive host**).

interstitial (in′tər-sti′shəl) (L. *inter,* among, + *sistere,* to stand) Situated in the interstices or spaces between structures such as cells, organs, or grains of sand.

intracellular (in-trə-sel′yə-lər) (L. *intra*, inside, + *cellula*, chamber). Occurring within a body cell or within body cells.

introvert (L. *intro*, inward, + *vertere*, to turn) The anterior narrow portion that can be withdrawn (introverted) into the trunk of a sipunculid worm.

J

juvenile hormone Hormone produced by the corpora allata of insects; among its effects are maintenance of larval or nymphal characteristics during development.

K

keratin (ker′ə-tən) (Gr. *kera*, horn, + *in*, suffix of proteins) A scleroprotein found in epidermal tissues and modified into hard structures such as horns, hair, and nails.

kinetosome (kin-et′ə-sōm) (Gr. *kinētos*, moving, + *sōma*, body) The self-duplicating granule at the base of the flagellum or cilium; similar to centriole, also called basal body or blepharoplast.

L

labium (lā′bē-əm) (L. a lip) The lower lip of the insect formed by fusion of the second pair of maxillae.

labrum (lā′brəm) (L. a lip) The upper lip of insects and crustaceans situated above or in front of the mandibles; also refers to the outer lip of a gastropod shell.

labyrinthodont (lab′ə-rin′thə-dänt) (Gr. *labyrinthos*, labyrinth, + *odous, odontos,* tooth) A group of Paleozoic amphibians containing the temnospondyls and the anthracosaurs.

Lamarckism Hypothesis, as expounded by Jean-Baptiste de Lamarck, of evolution by the acquisition during an organism's lifetime of characteristics that are transmitted to offspring.

lamella (lə-mel′ə) (L. dim. of *lamina*, plate) One of the two plates forming a gill in a bivalve mollusc. One of the thin layers of bone laid concentrically around an osteon (Haversian) canal. Any thin, platelike structure.

lateral (L. *latus*, the side, flank). Of or pertaining to the side of an animal; a *bilateral* animal has two sides.

larva (lar′və) pl. **larvae** (L. a ghost) An immature stage that is quite different from the adult.

lek (lek) (Sw. play, game) An area where animals assemble for communal courtship display and mating.

lemniscus (lem-nis′kəs) (L. ribbon). One of a pair of internal projections of the epidermis from the neck region of Acanthocephala, which functions in fluid control in the protrusion and invagination of the proboscis.

lepidosaurs (lep′ə-dō-sors) (L. *lepidos*, scale, + *sauros,* lizard) A lineage of diapsid reptiles that appeared in the Permian and that includes the modern snakes, lizards, amphisbaenids, and tuataras, and the extinct ichthyosaurs.

leptocephalus (lep′tə-sef′ə-ləs) pl. **leptocephali** (Gr. *leptos*, thin, + *kephalē*, head) Transparent, ribbonlike migratory larva of the European or American eel.

lobopodium (lō′bə-pō′dē-əm) (Gr. *lobos*, lobe, + *pous, podos,* foot). Blunt, lobelike pseudopodium.

lophophore (lōf′ə-fōr) (Gr. *lophos*, crest, + *phoros*, bearing) Tentacle-bearing ridge or arm within which is an extension of the coelomic cavity in lophophorate animals (ectoprocts, brachiopods, and phoronids).

lorica (lor′ə-kə) (L. *lorica*, corselet) A secreted, protective covering, as in phylum Loricifera.

lymph (limf) (L. *lympha*, water) The interstitial (intercellular) fluid in the body, also the fluid in the lymphatic space.

M

macroevolution (L. *makros,* long, large, + *evolvere*, to unfold) Evolutionary change on a grand scale, encompassing the origin of novel designs, evolutionary trends, adaptive radiation, and mass extinction.

macrogamete (mak′rə-gam′ēt) (Gr. *makros*, long, large, + *gamos*, marriage). The larger of the two gamete types in a heterogametic organism, considered the female gamete.

macronucleus (ma′krō-nū′klē-əs) (Gr. *makros,* long, large, + *nucleus,* kernel) The larger of the two kinds of nuclei in ciliate protozoa; controls all cell function except reproduction.

madreporite (ma′drə-pōr′īt) (Fr. *madrépore,* reef-building coral, + *ite,* suffix for some body parts) Sievelike structure that is the intake of the water-vascular system of echinoderms.

malacostracan (mal′ə-käs′trə-kən) (Gr. *malako,* soft, + *ostracon,* shell) Any member of the crustacean subclass Malacostraca, which includes both aquatic and terrestrial forms of crabs, lobsters, shrimps, pillbugs, sand fleas, and others.

malaria (mə-lar′ē-ə) (It. *malaria,* bad air) A disease marked by periodic chills, fever, anemia, and other symptoms, caused by *Plasmodium* spp.

Malpighian tubules (Mal-pig′ē-ən) (Marcello Malpighi, Italian anatomist, 1628–94) Blind tubules opening into the hindgut of nearly all insects and some myriapods and arachnids and functioning primarily as excretory organs.

mantle Soft extension of the body wall in certain invertebrates, for example, brachiopods and molluscs, which usually secretes a shell; thin body wall of tunicates.

manubrium (man-ū′brē-əm) (L. handle) The portion projecting from the oral side of a jellyfish medusa, bearing the mouth; oral cone; presternum or anterior part of sternum; handlelike part of malleus of ear.

marsupial (mär-sü′pē-əl) (Gr. *marsypion,* little pouch) One of the pouched mammals of the subclass Metatheria.

mass extinction A relatively short interval of geological time in which a large portion (75%–95%) of existing species or higher taxa are eliminated nearly simultaneously.

mastax (mas′təx) (Gr. jaws) Pharyngeal mill of rotifers.

matrix (mā′triks) (L. *mater,* mother) The intercellular substance of a tissue, or that part of a tissue into which an organ or process is set.

maxilla (mak-sil′ə) (L. dim. of *mala,* jaw) One of the upper jawbones in vertebrates; one of the head appendages in arthropods.

maxilliped (mak-sil′ə-ped) (L. *maxilla,* jaw, + *pes,* foot) One of the pairs of head appendages located just posterior to the maxilla in crustaceans; a thoracic appendage that has become incorporated into the feeding mouthparts.

medial (mē′dē-əl). Situated, or occurring, in the middle.

medulla (mə-dül′ə) (L. marrow) The inner portion of an organ in contrast to the cortex or outer portion. Also, hindbrain.

medusa (mə-dü′-sə) (Gr. mythology, female monster with snake-entwined hair) A jellyfish, or the free-swimming stage in the life cycle of cnidarians.

Mehlis' gland (me′ləs) Glands of uncertain function surrounding the junction of yolk duct, oviduct, and uterus in trematodes and cestodes.

meiosis (mī-ō′səs) (Gr. from *meioun,* to make small) The nuclear changes by means of which the chromosomes are reduced from the diploid to the haploid number; in animals, usually occurs in the last two divisions in the formation of the mature egg or sperm.

melanin (mel′ə-nin) (Gr. *melas,* black) Black or dark-brown pigment found in plant or animal structures.

membranelle A tiny membrane-like structure, may be formed by fused cilia.

merozoite (me′rə-zō′īt) (Gr. *meros*, part, + *zōon*, animal) A very small trophozoite at the stage just after cytokinesis has been completed in multiple fission of a protozoan.

mesenchyme (me′zn-kīm) (Gr. *mesos*, middle, + *enchyma*, infusion) Embryonic connective tissue; irregular or amebocytic cells often embedded in gelatinous matrix.

mesocoel (mez′ō-sēl) (Gr. *mesos*, middle, + *koilos*, hollow) Middle body coelomic compartment in some deuterostomes; anterior in lophophorates, corresponds to hydrocoel in echinoderms.

mesoderm (me′zə-dərm) (Gr. *mesos*, middle, + *derma*, skin) The third germ layer, formed in the gastrula between the ectoderm and endoderm; gives rise to connective tissues, muscle, urogenital and vascular systems, and the peritoneum.

mesoglea (mez′ō-glē′ə) (Gr. *mesos*, middle, + *glia*, glue) The layer of jellylike or cement material between the epidermis and gastrodermis in cnidarians and ctenophores; also may refer to jellylike matrix between epithelial layers in sponges.

mesohyl (me′sə-hil) (Gr. *mesos*, middle, + *hyle*, a wood) Gelatinous matrix surrounding sponge cells; mesoglea, mesenchyme.

mesonephros (me-zō-nef′rōs) (Gr. *mesos*, middle, + *nephros*, kidney) The middle of three pairs of embryonic renal organs in vertebrates. Functional kidney of fishes and amphibians; its collecting duct is a wolffian duct. Adj., **mesonephric.**

mesosome (mez′ə-sōm) (Gr. *mesos*, middle, + *sōma*, body). The portion of the body in lophophorates and some deuterostomes that contains the mesocoel.

metacercaria (me′tə-sər-ka′rē-ə) (Gr. *meta*, after, + *kerkos*, tail, + L. *aria*, connected with) Fluke juvenile (cercaria) that has lost its tail and has become encysted.

metacoel (met′ə-sēl) (Gr. *meta*, after, + *koilos*, hollow) Posterior coelomic compartment in some deuterostomes and lophophorates; corresponds to somatocoel in echinoderms.

metamere (met′ə-mēr) (Gr. *meta*, after, + *meros*, part) A repeated body unit along the longitudinal axis of an animal, a somite, or segment.

metamerism (mə-ta′-mə-ri′zəm) (Gr. *meta*, between, after, + *meros*, part) Condition of being made up of serially repeated parts (metameres); serial segmentation.

metamorphosis (Gr. *meta*, after, + *morphē*, form, + *osis*, state of) Sharp change in form during postembryonic development, for example, tadpole to frog or larval insect to adult.

bat/āpe/ärmadillo/herring/fēmale/finch/līce/ crocodile/crōw/duck/ūnicorn/tüna/ə indicates unaccented vowel sound "uh" as in mammal, fishes, cardinal, heron, vulture/stress as in bi-ol′o-gy, bi′o-log′i-cal

metanephridium (me′tə-nə-fri′di-əm) (Gr. *meta*, after, + *nephros*, kidney) A type of tubular nephridium with the inner open end draining the coelom and the outer open end discharging to the exterior.

metasome (met′ə-som) (Gr. *meta*, after, behind, + *sōma*, body). The portion of the body in lophophorates and some deuterostomes that contains the metacoel.

metazoa (met-ə-zō′ə) (Gr. *meta*, after, + *zōon*, animal) Multicellular animals.

microevolution (mī′krō-ev-ə-lü-shən) (L. *mikros*, small, + *evolvere*, to unfold) A change in the gene pool of a population across generations.

microfilariae (mīk′rə-fil-ar′ē-ē) (Gr. *mikros*, small, + L. *filum*, a thread). Partially developed juveniles bornealive by filarial worms (phylum Nematoda).

microgamete (mīk′rə-ga′-mēt) (Gr. *mikros*, small, + *gamos*, marriage). The smaller of the two gamete types in a heterogametic organism, considered the male gamete.

micron (µm) (mī′-krän) (Gr. neuter of *mikros*, small) One-thousandth of a millimeter; about 1/25,000 of an inch. Now largely replaced by micrometer (µm).

microneme (mī′krə-nēm) (Gr. *mikros*, small, + *nēma*, thread) One of the types of structures composing the apical complex in the Phylum Apicomplexa, slender and elongate, leading to the anterior and thought to function in host cell penetration.

micronucleus A small nucleus found in ciliate protozoa; controls the reproductive functions of these organisms.

microthrix See **microvillus.**

microtubule (Gr. *mikros*, small, + L. *tubule*, pipe) A long, tubular cytoskeletal element with an outside diameter of 20 to 27 nm. Microtubules influence cell shape and play important roles during cell division.

microvillus (Gr. *mikros*, small, + L. *villus*, shaggy hair) Narrow, cylindrical cytoplasmic projection from epithelial cells; microvilli form the brush border of several types of epithelial cells. Also, microvilli with unusual structure cover the surface of cestode tegument (also called **microthrix** [pl. **microtriches**]).

mictic (mik′tik) (Gr. *miktos*, mixed or blended) Pertaining to haploid egg of rotifers of the females that lay such eggs.

miracidium (mīr′ə-sid′ē-əm) (Gr. *meirakidion*, youthful person) A minute ciliated larval stage in the life of flukes.

mitochondrion (mīd′ə-kän′drē-ən) (Gr. *mitos*, a thread, + *chondrion*, dim. of *chondros*, corn, grain) An organelle in the cell in which aerobic metabolism takes place.

mitosis (mī-tō′səs) (Gr. *mitos*, thread, + *osis*, state of) Nuclear division in which there is an equal qualitative and quantitative division of the chromosomal material between the two resulting nuclei; ordinary cell division (indirect).

monoecious (mə-nē′shəs) (Gr. *monos*, single, + *oikos*, house) Having both male and female gonads in the same organism; hermaphroditic.

monogamy (mə-näg′ə-mē) adj. **monogamous** (Gr. *monos*, single, + *gamos*, marriage) The condition of having a single mate at any one time.

monophyletic (mä′nə-phī-le′tik) (Gr. *monos*, single, + *phyletikos*, pertaining to a phylum) Referring to a taxon whose units all evolved from a single parent stock; contrasts with **polyphyletic.**

monophyly (män′ō-fī′lē) (Gr. *monos*, single, + *phyle*, tribe) The condition that a taxon or other group of organisms contains the most recent common ancestor of the group and all of its descendants.

monotreme (mä′nō-trēm) (Gr. *monos*, single, + *trēma* hole) Egg-laying mammal of the order Monotremata.

morphogenesis (mor′fə-je′nə-səs) (Gr. *morphē*, form, + *genesis*, origin) Development of the architectural features of organisms; formation and differentiation of tissues and organs.

morphology (Gr. *morphē*, form, + *logos*, discourse) The science of structure. Includes cytology, the study of cell structure, histology, the study of tissue structure; and anatomy, the study of gross structure.

mosaic cleavage Type characterized by independent differentiation of each part of the embryo; determinate cleavage.

mucus (mū′kəs) (L. *mucus*, nasal mucus) Viscid, slippery secretion rich in mucins produced by secretory cells such as those in mucous membranes. Adj., **mucous.**

multiple fission A mode of asexual reproduction in some protistans in which the nuclei divide more than once before cytokinesis occurs.

multiplication of species The Darwinian theory that the evolutionary process generates new species through a branching of evolutionary lineages derived from an ancestral species.

mutation (mū-tā′shən) (L. *mutare*, to change) A stable and abrupt change of a gene; the heritable modification of a character.

mutualism (mü′chə-wə-li′zəm) (L. *mutuus*, lent, borrowed, reciprocal) A type of interaction in which two different species derive benefit from the association and in which the association is necessary to both; often symbiotic.

myocyte (mī′ə-sīt) (Gr. *mys*, muscle, + *kytos*, hollow vessel) Contractile cell (pinacocyte) in sponges.

myofibril (Gr. *mys*, muscle, + L. dim. of *fibra*, fiber) A contractile filament within muscle or muscle fiber.

myomere (mī′ə-mer) (Gr. *mys*, muscle, + *meros*, part) A muscle segment of successive segmental trunk musculature.

myotome (mī′ə-tōm) (Gr. *mys*, muscle, + *tomos*, cutting) A voluntary muscle segment in cephalochordates and vertebrates; that part of a somite destined to form muscles; the muscle group innervated by a single spinal nerve.

N

nacre (nā′kər) (F. mother-of-pearl) Innermost lustrous layer of mollusc shell, secreted by mantle epithelium. Adj., **nacreous.**

nares (na′rēz), sing. **naris** (L. nostrils) Openings into the nasal cavity, both internally and externally, in the head of a vertebrate.

natural selection The interactions between organismal character variation and the environment that cause differences in rates of survival and reproduction among varying organisms in a population; leads to evolutionary change if variation is heritable.

nauplius (naw′plē-əs) (L. a kind of shellfish) A free-swimming microscopic larval stage of certain crustaceans, with three pairs of appendages (antennules, antennae, and mandibles) and a median eye. Characteristic of ostracods, copepods, barnacles, and some others.

nekton (nek′tən) (Gr. neuter of *nēktos,* swimming). Term for actively swimming organisms, essentially independent of wave and current action. Compare with **plankton.**

nematocyst (ne-mad′ə-sist′) (Gr. *nēma,* thread, + *kystis,* bladder) Stinging organelle of cnidarians.

neo-Darwinism (nē′ō′där′wə-niz′əm) A modified version of Darwin's evolutionary theory that eliminates elements of the Lamarckian inheritance of acquired characteristics and pangenesis that were present in Darwin's formulation; this theory originated with August Weissmann in the late nineteenth century and, after incorporating Mendelian genetic principles, has become the currently favored version of Darwinian evolutionary theory.

neopterygian (nē-äp′tə-rij′ē-ən) (Gr. *neos,* new, + *pteryx,* fin) Any of a large group of bony fishes that includes most modern species.

neoteny (nē′ə-tē′nē, nē-ot′ə-nē) (Gr. *neos,* new, + *teinein,* to extend). An evolutionary process by which organismal development is retarded relative to sexual maturation; produces a descendant that reaches sexual maturity while retaining a morphology characteristic of the preadult or larval stage of an ancestor.

nephridium (nə-frid′ē-əm) (Gr. *nephridios,* of the kidney) One of the segmentally arranged, paired excretory tubules of many invertebrates, notably the annelids. In a broad sense, any tubule specialized for excretion and/or osmoregulation; with an external opening and with or without an internal opening.

nephron (ne′frän) (Gr. *nephros,* kidney) Functional unit of kidney structure of vertebrates, consisting of Bowman's capsule, an enclosed glomerulus, and the attached uriniferous tubule.

nephrostome (nef′rə-stōm) (Gr. *nephros,* kidney, + *stoma,* mouth) Ciliated, funnel-shaped opening of a nephridium.

nested hierarchy A pattern in which species are ordered into a series of increasingly more inclusive clades according to the taxonomic distribution of synapomorphies.

neuroglia (nü-räg′lē-ə) (Gr. *neuron,* nerve, + *glia,* glue) Tissue supporting and filling the spaces between the nerve cells of the central nervous system.

neuromast (Gr. *neuron,* sinew, nerve, + *mastos,* knoll) Cluster of sense cells on or near the surface of a fish or amphibian that is sensitive to vibratory stimuli and water current.

neuron (Gr. nerve) A nerve cell.

neuropodium (nü′rə-pō′dē-əm) (Gr. *neuron,* nerve, + *pous, podos,* foot) Lobe of parapodium nearer the ventral side in polychaete annelids.

neurosecretory cell (nü′rō-sə-krēd′ə-rē) Any cell (neuron) of the nervous system that produces a hormone.

niche The role of an organism in an ecological community; its unique way of life and its relationship to other biotic and abiotic factors.

notochord (nōd′ə-kord′) (Gr. *nōtos,* back, + *chorda,* cord) An elongated cellular cord, enclosed in a sheath, which forms the primitive axial skeleton of chordate embryos and adult cephalochordates.

notopodium (nō′tə-pō′dē-əm) (Gr. *nōtos,* back, + *pous, podos,* foot) Lobe of parapodium nearer the dorsal side in polychaete annelids.

nucleolus (nü-klē′ə-ləs) (dim. of L. *nucleus,* kernel) A deeply staining body within the nucleus of a cell and containing RNA; nucleoli are specialized portions of certain chromosomes that carry multiple copies of the information to synthesize ribosomal RNA.

nucleoplasm (nü′klē-ə-pla′zəm) (L. *nucleus,* kernel, + Gr. *plasma,* mold) Protoplasm of nucleus, as distinguished from cytoplasm.

nucleus (nü′klē-əs) (L. *nucleus,* a little nut, the kernel). The organelle in eukaryotes that contains the chromatin and which is bounded by a double membrane (nuclear envelope).

nurse cells Single cells or layers of cells surrounding or adjacent to other cells or structures for which the nurse cells provide nutrient or other molecules (for example, for insect oocytes or *Trichinella* spp. juveniles).

nymph (L. *nympha,* nymph, bride) An immature stage (following hatching) of a hemimetabolous insect that lacks a pupal stage.

O

ocellus (ō-sel′əs) (L. dim. of *oculus,* eye) A simple eye or eyespot in many types of invertebrates.

octomerous (ok-tom′ər-əs) (Gr. *oct,* eight, + *meros,* part) Eight parts, specifically, symmetry based on eight.

omasum (ō-mā′səm) (L. paunch) The third compartment of the stomach of a ruminant mammal.

ommatidium (ä′mə-tid′ē-əm) (Gr. *omma,* eye, + *idium,* small) One of the optical units of the compound eye of arthropods and molluscs.

omnivore (äm′nə-vōr) (L. *omnis,* all, + *vorare,* to devour) An animal that uses a variety of animal and plant material in its diet.

oncosphere (än′kəs-fər) (Gr. *onkinos,* a hook, + *sphaira,* ball) Rounded larva common to all cestodes; bears hooks.

ontogeny (än-tä′jə-nē) (Gr. *ontos,* being, + *geneia,* act of being born, from *genēs,* born) The course of development of an individual from egg to senescence.

oocyst (ō′ə-sist) (Gr. *ōion,* egg, + *kystis,* bladder) Cyst formed around zygote of malaria and related organisms.

oocyte (ō′ə-sīt) (Gr. *ōion,* egg, + *kytos,* hollow) Stage in formation of ovum, just preceding first meiotic division (primary oocyte) or just following first meiotic division (secondary oocyte).

ookinete (ō-ə-kī′nēt) (Gr. *ōion,* egg, + *kinein,* to move) The motile zygote of malaria organisms.

operculum (ō-per′kū-ləm) (L. cover) The gill cover in body fishes; horny plate in some snails.

opisthaptor (ō′pəs-thap′tər) (Gr. *opisthen,* behind, + *haptein,* to fasten). Posterior attachment organ of a monogenetic trematode.

organelle (Gr. *organon,* tool, organ, + L. *ella,* dimin. suffix) Specialized part of a cell; literally, a small organ that performs functions analogous to organs of multicellular animals.

osculum (os′kū-ləm) (L. *osculum,* a little mouth) Excurrent opening in a sponge.

osmoregulation Maintenance of proper internal salt and water concentrations in a cell or in the body of a living organism, active regulation of internal osmotic pressure.

osmosis (oz-mō′sis) (Gr. *ōsmos,* act of pushing, impulse) The flow of solvent (usually water) through a semipermeable membrane.

osmotroph (oz′mə-trōf) (Gr. *ōsmos,* a thrusting, impulse, + *trophē,* to eat) A heterotrophic organism that absorbs dissolved nutrients.

osphradium (äs-frā′dē-əm) (Gr. *osphradion,* small bouquet, dim., of *osphra,* smell) A sense organ in aquatic snails and bivalves that tests incoming water.

ossicles (L. *ossiculum,* small bone) Small separate pieces of echinoderm endoskeleton. Also, tiny bones of the middle ear of vertebrates.

ostium (L. door) Opening.

otolith (ōd′ə-lith′) (Gr. *ous, otos,* ear, + *lithos,* stone) Calcerous concretions in the membranous labyrinth of the inner ear of lower vertebrates or in the auditory organ of certain invertebrates.

outgroup In phylogenetic systematic studies, a species or group of species closely related to but not included within a taxon whose phylogeny is being studied, and used to polarize variation of characters and to root the phylogenetic tree.

outgroup comparison A method for determining the polarity of a character in cladistic analysis of a taxonomic group. Character states found within the group being studied are judged ancestral if they occur also in related taxa outside the study group (= outgroups); character states that occur only within the taxon being studied but not in outgroups are judged to have been derived evolutionarily within the group being studied.

oviger (ō′vi-jər) (L. *ovum*, egg, + *gerere*, to bear) Leg that carries eggs in pycnogonids.

oviparity (ō′və-pa′rəd-ē) (L. *ovum*, egg, + *parere*, to bring forth) Reproduction in which eggs are released by the female; development of offspring occurs outside the maternal body. Adj., **oviparous** (ō-vip′ə-rəs).

ovipositor (ō′ve-päz′əd-ər) (L. *ovum*, egg, + *positor*, builder, placer, + *or*, suffix denoting agent or doer) In many female insects a structure at the posterior end of the abdomen for laying eggs.

ovoviviparity (ō′vo-vī-və-par′ə-dē) (L. *ovum*, egg, + *vivere*, to live, + *parere*, to bring forth) Reproduction in which eggs develop within the maternal body without additional nourishment from the parent and hatch within the parent or immediately after laying. Adj., **ovoviviparous** (ō-vo-vī-vip′ə-rəs).

ovum (L. *ovum*, egg) Mature female germ cell (egg).

P

paedogenesis (pē-dō-jen′ə-sis) (Gr. *pais*, child, + *genēs*, born) Reproduction by immature or larval animals caused by acceleration of maturation. Progenesis.

paedomorphosis (pē-dō-mor′fə-səs) (Gr. *pais*, child, + *morphē*, form) Displacement of ancestral juvenile features to later stages of the ontogeny of descendants.

pangenesis (pan-jen′ə-sis) (Gr. *pan*, all, + *genesis*, descent) Darwin's hypothesis that hereditary characteristics are carried by individual body cells that produce particles collecting in the germ cells.

papilla (pə-pil′ə) pl. **papillae** (L. nipple) A small nipplelike projection. A vascular process that nourishes the root of a hair, feather, or developing tooth.

papula (pa′pü-lə) pl. **papulae** (L. pimple) Respiratory processes on skin of sea stars; also, pustules on skin.

parabronchi (par-ə-brong′kī) (Gr. *para*, beside, + *bronchos*, windpipe) Fine air-conduction pathways of the bird lung.

paraphyly (par′ə-fī′lē) (Gr. *para*, before, + *phyle*, tribe) The condition that a taxon or other group of organisms contains the most recent common ancestor of all members of the group but excludes some descendants of that ancestor.

parapodium (pa′rə-pō′dē-əm) (Gr. *para*, beside, + *pous*, *podos*, foot) One of the paired lateral processes on each side of most segments in polychaete annelids; variously modified for locomotion, respiration, or feeding.

parasitism (par′ə-sit′iz-əm) (Gr. *parasitos*, from *para*, beside, + *sitos*, food) The condition of an organism living in or on another organism (host) at whose expense the parasite is maintained; destructive symbiosis.

parenchyma (pə-ren′kə-mə) (Gr. anything poured in beside) In simpler animals, a spongy mass of vacuolated mesenchyme cells filling spaces between viscera, muscles, or epithelia; in some, the cells are cell bodies of muscle cells. Also, the specialized tissue of an organ as distinguished from the supporting connective tissue.

parenchymula (pa′rən-kīm′yə-lə) (Gr. *para*, beside, + *enchyma*, infusion) Flagellated, solid-bodied larva of some sponges.

parietal (pä-rī′ə-təl) (L. *paries*, wall) Something next to, or forming part of, a wall of a structure.

parthenogenesis (pär′thə-nō-gen′ə-sis) (Gr. *parthenos*, virgin, + L. from Gr. *genesis*, origin) Unisexual reproduction involving the production of young by females not fertilized by males; common in rotifers, cladocerans, aphids, bees, ants, and wasps. A parthenogenetic egg may be diploid or haploid.

pecten (L. comb) Any of several types of comblike structures on various organisms, for example, a pigmented, vascular, and comblike process that projects into the vitreous humor from the retina at a point of entrance of the optic nerve in the eyes of all birds and many reptiles.

pectoral (pek′tə-rəl) (L. *pectoralis*, from *pectus*, the breast) Of or pertaining to the breast or chest; to the pectoral girdle; or to a pair of horny shields of the plastron of certain turtles.

pedalium (pə-dal′ē-əm) (Gr. *pedalion*, a prop, rudder) The flattened, bladelike base of a tentacle or group of tentacles in the cnidarian class Cubozoa.

pedal laceration Asexual reproduction found in sea anemones, a form of fission.

pedicel (ped′ə-sel) (L. *pediculus*, little foot) A small or short stalk or stem. In insects, the second segment of an antenna or the waist of an ant.

pedicellaria (ped′ə-sə-lar′ē-ə) (L. *pediculus*, little foot, + *aria*, like or connected with) One of many minute pincerlike organs on the surface of certain echinoderms.

pedipalps (ped′ə-palps′) (L. *pes*, *pedis*, foot, + *palpus*, stroking, caress) Second pair of appendages of arachnids.

pedogenesis See **paedogenesis.**

peduncle (pē-dun′kəl) (L. *pedunculus*, dim. of *pes*, foot) A stalk. Also, a band of white matter joining different parts of the brain.

pelage (pel′ij) (Fr. fur) Hairy covering of mammals.

pelagic (pə-laj′ik) (Gr. *pelagos*, the open sea) Pertaining to the open ocean.

pellicle (pel′ə-kəl) (L. *pellicula*, dim. of *pelis*, skin) Thin, translucent, secreted envelope covering many protozoa.

pentadactyl (pen-tə-dak′təl) (Gr. *pente*, five, + *daktylos*, finger) With five digits, or five fingerlike parts, to the hand or foot.

periostracum (pe-rē-äs′trə-kəm) (Gr. *peri*, around, + *ostrakon*, shell) Outer horny layer of a mollusc shell.

peripheral (pə-ri′fər-əl) (Gr. *peripherein*, to move around) Structure or location distant from center, near outer boundaries.

periproct (per′ə-präkt) (Gr. *peri*, around, + *prōktos*, anus) Region of aboral plates around the anus of echinoids.

perisarc (per′ə-särk) (Gr. *peri*, around, + *sarx*, flesh) Sheath covering the stalk and branches of a hydroid.

perissodactyl (pə-ris′ə-dak′təl) (Gr. *perissos*, odd, + *daktylos*, finger, toe) Pertaining to an order of ungulate mammals with an odd number of digits.

peristomium (per′ə-stō′mē-əm) (Gr. *peri*, around, + *stoma*, mouth) Foremost true segment of an annelid; it bears the mouth.

peritoneum (per′ə-tə-nē′əm) (Gr. *peritonaios*, stretched around) The membrane that lines the coelom and covers the coelomic viscera.

Permian extinction A mass extinction that occurred 245 million years ago in which 96% of existing species became extinct, marking the end of the Paleozoic era.

perpetual change The most basic theory of evolution, that the living world is neither constant nor cycling, but is always undergoing irreversible modification through time.

phagocyte (fag′ə-sīt) (Gr. *phagein*, to eat, + *kytos*, hollow vessel) Any cell that engulfs and devours microorganisms or other particles.

phagocytosis (fag′ə-sī-tō′səs) (Gr. *phagein*, to eat, + *kytos*, hollow vessel) The engulfment of a particle by a phagocyte or a protozoan.

phagosome (fa′gə-sōm) (Gr. *phagein*, to eat, + *sōma*, body) Membrane-bound vessel in cytoplasm containing food material engulfed by phagocytosis.

phagotroph (fag′ə-trōf) (Gr. *phagein*, to eat, + *trophē*, food) A heterotrophic organism that ingests solid particles for food.

pharynx (far′inks) pl. **pharynges** (Gr. *pharynx*, gullet) The part of the digestive tract between the mouth cavity and the esophagus that, in vertebrates, is common to both digestive and respiratory tracts. In cephalochordates the gill slits open from it.

phenetic taxonomy (fə-ne′tik) (Gr. *phaneros*, visible, evident) Refers to the use of a criterion of overall similarity to classify organisms into taxa; contrasts with classification based explicitly on a reconstruction of phylogeny.

phenotype (fē′nə-tīp) (Gr. *phainein*, to show) The visible or expressed characteristics of an organism, controlled by the genotype, but not all genes in the genotype are expressed.

pheromone (fer′ə-mōn) (Gr. *pherein*, to carry, + *hormōn*, exciting, stirring up) Chemical substance released by one organism that influences the behavior or physiological processes of another organism.

photoautotroph (fōd-ə-aw′-tō-trōf) (Gr. *phōtos*, light, + *autos*, self, + *trophos*, feeder) An organism requiring light as a source of energy for making organic nutrients from inorganic raw materials.

phototaxis (fōd′ō-tak′sis) (Gr. *phōtos*, light, + *taxis*, arranging, order) A taxis in which light is the orienting stimulus. An involuntary tendency for an organism to turn toward (positive) or away from (negative) light.

phyletic gradualism A model of evolution in which morphological evolutionary change is continuous and incremental and occurs mainly within unbranched species or lineages over long periods of geological time; contrasts with **punctuated equilibrium.**

phylogenetic species concept An irreducible (basal) cluster of organisms, diagnosably distinct from other such clusters, and within which there is a parental pattern of ancestry and descent.

phylogenetic systematics See **cladistics.**

phylogenetic tree A branching diagram whose branches represent evolutionary lineages and depicts the common descent of species or higher taxa.

phylogeny (fī′läj′ə-nē) (Gr. *phylon*, tribe, race, + *geneia*, origin) The origin and diversification of any taxon, or the evolutionary history of its origin and diversification, usually presented in the form of dendrogram.

phylum (fī′ləm) pl. **phyla** (N.L. from Gr. *phylon*, race, tribe) A chief category, between kingdom and class, of taxonomic classifications into which are grouped organisms of common descent that share a fundamental pattern of organization.

physiology (L. *physiologia*, natural science) A branch of biology dealing with the organic processes and phenomena of an organism or any of its parts or a particular bodily process.

phytoflagellates (fī-tə-fla′jə-lāts) Members of the class Phytomastigophorea, plantlike flagellates.

pinacocyte (pin′ə-kō-sīt′) (Gr. *pinax*, tablet, + *kytos*, hollow vessel) Flattened cells comprising dermal epithelium in sponges.

pinna (pin′ə) (L. feather, sharp point) The external ear. Also a feather, wing, or fin or similar part.

pinocytosis (pin′o-sī-tō′sis, pīn′o-sī-tō′sis) (Gr. *pinein*, to drink, + *kytos*, hollow vessel, + *osis*, condition) Taking up of fluid by endocytosis; cell drinking.

placenta (plə-sen′tə) (L. flat cake) The vascular structure, embryonic and maternal, through which the embryo and fetus are nourished while in the uterus.

placoderms (plak′ə-dərmz) (Gr. *plax*, plate, + *derma,* skin) A group of heavily armored jawed fishes of the Lower Devonian to Lower Carboniferous.

placoid scale (pla′koyd) (Gr. *plax, plakos*, tablet, plate) Type of scale found in cartilaginous fishes, with basal plate of dentin embedded in the skin and a backward-pointing spine tipped with enamel.

plankton (plank′tən) (Gr. neuter of *planktos,* wandering) The passively floating animal and plant life of a body of water; compares with **nekton.**

plantigrade (plan′tə-grād′) (L. *planta*, sole, + *gradus*, step, degree) Pertaining to animals that walk on the whole surface of the foot (for example, humans and bears); compares with **digitigrade.**

planula (plan′yə-lə) (N.L. dim. from L. *planus,* flat) Free-swimming, ciliated larval type of cnidarians; usually flattened and ovoid, with an outer layer of ectodermal cells and an inner mass of endodermal cells.

planuloid ancestor (plan′yə-loid) (L. *planus,* flat, + Gr. *eidos*, form) Hypothetical form representing ancestor of Cnidaria and Platyhelminthes.

plasma membrane (plaz′mə) (Gr. *plasma*, a form, mold) A living, external, limiting, protoplasmic structure that functions to regulate exchange of nutrients across the cell surface.

plastron (plast′trən) (Fr. *plastron*, breast plate) Ventral body shield of turtles; structure in corresponding position in certain arthropods; thin film of gas retained by epicuticle hairs of aquatic insects.

pleura (plü′rə) (Gr. side, rib) The membrane that lines each half of the thorax and covers the lungs.

podium (pō′dē-əm) (Gr. *pous, podos*, foot) A footlike structure, for example, the tube foot of echinoderms.

poikilothermic (poi-ki′lə-thər′mik) (Gr. *poikilos*, variable, + thermal) Pertaining to animals whose body temperature is variable and fluctuates with that of the environment; cold-blooded; compares with **ectothermic.**

polarity (Gr. *polos*, axis) In systematics, the ordering of alternative states of a taxonomic character from ancestral to successively derived conditions in an evolutionary transformation series. In developmental biology, the tendency for the axis of an ovum to orient corresponding to the axis of the mother. Also, condition of having opposite poles; differential distribution of gradation along an axis.

Polian vesicles (pō′le-ən) (From G. S. Poli, 1746–1825, Italian naturalist) Vesicles opening into ring canal in most asteroids and holothuroids.

polyandry (pol′y-an′drē) (Gr. *polys*, many, + *anēr*, man) Condition of having more than one male mate at one time.

polygamy (pə-lig′ə-mē) (Gr. *polys*, many, + *gamos*, marriage) Condition of having more than one mate at one time.

polygyny (pə-lij′ə-nē) (Gr. *polys*, many + *gynē*, woman) Condition of having more than one female mate at one time.

polymorphism (pä′lē-mor′fi-zəm) (Gr. *polys*, many, + *morphē*, form) The presence in a species of more than one structural type of individual.

polyp (päl′əp) (Fr. *polype*, octopus, from L. *polypus*, many footed) The sessile stage in the life cycle of cnidarians.

polyphyletic (pä′lē-fī-led′-ik) (Gr. *polys*, many, + *phylon*, tribe) Derived from more than one ancestral source; opposed to monophyletic.

polyphyly (pä′lē-fī′lē) (Gr. *polys*, full + *phylon*, tribe) The condition that a taxon or other group of organisms does not contain the most recent common ancestor of all members of the group, implying that it has multiple evolutionary origins; such groups are not valid as formal taxa and are recognized as such only through error.

polyphyodont (pä-lē-fī′ə-dänt) (Gr. *polyphyes*, manifold, + *odous*, tooth) Having several sets of teeth in succession.

polypide (pä′lē-pīd) (L. *polypus*, polyp) An individual or zooid in a colony, specifically in ectroprocts, which has a lophophore, digestive tract, muscles, and nerve centers.

pongid (pän′jəd) (L. *Pongo*, type genus of orangutan) Of or relating to the primate family Pongidae, comprising the anthropoid apes (gorillas, chimpanzees, gibbons, orangutans).

population (L. *populus*, people) A group of organisms of the same species inhabiting a specific geographical locality.

porocyte (pō′rə-sīt) (Gr. *porus*, passage, pore, + *kytos*, hollow vessel) Type of cell found in asconoid sponges through which water enters the spongocoel.

portal system (L. *porta*, gate) System of large veins beginning and ending with a bed of capillaries; for example, hepatic portal and renal portal system in vertebrates.

positive assortative mating A tendency of an individual to mate preferentially with others whose phenotypes are similar to its own.

posterior (L. latter). Situated at or toward the rear of the body; situated toward the back; in human anatomy the upright posture makes posterior and dorsal identical.

preadaptation The possession of a trait that coincidentally predisposes an organism for survival in an environment different from those encountered in its evolutionary history.

precocial (prē-kō′shəl) (L. *praecoquere*, to ripen beforehand) Referring (especially) to birds whose young are covered with down and are able to run about when newly hatched.

predaceous, predacious (prē-dā′shəs) (L. *praedator*, a plunderer, *praeda*, prey) Living by killing and consuming other animals; predatory.

predator (pred′ə-tər) (L. *praedator*, a plunderer, *praeda*, prey) An organism that preys on other organisms for its food.

prehensile (prē-hen′səl) (L. *prehendere*, to seize) Adapted for grasping.

primate (prī-māt) (L. *primus*, first) Any mammal of the order Primates, which includes the tarsiers, lemurs, marmosets, monkeys, apes, and humans.

primitive (L. *primus*, first) Primordial; ancient; little evolved; said of characteristics closely approximating those possessed by early ancestral types.

proboscis (prō-bäs′əs) (Gr. *pro*, before, + *boskein*, feed) A snout or trunk. Also, tubular sucking or feeding organ with the mouth at the end as in planarians, leeches, and insects. Also, the sensory and defensive organ at the anterior end of certain invertebrates.

producers (L. *producere*, to bring forth) Organisms, such as plants, able to produce their own food from inorganic substances.

proglottid (prō-gläd′əd) (Gr. *proglōttis*, tongue tip, from *pro*, before, + *glōtta*, tongue, + *id*, suffix) Portion of a tapeworm containing a set of reproductive organs; usually corresponds to a segment.

prokaryotic, procaryotic (pro-kar′ē-ät′ik) (Gr. *pro*, before, + *karyon*, kernel, nut) Not having a membrane-bound nucleus or nuclei. Prokaryotic cells are more primitive than eukaryotic cells and persist today in the bacteria and cyanobacteria.

pronephros (prō-nef′rəs) (Gr. *pro*, before, + *nephros*, kidney) Most anterior of three pairs of embryonic renal organs of vertebrates; functional only in adult hagfishes and larval fishes and amphibians; vestigial in mammalian embryos. Adj., **pronephric.**

prosimian (prō-sim′ē-ən) (Gr. *pro*, before, + L. *simia*, ape) Any member of a group of primitive, arboreal primates: lemurs, tarsiers, lorises, and so on.

prosopyle (präs′-ə-pīl) (Gr. *prosō*, forward, + *pylē*, gate) Connections between the incurrent and radial canals in some sponges.

prostomium (prō-stō′mē-əm) (Gr. *pro*, before, + *stoma*, mouth) In most annelids and some molluscs, that part of the head located in front of the mouth.

protein (prō′tēn, prō′tē-ən) (Gr. *protein*, from *proteios*, primary) A macromolecule of carbon, hydrogen, oxygen, and nitrogen and sometimes sulfur and phosphorus; composed of chains of amino acids joined by peptide bonds; present in all cells.

prothoracic glands Glands in the prothorax of insects that secrete the hormone ecdysone.

prothoracicotropic hormone See **ecdysiotropin.**

protist (prō′-tist) (Gr. *prōtos*, first) A member of the kingdom Protista, generally considered to include the protozoa and eukaryotic algae.

protocoel (prō′tə-sēl) (Gr. *prōtos*, first, + *koilos*, hollow) The anterior coelomic compartment in some deuterostomes, corresponds to the axocoel in echinoderms.

protonephridium (prō-tō-nə-frid′ē-əm) (Gr. *prōtos*, first, + *nephros*, kidney) Primitive osmoregulatory or excretory organ consisting of a tubule terminating internally with flame bulb or solenocyte; the unit of a flame bulb system.

protopod, protopodite (prō′tə-päd, prō-top′ə-dīt) (Gr. *prōtos*, first, + *pous, podos*, foot) Basal portion of crustacean appendage, containing coxa and basis.

Protostomia (prō′də-stō′mē-ə) (Gr. *prōtos*, first, + *stoma*, mouth) A group of phyla in which cleavage is determinate, the coelom (in coelomate forms) is formed by proliferation of mesodermal bands (schizocoelic formation), the mesoderm is formed from a particular blastomere (called 4d), and the mouth is derived from or near the blastopore. Includes the Annelida, Arthropoda, Mollusca, and a number of minor phyla. Compares with **Deuterostomia.**

proventriculus (pro′ven-trik′ū-ləs) (L. *pro*, before, + *ventriculum*, ventricle) In birds the glandular stomach between the crop and gizzard. In insects, a muscular dilation of foregut armed internally with chitinous teeth.

proximal (L. *proximus*, nearest) Situated toward or near the point of attachment; opposite of distal, distant.

proximate cause (L. *proximus*, nearest, + *causa*) The factors that underlie the functioning of a biological system at a particular place and time, including those responsible for metabolic, physiological, and behavioral functions at the molecular, cellular, organismal, and population levels. Immediate cause.

pseudocoel (sü′də-sēl) (Gr. *pseudēs*, false, + *koilos*, hollow) A body cavity not lined with peritoneum and not a part of the blood or digestive systems, embryonically derived from the blastocoel.

pseudopodium (sü′də-pō′dē-əm) (Gr. *pseudēs*, false, + *podion*, small foot, + *eidos*, form) A temporary cytoplasmic protrusion extended out from a protozoan or ameboid cell and serving for locomotion or for taking up food.

punctuated equilibrium A model of evolution in which morphological evolutionary change is discontinuous, being associated primarily with discrete, geologically instantaneous events of speciation leading to phylogenetic branching; morphological evolutionary stasis characterizes species between episodes of speciation; contrasts with **phyletic gradualism.**

pupa (pū′pə) (L. girl, doll, puppet) Inactive quiescent state of the holometabolous insects. It follows the larval stages and precedes the adult stage.

Q

queen In entomology, the single fully developed female in a colony of social insects such as bees, ants, and termites, distinguished from workers, nonreproductive females, and soldiers.

R

radial canals Canals along the ambulacra radiating from the ring canal of echinoderms; also choanocyte-lined canals in syconoid sponges.

radial cleavage Type in which early cleavage planes are symmetrical to the polar axis, each blastomere of one tier lying directly above the corresponding blastomere of the next layer; indeterminate cleavage.

radial symmetry A morphological condition in which the parts of an animal are arranged concentrically around an oral-aboral axis, and more than one imaginary plane through this axis yields halves that are mirror images of each other.

radiata (rā′dē-ä′tə (L. *radius*, ray) Phyla showing radial symmetry, specifically Cnidaria and Ctenophora.

radiole (rā′dē-ōl) (L. *radiolus*, dim. of *radius*, ray, spoke of a wheel) Structure extending from head of some sedentary polychaetes used in feeding on suspended particles.

radula (ra′jə-lə) (L. scraper) Rasping tongue found in most molluscs.

ratite (ra′tīt) (L. *ratis*, raft) Having an unkeeled sternum; compares with **carinate.**

recapitulation Summing up or repeating; hypothesis that an individual repeats its phylogenetic history in its development.

redia (rē′dē-ə) pl. **rediae** (rē′dē-ē) (from Francesco Redi, 1626–97, Italian biologist) A larval stage in the life cycle of flukes; it is produced by a sporocyst larva, and in turn gives rise to many cercariae.

regulative development Progressive determination and restriction of initially totipotent embryonic material.

reproductive barrier (L. *re*, + *producere*, to lead forward; M.F. *barriere*, bar) The factors that prevent one sexually propagating population from interbreeding and exchanging genes with another population.

rete mirabile (rē′tē mə-rab′ə-lē) (L. wonderful net) A network of small blood vessels so arranged that the incoming blood runs countercurrent to the outgoing blood and thus makes possible efficient exchange between the two bloodstreams. Such a mechanism serves to maintain the high concentration of gases in the fish swim bladder.

reticulopodia (rə-tik′ū-lə-pō′dē-ə) (L. *reticulum*, dim. of *rete*, net, + *podos, pous*, foot) Pseudopodia that branch and rejoin extensively.

reticulum (rə-tik′yə-ləm) (L. *rete*, dim. *reticulum*, a net) Second stomach of ruminants; a netlike structure.

bat/āpe/ärmadillo/herring/fēmale/finch/līce/ crocodile/crōw/duck/ūnicorn/tüna/ə indicates unaccented vowel sound "uh" as in mammal, fishes, cardinal, heron, vulture/stress as in bi-ol′o-gy, bi′o-log′i-cal

rhabdite (rab′dīt) (Gr. *rhabdos*, rod) Rodlike structures in the cells of the epidermis or underlying parenchyma in certain turbellarians. They are discharged in mucous secretions.

rhinophore (rī′nə-fōr) (Gr. *rhis*, nose, + *pherein*, to carry). Chemoreceptive tentacles in some molluscs (opisthobranch gastropods).

rhipidistian (rip-ə-dis′tē-ən) (Gr. *rhipis*, fan, + *histion*, sail, web) Member of a group of Paleozoic lobe-finned fishes.

rhopalium (rō-pā′lē-əm) (N.L. from Gr. *rhopalon*, a club) One of the marginal, club-shaped sense organs of certain jellyfishes; tentaculocyst.

rhoptries (rōp′trēz) (Gr. *thopalon*, club, + *tryō*, to rub, wear out) Club-shaped bodies in Apicomplexa making up one of the structures of the apical complex; open at anterior and apparently functioning in penetration of host cell.

rhynchocoel (ring′kō-sēl) (Gr. *rhynchos*, snout, + *koilos*, hollow) In nemertines, the dorsal tubular cavity that contains the inverted proboscis. It has no opening to the outside.

rostrum (räs′trəm) (L. ship's beak) A snoutlike projection on the head.

rumen (rü′mən) (L. cud) The large first compartment of the stomach of ruminant mammals.

ruminant (rüm′ə-nənt) (L. *ruminare*, to chew the cud) Cud-chewing artiodactyl mammals with a complex four-chambered stomach.

S

sagittal (saj′ə-dəl) (L. *sagitta*, arrow) Pertaining to the median anteroposterior plane that divides a bilaterally symmetrical organism into right and left halves.

saprophagous (sə-präf′ə-gəs) (Gr. *sapros*, rotten, + *phagos*, from *phagein*, to eat) Feeding on decaying matter; saprobic; saprozoic.

saprophyte (sap′rə-fīt) (Gr. *sapros*, rotten, + *phyton*, plant) A plant living on dead or decaying organic matter.

saprozoic nutrition (sap′rə-zō′ik) (Gr. *sapros*, rotten + *zōon*, animal) Animal nutrition by absorption of dissolved salts and simple organic nutrients from surrounding medium; also refers to feeding on decaying matter.

sauropterygians (so-räp′tə-rij′ē-əns) (Gr. *sauros*, lizard, + *pteryginos*, winged) Mesozoic marine reptiles.

schizocoel (skiz′ə-sēl) (Gr. *schizo*, from *schizein*, to split, + *koilos*, hollow) A coelom formed by the splitting of embryonic mesoderm. Noun, **schizocoelomate**, an animal with a schizocoel, such as an arthropod or mollusc. Adj., **schizocoelous.**

schizocoelous mesoderm formation (skiz′ō-sē-ləs) Embryonic formation of the mesoderm as cords of cells between ectoderm and endoderm; splitting of these cords results in the coelomic space.

schizogony (ski′-zä-gə-nē) (Gr. *schizein*, to split, + *gonos*, seed) Multiple asexual fission.

sclerite (skle′rēt) (Gr. *skleros*, hard) A hard chitinous or calcareous plate or spicule; one of the plates making up the exoskeleton of arthropods, especially insects.

scleroblast (skler′ə-blast) (Gr. *skleros*, hard, + *blastos*, germ) An amebocyte specialized to secrete a spicule, found in sponges.

sclerocyte (skler′ə-sīt) (Gr. *skleros*, hard, + *kytos*, hollow vessel). An amebocyte in sponges that secretes spicules.

sclerotin (skler′-ə-tən) (Gr. *sklerotēs*, hardness) Insoluble, tanned protein permeating the cuticle of arthropods.

sclerotization (skle′rə-tə-zā′shən) Process of hardening of the cuticle of arthropods by the formation of stabilizing cross linkages between peptide chains of adjacent protein molecules.

scolex (skō′leks) (Gr. *skōlex*, worm, grub) The holdfast, or so-called head, of a tapeworm; bears suckers and, in some, hooks; posterior to it new proglottids are differentiated.

scyphistoma (sī-fis′tə-mə) (Gr. *skyphos*, cup, + *stoma*, mouth) A stage in the development of scyphozoan jellyfishes just after the larva becomes attached; the polyp form of a scyphozoan.

sebaceous (sə-bāsh′əs) (L. *sebaceus*, made of tallow) A type of mammalian epidermal gland that produces a fatty substance.

sebum (sē′bəm) (L. grease, tallow) Oily secretion of the sebaceous glands of the skin.

sedentary (sed′ən-ter-ē) Stationary, sitting, inactive; staying in one place.

sensillum pl. **sensilla** (sin-si′ləm) (L. *sensus*, sense) A small sense organ, especially in the arthropods.

septum pl. **septa** (L. fence) A wall between two cavities.

serial homology See **homology.**

serosa (sə-rō′sə) (N.L. from L. *serum*, serum) The outer embryonic membrane of birds and reptiles; chorion. Also, the peritoneal lining of the body cavity.

serous (sir′əs) (L. *serum*, serum) Watery, resembling serum; applied to glands, tissue, cells, fluid.

serum (sir′əm) (L. whey, serum) The liquid that separates from the blood after coagulation; blood plasma from which fibrinogen has been removed. Also, the clear portion of a biological fluid separated from its particular elements.

sessile (ses′əl) (L. *sessilis*, low, dwarf) Attached at the base; fixed to one spot, not able to move about.

seta (sēd′ə), pl. **setae** (sē′tē) (L. bristle) A needlelike chitinous structure of the integument of annelids, arthropods, and others.

sexual selection Charles Darwin's theory that there exists a struggle among males for mates and that characteristics favorable for mating may prevail through reproductive success even if they are not advantageous in the struggle for survival.

siliceous (sə-li′shəs) (L. *silex*, flint) Containing silica.

simian (sim′ē-ən) (L. *simia*, ape) Pertaining to monkeys or apes.

sinus (sī′nəs) (L. curve) A cavity or space in tissues or in bone.

siphonoglyph (sī′fan′ə-glif) (Gr. *siphon*, reed, tube, siphon, + *glyphē*, carving) Ciliated furrow in the gullet of sea anemones.

siphuncle (sī′fun-kəl) (L. *siphunculus*, small tube) Cord of tissue running through the shell of a nautiloid, connecting all chambers with the body of animal.

sister taxon (=“sister group”) The relationship between a pair of species or higher taxa that are each other's closest phylogenetic relatives.

soma (sō′mə) (Gr. body) The whole of an organism except the germ cells (germ plasm).

somatic (sō-mat′ik) (Gr. *sōma*, body) Refers to the body, for example, somatic cells in contrast to germ cells.

somatocoel (sə-mat′ə-sēl) (Gr. *sōma*, the body, + *koilos*, hollow) Posterior coelomic compartment of echinoderms; left somatocoel gives rise to oral coelom, and right somatocoel becomes aboral coelom.

somite (sō′mīt) (Gr. *sōma*, body) One of the blocklike masses of mesoderm arranged segmentally (metamerically) in a longitudinal series beside the digestive tube of the embryo; metamere.

sorting Differential survival and reproduction among varying individuals; often confused with natural selection which is one possible cause of sorting.

speciation (spē′sē-ā′shən) (L. *species*, kind) The evolutionary process or event by which new species arise.

species (spē′shez, spē′sēz) sing. and pl. (L. particular kind) A group of interbreeding individuals of common ancestry that are reproductively isolated from all other such groups; a taxonomic unit ranking below a genus and designated by a binomen consisting of its genus and the species name.

species selection Differential rates of speciation and/or extinction among varying evolutionary lineages caused by interactions among species-level characteristics and the environment.

specific epithet The second, uncapitalized word in the binomial name of a species. It is usually an adjective modifying the first word, which identifies the genus into which the species is placed.

spermatheca (spər′mə-thē′kə) (Gr. *sperma*, seed, + *thēkē*, case) A sac in the female reproductive organs for the reception and storage of sperm.

spermatophore (spər-mad′ə-fōr′) (Gr. *sperma*, *spermatos*, seed, + *pherein*, to bear) Capsule or packet enclosing sperm, produced by males of several invertebrate groups and a few vertebrates.

spicule (spi′kyul) (L. dim. of *spica*, point) One of the minute calcareous or siliceous skeletal bodies found in sponges, radiolarians, soft corals, and sea cucumbers.

spiracle (spi′rǝ-kǝl) (L. *spiraculum*, from *spirare*, to breathe) External opening of a trachea in arthropods. One of a pair of openings on the head of elasmobranchs for passage of water. Exhalent aperture of tadpole gill chamber.

spiral cleavage A type of early embryonic cleavage in which cleavage planes are diagonal to the polar axis and unequal cells are produced by the alternate clockwise and counterclockwise cleavage around the axis of polarity; determinate cleavage.

spongin (spun′jin) (L. *spongia*, sponge) Fibrous, collagenous material making up the skeletal network of horny sponges.

spongocoel (spun′jō-sēl) (Gr. *spongos*, sponge, + *koilos*, hollow) Central cavity in sponges.

spongocyte (spun′jō-sīt) (Gr. *spongos*, sponge, + *kytos*, hollow vessel) A cell in sponges that secretes spongin.

sporocyst (spō′rǝ-sist) (Gr. *sporos*, seed, + *kystis*, pouch) A larval stage in the life cycle of flukes; it originates from a miracidium.

sporogony (spor-äg′ǝ-nē) (Gr. *sporos*, seed, + *gonos*, birth) Multiple fission to produce sporozoites after zygote formation.

sporozoite (spō′rǝ-zō′īt) (Gr. *sporos*, seed, + *zōon*, animal, + *ite*, suffix for body part) A stage in the life history of many sporozoan protozoa; released from oocysts.

stabilizing selection Natural selection that favors average values of a continuously varying trait and disfavors extreme values.

statoblast (stad′ǝ-blast) (Gr. *statos*, standing, fixed, + *blastos*, germ) Biconvex capsule containing germinative cells and produced by most freshwater ectoprocts by asexual budding. Under favorable conditions it germinates to give rise to a new zooid.

statocyst (Gr. *statos*, standing, + *kystis*, bladder) Sense organs of equilibrium; a fluid-filled cellular cyst containing one or more granules (statoliths) used to sense direction of gravity.

stenohaline (sten-ǝ-ha′līn, -lǝn) (Gr. *stenos*, narrow, + *hals*, salt) Pertaining to aquatic organisms that have restricted tolerance to changes in environmental saltwater concentration.

stereom (ster′ē-ōm) (Gr. *stereos*, solid, hard, firm) Meshwork structure of endoskeletal ossicles of echinoderms.

sternum (ster′nǝm) (L. breastbone) Ventral plate of an arthropod body segment; breastbone of vertebrates.

stigma (Gr. *stigma*, mark, tattoo mark) Eyespot in certain protozoa. Spiracle of certain terrestrial arthropods.

stolon (stō′lǝn) (L. *stolō, stolonis*, a shoot, or sucker of a plant) A rootlike extension of the body wall giving rise to buds that may develop

into new zooids, thus forming a compound animal in which the zooids remain united by the stolon. Found in some colonial anthozoans, hydrozoans, ectoprocts, and ascidians.

stoma (stō′mǝ) (Gr. mouth) A mouthlike opening.

strobila (strō′bǝ-lǝ) (Gr. *strobile*, lint plug like a pine cone [*strobilos*]) A stage in the development of the scyphozoan jellyfishes. Also, the chain of proglottids of a tapeworm.

subnivean (sǝb-ni′vē-ǝn) (L. *sub*, under, below, + *nivis*, snow) Applied to environments beneath snow, in which snow insulates against a colder atmospheric temperature.

sycon (sī′kon) (Gr. *sykon*, fig) A type of canal system in certain sponges. Sometimes called syconoid.

symbiosis (sim-bī-ōs′ǝs, sim′bē-ōs′ǝs) (Gr. *syn*, with, + *bios*, life) The living together of two different species in an intimate relationship. Symbiont always benefits; host may benefit, may be unaffected, or may be harmed (mutualism, commensalism, and parasitism).

synapomorphy (sin-ap′o-mor′fē) (Gr. *syn*, together with, + *apo*, of, + *morphē*, form) Shared, evolutionarily derived character states that are used to recover patterns of common descent among two or more species.

synapsids (si-nap′sǝdz) (Gr. *synapsis*, contact, union) An amniote lineage comprising the mammals and the ancestral mammal-like reptiles, having a skull with a single pair of temporal openings.

syncytium (sin-sish′ē-ǝm) (Gr. *syn*, with, + *kytos*, hollow vessel) A mass of protoplasm containing many nuclei and not divided into cells.

syngamy (sin′gǝ-mē) (Gr. *syn*, with, + *gamos*, marriage) Fertilization of one gamete with another individual gamete to form a zygote, found in most animals with sexual reproduction.

syrinx (sir′inks) (Gr. shepherd's pipe) The vocal organ of birds located at the base of the trachea.

systematics (sis-tǝ-mad′ iks) Science of classification and reconstruction of phylogeny.

T

tactile (tak′til) (L. *tactilis*, able to be touched, from *tangere*, to touch) Pertaining to touch.

tagma pl. **tagmata** (Gr. *tagma*, arrangement, order, row) A compound body section of an arthropod resulting from embryonic fusion of two or more segments; for example, head, thorax, abdomen.

tagmatization, tagmosis Organization of the arthropod body into tagmata.

taxon (tak′son) pl. **taxa** (Gr. *taxis*, order, arrangement) Any taxonomic group or entity.

taxonomy (tak-sän′ǝ-mi) (Gr. *taxis*, order, arrangement, + *nomas*, law) Study of the principles of scientific classification; systematic ordering and naming of organisms.

tegument (teg′ū-ment) (L. *tegumentum*, from *tegere*, to cover) An integument; specifically external covering in cestodes and trematodes, formerly believed to be a cuticle.

teleology (tel′ē-äl′ǝ-jē) (Gr. *telos*, end, + L. *logia*, study of, from Gr. *logos*, word) The philosophical view that natural events are goal-directed and are preordained, as opposed to the scientific view of mechanical determinism.

telson (tel′sǝn) (Gr. extremity) Posterior projection of the last body segment in many crustaceans.

tergum (ter′gǝm) (L. back) Dorsal part of an arthropod body segment.

test (L. *testa*, shell) A shell or hardened outer covering.

tetrapods (te′trǝ-päds) (Gr. *tetras*, four, + *pous, podos*, foot) Four-footed vertebrates; the group includes amphibians, reptiles, birds, and mammals.

theory A scientific hypothesis or set of related hypotheses that offer very powerful explanations for a wide variety of related phenomena and serve to organize scientific investigation of those phenomena.

therapsid (thǝ-rap′sid) (Gr. *theraps*, an attendant) Extinct Mesozoic mammal-like reptiles from which true mammals evolved.

thoracic (thō-ra′sǝk) (L. *thōrax*, chest) Pertaining to the thorax or chest.

Tiedemann's bodies (tēd′ǝ-mǝnz) (from F. Tiedemann, German anatomist) Four or five pairs of pouchlike bodies attached to the ring canal of sea stars, apparently functioning in production of coelomocytes.

tissue (ti′shü) (M.E. *tissu*, tissue) An aggregation of cells, usually of the same kind, organized to perform a common function.

torsion (L. *torquere*, to twist) A twisting phenomenon in gastropod development that alters the position of the visceral and pallial organs by 180 degrees.

toxicyst (tox′i-sist) (Gr. *toxikon*, poison, + *kystis*, bladder) Structures possessed by predatory ciliate protozoa, which on stimulation expel a poison to subdue the prey.

trachea (trā′kē-ǝ) (M.L. windpipe) The windpipe. Also, any of the air tubes of insects.

trend A directional change in the characteristic features or patterns of diversity in a group of organisms when viewed over long periods of evolutionary time in the fossil record.

trichinosis (trik-ǝn-o′sǝs) Disease caused by infection with the nematode *Trichinella spiralis*.

trichocyst (trik′ǝ-sist) (Gr. *thrix*, hair, + *kystis*, bladder) Saclike protrusible organelle in the ectoplasm of ciliates, which discharges as a threadlike weapon of defense.

triploblastic (trip′lō-blas′tik) (Gr. *triploos*, triple, + *blastos*, germ) Pertaining to metazoa in which the embryo has three primary term layers—ectoderm, mesoderm, and endoderm.

trochophore (trōk′ǝ-for) (Gr. *trochos*, wheel, + *pherein*, to bear) A free-swimming ciliated marine larva characteristic of most molluscs

bat/āpe/ärmadillo/herring/fēmale/finch/līce/crocodile/crōw/duck/ūnicorn/tüna/ǝ indicates unaccented vowel sound "uh" as in mammal, fishes, cardinal, heron, vulture/stress as in bi-ol′o-gy, bi′o-log′i-cal

and certain ectoprocts, brachiopods, and marine worms; an ovoid or pyriform body with preoral circlet of cilia and sometimes a secondary circlet behind the mouth.

trophallaxis (trŏf′ə-lak′səs) (Gr. *trophē*, food, + *allaxis*, barter, exchange) Exchange of food between young and adults, especially certain social insects.

trophic (trō′fək) (Gr. *trophē*, food) Pertaining to nutrition; or pertaining to hormones that influence other endocrine glands or growth.

trophosome (trŏf′ə-sōm) (Gr. *trophē*, food, + *soma*, body) Organ in poganophorans bearing mutualistic bacteria; derived from midgut.

trophozoite (trŏf′ə-zō′īt) (Gr. *trophē*, food, + *zōon*, animal) Adult stage in the life cycle of a protozoan in which it is actively absorbing nourishment.

tube feet (podia) Numerous small, muscular, fluid-filled tubes projecting from body of echinoderms; part of water-vascular system; used in locomotion, clinging, food handling, and respiration.

tubulin (tü′bū-lən) (L. *tubulus*, small tube, + *in*, belonging to) Globular protein forming the hollow cylinder of microtubules.

tunic (L. *tunica*, tunic, coat) In tunicates, a cuticular, cellulose-containing covering of the body secreted by the underlying body wall.

typhlosole (tif′lə-sōl′) (Gr. *typhlos*, blind, + *sōlēn*, channel, pipe) A longitudinal fold projecting into the intestine in certain invertebrates such as the earthworm.

typological species concept The discredited, pre-Darwinian notion that species are classes defined by the presence of fixed, unchanging characters (= "essence") shared by all members.

U

ultimate cause (L. *ultimatus*, last, + *causa*) The evolutionary factors responsible for the origin, state of being, or purpose of a biological system.

umbo (um′bō) pl. **umbones** (əm-bō′nēz) (L. boss of a shield) One of the prominences on either side of the hinge region in a bivalve mollusc shell. Also, the "beak" of a brachiopod shell.

undulating membrane A membranous structure on a protozoan associated with a flagellum; on other protozoa may be formed from fused cilia.

ungulate (un′gū-lət) (L. *ungula*, hoof) Hooved. Noun, any hooved mammal.

uniformitarianism (ū′nə-fōr′mə-ter′ē-ə-niz′əm) Methodological assumptions that the laws of chemistry and physics have remained constant throughout the history of the earth, and that past geological events occurred by processes that can be observed today.

uropod (ū′rə-pod) (Gr. *oura*, tail, + *pous*, *podos*, foot) Posteriormost appendage of many crustaceans.

V

vacuole (vak′yə-wōl) (L. *vacuus*, empty, + Fr. *ole*; dimin. suffix) A membrane-bounded, fluid-filled space in a cell.

valve (L. *valva*, leaf of a double door) One of the two shells of a typical bivalve mollusc or brachiopod.

variation (L. *varius*, various) Differences among individuals of a group or species that cannot be ascribed to age, sex, or position in the life cycle.

velarium (və-la′rē-əm) (L. *velum*, veil, covering). Shelf-like extension of the subumbrella edge in cubozoans (phylum Cnidaria).

veliger (vēl′ə-jər, vel-) (L. *velum*, veil, covering) Larval form of certain molluscs; develops from the trochophore and has the beginning of a foot, mantle, shell, and so on.

velum (vē′ləm) (L. veil, covering) A membrane on the subumbrella surface of jellyfishes of class Hydrozoa. Also, a ciliated swimming organ of the veliger larva.

ventral (ven′trəl) (L. *venter*, belly). Situated on the lower or abdominal surface.

vestige (ves′tij) (L. *vestigium*, footprint) A rudimentary organ that may have been well developed in some ancestor or in the embryo.

vibrissa (vī′bris′ə), pl. **vibrissae** (L. nostril-hair) Stiff hairs that grow from the nostrils or other parts of the face of many mammals and that serve as tactile organs; "whiskers."

villus (vil′əs) pl. **villi** (L. tuft of hair) A small fingerlike, vascular process on the wall of the small intestine. Also one of the branching, vascular processes on the embryonic portion of the placenta.

viscera (vis′-ər-ə) (L. pl. of *viscus*, internal organ) Internal organs in the body cavity.

visceral (vis′ər-əl) Pertaining to viscera.

viviparity (vī′və-par′ə-dē) (L. *vivus*, alive, + *parere*, to bring forth) Reproduction in which eggs develop within the female body, with nutritional aid of maternal parent as in therian mammals, many reptiles, and some fishes; offspring are born as juveniles. Adj., **viviparous** (vī-vip′ə-rəs).

W

water-vascular system System of fluid-filled closed tubes and ducts peculiar to echinoderms; used to move tentacles and tube feet that serve variously for clinging, food handling, locomotion, and respiration.

weir (wēr) (Old English *wer*, a fence placed in a stream to catch fish). Interlocking extensions of a flame cell and a collecting tubule cell in some protonephridia.

X

X-organ Neurosecretory organ in eyestalk of crustaceans that secretes molt-inhibiting hormone.

Y

Y-organ Gland in the antennal or maxillary segment of some crustaceans that secretes molting hormone.

Z

zoecium, zooecium (zō-ē′shē-əm) (Gr. *zō-on*, animal, + *oikos*, house) Cuticular sheath or shell of Ectoprocta.

zoochlorella (zō′ə-klōr-el′ə) (Gr. *zōon*, animal, + *Chlorella*) Any of various minute green algae (usually *Chlorella*) that live symbiotically within the cytoplasm of some protozoa and other invertebrates.

zooflagellates (zō′ə-fla′jə-lāts) Members of the Zoomastigophora, the animal-like flagellates (phylum Sarcomastigophora).

zooid (zō-oid) (Gr. *zōon*, animal) An individual member of a colony of animals, such as colonial cnidarians and ectoprocts.

zooxanthella (zo′ə-zan-thəl′ə) (Gr. *zōon*, animal, + *xanthos*, yellow) A minute dinoflagellate alga living in the tissues of many types of marine invertebrates.

zygote (Gr. *zygōtos*, yoked) The fertilized egg.

zygotic meiosis Meiosis that takes place within the first few divisions after zygote formation; thus all stages in the life cycle other than the zygote are haploid.

Credits

Photographs

Photo Research by Connie Mueller

Chapter 1

Figure1.1a: © William Ober; **Figure 1.1b:** Cleveland P. Hickman, Jr.; **Figure 1.1c:** Courtesy Duke University Marine Laboratory; **Figure 1.1d–e:** Cleveland P. Hickman, Jr.; **TA 1.1 Page 4:** Courtesy Foundation for Biomedical Research; **Figure 1.2a:** Courtesy American Museum of Natural History, Neg. #326662; **Figure 1.2b:** Courtesy The Natural History Museum, London; **Figure 1.3:** Courtesy The Natural History Museum, London; **Figure 1.4:** Courtesy The Natural History Museum, London; **Figure 1.6a:** BAL4391 Portrait of Charles Darwin, 1840 by George Richmond (1809-96) Down House, Downe, Kent, UK/Bridgeman Art Library, London/New York; **Figure 1.6b:** © Stock Montage; **Figure 1.7:** Cleveland P. Hickman, Jr.; **Figure 1.8:** Cleveland P. Hickman, Jr.; **TA 1.2 Page 10:** Courtesy The Natural History Museum, London; **Figure 1.10a:** © A.J. Copley/Visuals Unlimited; **Figure 1.10b:** © G.O. Poinar, Oregon State University, Corvallis; **Figure 1.10c:** © Ken Lucas/Biological Photo Service; **Figure 1.10d:** © Roberta Hess Poinar; **Figure 1.11a:** © Boehm Photography; **Figure 1.12:** Cleveland P. Hickman, Jr.; **Figure 1.16:** Courtesy Library of Congress; **Figure 1.21b:** Cleveland P. Hickman, Jr.; **Figure 1.22:** Courtesy of Storrs Agricultural Experiment Station, University of Connecticut at Storrs; **Figure 1.25a–b:** © Michael Tweedie/Photo Researchers, Inc.; **Figure 1.27:** © Timothy W. Ranson/Biological Photo Service; **Figure 1.28:** © S. Krasemann/Photo Researchers, Inc.; **Figure 1.31:** Courtesy Natural Resources Canada.

Chapter 2

Opener: © Stephen Dalton/Photo Researchers, Inc.; **Figure 2.4a-c:** © Ed Reschke; **Figure 2.5a-b:** © Ed Reschke; **Figure 2.6a-b:** © Ed Reschke; **Figure 2.6c:** Cleveland P. Hickman, Jr.; **Figure 2.6d:** © Ed Reschke; **Figure 2.7a-c:** © Ed Reschke.

Chapter 3

Opener: Cleveland P. Hickman, Jr.; **Figure 3.1:** Courtesy Library of Congress; **Figure 3.5:** Courtesy American Museum of Natural History, Neg. #334101; **Figure 3.6a:** © M. Cole/Animals Animals/Earth Scenes; **Figure 3.6b:** © D. Allen/Animals Animals/Earth Scenes; **Figure 3.8:** Courtesy of Dr. George W. Byers, University of Kansas.

Chapter 4

Opener: © Michael Abbey/Visuals Unlimited; **Figure 4.2:** Courtesy Dr. Ian R. Gibboons; **Figure 4.4a–c:** © M. Abbey/Visuals Unlimited; **Figure 4.5:** © L. Evans Roth/Biological Photo Service; **Figure 4.15a:** Courtesy Gustaaf M. Hallegraeff; **Figure 4.15b:** © A.M. Siegelman/Visuals Unlimited; **Figure 4.16:** Photo courtesy J. and M. Cachon. From Lee, J. J., Hunter, S.H. and Boves, E.C. (editors). 1985. "An Illustrated Guide to the Protozoa", Society of Protozoologists. Permission by Edna Kaneshiro.

Chapter 5

Opener: Larry S. Roberts; **Figure 5.2:** Larry S. Roberts; **Figure 5.4:** Larry S. Roberts; **Figure 5.11a:** Larry S. Roberts, **Figure 5.11b:** Larry S. Roberts; **Figure 5.11c:** Larry S. Roberts.

Chapter 6

Opener: Larry S. Roberts; **Figure 6.1a:** © William Ober; **Figure 6.6:** © Rick Harbo; **Figure 6.7:** © CABISCO/Phototake; **Figure 6.8:** © CABISCO/Visuals Unlimited; **Figure 6.10:** © Daniel W. Gotshall; **Figure 6.12:** © Peter Parks/OSF/Animals Animals/Earth Scenes; **Figure 6.13a–b:** Larry S. Roberts; **Figure 6.14:** © Rick Harbo; **Figure 6.15:** © Rick Harbo; **Figure 6.18:** © Daniel W. Gotshall; **Figure 6.19a:** © Jeff Rotman Photography; **Figure 6.19b:** Larry S. Roberts; **Figure 6.21:** Larry S. Roberts; **Figure 6.22a–b:** © Rick Harbo; **Figure 6.23a:** Cleveland P. Hickman, Jr.; **Figure 6.23b–c:** Larry S. Roberts; **Figure 6.26:** Larry S. Roberts; **Figure 6.28a:** © Jeff Rotman Photography; **Figure 6.28b:** © Kjell Sandved/Butterfly Alphabet.

Chapter 7

Opener: Larry S. Roberts; **Figure 7.2:** © CABISCO/Visuals Unlimited; **Figure 7.3:** © James L. Castner; **Figure 7.9a:** Larry S. Roberts; **Figure 7.10:** Photo by R.E. Kuntz from "A Pictorial Presentation of Parasites," edited by H. Zaiman, M.D.; **Figure 7.12:** © CABISCO/Visuals Unlimited; **Figure 7.16:** Larry S. Roberts; **Figure 7.17:** © Stan Elems/Visuals Unlimited; **Figure 7.19:** Cleveland P. Hickman, Jr.

Chapter 8

Opener: Photo #1206 by D. Despommier from "A Pictorial Presentation of Parasites," edited by H. Zaiman, M.D.; **Figure 8.9a:** Courtesy Frances M. Hickman; **Figure 8.9b:** Photo by G.W. Kelley, Jr. from "A Pictorial Presentation of Parasites," edited by H. Zaiman, M.D.; **Figure 8.10:** Photo by E. Pike from "A Pictorial Presentation of Parasites," edited by H. Zaiman, M.D.; **Figure 8.11:** From "A Pictorial Presentation of Parasites," edited by H. Zaiman, M.D.; **Figure 8.12a:** © R. Calentine/Visuals Unlimited; **Figure 8.12b:** From "A Pictorial Presentation of Parasites," edited by H. Zaiman, M.D.; **Figure 8.13:** Contributed by E.L. Schiller, Armed Forces Institute of Pathology; **Figure 8.14:** Larry S. Roberts.

Chapter 9

Opener: Larry S. Roberts; **Figure 9.1a-c:** © Rick Harbo; **Figure 9.1d:** © Daniel W. Gotshall; **Figure 9.1e:** Larry S. Roberts; **Figure 9.3:** Larry S. Roberts; **Figure 9.6:** © Kjell Sandved/Butterfly Alphabet; **Figure 9.8:** © Rick Harbo; **Figure 9.14a-b:** © Daniel W. Gotshall; **Figure 9.15a-b:** © Alex Kerstitch/Sea of Cortez Enterprises; **Figure 9.16a-b:** Cleveland P. Hickman, Jr., **Figure 9.18a:** © Rick Harbo; **Figure 9.18b:** Larry S. Roberts; **Figure 9.19:** © Tom Phillipp Underwater Photography; **Figure 9.20a:** Larry S. Roberts; **Figure 9.20b:** Cleveland P. Hickman, Jr.; **Figure 9.21:** Larry S. Roberts; **Figure 9.24a–b:** Larry S. Roberts; **Figure 9.25:** Larry S. Roberts; **Figure 9.26:** © Rick Harbo; **Figure 9.27:** © Rick Harbo; **Figure 9.29b:** Courtesy Richard J. Neves; **Figure 9.30:** © Dave Fleetham/Tom Stack & Associates; **Figure 9.31:** Courtesy M. Butschler, Vancouver Public Aquarium; **Figure 9.32:** Courtesy Captain Louis Usie.

Chapter 10

Opener: Photo gear, #CRAB 02.TIF; **Figure 10.3a–b:** Larry S. Roberts; **Figure 10.5:** © Stan Elems/Visuals Unlimited; **Figure 10.6:** Larry S. Roberts; **Figure 10.14:** © G.L. Twiest/Visuals Unlimited; **Figure 10.17:** © Timothy Branning; **Figure 10.19:** Cleveland P. Hickman, Jr.

Chapter 11

Opener: Larry S. Roberts; **Figure 11.1a–b:** © A.J. Copley/Visuals Unlimited; **Figure 11.5a:** © James L. Castner; **Figure 11.5b:** © John H. Gerard/Nature Press; **Figure 11.6:** © John H. Gerard/Nature Press; **Figure 11.7:** © John H. Gerard/Nature Press; **Figure 11.8:** © John H. Gerard/Nature Press; **Figure 11.9a–b:** © James L. Castner; **Figure 11.10a:** © J.A. Alcock/Visuals Unlimited; **Figure 11.10b:** © James L. Castner; **Figure 11.11:** © John H. Gerard/Nature Press; **Figure 11.12:** Larry S. Roberts; **Figure 11.13:** © D.S. Snyder/Visuals Unlimited; **Figure 11.22:** ©

CABISCO/Phototake; **Figure 11.26a–b:** © Rick Harbo; **Figure 11.27:** Cleveland P. Hickman, Jr.; **Figure 11.28:** Larry S. Roberts; **Figure 11.29a:** © Rick Harbo; **Figure 11.29b:** © Kjell Sandved/Butterfly Alphabet; **Figure 11.29c:** © Kjell Sandved/Butterfly Alphabet; **Figure 11.31a:** Cleveland P. Hickman, Jr.; **Figure 11.31b:** © Rick Harbo; **Figure 11.31c:** Cleveland P. Hickman, Jr., **Figure 11.31d–e:** Larry S. Roberts; **Figure 11.32a:** © James L. Castner; **Figure 11.33a:** © James L. Castner; **Figure 11.35b:** © James L. Castner; **Figure 11.36a–b:** © Ron West/Nature Photography; **Figure 11.37:** © James L. Castner; **Figure 11.41c:** © James L. Castner; **Figure 11.43a:** Cleveland P. Hickman, Jr.; **Figure 11.43b:** © John H. Gerard/Nature Press; **Figure 11.44:** Cleveland P. Hickman, Jr.; **Figure 11.46a:** Cleveland P. Hickman, Jr.; **Figure 11.46b:** © John H. Gerard/Nature Press; **Figure 11.46c:** © CABISCO/Phototake; **Figure 11.47a–b:** © James L. Castner; **Figure 11.48:** © John H. Gerard/Nature Press; **Figure 11.49:** Courtesy James E. Lloyd; **Figure 11.50:** © James L. Castner; **Figure 11.51:** © K. Lorenzen/Andromeda Productions; **Figure 11.52a:** © John H. Gerard/Nature Press; **Figure 11.52b:** © James L. Castner; **Figure 11.53:** Larry S. Roberts; **Figure 11.54a–c:** © James L. Castner; **Figure 11.55a:** © Leonard Lee Rue, III; **Figure 11.55b:** © James L. Castner; **Figure 11.55c:** © John H. Gerard/Nature Press; **Figure 11.56:** © James L. Castner; **Figure 11.57:** © James L. Castner; **Figure 11.58b:** © James L. Castner; **Figure 11.59:** Courtesy Jay Georgi; **Figure 11.60:** © James L. Castner; **Figure 11.61:** © James L. Castner; **Figure 11.62:** © James L. Castner; **Figure 11.63:** © John D. Cunningham/Visuals Unlimited.

Chapter 12

Opener: © William Ober; **Figure 12.5:** © James L. Castner; **Figure 12.6:** Courtesy Diane R. Nelson; **Figure 12.8b:** Larry S. Roberts; **Figure 12.9:** © Ken Lucas/Biological Photo Service; **Figure 12.10a:** © Robert Brons/Biological Photo Service; **Figure 12.10b:** © Robert Brons/Biological Photo Service.

Chapter 13

Opener: © Paul Gier/Visuals Unlimited; **Figure 13.1a:** © Rick Harbo; **Figure 13.1b–c:** Larry S. Roberts; **Figure 13.1d:** © Rick Harbo; **Figure 13.4f:** © Rick Harbo; **Figure 13.5:** Larry S. Roberts; **Figure 13.7a:** © Rick Harbo; **Figure 13.7b:** Larry S. Roberts; **Figure 13.10:** © Rick Harbo; **Figure 13.11a–b:** Larry S. Roberts; **Figure 13.14:** © Rick Harbo; **Figure 13.17:** Larry S. Roberts; **Figure 13.26b:** From Theusen, E.V. and R. Bert, 1987. Canadian Journal of Zoology, 65:181-87/NRC Canada.

Chapter 14

Opener: © Heather Angel/Biofoto; **Figure 14.4:** Courtesy R.P.S. Jeffries, © The Natural History Museum, London; **Figure 14.7:** © David B. Fleetham/Visuals Unlimited.

Chapter 15

Opener: Larry S. Roberts; **Figure 15.9:** © Jeff Rotman Photography; **Figure 15.18a–b:** © John G. Shedd Aquarium 1997/Patrice Ceisel; **Figure 15.27:** © Will Troyer/Visuals Unlimited; **Figure 15.29:** © Daniel W. Gotshall; **Figure 15.30:** © Frederick R. McConnaughey.

Chapter 16

Opener: © Gary Meszaros/Visuals Unlimited; **Figure 16.6a–b:** Courtesy L. Houck; **Figure 16.9:** Cleveland P. Hickman, Jr.; **Figure 16.11a:** © Ken Lucas/Biological Photo Service; **Figure 16.11b:** Cleveland P. Hickman, Jr.; **Figure 16.12:** Cleveland P. Hickman, Jr.; **Figure 16.13:** Courtesy American Museum of Natural History, Neg.# 125617; **Figure 16.14:** Cleveland P. Hickman, Jr.; **Figure 16.15:** Cleveland P. Hickman, Jr.

Chapter 17

Opener: Courtesy National Zoological Park, Smithsonian Institution, Jessie Cohen, photographer; **Figure 17.7:** Cleveland P. Hickman, Jr.; **Figure 17.8:** Cleveland P. Hickman, Jr.; **Figure 17.10:** © John Mitchell/Photo Researchers, Inc.; **Figure 17.11:** Cleveland P. Hickman, Jr.; **Figure 17.12:** © Javier Andrada; **Figure 17.13:** © Leonard Lee Rue, III; **Figure 17.14:** © Leonard Lee Rue, III; **Figure 17.16:** © Leonard Lee Rue, III; **Figure 17.17:** Cleveland P. Hickman, Jr.; **Figure 17.18:** © Leonard Lee Rue, III; **Figure 17.21:** Cleveland P. Hickman, Jr.; **Figure 17.22:** © Renee Lynn/Tony Stone Images, Inc.; **Figure 17.24a–b:** Cleveland P. Hickman, Jr.

Chapter 18

Opener: PhotoDisc, Vol. 6, #6321; **Figure 18.1:** Courtesy American Museum of Natural History, Neg. # 125065; **Figure 18.15:** © J.L. McAlonay/Visuals Unlimited; **Figure 18.19a–b:** © Leonard Lee Rue, III; **Figure 18.22:** © John Gerlach/Visuals Unlimited; **Figure 18.23:** © Richard R. Hansen/Photo Researchers, Inc.; **Figure 18.25:** © Leonard Lee Rue, III; **Figure 18.27:** Cleveland P. Hickman, Jr.; **Figure 18.28:** Cleveland P. Hickman, Jr.; **Figure 18.29:** Cleveland P. Hickman, Jr.; **Figure 18.30:** Cleveland P. Hickman, Jr.

Chapter 19

Opener: © Jeff Lepore/Photo Researchers, Inc.; **Figure 19.5:** © Leonard Lee Rue, III; **Figure 19.6a–b:** © Leonard Lee Rue, III; **Figure 19.7:** Courtesy Robert E. Treat; **Figure 19.11:** © Leonard Lee Rue, III; **Figure 19.12:** © John Gerlach/Visuals Unlimited; **Figure 19.13:** Cleveland P. Hickman, Jr.; **Figure 19.15:** © S. Maslowski/Visuals Unlimited; **Figure 19.17:** © Kjell Sandved/Visuals Unlimited; **Figure 19.18:** © Leonard Lee Rue, III; **Figure 19.20:** © G. Herben/Visuals Unlimited; **Figure 19.22:** Courtesy Zoological Society of San Diego; **Figure 19.23a:** Courtesy Zoological Society of San Diego; **Figure 19.23b:** Cleveland P. Hickman, Jr.; **Figure 19.24:** © Joe McDonald/Visuals Unlimited; **Figure 19.25:** © John Reader; **Figure 19.28:** © John Gerlach/Visuals Unlimited; **Figure 19.29:** Cleveland P. Hickman, Jr.; **Figure 19.30:** Cleveland P. Hickman, Jr.; **Figure 19.31:** © William Ober; **Figure 19.32:** Cleveland P. Hickman, Jr.

Line Art and Extracts

Chapter 1

Figure 1.5: Source: A. Moorehead, *Darwin and the Beagle*, 1969, Harper & Row, New York, NY. **Figure 1.14:** Source: J. J. Sepkoski, Jr., *Paleobiology*, 7:36-53, 1981. **Figure 1.15:** From Peter H. Raven and George B. Johnson, *Biology*, 4th edition. Copyright © 1996 McGraw-Hill Company, Inc., Dubuque, IA. All Rights Reserved. Reprinted by permission. **Figure 1.17:** Source: J. Cracraft in *IBIS*, 116:294-521, 1974. **Figure 1.20:** Source: P.R. Grant, "Speciation and Adaptive Radiation of Darwin's Finches" in *American Scientist*, 69:653-663, 1981. **Figure 1.21:** From Peter H. Raven and George B. Johnson, *Biology*, 4th edition. Copyright © 1996 McGraw-Hill Company, Inc., Dubuque, IA. All Rights Reserved. Reprinted by permission. **Figure 1.25c:** Source: P.M. Brakefield, "Industrial Melanism: Do We Have the Answers?" in *Trends in Ecology and Evolution*, 2:117-122, 1987. **Figure 1.26:** Source: A.E. Mourant, *The Distribution of Human Blood*, 1954, Ryerson Press, Toronto, Ontario, Canada. **Figure 1.30:** Source: D.M. Raup and J.J. Sepkoski, Jr., "Mass Extinctions in the Marine Fossil Record" in *Science*, 215:1502-1504, 1982.

Chapter 2

Figure 2.1: Source: J.T. Bonner, *The Evolution of Complexity*, 1988, Princeton University Press. **Figure 2.2:** Source: C.R. Taylor, et al., "Scaling of Energetic Costs of Running to Body Size in Animals" in *American Journal of Physiology*, 219(4): 1104-1107, October 1970.

Figure 2.8: From Kent M. Van De Graaff and Stuart Ira Fox, *Concepts of Human Anatomy & Physiology*, 4th edition. Copyright © 1995 McGraw-Hill Company, Inc., Dubuque, IA. All Rights Reserved. Reprinted by permission.

Chapter 3

Figure 3.3: Source: E.O. Wiley, *Phylogenetics*, 1981, John Wiley & Sons, New York, NY.

Chapter 4

Figure 4.1: Source: J. Lasman in *Journal of Parasitology*, 24:244-248, 1977.
Figure 4.2: From Peter H. Raven and George B. Johnson, *Biology*, 4th edition. Copyright © 1996 McGraw-Hill Company, Inc., Dubuque, IA. All Rights Reserved. Reprinted by permission.

Chapter 7

Figure 7.5: Source: Based on drawing by L.T. Threadgold from Larry S. Roberts and J. Janovy, Jr., *Foundations of Parasitology*, 5th edition, 1996, McGraw-Hill Company, Inc., Dubuque, IA.
Figure 7.11: Source: J.F. Mueller and H.J. Van Cleave, *Roosevelt Wildlife Annals*, 1932.
Figure 7.13: Source: D.J. Morseth, *Journal of Parasitology*, 53:492-500, 1967.
Figure 7.20: Source: W.E. Sterrer, "Systematics and Evolution Within the Gnathostomulida" in *Systematic Zoology*, 21:151, 1972.

Chapter 8

Figure 8.5: Source: R.M. Kristensen, "Loricifera, a New Phylum with Aschelminthes Characters from the Meiobenthos" in *Zeitsch. Zool. Syst. Evol.*, 21:163, 1983.
Figure 8.16: Source: C. Con, "Kamptozoa" in H.G. Bronn, ed., *Klassen und Ordnungen des Tier-Reichs*, Vol. 4, Part 2, 1936, Akademische Verlagsgesselschaft, Leipzig.
Figure 8.17: Source: C. Nielsen, *Animal Evolution. Interrelationships of the Living Phyla*, 1995, Oxford University Press.

Chapter 10

Figure 10.7: Source: P. Fauvel, "Annelides Polychetes. Reproduction" in P.P. Grasse, ed., *Traite de Zoologie*, Vol. 5, Part 1, 1959, Masson et Cie, Paris; modified from W.M. Woodworth, 1907.

Chapter 11

Figure 11.3: From *Synopsis and Classification of Living Organisms* edited by S.P. Parker, 1982, Volume 2. Copyright © 1982 McGraw-Hill Company, Inc., New York, NY. All Rights Reserved. Reprinted by permission.
Figure 11.40: From Peter H. Raven and George B. Johnson, *Biology*, 4th edition. Copyright © 1996 McGraw-Hill Company, Inc., Dubuque, IA. All Rights Reserved. Reprinted by permission.

Chapter 12

Figure 12.12b: Source: W.D. Russell-Hunger, *A Biology of Higher Invertebrates*, 1969, Macmillan Publishing Company, New York, NY.

Chapter 13

Figure 13.4: Source: Drawing by Tim Doyle.
Figure 13.19: Source: A.N. Baker, et al., "A New Class of Echinodermata from New Zealand" in *Nature,* 321:862-864, 1986.
Figure 13.22: Source: W.D. Russell-Hunger, *A Biology of Higher Invertebrates*, 1969, Macmillan Publishing Company, New York, NY.

Chapter 14

Lyrics page 268: Reprinted by permission of Alpha Music, Inc.
Figure 14.9: Source: J.J. Gould, *Wonderful Life*, 1989;, W.W. Norton & Company.
Figure 14.12: Source: R.J. Aldridge, et al., "The Anatomy of Conodonts" in *Phil. Trans. Roy. Soc. London B*, 340:405-421, 1993.
Figure 14.13: Source: R. Zangerl and M.E. Williams in *Paleontology,* 18:333-341, 1975.

Chapter 15

Figure 15.4a: Source: R. Conniff, *Audubon*, March, 1991.
Figure 15.4b: Source: F.H. Pough, et al., *Vertebrate Life*, 1966, Macmillan, NY; and D. Jensen, *Scientific American,* 214(2): 82-90, 1966.
Figure 15.20: From P. Castro and M.E. Huber, *Marine Biology*, 2nd edition. Copyright © 1997 McGraw-Hill Company, Inc., Dubuque, IA. All Rights Reserved. Reprinted by permission.

Chapter 16

Figure 16.5: Source: W.E. Duellmann and L. Trueb, *Biology of Amphibians*, 1986, McGraw-Hill Company, Inc., New York, NY.

Chapter 17

Figure 17.9: Source: R.M. Alexander, *The Chordates*, 1975, Cambridge University Press, England.

Chapter 18

Figure 18.5b: Source: P. Wellenhofer, "Archaeopteryx" in *Scientific American*, 262:70-77, May 1990.
Figure 18.10b: Source: K. Schmidt-Nielsen, *Animal Physiology*, 4th edition, 1990, Cambridge University Press.
Figure 18.25b: Source: S.R. Johnson and I.T. McCowan, "Thermal Adaptation as a Factor Affecting Colonizing Success of Introduced Sturdinae (Aves) in North America" in *Canadian Journal of Zoology*, 52:1559-1576, 1974.

Chapter 19

Figure 19.1: Source: R.J. Carroll, *Vertebrate Paleontology and Evolution*, 1988, W.H. Freeman & Company, New York, NY.
Figure 19.3: Source: J.Z. Young, *The Life of Mammals*, 1975, Oxford University Press.
Figure 19.16: Source: N. Suga, "Biosonar and Neural Computation in Bats" in *Scientific America*, 262:60-68, June 1990.
Figure 19.19: Source: J.A. Lillegraven, et al., "The Origin of Eutherian Mammals" in *Biological Journal of Linnean Society*, 32:281-336, 1987.

Index

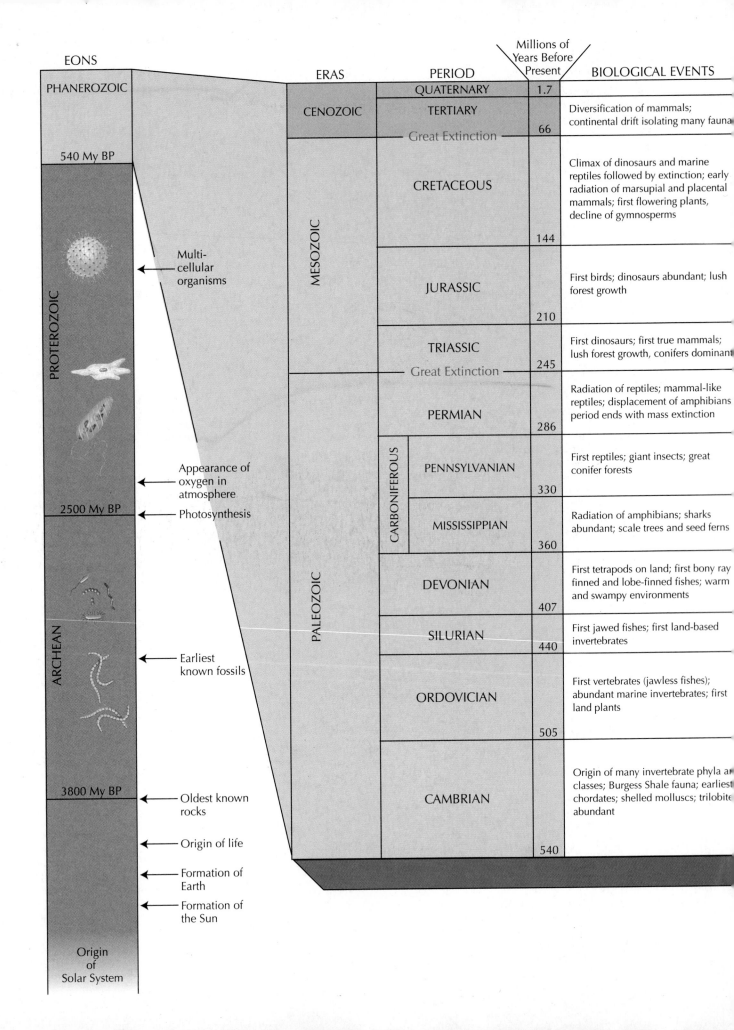

EONS		ERAS	PERIOD	Millions of Years Before Present	BIOLOGICAL EVENTS
PHANEROZOIC		CENOZOIC	QUATERNARY	1.7	
			TERTIARY		Diversification of mammals; continental drift isolating many fauna
			Great Extinction	66	
540 My BP		MESOZOIC	CRETACEOUS		Climax of dinosaurs and marine reptiles followed by extinction; early radiation of marsupial and placental mammals; first flowering plants, decline of gymnosperms
				144	
PROTEROZOIC	Multi-cellular organisms		JURASSIC		First birds; dinosaurs abundant; lush forest growth
				210	
			TRIASSIC		First dinosaurs; first true mammals; lush forest growth, conifers dominant
			Great Extinction	245	
		PALEOZOIC	PERMIAN		Radiation of reptiles; mammal-like reptiles; displacement of amphibians period ends with mass extinction
	Appearance of oxygen in atmosphere			286	
2500 My BP	Photosynthesis		CARBONIFEROUS PENNSYLVANIAN		First reptiles; giant insects; great conifer forests
				330	
			CARBONIFEROUS MISSISSIPPIAN		Radiation of amphibians; sharks abundant; scale trees and seed ferns
				360	
			DEVONIAN		First tetrapods on land; first bony ray finned and lobe-finned fishes; warm and swampy environments
				407	
ARCHEAN	Earliest known fossils		SILURIAN	440	First jawed fishes; first land-based invertebrates
			ORDOVICIAN		First vertebrates (jawless fishes); abundant marine invertebrates; first land plants
				505	
3800 My BP	Oldest known rocks		CAMBRIAN		Origin of many invertebrate phyla and classes; Burgess Shale fauna; earliest chordates; shelled molluscs; trilobite abundant
	Origin of life			540	
	Formation of Earth				
	Formation of the Sun				
Origin of Solar System					